Brucella

Molecular Microbiology and Genomics

Edited by

Ignacio López-Goñi

Departamento de Microbiología y Parasitología
Instituto de Salud Tropical
Universidad de Navarra
Pamplona
Spain

and

David O'Callaghan

INSERM U1047
Universite de Montpellier 1
UFR Médecine
Nimes
France

Caister Academic Press

Caister Academic Press
Norfolk, UK

www.caister.com

British Library Cataloguing-in-Publication Data
A catalogue record for this book is available from the British Library

ISBN: 978-1-904455-93-6

Cover design adapted from Figure 10.2

Printed and bound in Great Britain

Contents

Contributors

Eric Anderson
Department of Microbiology and Immunology
Brody School of Medicine
East Carolina University
Greenville, NC
USA

andersone@ecu.edu

Christian Baron
Département de Biochimie
Université de Montréal
Montréal, QC
Canada

christian.baron@umontreal.ca

Sophie Bonnot
Laboratoire de Parasitologie Moléculaire
ULB-IBMM
Charleroi
Belgium

sbonnot@ulb.ac.be

Esteban Chaves-Olarte
Programa de Investigación en Enfermedades Tropicales
Escuela de Medicina Veterinaria
Universidad Nacional
Heredia
Costa Rica;
Centro de Investigación en Enfermedades Tropicales
Facultad de Microbiología
Universidad de Costa Rica
San José
Costa Rica

esteban.chaves@ucr.ac.cr

Axel Cloeckaert
INRA, UR1282
Infectiologie Animale et Santé Publique, IASP
Nouzilly
France

axel.cloeckaert@tours.inra.fr

Thomas A. Ficht
Veterinary Pathobiology
College of Veterinary Medicine and Biomedical Science
Texas Agrilife Research
Faculty of Genetics
Texas A&M University
College Station, TX
USA

tficht@cvm.tamu.edu

Juan M. García-Lobo
Departamento de Biología Molecular
Universidad de Cantabria
Instituto de Biomedicina y Biotecnología de Cantabria (IBBTEC)
UC/CSIC/SODERCAN
Santander
Spain

jmglobo@unican.es

Krishna K. Gopaul
FAO/WHO Collaborating Centre for Brucellosis, OIE Brucellosis Reference Centre
Department of Bacteriology
Animal Health and Veterinary Laboratories Agency (AHVLA)
Addlestone, Surrey
UK

krishna.gopaul@ahvla.gsi.gov.uk

Caterina Guzmán-Verri
Programa de Investigación en Enfermedades Tropicales
Escuela de Medicina Veterinaria
Universidad Nacional
Heredia
Costa Rica

catguz@medvet.una.ac.cr

Peter Kämpfer
Institut für Angewandte Mikrobiologie
Justus-Liebig Universität Giessen, IFZ
Giessen
Germany

peter.kaempfer@agrar.uni-glessen.de

Stephan Köhler
Centre d'études d'agents Pathogènes et Biotechnologies pour la Santé (CPBS), UMR 5236 CNRS
Université Montpellier I–Université Montpellier II
Montpellier
France

stephan.kohler@cpbs.cnrs.fr

Jean-Jacques Letesson
Research Unit in Molecular Biology
Department of Biology, NARILIS
University of Namur (FUNDP)
Namur
Belgium

jean-jacques.letesson@fundp.ac.be

Ignacio López-Goñi
Departamento de Microbiología y Parasitología
Instituto de Salud Tropical
Universidad de Navarra
Pamplona
Spain

ilgoni@unav.es

Marcos Mancilla
Department of Microbiology and Parasitology
Institute of Tropical Health
University of Navarra
Pamplona
Spain

mmancillae@unav.es

Daniel W. Martin
Department of Microbiology and Immunology
Brody School of Medicine
East Carolina University
Greenville, NC
USA

martinda@ecu.edu

David Martinson
Department of Microbiology and Immunology
Brody School of Medicine
East Carolina University
Greenville, NC
USA

martinsond08@students.ecu.edu

Evan Menscher
Department of Microbiology and Immunology
Brody School of Medicine
East Carolina University
Greenville, NC
USA

menschere@ecu.edu

Edgardo Moreno
Programa de Investigación en Enfermedades Tropicales
Escuela de Medicina Veterinaria
Universidad Nacional
Heredia
Costa Rica;
Instituto Clodomiro Picado
Universidad de Costa Rica
San José
Costa Rica

emoreno@racsa.co.cr

David O'Callaghan
INSERM U1047
Universite de Montpellier 1
UFR Médecine
Nimes
France

david.ocallaghan@univ-montp1.fr

Jenifer Ojeda
Department of Microbiology and Immunology
Brody School of Medicine
East Carolina University
Greenville, NC
USA

ojeda03@students.ecu.edu

Eustache Paramithiotis
Caprion Proteomics, Inc.
Montreal
Quebec
Canada

eparamithiotis@caprion.com

Allison C. Rice-Ficht
Molecular and Cellular Medicine
College of Medicine
Faculty of Genetics
Texas A&M University
College Station, TX
USA

a-ficht@tamu.edu

María C. Rodríguez
Departamento de Biología Molecular
Universidad de Cantabria
Instituto de Biomedicina y Biotecnología de Cantabria (IBBTEC)
UC/CSIC/SODERCAN
Santander
Spain

mariacruz.rodriguez@gestion.unican.es

R. Martin Roop II
Department of Microbiology and Immunology
Brody School of Medicine
East Carolina University
Greenville, NC
USA

roopr@ecu.edu

Félix J. Sangari
Departamento de Biología Molecular
Universidad de Cantabria
Instituto de Biomedicina y Biotecnología de Cantabria (IBBTEC)
UC/CSIC/SODERCAN
Santander
Spain

sangarif@unican.es

Holger C. Scholz
Bundeswehr Institute of Microbiology
Department of Bacteriology and Toxinology
Munich
Germany

holger1scholz@bundeswehr.org

Asunción Seoane
Departamento de Biología Molecular
Universidad de Cantabria
Instituto de Biomedicina y Biotecnología de Cantabria (IBBTEC)
UC/CSIC/SODERCAN
Santander
Spain

seoanea@unican.es

Bruno W. Sobral
Virginia Bioinformatics Institute
Virginia Tech
Blacksburg, VA
USA

sobral@vbi.vt.edu

Matthieu Terwagne
Research Unit in Molecular Biology
Department of Biology, NARILIS
University of Namur (FUNDP)
Namur
Belgium

matthieu.terwange@fundp.ac.be

Renee M. Tsolis
University of California, Davis
Medical Microbiology & Immunology
Davis, CA
USA

rmtsolis@ucdavis.edu

Nieves Vizcaíno
Departamento de Microbiología y Genética
Universidad de Salamanca
Salamanca
Spain

vizcaino@usal.es

Alice R. Wattam
Virginia Bioinformatics Institute
Virginia Tech
Blacksburg, VA
USA

wattam@vbi.vt.edu

Adrian M. Whatmore
FAO/WHO Collaborating Centre for Brucellosis
OIE Brucellosis Reference Centre
Department of Bacteriology
Animal Health and Veterinary Laboratories Agency (AHVLA)
Addlestone, Surrey
UK

adrian.whatmore@ahvla.gsi.gov.uk

Jean-Yves Winum
Institut des Biomolécules Max Mousseron (IBMM), UMR 5247 CNRS
Université Montpellier I–Université Montpellier II
Ecole Nationale Supérieure de Chimie de Montpellier
Montpellier
France

jean-yves.winum@univ-montp2.fr

Preface

Brucellosis is the term used to describe a disease caused by the members of the genus *Brucella* in animals and humans. The range of primary hosts of *Brucella* is large and encompasses several domestic or semi-domestic animals, pinnipeds and cetaceans, certain wild rodents and, to a lesser extent, other species of warm-blooded vertebrates. Recently, *Brucella* have been isolated from fish and a frog. In domestic ruminants, the disease is manifested mostly as abortion and infertility. Humans acquire brucellosis from domestic animals and are not themselves a source of contagion. Although rarely fatal, the human disease is insidious and debilitating, and may leave severe sequelae if untreated.

Brucellosis has a world wide impact in terms of its epidemiology, human health risks and its impact on trade. Brucellosis should be considered as a disease of poverty; the greatest impact of brucellosis is in the poorer rural areas of the world, without the funds or the infrastructure to establish surveillance and eradication programs. The World Health Organization considers brucellosis to be the world's most widespread bacterial zoonosis and one of the neglected zoonotic diseases.

During the infection *Brucella* spends most of its time inside phagocytic cells. A major aspect of the pathogenicity of *Brucella* is its capacity to survive and replicate within host macrophages. This impressive ability has led to the description of *Brucella* as a 'facultatively extracellular intracellular parasite'. However, the *Brucella* genome does not seem to encode classical virulence factors found in other well-known bacterial pathogens, and the basis of this pathogenic nature is still an area of intense research. From this perspective, the use of comprehensive system biology approaches, such as genomics, proteomics and transcriptomics, has contributed to the improvement of our understanding of the biology, including pathogenicity, and evolution of members of the genus *Brucella*. Written by highly acclaimed *Brucella* scientists, this book comprehensively reviews the most important advances in the field.

Brucella belongs to the class α-proteobacteria which is one of the largest and diverse groups within the phylum Proteobacteria. Comparative genome analysis revealed that *Brucella* is genetically related to plant-associated symbionts and pathogens. For many years, the taxonomy of *Brucella* remained unchanged. Already at the time of writing around 40 *Brucella* genome sequences are available and continuing advances in next generation sequencing approaches promise to provide the next major wave of molecular data for bacteria. The recent development of new molecular typing methods and comparative genome analysis has resulted in rapid advances in the understanding of *Brucella* diversity and, after many years of stagnancy, a process of expansion of the genus is in progress. The recent identification of novel atypical *Brucella* strains and novel species poses a new challenge to the characterization, classification and nomenclature of the genus. Genome sequencing projects of these new isolates have been initiated to obtain a better understanding of biology and evolution of members of the genus *Brucella*, which is reviewed in Chapter 1. Likewise, Chapter 2 summarizes the current state of knowledge of the genetic diversity within *Brucella*, including both the classical and new species, with particular emphasis on comparing the genome sequences, the phylogenies produced by a variety of methods and specific families of genes in particular functions and pathways.

The sequence analysis of *Brucella* has also revealed the presence of various genomic islands scattered through its genome, a fact that strongly suggests the acquisition of DNA by horizontal transfer events. The analysis of the genetic content of these regions indicates that they might represent a source of virulence factors. Recently, it has been demonstrated that some predicted genomic islands are unstable and can be excised from the chromosome by recombination. This experimental evidence points out the foreign origin of such loci and suggests an explanation for the polymorphism related to these large chromosomal regions exhibited by *Brucella*. Chapter 3 summarizes the advances in the identification and characterization of *Brucella* genomic islands, and its relation to pathogenicity in the context of genome plasticity and virulence attenuation.

For many years, the diagnosis of brucellosis depended on tried and tested cultural and serological methods. New molecular approaches, based on the ever increasing genomic data available for the genus, are revolutionizing our ability to understand relationships between strains, within a group of organisms traditionally described as homogeneous. Chapter 4 presents a review of the recent advances in molecular approaches to *Brucella* diagnostics and epidemiology.

The pathogenesis of brucellosis is dependent on the ability of the bacterium to enter epithelial and phagocytic cells, to avoid host cell defences and create a novel intracellular replication niche by capturing vesicles from endoplasmic reticulum exit sites. From this perspective, the use of comprehensive system biology approaches has contributed to the dissection and unravelling of aspects of the life cycle of *Brucella*. In Chapters 5 and 6, we describe the advantages and limitations of transcriptomic and proteomics approaches employed in the field of brucellosis to follow, at the cellular and the molecular level, the pathogenesis of brucellosis. The emerging picture is one of a pathogen that does not rely on discrete virulence determinants to establish successful infections, but rather of a microbe that utilizes synchronized complex metabolic and signalling systems to reach its replicating niche. These systems include secretion machineries including the type IV secretion VirB system (Chapter 11) and signalling complexes such as the two-component regulatory BvrR/BvrS system (Chapter 10) and the quorum sensing transcriptional regulator VjbR (Chapter 8), all of which have a profound impact on the metabolic state of the bacterium and particularly on the homeostasis of the cell envelope. The particular characteristics of the *Brucella* outer membrane that could affect the interaction with the host and be related to the differences in pathogenicity and host tropism among the brucellae, is reviewed in Chapter 7. Moreover, Chapter 9 discusses what is presently known about the metal acquisition and homeostasis systems in *Brucella* strains. Finally, approaches to develop improve vaccines and novel strategies for the treatment of this infectious disease are also discussed in Chapters 12 and 13.

It is our duty as Editors of this book to thanks to all co-authors for their willingness to share openly with the readers many interesting and challenging new ideas. Undoubtedly, these will have an impact not only in everyone with an interest in *Brucella* and brucellosis but also in those interested in microbial pathogenesis and host–parasite relationship, cellular microbiologists and immunologists, and vaccine development. The timely initiative of the Publishers of the book deserves also our gratitude.

Ignacio López-Goñi and
David O'Callaghan

Brucella: Relationship to Other *Alphaproteobacteria*, Current Taxonomy and the Emergence of New Species

1

Holger C. Scholz, Peter Kämpfer and Axel Cloeckaert

Abstract

The genus *Brucella* belongs to the class *Alphaproteobacteria,* which is one of the largest and most diverse groups within the phylum Proteobacteria. Comparative genome analysis revealed that *Brucella* is genetically related to plant-associated symbionts and pathogens. The genus *Ochrobactrum,* consisting of saprophytes that occasionally infect humans, is the closest phylogenetic neighbour. For many years, the taxonomy of *Brucella* remained unchanged. The recent development of new molecular typing methods and comparative genome analysis however, has resulted in rapid advances in the understanding of *Brucella* diversity and, after many years of stagnancy, a process of expansion of the genus is in progress. The recent increased occurrence of novel atypical *Brucella* strains and novel species poses a new challenge to the characterization, classification and nomenclature of the genus *Brucella.* Genome sequencing projects of these new isolates have been initiated to obtain a better understanding of biology (including pathogenicity) and evolution of members of the genus *Brucella.*

Relationship to other *Alphaproteobacteria*

The genus *Brucella* (Meyer and Shaw, 1920) belongs to the family *Brucellaceae* within the order *Rhizobiales** (see Remarks on p. 9) of the class *Alphaproteobacteria.* This class is one of the largest and diverse groups within the phylum *Proteobacteria.* Of central importance both for medical and agricultural perspectives, is the intimate association of many *Alphaproteobacteria* with higher eukaryotes, either as plant mutualists or plant, animal or human pathogens (Batut *et al.,* 2004). The interactions range from pericellular colonization through facultative intracellular multiplication to obligate intracellular lifestyles (Batut *et al.,* 2004). Interestingly, common strategies necessary for the invasion and survival in association with cells have been preserved on chromosomes of both the animal and plant-associated *Alphaproteobacteria* (Moreno, 1998; Batut *et al.,* 2004). Comparative genome sequence analysis of a large number of alphaproteobacterial genomes which became available within the last years has provided new insight into the evolution and microevolution of these bacteria. Based on 104 protein families of 72 *Alphaproteobacteria,* Williams *et al.* (2007) recently calculated a phylogenetic tree with robust phylogenetic inference. A robust multigene phylogenetic tree for the order *Rhizobiales* can also be found at the PathoSystems Resource Integration Centre (PATRIC), accessible via the internet (http://www.patricbrc.org). Both phylogenetic reconstructions demonstrate the genetic relatedness of *Brucella* to plant symbionts and plant pathogens as well as to *Bartonella,* another facultative intracellular pathogen of animal and man.

Relationship to plant symbionts and plant pathogens

From recent genome sequencing analysis, in particular the genome of *Brucella suis* 1330, it became evident that *Brucella* shares extensive genetic similarity with some plant pathogens and microbes that live symbiotically with plants, such as *Agrobacterium, Rhizobium* (note that *Agrobacterium* species been transferred to the genus *Rhizobium,* however this is strongly contested by the *Agrobacterium* research community) and *Mesorhizobium* (Paulsen *et al.*, 2002). Paulsen *et al.* found that a total of 1902 *B. suis* ORFs were conserved in all three genomes of *M. loti, S. meliloti*, and *A. tumefaciens*, and 2408 *B. suis* 1330 ORFs were conserved in at least one of these three genomes. In addition to gene sequence conservation, *B. suis* 1330 chromosome I shares extensive regions of gene synteny (gene order) with the *M. loti* chromosome, whereas chromosome II does not. Instead, chromosome II possesses a cluster of plasmid-like replication genes similar to plasmid replication genes from *Agrobacterium* Ti plasmids, and plasmids from other organisms including *Rhizobium* spp. (Paulsen *et al.* 2002).

Detailed sequence analysis revealed the existence of an intact beta-ketoadipate pathway in *B. suis* 1330 (Paulsen *et al.*, 2005) and also in *B. microti* (Audic *et al.*, 2009). This pathway, which enables the bacterium to break down plant-derived aromatic compounds and lignin monomers, is widely distributed and highly conserved in soil bacteria and fungi, including *Pseudomonas putida, Acinetobacter calcoaceticus, Agrobacterium tumefaciens, Rhodococcus erythropolis* (Harwood and Parales, 1996). Therefore, it seems probable that some *B. suis* strains can persist and even might multiply in the environment. Various studies have been performed to confirm this hypothesis. In a study, carried out by the US Environmental Protection Agency (EPA), the persistence of *B. suis* biotype I strain Battelle BRU163 on various outdoor materials, was evaluated. *B. suis* persisted as long as 28 days (longest period tested) on various matrices (aluminium, wood, glass, soil) under low and moderate temperatures without UV light (report see: EPA/600/R-10/026; April 2010; www.epa.gov/ord).

Besides *B. suis*, the recently described novel species *B. microti* (see below) that is metabolically highly active, could be recovered directly from soil after long-time storage (6 months) at 4°C and from soil samples collected at the same site seven years after primary isolation of this novel species from common voles (Scholz *et al.*, 2008c). In summary, it seems probable that *Brucella* has evolved from a soil/plant-associated ancestral bacterium and speculatively, at least some strains of *B. suis* and *B. microti* may have a replicative life cycle outside of a mammalian host.

Relationship to animal and human pathogens

Ochrobactrum is the closest phylogenetic neighbour

In 1987 De Ley and co-authors (De Ley *et al.*, 1987) investigated the taxonomic neighbourhood of *Brucella* by comparative ribosomal ribonucleic acid cistron analysis (DNA/RNA hybridizations). At his time, the taxonomic neighbourhood of *Brucella* was unknown and even the taxonomic position of *Brucella* was uncertain. From their results De Ley *et al.* concluded that *Brucella* and *Achromobacter* CDC Group Vd strains are closely related, belonging to the same rRNA branch within rRNA superfamily IV.

One year later Holmes *et al.* constituted the genus *Ochrobactrum,* by outgrouping of *Achromobacter* species biotypes 1 and 2, type A strains, and *Achromobacter* CDC-groups Vd-1 and Vd-2 (Holmes *et al.*, 1988). At this time *O. anthropi* was constituted as the sole species of this novel genus. The name 'anthropi' (an. thro'pi. Gr. n. *anthropos,* a human being; N.L. gen. n. *anthropi,* of a human being) was given because virtually all strains at this time had been recovered from human clinical specimens. Ten years later (1998) Velasco *et al.* studied the relatedness of *Brucella* spp. and multiple strains of *O. anthropi* by protein profiling, Western blot analysis, immunoelectrophoresis and 16S rRNA gene analysis (Velasco *et al.*, 1998). They recognized an intense serological cross-reactivity between *Brucella* spp. and strains included in *O. anthropi.* They also identified a subset of strains among *O. anthropi* with even a higher relatedness to *Brucella.* Ribosomal RNA gene analysis revealed that these strains share similarities of > 98.5% with the consensus 16S rRNA gene sequence of the classical *Brucella* species. Based on their findings, they established the second *Ochrobactrum* species, *O. intermedium* which name should reflect the intermediate position between *Ochrobactrum* and *Brucella.* It has to be mentioned here, that *Brucella* and *Ochrobactrum* species share more than 97% 16S rRNA gene sequence similarity and several other phenotypic traits, so that it could be argued, that they constitute a single genus. Furthermore, from comparative genome analysis it became evident that *Ochrobactrum* shares various other genes with *Brucella,* such as genes encoding the outer membrane porin proteins Omp2a and Omp2b, the Omp25/Omp31 family of proteins, the outer membrane lipoproteins Omp10, Omp16, and Omp19, and the periplasmic BP26 protein (Cloeckaert *et al.,* 1999 and unpublished results). The close genetic relatedness of *Ochrobactrum* and *Brucella* was confirmed by *recA* and 16S rRNA gene analysis of a large collection of *Ochrobactrum* strains belonging to all currently known species (Scholz *et al.,* 2008a). Similarities of both *recA* and 16S rRNA genes were found to be markedly lower among the various *Ochrobactrum* species than to members of the genus *Brucella.* In other words, *O. anthropi* and *O. intermedium* were more closely related to members of the genus *Brucella* than to species of their own genus.

In a recent study, the genomes of *Brucella* spp. were compared with the genomes of three agrobacterial species and the type strains of *O. anthropi* and *O. intermedium* by using base compositional and proteome based methods (Bohlin *et al.,* 2010). The authors found that at the protein level the similarity between *Brucella* species and *Ochrobactrum* species was higher than between the various species within the genus *Agrobacterium.* Thus the diversity among species of the genus *Agrobacterium* was found to be higher than between species of the genus *Brucella* and *Ochrobactrum.* Besides the confirmation of the close genetic relatedness of *Ochrobactrum* and *Brucella,* the authors again confirmed the high genetic relatedness of the various *Brucella* species as demonstrated by previous studies (Hoyer and McCullough, 1968a,b; Verger *et al.,* 1985; De Ley *et al.,* 1987; Paulsen *et al.,* 2002). From a taxonomic point of view, this again raises the question whether the currently recognized species in the genus *Brucella* should be combined to a single species (see below). The authors discuss that such a scenario would imply that the members of the genus *Ochrobactrum* would join the genus *Brucella,* while the present species in the genus *Brucella* would all be one species with different subspecies. Whereas the discussions to combine the various *Brucella* species to a single species (*B. melitensis*) is not new, the question about unification of *Ochrobactrum* and *Brucella* has only recently become an issue, as a result of whole genome comparison.

Indeed in more recent publications such a scenario has been discussed by different authors (Whatmore, 2009; Ficht, 2010). Most of the authors argue that both the single species concept and the unification of the two genera would be dangerous to scientists that are not familiar with the varying pathogenicity and transmissibility of the classical species and biovars. However, a unified genus would not be under the umbrella of *Ochrobactrum* (Ficht, 2010) but, because of priority rule 38 of the *International Code of Nomenclature of Bacteria* (available at: http://www.ncbi.nlm.nih.gov/bookshelf/br.fcgi?book=icnb), has to be named *Brucella.* This would imply that the current *Ochrobactrum* species would have to be transferred to the genus *Brucella.* Other concerns led to the argument that according to rule 56a of the *International Code of Nomenclature of Bacteria* the *nomen periculosum,* i.e. a name whose application probably leads to accidents endangering health or life or both or of serious economic consequences, should be applied to the *Brucella* genus (Whatmore, 2009). An example for a *nomen periculosum* has been the suggestion to reclassify *Yersinia pestis,* the causative agent of plague, as *Yersinia pseudotuberculosis* subsp. *pestis* (Opinion 60) which has been rejected. However, also this rule can not be applied to *Brucella,* because according to rule 23b the combined species name would be *B. melitensis* with presumably the other species recognized as subspecies with their biovars, i.e. *Brucella melitensis* subsp. *abortus* bivovar 1. Since *B. melitensis* is the most dangerous human pathogenic species, no additional danger would be caused. In fact all *Brucella* species despite their different human pathogenic potential are currently classified as Biosafety Level 3 (BSL 3) pathogens, and therefore *per se* have to be handled in a Containment level 3 laboratory.

The above-mentioned species/subspecies concept is also used for other clinically important genera, like for example *Francisella,* which includes highly pathogenic species/subspecies (*F.*

tularensis subsp. *tularensis*) and at the same time less pathogenic subspecies (i.e. *F. tularensis* subsp. *holarctica*). In fact, by the unification of *Ochrobactrum* and *Brucella* one would end up with a structure similar to the genera *Burkholderia* (*B.*), and *Yersinia* (*Y.*), which besides the highly pathogenic species (*B. mallei/pseudomallei* and *Y. pestis*) also comprise facultative pathogens and non-pathogenic members. Thus genera including both highly pathogenic members and non-pathogenic members are much more common than genera comprising only of pathogenic species, like in the case of *Brucella*.

Although *Ochrobactrum* and *Brucella* are genetically highly related, they can be differentiated from each other by the presence of the insertion sequence IS*711* and *bcsp31,* a gene encoding a surface protein of unknown function in *Brucella*. These genetic elements are absent from all *Ochrobactrum* strains isolated so far or at least can not be detected by commonly used diagnostic PCR-assays (Cloeckaert, Scholz, unpublished). This is of major importance in the diagnoses of brucellosis, which is a notifiable disease.

Genetic relatedness to *Bartonella*

Like *Brucella,* pathogenic *Bartonella* species are facultative intracellular pathogens. However, whereas *Bartonella* represent arthropod-borne pathogens that typically parasitize erythrocytes in their mammalian reservoir hosts (Seubert *et al.*, 2002), *Brucella* replicates in macrophages and is not transmitted by arthropod vectors.

Comparative 16S rRNA gene analysis (O'Connor *et al.*, 1991) and DNA–DNA hybridization studies (Brenner *et al.*, 1993) led to the unification of the genera *Rochalimaea* and *Bartonella* to the combined genus *Bartonella* and the removal of the family *Bartonellaceae* from the order *Rickettsiales*. Based on these studies it became obvious that the closest relative to *Bartonella* is the genus *Brucella*. Phylogenetic reconstructions based on other loci (Birtles and Raoult, 1996; Sweger *et al.*, 2000) and more recently also on whole genome comparisons (http://www.patricbrc.org) confirmed that *Bartonella* is genetically most closely related to *Brucella*. Indeed, analysis of the genomes from *Bartonella quintana* and *Bartonella henselae* revealed that both genomes are reduced versions of chromosome I of *B. melitensis* (Alsmark *et al.*, 2004). In addition a segment with similarity to genes located on chromosome II of *B. melitensis* is present in the two *Bartonella* species which is flanked by two rRNA operons. This suggests that the *B. melitensis*-specific fragment was acquired by integration of megareplicon DNA in a common ancestor of the two *Bartonella* species (Alsmark *et al.*, 2004). As a result of these integration and recombination events, *Brucella* and *Bartonella* also share common virulence-associated genes, such as, for example, the BvrR/BvrS two-component regulatory system (Guzman-Verri *et al.*, 2002) or a VirB type IV secretion system (O'Callaghan *et al.*, 1999).

Classification and current taxonomy

The genus *Brucella* (Meyer and Shaw, 1920) belongs to the family *Brucellaceae* within the order *Rhizobiales** of the class *Alphaproteobacteria*. It was named in honour to Sir David Bruce, a military physician who first discovered the aetiological agent of brucellosis (Malta fever).

In addition to the type genus *Brucella,* six further genera, namely *Crabtreella, Daeguia, Mycoplana, Ochrobactrum, Paenochrobactrum* and *Pseudochrobactrum,* are currently recognized within the family *Brucellaceae* (for details see http://www.bacterio.cict.fr/classifgenerafamilies.html#*Brucell*aceae). The closest phylogenetic (i.e. sequenced-based evaluation) neighbour of *Brucella* is *Ochrobactrum,* a saprophyte that occasionally infects humans (see above).

Initially, Meyer and Shaw (1920) placed two species in the genus *Brucella,* the causative agent of Malta-fever, *B. melitensis,* as the type species of the genus and *B. abortus,* the aetiological agent of bovine brucellosis characterized by abortion (both were initially described in the nineteenth century). *B. melitensis* was named after the region where the disease has been described for the first time (melitensis is the Latin name for the island Malta). *B. abortus* was named after the discovery that the organism was responsible for abortion in cattle. Later names were given to the main source of isolation (preferred natural host).

The third *Brucella* species, *B. suis*, was added in 1929 (Huddleson, 1929), followed by *B. ovis* in 1956 (Buddle, 1956), *B. neotomae* in 1957 (Stoenner and Lackman, 1957) and by *B. canis* in 1968 (Carmichael and Brunner, 1968). All six species names were included on the Approved Lists of Bacterial Names, thus retaining them in a modern context. These species are today known as the 'classical' *Brucella* species. Because the extensive battery of modern molecular methods for species delineation were not available at the time these species were originally described, delineation and classification was solely based on phenotypical traits and the apparent preferred natural host. However, early in the history of research on members of the genus *Brucella*, it was recognized that differences among the various species were only marginal and doubts rose whether the various *Brucella* species should be taxonomically regarded as separate species or as a single species. With the advent of molecular methods like DNA–DNA reassociation and later 16S rRNA gene sequencing it became obvious that *Brucella* consists of genetically highly related strains and it would be appropriate to consider them to represent a single species (i.e. a monospecific genus). Based on the knowledge that all *Brucella* species were indistinguishable by means of their 16S rRNA genes and that all species share DNA–DNA reassociation values >90% (a widely accepted borderline for bacterial species delineation is 70% DNA–DNA similarity), Verger *et al.* 1985 consequently proposed to combine all *Brucella* species into a single species, *B. melitensis*, and the others to be recognized as biovars (e.g. *Brucella melitensis* biovar Abortus 1). They also provided an amended species description of *B. melitensis*. Verger's results were an extension of earlier work (Hoyer and McCullough., 1968a,b) that was later confirmed by another study (De Ley *et al.*, 1987). However, in the Manchester Report of 1988, the Subcommittee on the Taxonomy of *Brucella* noted that the existing vernacular names for the nomen species *Brucella melitensis, Brucella abortus*, etc., could be retained for non-taxonomic purposes to avoid confusion (Corbel, 1988). This suggestion was made mainly for medical-clinical issues, because the various species and biovars differ significantly in their ability to cause disease in humans (or other animals). In subsequent years the subcommittee confirmed the acceptance of the single species concept in its interim report of the years 1991–1993, although there appeared to be problems associated with the proposed nomenclature (Gargani and López-Merino, 2006). On reflection the use of a single species name together with six vernacular names that were not distinguishable from the scientific names they were meant to replace was a recipe for confusion and misunderstanding.

Over the course of time the suggestion to retain the vernacular names for non-taxonomic purposes led to confusion in terms of a stable and unambiguous nomenclature and taxonomy. This is probably the reason why in 2003 the Subcommittee on the Taxonomy of *Brucella* unanimously agreed on a return to pre-1986 taxonomic interpretation of the genus *Brucella*, implying the division of the genus into the six classical *Brucella* nomenspecies with their biovars (Osterman and Moriyón, 2006). However, the names of prokaryotes once validly published always remain validly published, even if they are considered to be synonyms of other names. Consequently the use of either a single species concept or a six species concept for the classical *Brucella* species only requires that different taxonomic opinions are expressed, together with the use of appropriate species descriptions. At present, although there is on going debate with respect to the status of both the classical species and novel species within the genus *Brucella*, a modern and more comprehensive description/delineation of the genus is still lacking. This is especially relevant in terms of clearly distinguishing between members of the genera *Brucella, Ochrobactrum, Paenochrobactrum* and *Pseudochrobactrum*.

At the time of writing the genus *Brucella* consists of ten species with validly published names, *B. melitensis, B. abortus, B. suis, B. ovis, B. neotomae* and *B. ovis*, known as the six classical species, two species from marine mammals, *B. ceti* and *B. pinnipedialis* and two novel *species B. microti* and *B. inopinata*.

The correct names of the species within the genus *Brucella* with details of type and reference strains can be found on the internet site of the ICSB Subcommittee on Taxonomy of *Brucella* (http://www.the-icsp.org/taxa/*Brucella*list.htm).

New *Brucella* species

Because of the close genetic and phenotypic relatedness among the various *Brucella* species and the continuous uncertainty in taxonomy, the description of novel species is a challenge. In addition, the gold standard for bacterial species delineation can not be applied, because all species, including the novel species have a DNA–DNA relatedness >70% compared with the type species *B. melitensis* and therefore would be placed into a single species. Historically, classification of *Brucella* species was based on phenotypic traits, i.e. CO_2 requirement, H_2S production, dye-sensitivity, oxidative metabolic profiles, lysis by *Brucella*-specific bacteriophages, and agglutination with monospecific antisera. Another important criterion for species delineation was the observed preference for a natural host. However, this requires that (i) a large number of isolates is available and (ii) that a preferred host does exist. Although the inclusion of a large number of isolates is highly appreciated when a novel species is described, it has to be mentioned that that this is no mandatory prerequisite.

Furthermore, although, the isolation frequencies of different *Brucella* species from infected hosts are consistent with some type of host preference (Verger *et al.*, 1982), the range of hosts infected by a specific *Brucella* species can be enormous. As an example, cattle have been described as the natural or primary hosts for *B. abortus*. However, *B. abortus* could meanwhile be isolated from various orders of mammalian hosts, such as horses, pigs, sheep, goats, Bactrian camels, dromedary camels, water buffalo, yaks, elk, dogs (Crawford *et al.*, 1990), and humans (Arimi *et al.*, 2005; Swai and Schoonman, 2009).

Minimal standards for the description of novel *Brucella* species were proposed in 1975 before molecular methods had been developed (Corbel and Brinley Morgan, 1975) and can be used as the basis for the description of novel species. However, classical methods such as the comparison of the electrophoretic patterns of acid-phenol-soluble proteins or the measurement of the absorption spectra of the major cytochrome bands are no longer practiced. Hence, the existing minimal standards for the description of novel *Brucella* species require a fundamental revision in respect to the developed molecular typing techniques and evolutionary findings.

The description of novel *Brucella* species is also hampered by the genus description itself. Here commonly variable phenotypic features, like motility, certain biochemical reactions (e.g. Voges-Proskauer) are part of the description. This implies that for example the description of a motile *Brucella* species would be in conflict with the current genus description and basically such an isolate could not be a member of the genus unless an amended genus description has been proposed. On the other hand it is well known that even at the species level motility is a variable marker, e.g. *Y. enterocolitica* is motile at 20–25°C but not at 30°C (Bottone, 1998).

Indeed, already the description of the novel fast growing and Voges-Proskauer reaction positive species *B. microti* and *B. inopinata* correctly would have required an amendment of the current genus description. As a conclusion such variable criteria should be used at the species level rather than at the genus level.

The classification of the most recently described species *B. ceti*, *B. pinnipedialis*, *B. microti*, and *B. inopinata* followed a modern polyphasic approach consisting of a battery of classical phenotypical and modern molecular methods (Tindal *et al.*, 2010). This approach can be used as a new standard for the description of novel *Brucella* species.

Brucella ceti and *Brucella pinnipedialis*

After the addition of *B. canis* in 1968 no new species was added to the genus until 2007, presumably because of the above mentioned difficulties. In 1994, Ewalt *et al.* (1994) described an atypical[+] (see Remarks on p. 9) novel *Brucella* strain isolated from a bottlenose dolphin. In 1997, following the isolation of a series of novel strains from sea mammals, Jahans *et al.* (1997) proposed that they comprised a new nomen species which they called *B. maris*. The name *B. maris* never officially recognized and after 10 years of further research, the sea mammal isolates were subdivided into two novel *Brucella* species, according to specific and phenotypic molecular markers (Cloeckaert *et al.*, 2001; Jacques *et al.*, 2007), named *B. ceti* and *B. pinnipedialis* with cetaceans and seals as their preferred hosts (Foster *et al.*, 2007). More recent molecular analyses indicate that both *B. ceti* and *B. pinnipedialis* can further be subdivided into separate subclusters (Bourg *et al.*, 2007, Dawson *et al.*, 2008; Maquart *et al.*, 2009; Zygmunt *et al.*, 2010). One of these subclusters seems specific of dolphin isolates and a novel species name, i.e. *B. delphini*, has been proposed for them (Groussaud *et al.*, 2007). Most marine mammal isolates described have been isolated from the Atlantic and interestingly marine mammal strains, mostly from cetaceans, have also been isolated from the Pacific. These isolates from the Pacific may also constitute a separate species (Maquart *et al.*, 2009). Three human cases with naturally acquired infection by *Brucella* strains presumably of marine origin have been reported, one case of spinal osteomyelitis from a patient in New Zealand (McDonald *et al.*, 2006) and two neurobrucellosis cases from Peruvian patients (Sohn *et al.*, 2003). Interestingly, these human isolates presented the same genotype as strains from cetaceans from the Pacific (Whatmore *et al.*, 2008).

Brucella microti

Brucella microti was initially isolated from systemically diseased common voles (*Microtus arvalis*) in South Moravia in the Czech Republic (Hubalek *et al.*, 2007) and was published as a novel species in 2008 (Scholz *et al.*, 2008b). Meanwhile this novel species has also been isolated from mandibular lymph nodes of red foxes in Lower Austria (Scholz *et al.*, 2009) and even directly from soil (Scholz *et al.*, 2008c). Long-time persistence in the soil suggests that *B. microti* might at least have a transient reservoir in the environment and that common voles (and other animals) become infected by (aerolized) contaminated soil. Foxes are presumably infected by the ingestion of diseased rodents. Currently there is an ongoing project to study the prevalence of *B. microti* in various animal species.

At the time of writing, 12 isolates of *B. microti* (two from common voles, three from soil and seven from foxes) exist (Al Dahouk *et al.*, 2010). All isolates carry the phage-related genomic insertion of 11 kb (Scholz, unpublished) and can therefore be differentiated from other species by the use of a single PCR (Scholz *et al.*, 2008c) or in a modified multiplex PCR (Mayer-Scholl *et al.*, 2010). Although *B. microti* shares most of the phenotypic characteristics of the genus *Brucella* (e.g. agglutination with anti *Brucella* A and M sera), it can be clearly distinguished from the classical *Brucella* species by its rapid growth on standard culture media and its high metabolic activity, which is comparable to member of the soil bacterial genus *Ochrobactrum* (Scholz *et al.*, 2008b, Al Dahouk *et al.*, 2010). However, genome sequencing, surprisingly revealed 99.4% identity compared with *B. suis* 1330 in aligned regions (Audic *et al.*, 2009). Only a limited number of *B. microti* specific genes were identified and additional *Ochrobactrum*-related genes could not be detected. Thus, the basis for the differences in growth behaviour remains to be elucidated. *B. microti* will therefore be an interesting model organism to study regulatory effects on gene expression. An important outcome of genome analysis was the finding that the 23S rRNA gene was markedly different from all other *Brucella* species in terms of (i) an altered intervening sequence (IVS) and (ii) alterations in the large subunit (Audic *et al.*, 2010). Currently it is unknown whether these alterations have a positive effect on the translation efficiency which might explain the fast growth and high metabolic activity. Interestingly, another fast-growing *Brucella* species, *B. inopinata,* isolated in the USA (details see below) has an identical atypical IVS. This IVS can also be found in some other atypical and fast growing *Brucella* strains, isolated from rodents in Australia (Scholz, unpublished).

Recently, *B. microti* was shown to replicate in human macrophages and in human and murine macrophage-like cells (Jiménez de Bagüés *et al.*, 2010). In contrast to other species of *Brucella*, *B. microti* is highly virulent in murine models of infection in which 10^5 colony-forming units of *B. microti* killed 82% of Balb/c mice within 7 days (Jiménez de Bagüés *et al.*, 2010). No human infections with *B. microti* have been reported so far.

Brucella inopinata

Brucella inopinata (strain BO1) was isolated from an inflamed breast implant of a 71-year-old patient in the USA with clinical sings of brucellosis and was originally described as a *Brucella*-like organism (De *et al.*, 2008). By applying a polyphasic approach this *Brucella*-like organism could be clearly classified as a novel *Brucella* species (Scholz *et al.*, 2010). Because of its unexpected isolation, the name *Brucella inopinata* (in.o.pi.na`ta. L. fem. adj. inopinata, unexpected) was given.

Like *B. microti*, *B. inopinata* is characterized by rapid growth and high metabolic activity. But unlike the classical *Brucella* species, including *B. microti*, *B. inopinata* is characterized by a unique 16S rRNA gene sequence differing in five nucleotides when compared with the 16S rRNA consensus sequence of all other *Brucella* species (Scholz *et al.*, 2010). *B. inopinata* also exhibits a significantly lower level of sequence similarity in various housekeeping genes (De *et al.*, 2008) compared with other *Brucella* species. Consequently it is the most divergent species within the genus *Brucella*. Although *B. inopinata* represents the most diverse species it shares a DNA–DNA relatedness of >80% with *B. melitensis* and therefore would still belong to the same species if the commonly accepted borderline for bacterial species delineation would be applied. Currently this species consists of a single strain.

Other atypical *Brucella* strains and potential novel species

For unknown reasons atypical isolates have historically been attributed to *B. suis*, although none of these had been isolated from swine. Some of these isolates, isolated from rodents in Australia, originally classified as *B. suis* biovar 3, were recently re-classified as a potential novel *Brucella* species (see below). In several molecular analyses, *B. suis* biovar 5 appears only distantly related to the other *B. suis* biovars and certainly needs to be re-classified.

Beside these historically falsely classified isolates, more recently various *Brucella* researches repeatedly reported the occurrence of new atypical *Brucella* isolates from different hosts that could not be attributed to one of the existing species.

One possible reason for this recent emergence of new isolates is the improvement of molecular identification and typing techniques combined with a better awareness of the existence of atypical *Brucellae* in unexpected hosts or in the environment.

The following text passages summarize what is currently known about the presence of potential novel *Brucella* species.

An unusual *Brucella* strain (BO2) was isolated from a lung biopsy in a 52-year-old patient from Australia with chronic destructive pneumonia (Tiller *et al.*, 2010a). Phenotypical- and molecular analyses placed strain BO2 in close proximity to *B. inopinata*. Hence, the BO2 strain probably represents a novel lineage of *B. inopinata* (Tiller *et al.*, 2010a). However, extended molecular analyses are necessary to confirm this conclusion.

Seven strains, isolated in 1964 from three native rodent species in North Queensland, Australia, and previously reported as *B. suis* biovar 3, were reinvestigated for their exact classification (Tiller *et al.*, 2010b). Molecular analysis indicated that the seven strains are unique and distinct from all known *Brucella* species and therefore might represent a novel species. Molecular data also indicate that the seven strains are related to *B. inopinata* and strain BO2 (Tiller *et al.*, 2010b).

Atypical *Brucella* strains were recently isolated for the first time from non-human primates in association with two cases of stillbirth (Schlabritz-Loutsevitch *et al.*, 2009). Preliminary phenotypical and molecular analyses indicate that the two isolates are different from all known *Brucella* species (Schlabritz-Loutsevitch *et al.*, 2009).

Novel atypical *Brucella* strains could also be isolated from red foxes in Austria (Scholz, Hofer, unpublished). These strains are slowly growing and exhibit a large number of IS*711* elements comparable to *B. ceti* and *B. pinnipedialis* (Scholz, Cloeckaert, unpublished). They show very low metabolic activity using the Taxa Profile™ assay to measure substrate utilization and do not cluster with any known *Brucella* species (Al Dahouk *et al.*, 2010). Deeper molecular analysis by MLST and MLVA confirmed their separate position within the genus *Brucella* (Scholz, Whatmore, Vergnaud, unpublished).

A potential novel *Brucella* species has been isolated from African bullfrogs In contrast to all other *Brucella* species, the isolates are motile and equipped with a single laterally attached flagellum as it has been described for *O. anthropi* (Eisenberg, Scholz, Noeckler, unpublished). Genome sequencing confirmed that it belongs to the genus *Brucella*. However, various additional genes related to genes of soil bacteria, i.e. *Agrobacterium* could be identified, of which most encode ABC transporters (Scholz, Binnewies, unpublished).

Remarks

*According to Rule 51b of the International Code of Nomenclature of Bacteria, the order name *Rhizobiales* Kuykendall 2006 is illegitimate because it contains the genus *Hyphomicrobium* (Stutzer and Hartleb, 1899; Approved Lists 1980) which is the type of the order *Hyphomicrobiales* (Douglas, 1957; Approved Lists 1980). For details see LISTS EDITOR, IJSEM: Validation list n 107 (footnote). Int. J. Syst. Evol. Microbiol., 2006, *56*, 1–6.

+The use of the term 'atypical' is used in the context of the delineation of members of the genus *Brucella* that dates back to the minimal standards paper of Corbel and Brinely Morgan (1975). However, the discovery of additional isolates that may be considered to be members of the genus *Brucella* that do not conform that definition as well as the discovery of novel taxa that bear and evolutionary relationship to members of this genus (e.g. *Ochrobactrum, Paenochrobactrum* and *Pseudochrobactrum)* clearly indicate the need for an in depth, impartial evaluation of the facts, both in relation to the classical species and new species of the genus *Brucella* as well as members of 'closely related' genera. This also calls for a re-examination of both the taxonomy and nomenclature of the organisms concerned with a view to providing future stability preferably on the basis of the recommendations outlined by Tindall *et al.* (2010). This is in the interest of science in general as well as in the interest of those with an interest in solving longstanding taxonomic/nomenclatural problems with a group of organisms that were of concern to the WHO 60 years ago, and on who's initiative the ICSP subcommittee on the taxonomy of the genus *Brucella* was originally established.

Acknowledgements

We would like to thank Dr. Brian J. Tindall, DSMZ Braunschweig, for comments relating to the interrelationship between taxonomy, nomenclature and the Bacteriological Code.

References

Al Dahouk, S., Scholz, H.C., Tomaso, H., Bahn, P., Goellner, C., Karges, W., Appel, B., Hensel, A., Neubauer, H., and Noeckler, K. (2010). Differential phenotyping of *Brucella* species using a newly developed semi-automated metabolic system. BMC Microbiol. *23*, 10, 269.

Alsmark, C.M, Frank, A.C., Karlberg, E.O., Legault, B.A., Ardell, D.H., Canbäck, B., Eriksson, A.S., Näslund, A.K., Handley, S.A., Huvet, M., *et al.* (2002). The louse-borne human pathogen *Bartonella quintana* is a genomic derivative of the zoonotic agent *Bartonella henselae*. Int. J. Med. Microbiol. *291*, 555–560.

Arimi, S.M., Koroti, E., Kang'ethe, E.K., Omore, A.O., and McDermott. J.J. (2005). Risk of infection with *Brucella abortus* and *Escherichia coli* O157, H7 associated with marketing of unpasteurized milk in Kenya. Acta Trop. *96*, 1–8.

Audic, S., Lescot, M., Claverie, J.M., and Scholz, H.C. (2009). *Brucella microti*: the genome sequence of an emerging pathogen. BMC Genomics *4*, 10, 352.

Batut, J., Andersson, S.G., and O'Callaghan, D. (2004). The evolution of chronic infection strategies in the alpha-proteobacteria. Nat. Rev. Microbiol. *2*, 933–945.

Birtles, R.J, and Raoult, D. (1996). Comparison of partial citrate synthase gene (*gltA*) sequences for phylogenetic analysis of *Bartonella* species. Int. J. Syst. Bacteriol. *46*, 891–897.

Bohlin, J., Snipen, L., Cloeckaert, A., Lagesen, K., Ussery, D., Kristoffersen, A.B., and Godfroid, J. (2010). Genomic comparisons of *Brucella* spp. and closely related bacteria using base compositional and proteome based methods. BMC Evol. Biol. *13*, 10, 249.

Bottone, E.J. (1998). *Yersinia enterocolitica* infections. In Infectious diseases, 2nd ed., Gorbach, S.L., Bartlett, J.G, Blacklow, N.R, eds. (WB Saunders Company: Philadelphia), p. 733.

Bourg, G., O'Callaghan, D., and Boschiroli, M.L. (2007). The genomic structure of Brucella strains isolated from marine mammals gives clues to evolutionary history within the genus. Vet. Microbiol. *125*, 375–380.

Brenner, D.J., O'Connor, S.P., Winkler, H.H., and Steigerwalt, A.G. (1993). Proposals to unify the genera *Bartonella* and *Rochalimaea,* with descriptions of *Bartonella quintana* comb. nov., *Bartonella vinsonii* comb. nov., *Bartonella henselae* comb. nov., and *Bartonella elizabethae* comb. nov., and to remove the family *Bartonellaceae* from the order *Rickettsiales*. Int. J. Syst. Bacteriol. 43, 777–786.

Buddle, M.B. (1956). Studies on *Brucella ovis n.* sp., a cause of genital disease of sheep in New Zealand and Australia. J. Hyg. *54*, 351–364.

Carmichael, L.E., and Brunner, D.W. (1968). Characteristics of a newly recognized species of *Brucella* responsible for infectious canine abortions. Cornell Vet. *58*, 579–592.

Cloeckaert, A., Tibor, A., and Zygmunt, M.S. (1999). *Brucella* outer membrane lipoproteins share antigenic determinants with bacteria of the family *Rhizobiaceae*. Clin. Diagn. Lab. Immunol. *6*, 627–629.

Cloeckaert, A., Verger, J.M., Grayon, M., Paquet, J.Y., Garin-Bastuji, B., Foster, G., and Godfroid, J. (2001). Classification of *Brucella* spp. isolated from marine mammals by DNA polymorphism at the omp2 locus. Microbes Infect. *3*, 729–738.

Corbel, M.J. (1988). Recommendation on a change in taxonomic opinion on genus *Brucella* as being monospecific, harbouring *Brucella melitenis* as the single species with 18 biovars. Int. J. Syst. Bact. *38*, 450–452.

Corbel, M.J. and Brinley Morgan, W.J. (1975). Proposal for minimal standards for descriptions of new species and biotypes of the genus *Brucella*. Int. J. Syst. Bacteriol. *25*, 83–89.

Crawford, R.P., Huber, J.D., and Adams, B.S. (1990). Epidemiology and surveillance. In Animal brucellosis, Klaus, N., and Robert, D., eds. (CRC Press: Boca Raton, FL), pp. 131–151.

Dawson, C.E., Stubberfield, E.J., Perrett, L.L., King, A.C., Whatmore, A.M, Bashiruddin, J.B., Stack, J.A., and Macmillan, A.P. (2008). Phenotypic and molecular characterisation of *Brucella* isolates from marine mammals. BMC Microbiol. *17*, 8, 224.

De, B.K., Stauffer, L., Koylass, M.S., Sharp, S.E., Gee, J.E., Helsel, L.O., Steigerwalt, A.G., Vega, R., Clark, T.A., Daneshvar, M.I., *et al.* (2008). Novel *Brucella* strain (BO1) associated with a prosthetic breast implant infection. J. Clin. Microbiol. *46*, 43–49.

De Ley, J., Mannheim, W., Segers, P., Lievens, A., Denijn, M., Vanhoucke, M., and Gillis, M. (1987). Ribosomal ribonucleic acidcistron similarities and taxonomic neighborhood of *Brucella* andCDC group Vd. Int. J. Syst. Bacteriol. *37*, 3542.

Ewalt, D.R., Payeur, J.B., Martin, B.M., Cummins, D.R., Miller, W.G. (1994). Characteristics of a *Brucella* species from a bottlenose dolphin (Tursiops truncatus). J. Vet. Diagn. Invest. *6*, 448–452.

Ficht, T. (2010). *Brucella* taxonomy and evolution. Future Microbiol. *5*, 859–866.

Foster, G., Osterman, B., Godfroid, J., Jacques, I., and Cloeckaert, A. (2007). *Brucella ceti* sp. nov. and *Brucella pinnipedialis* sp. nov. for *Brucella* strains with cetaceans and seals as their preferred hosts. Int. Syst. Evol. Microbiol. *57*, 2688–2693.

Gargani, G., and López-Merino, A. (2006). International Committee on Systematic Bacteriology; Subcommittee on the taxonomy of *Brucella*: Correspondence Report (Interim Report), 1991–1993. Int. J. Syst. Evol. Microbiol. *56*, 1167–1168.

Groussaud, P., Shankster, S.J., Koylass, M.S., and Whatmore, A.M. (2007). Molecular typing divides marine mammal strains of *Brucella* into at least three groups with distinct host preferences. J. Med. Microbiol. *56*, 1512–1518.

Guzmán-Verri, C., Manterola, L., Sola-Landa, A., Parra, A., Cloeckaert, A., Garin, J., Gorvel, J.P., Moriyón, I., Moreno, E., and López-Goñi, I. (2002). The two-component system BvrR/BvrS essential for *Brucella abortus* virulence regulates the expression of outer membrane proteins with counterparts in members of the *Rhizobiaceae*. Proc. Natl. Acad. Sci. U.S.A. *99*, 12375–12380.

Harwood, C.S., Parales, R.E. (1996). The beta-ketoadipate pathway and the biology of self-identity. Annu. Rev. Microbiol. *50*, 553–590.

Holmes, B., Popoff, M., Kiredjian, M., and Kersters, K. (1988). *Ochrobactrum anthropi* gen. nov., sp. nov. from human clinical specimens and previously known as Group Vd. Int. J. Syst. Bacteriol. *38*, 406–416.

Hoyer, B.H., and McCullough, N.B. (1968a). Polynucleotide homologies of *Brucella* deoxyribonucleic acids. J. Bacteriol. *95*, 444–448.

Hoyer, B.H., and McCullough, N.B. (1968b). Homologies of deoxyribonucleic acids from *Brucella ovis*, canine abortion organisms, and other *Brucella* species. J. Bacteriol. *96*, 1783–1790.

Hubalek, Z., Scholz, H.C., Sedlaceck, I., Melzer, F., Sanogo, Y. and Nesvadbova, J. (2007). Brucellosis of the Common Vole (*Microtus arvalis*). Vector Borne Zoonotic Dis. *7*, 679–687.

Huddleson, I.F. (1929). The differentiation of the species of the genus *Brucella*. Michigan State College Agricultural Experimental Station Technical Bulletin *100*, 1–16.

Jacques, I., Grayon, M., and Verger, J.M. (2007). Oxidative metabolic profiles of *Brucella* strains isolated from marine mammals: contribution to their species classification. FEMS Microbiol. Lett. *270*, 245–249.

Jahans, K.L., Foster, G., and Broughton, E.S. (1997). The characterisation of *Brucella* strains isolated from marine mammals. Vet. Microbiol. *57*, 373–382.

Jiménez de Bagüés, M.P., Ouahrani-Bettache, S., Quintana, J.F., Mitjana, O., Hanna, N., Bessoles, S., Sanchez, F., Scholz, H.C., Lafont, V., Köhler, S., *et al.* (2010). The new species *Brucella microti* replicates in macrophages and causes death in murine models of infection. J. Infect. Dis. *202*, 3–10.

McDonald, W.L., Jamaludin, R., Mackereth, G., Hansen, M., Humphrey, S., Short, P., Taylor, T., Swingler, J., Dawson, C.E., Whatmore, A.M., *et al.* (2006). Characterization of a *Brucella* sp. strain as a marine-mammal type despite isolation from a patient with spinal osteomyelitis in New Zealand. J. Clin. Microbiol. *44*, 4363–4370.

Maquart, M., Le Flèche, P., Foster, G., Tryland, M., Ramisse, F., Djønne, B., Al Dahouk, S., Jacques, I., Neubauer, H., Walravens, K., *et al.* (2009). MLVA-16 typing of 295 marine mammal *Brucella* isolates from different animal and geographic origins identifies

7 major groups within *Brucella ceti* and *Brucella pinnipedialis*. BMC Microbiol. *20, 9,* 145.

Mayer-Scholl, A., Draeger, A., Göllner, C., Scholz, H.C., and Nöckler, K. (2010). Advancement of a multiplex PCR for the differentiation of all currently described *Brucella* species. J. Microbiol. Methods. *80,* 112–114.

Meyer, K.F., and Shaw, E.B. (1920). A comparison of the morphologic, cultural and biochemical characteristics of *B. abortus* and *B. melitensis* from cattle. Studies on the genus *Brucella* nov. gen. J. Inf. Dis. *27,* 173–184.

Moreno E. (1998). Genome evolution within the alpha Proteobacteria: why do some bacteria not possess plasmids and others exhibit more than one different chromosome? FEMS Microbiol. Rev. *22,* 255–275.

O'Callaghan, D., Cazevieille, C., Allardet-Servent, A., Boschiroli, M. L., Bourg, G., Foulongne, V., Frutos, P., Kulakov, Y., and Ramuz, M. (1999). A homologue of the Agrobacterium tumefaciens VirB and Bordetella pertussis Ptl type IV secretion systems is essential for intracellular survival of *Brucella suis.* Mol. Microbiol. *33,* 1210–1220.

O'Connor, S.P., Dorsch, M., Steigerwalt, A.G., Brenner, D.J., and Stackebrandt, E. (1991). 16S rRNA sequences of *Bartonella bacilliformis* and cat scratch disease bacillus reveal phylogenetic relationships with the alpha-2 subgroup of the class *Proteobacteria*. J. Clin. Microbiol. *29,* 2144–2150.

Osterman, B., and Moriyón, I. (2006). International Committee on Systematics of Prokaryotes Subcommittee on the taxonomy of *Brucella.* Minutes of the meeting, 17 September 2003, Pamplona, Spain. Int. J. Syst. Evol. Microbiol. *56,* 1173–1175.

Paulsen, I.T., Seshadri, R., Nelson, K.E., Eisen, J.A., Heidelberg, J.F., Read, T.D., Dodson, R.J., Umayam, L., Brinkac, L.M., Beanan, M.J., *et al.* (2002). The *Brucella suis* genome reveals fundamental similarities between animal and plant pathogens and symbionts. Proc. Natl. Acad. Sci. U.S.A. *99,* 13148–13153.

Schlabritz-Loutsevitch, N.E., Whatmore, A.M., Quance, C.R., Koylass, M.S., Cummins, L.B., Dick, E.J. Jr., Snider, C.L., Cappelli, D., Ebersole, J.L., Nathanielsz, P.W., *et al.* (2009). A novel *Brucella* isolate in association with two cases of stillbirth in non-human primates – first report. J. Med. Primatol. *38,* 70–73.

Scholz, H.C., Al Dahouk, S., Tomaso, H., Neubauer, H., Witte, A., Schloter, M., Kämpfer, P., Falsen, E., Pfeffer, M., and Engel, M. (2008a). Genetic diversity and phylogenetic relationships of bacteria belonging to the *Ochrobactrum-Brucella* group by *recA* and 16S rRNA gene-based comparative sequence analysis. Syst. Appl. Microbiol. 31, 1–16.

Scholz, H.C., Hubalek, Z., Sedlácek, I., Vergnaud, G., Tomaso, H., Al Dahouk, S., Melzer, F., Kämpfer, P., Neubauer, H., *et al.* (2008b). *Brucella microti* sp. nov., isolated from the common vole, *Microtus arvalis*. Int. J. Syst. Evol. Microbiol. *58,* 375–382.

Scholz, H.C., Hubalek, Z., Nesvadbova, J., Tomaso, H., Vergnaud, G., Le Flèche, P., Whatmore, A. M., Al Dahouk, S., Krüger, M., Lodri, C., *et al.* (2008c). Isolation of *Brucella microti* from soil. Emerg. Infect. Dis. *14,* 1316–1317.

Scholz, H.C., Hofer E., Vergnaud, G., Le Fleche, P., Whatmore, A.M., Al Dahouk, S., Pfeffer M, Krüger M, Cloeckaert A, and Tomaso H. (2009). Isolation of *Brucella microti* from mandibular lymph nodes of red foxes, Vulpes vulpes, in lower Austria. Vector Borne Zoonotic Dis. *2,* 153–156.

Scholz, H.C., Nöckler, K., Göllner, C., Bahn, P., Vergnaud, G., Tomaso, H., Al Dahouk, S., Kämpfer, P., Cloeckaert, A., Maquart, M., *et al.* (2010). *Brucella inopinata* sp. nov., isolated from a breast implant infection. Int. J. Syst. Evol. Microbiol. *60,* 801–808.

Seubert, A., Schulein, R., and Dehio, C. (2002). Bacterial persistence within erythrocytes: a unique pathogenic strategy of *Bartonella* spp. Vet. Microbiol. *90,* 329–339.

Sohn, A.H., Probert, W.S., Glaser, C.A., Gupta, N., Bollen, A.W., Wong, J.D., Grace, E.M., and McDonald, W.C. (2003). Human neurobrucellosis with intracerebral granuloma caused by a marine mammal *Brucella* spp. Emerg. Infect. Dis. *9,* 485–488.

Stoenner, H.G., and Lackman, D.B. (1957). A new species of *Brucella* isolated from the desert wood rat, *Neotoma lepida* Thomas. Am. J. Vet. Res. *18,* 947–951.

Swai, E.S., and Schoonman, L. (2009). Human brucellosis: seroprevalence and risk factors related to high risk occupational groups in Tanga municipality, Tanzania. Zoonoses Publ. Health *56,* 183–187.

Sweger, D., Resto-Ruiz, S., Johnson, D.P., Schmiederer, M., Hawke, N., and Anderson, B. (2000). Conservation of the 17-kilodalton antigen gene within the genus *Bartonella*. Clin. Diagn. Lab. Immunol. *7,* 251–257.

Tiller, R.V., Gee, J.E., Lonsway, D.R., Gribble, S., Bell, S.C., Jennison, A.V., Bates, J., Coulter, C., Hoffmaster, A.R., and De, B.K. (2010). Identification of an unusual *Brucella* strain (BO2) from a lung biopsy in a 52 year-old patient with chronic destructive pneumonia. BMC Microbiol. *27,* 10, 23.

Tiller, R.V., Gee, J.E., Frace, M.A., Taylor, T.K., Setubal, J.C., Hoffmaster, A.R., De, B.K. (2010b). Characterization of novel *Brucella* strains originating from wild native rodent species in North Queensland, Australia. Appl. Environ. Microbiol. *76,* 5837–5845.

Tindall, B.J., Rosselló-Móra, R., Busse, H.J., Ludwig, W., and Kämpfer, P. (2010) Notes on the characterization of prokaryote strains for taxonomic purposes. Int. J. Syst. Evol. Microbiol. *60,* 249–266.

Velasco, J., Romero, C., López-Goñi, I., Leiva, J., Díaz, R., and Moriyón, I. (1998). Evaluation of the relatedness of *Brucella* spp. and *Ochrobactrum anthropi* and description of Ochrobactrum intermedium sp. nov., a new species with a closer relationship to *Brucella* spp. Int. J. Syst. Bacteriol. 48, 759–768.

Verger, J.M., Due´e, J.P., and Grayon, M. (1982). *Brucella* isolates in France: evaluation of ten years of typing. Ann. Microbiol. (Paris) *133,* 433–447.

Verger, J.M., Grimont, F., Grimont, P.A.D., and Grayon, M. (1985). *Brucella,* a monospecific genus as shown by deoxyribonucleic acid hybridization. Int. J. Syst. Bacteriol. 35, 292–295.

Whatmore, A.M. (2009). Current understanding of the genetic diversity of *Brucella,* an expanding genus of zoonotic pathogens. Infect. Genet. Evol. 9, 1168–1184.

Whatmore, A.M., Dawson, C.E., Groussaud, P., Koylass, M.S., King, A.C., Shankster, S.J., Sohn, A.H., Probert, W.S., and McDonald, W.L. (2008). Marine mammal *Brucella* genotype associated with zoonotic infection. Emerg. Infect. Dis. *14*, 517–518.

Williams, K.P., Sobral, B.W., and Dickerman, A.W. (2007). A robust species tree for the alphaproteobacteria. J. Bacteriol. *189*, 4578–4586.

Zygmunt, M.S., Maquart, M., Bernardet, N., Doublet, B., and Cloeckaert, A. (2010). Novel IS711-specific chromosomal locations useful for identification and classification of marine mammal *Brucella* strains. J. Clin. Microbiol. *48*, 3765–3769.

Comparative Genomics and Phylogenomics of *Brucella*

2

Bruno W. Sobral and Alice R. Wattam

Abstract

Brucella species are characterized by extremely high levels of nucleotide similarity and yet vary in microbial and disease phenotypes, as well as in pathogenicity and host preference. These variations initially resulted in classification of six species; *B. abortus, B. canis, B. melitensis, B. neotomae, B. ovis* and *B. suis*. The lack of sequence diversity has inhibited molecular studies, but the development of new techniques and the recent availability of genome sequences have revealed interesting differences, including the expansion of the known *Brucella* species.

Recently, the *Brucella* genus has expanded by the discovery of four new species, *B. ceti* and *B. pinnipedialis* from marine mammals, *B. inopinata* from new human isolates, and *B. microti*, isolated from a rodent in the Czech Republic. An as yet unnamed isolate from Australian rodents might also be classified as a new species. This review summarizes the current state of knowledge of the genetic diversity within *Brucella*, including both the classical and new species, with particular emphasis on comparing the genome sequences, the phylogenies produced by a variety of methods and specific families of genes in particular functions and pathways.

The genome sequences

The first sequences: *B. melitensis* 16M and *B. suis* 1330

In 2002, two *Brucella* genomes were sequenced and analysed. The genome of *B. melitensis* 16M, which infects goats and sheep, was the first sequence and provided information on the basic nature of the organism with some comparisons to *Sinorhizobium meliloti* (DelVecchio *et al.*, 2002). *B. melitensis* 16M contains 3,294,935 bp distributed over two circular chromosomes of 2,117,144 bp and 1,177,787 bp with a 57% GC content; 3197 ORFs were identified, and several genes, operons and pathways of interest were noted, among them genes necessary to produce a Type IV secretion system and, despite the fact *Brucella* are immobile, they contain flagellar genes. They also noted that *B. melitensis* 16M has the genes enabling it to use erythritol, a key sugar produced by the placental of pregnant hosts (Smith *et al.*, 1962). Previously, a correlation between erythritol oxidation and virulence in different animal models had been suggested (Meyer, 1966).

The opportunity to compare 16M to another *Brucella* genome came with the release of the *B. suis* 1330 genome later that same year (Paulsen *et al.*, 2002). *B. suis* is typically isolated from swine and strain 1330 is the standard reference strain for *B. suis* biovar 1 (Morgan and Corbel, 1976). The *B. suis* 1330 genome was found to consist of two circular chromosomes of 2,107,792 bp (Chr I) and 1,207,381 bp (Chr II). A total of 2185 and 1203 ORFs were identified on Chr I and II, respectively. They noted differences between the genes found on the two chromosomes, with Chr I encoding the majority of the core metabolic machinery for processes such as transcription, translation, and protein synthesis. Chr II appeared to have genes more relevant to cellular processes and plasmid functions, primarily because of the presence of three clusters of flagellar biosynthesis and secretion genes and the clusters of conjugation-associated and plasmid-like replication genes.

The comparison of *B. melitensis* 16M and *B. suis* 1330 genomes showed extensive similarity and gene synteny between the two genomes, with more than 90% of the genes sharing 98–100% identity (Paulsen *et al.*, 2002). Genes with more variability were the hypothetical genes and the genes thought to be surface-exposed proteins (Paulsen *et al.*, 2002). There were also regions unique to each genome ranging in size from 108 to 20,883 bp that seemed to have entered as phage-mediated integration events. Some of these regions encoded proteins predicted to be exposed at the bacterial surface and it was suggested that these areas might contain genes responsible for the differing degrees of host pathogenicity seen among the different species, and also might contain the genetic basis for host preference. They also found a total of 7307 SNPs within a shared backbone of 3,237,820 bp, with the level of intragenic SNPs being lower in *Brucella* than described in comparisons within other bacterial species at the time, further substantiating the high degree of genomic similarity between these two *Brucella* species. A number of known and suggested virulence factors were also identified in the two genomes, including urease gene clusters.

Three *B. abortus* genomes

Three years later, two *B. abortus* genome sequences were released (9-941 and 2308) providing information for three of the pathogenic species of the genus. Each additional sequence allowed for interspecific as well as intraspecific comparisons. *Brucella abortus* 9-941 was originally isolated from an infected cattle herd in north-western Wyoming and is a member of biovar 1 (Halling *et al.*, 2005). The genome was found to be 3.3 Mb and composed of two circular chromosomes of 2,124,242 (Chr I) and 1,162,780 bp (Chr II). The G+C contents of Chr I and Chr II are 57.2 and 57.3%, respectively, identical to that found for the two chromosomes of both *B. suis* 1330 and *B. melitensis* 16M. The *B. abortus* 9-941 genome was annotated as having 3296 ORFs annotated as genes, 2158 on Chr I and 1138 on Chr II, similar to finding for the other genomes (*B. suis* with 3388 and *B. melitensis* with 3197). It was noted that differences in the number of ORFs found among the three *Brucella* genomes derived primarily from differences in annotation of short ORFs and from large insertions–deletions (indels). Similarity among the *Brucella* orthologues was high, with an average amino acid sequence identity of greater than 99% (Halling *et al.*, 2005). What initially looked to be unique to *B. melitensis* and *B. suis* was now viewed in a new light, with many of the fragments present in *B. abortus* 9-941 (Halling *et al.*, 2005).

The next *B. abortus* genome sequence came from was strain 2308, a standard laboratory biovar I strain virulent for humans, cattle, and certain other domestic animals (Chain *et al.*, 2005). Like all other *Brucella* genomes sequenced before it, the genome of *B. abortus* strain 2308 consisted of two circular chromosomes, Chr I (2,121,359 bp) and Chr II (1,156,950 bp), coding for 2280 and 1214 annotated open reading frames. Comparisons of *B. melitensis* 16M, *B. suis* 1330 and *B. abortus* 2308 showed an average of >94% identity at the nucleotide level. Comparisons between the 9-941 and 2308 *B. abortus* showed that these two strains are very close, with the exception of a small number of strain-specific deletions and polymorphisms (Chain *et al.*, 2005).

A 25-kb sequence that had previously been identified as missing in *B. abortus* 544, the biovar 1 type strain (Vizcaíno *et al.*, 2000, 2001), shown to encode a number of ORFs that may be involved in polysaccharide synthesis (Halling *et al.*, 2005), was also found to be missing in both *B. abortus* genomes. In addition, a *B. suis*-specific fragment was found to be missing from the two *B. abortus* genomes and also from *B. melitensis* 16M. This 2774-bp fragment encoding a predicted surface protein and two partial ORFs with homology to the insertion sequences IS711 and ISBm1. Chain *et al.* (2005) noted that this fragment contained an amino acid ABC-type transporter and a glutathione *S*-transferase and they concluded that this amino acid transporter is probably involved in the transport of arginine, ornithine, and lysine, and its deletion may be responsible for the observed inability of *B. melitensis* and *B. abortus* to oxidize these amino acids.

B. abortus S19 was the first vaccine strain to be sequenced (Crasta *et al.*, 2008). S19 is a spontaneously attenuated vaccine, and underlying molecular mechanisms that cause the loss of virulence are not well understood (Crasta *et al.*, 2008). The 3.2 Mb genome was in two chromosomes with one of 2,122,487 bp and the second of 1,161,449 bp, had 3062 ORFs. The natural comparison between this strain and the two virulent *B. abortus* sequences (2308 and 9-941) revealed that more than 96% of the ORFs were identical to the ORFs of 9-941 and 2308. There were only 263 ORFs that had less than 100% homology between S19 and both 9-941 and 2308. Of these, 45 had the same differences between S19 and the other two genomes, with these genes being identical in the two virulent strains (Crasta *et al.*, 2008). The authors described these proteins as having 'major virulence-associated differences' and one protein in this group was noted as having structural similarity to virulence factors YadA of *Yersinia enterocolitica* and the Hia adhesins from *Haemophilus influenzae* (Crasta *et al.*, 2008).

The *B. ovis* genome

Brucella ovis is a rough lipopolysaccharide (LPS) isolate that retains virulence and was recently sequenced (Tsolis *et al.*, 2009). Associated with epididymitis and abortion in sheep (Buddle, 1956), it is not a human zoonotic. The genome of this strain (3.29 Mb) has two circular chromosomes of 2,111,370 bp and 1,164,220 bp that were predicted to encode a total of 2890 proteins. The analysis revealed a high percentage of pseudogenes and increased numbers of transposable elements compared with the *Brucella* species that can cause human disease. The authors suggested that genome degradation has occurred concomitant with narrowing of the host range of *B. ovis*.

Tsolis *et al.* (2009) made a detailed effort to link *Brucella* phenotypes to underlying biochemistry. They noted a study showing that the four *Brucella* species capable of causing human disease are capable of the oxidative metabolism of arabinose, galactose, ribose, xylose, glucose and erythritol, while *B. ovis* is not (Meyer, 1969). Analysis of the *B. ovis* genome showed that two genes important in erythritol oxidation, *eryA* and *eryD*, appear to be inactivated by mutation (Tsolis *et al.*, 2009). Two genes within in the erythritol operon, *eryB* and *eryC*, have been found to be essential for growth inside the macrophage, although erythritol has not been described in human macrophages (Kohler *et al.*, 2002). Tsolis *et al.* (2009) suggested that these two mutations in *B. ovis*, as well as nonsense codons in *eryF* and *eryG*, might explain its reduced virulence. In addition, it was also noted that despite the fact that *B. ovis* is urease negative, it contains the two urease clusters that all *Brucella* genomes have (Tsolis *et al.*, 2009). This may play a part in transmission, as urease has been found to be important for other *Brucella* species to survive passage through the stomach in the mouse model of infection (Bandara *et al.*, 2007; Sangari *et al.*, 2007). Oral transmission is not the main infection route for *B. ovis* (Blasco, 1990) and the fact that it is urease-negative may explain this (Tsolis *et al.*, 2009). A detailed examination of this operon in all *Brucella* genomes is warranted to see whether these specific differences are found in related genomes.

The *B. microti* genome

A strain of *Brucella* isolated during an epizootic in the Czech Republic (Hubalek *et al.*, 2007) from the common vole (*Microtus arvalis*) has been described as a new species (Scholz *et al.*, 2008a) and recently sequenced (Audic *et al.*, 2009). Isolates of this new species (CCM 4915 and CCM 4916) were clearly distinguished from all known *Brucella* species and their biovars by means of both their phenotypic and molecular properties (Scholz *et al.*, 2008a), thus validating the new species designation. A recent finding shows that *B. microti* has an enhanced capacity for intramacrophagic replication in human macrophages compared with that of *B. suis*, exhibiting a high pathogenic potential (Jiménez de Bagüés *et al.*, 2010).

An earlier phenotypic analysis (Scholz *et al.*, 2008A) suggested that a sequence from this species would show that it was closer to *Ochrobactrum* then to *Brucella*. Analysis of the newly sequenced *B. microti* genome showed that this was not the case (Audic *et al.*, 2009). The genome sequence of *B. microti* was found to be highly similar to the known *Brucella* species, and more remotely related to *O. anthropi*. In fact, the overall sequence similarity between *B. microti* and *B. suis* 1330 was 99.84% (Audic *et al.*, 2009) with a notable difference being an 11.7 kb insert in the *microti* genome that appears to be a unique island of probable phage origin but with unknown functionality. This genome, with 3291 protein coding genes, is composed of 3.3 Mb over two chromosomes, 2117 and 1220 kb in size.

One similarity that *B. microti* had with *O. anthropi* was the intervening sequence (IVS) of the 23*S* rRNA. These are highly conserved among *Brucella* species and correspond to inverted DNA repeats characteristic of mobile genetic elements (Evguenieva-Hackenberg, 2005). The IVS of *B. microti* contains sequence similarity to both *O. anthropi* and to *B. inopinata* (Audic *et al.*, 2009), but in different areas. Neither of these is like the highly conserved IVS that the other *Brucella* species share. Fragmentation of 23S rRNA during post-transcriptional processing of precursor rRNA has been reported in bacteria (Hsu *et al.*, 1992) and it is speculated that in *B. microti* the 23S maturation leading to the IVS removal may not occur because of these sequence variations (Audic *et al.*, 2009).

It was noted that the biochemical profile of *B. microti* was more similar to *Ochrobactrum* than to *Brucella*, so to examine longevity of this new isolate in soil, samples were taken from the original collection site. This soil, stored at 4°C, was positive for *B. microti* seven years after the soil samples were taken (Scholz *et al.*, 2008c). This long-term survival in soil led these authors to hypothesize that soil could be a reservoir for this species. A previous study noted that *Brucella* could survive for up to 66 days in moist soil and up to 185 days in cold soil (Crawford *et al.*, 1990). It could be that all *Brucella* have this extended capability for soil survival and that previous studies neglected to lengthen the time of analysis, or it could be that *B. microti* is unique among the known *Brucella* in its ability to survive long term in this type of environment. Certainly survival and even thriving in soil is not unique to the other *Brucellaceae* genera, *Ochrobactrum* and *Mycoplana*.

New sequences, multiple genomes

A comparative analysis of 10 *Brucella* genomes, including new genomes *B. canis*, *B. ceti* Cudo, *B. melitensis* ATCC 23457, and *B. suis* ATCC 23445, was conducted recently (Wattam *et al.*, 2009). The *B. ceti* Cudo strain is an incomplete genome in seven contigs, and the new *suis* and *melitensis* genomes represent sequences from new biovars, both representing biovar 2 of their individual species. All genomes were re-annotated using the same method and retrieved from the Pathosystems Resource Integration Centre (PATRIC, Snyder *et al.*, 2007) prior to analysis to ensure uniformity.

Of the newly described genomes, *B. canis* had 3.3 Mb split over two chromosomes of 2105 and 1206 kb, with 3500 genes annotated. The *B. melitensis* ATCC 23457 genome had 3.31 Mb in two chromosomes of 2125 and 1185 kb with 3425 genes. The *B. suis* ATCC 23445 (total length 3.32 Mb) chromosomes had different lengths from what is normally found in *Brucella* and this was due to a 210 kb translocation from Chr I (1, 923 kb) to Chr II (1400 kb). The incomplete *B. ceti* Cudo genome, in seven contigs, was 3.31 Mb with 3501 genes. It was also found that three *abortus* genomes examined (9-941, 2308 and S19) all share the 640 bp inversion on the second chromosome compared with all other *Brucella* genomes previously identified in the physical mapping studies (Michaux Charachon *et al* 1997; Tsoktouridis *et al.*, 2003). It was also noted that several areas of the genome that contain proteins of interest, like the Type IV secretion system, *tra* genes, and genes involved in LPS synthesis appear to have been acquired horizontally.

Forty *Brucella* genomes

The Broad Institute has recently sequenced 25 *Brucella* genomes, which along with the pre-existing genomes provides a sequence from every biovar, as well as other isolates of interest (http://www.broadinstitute.org/annotation/genome/brucella_group/MultiHome.html). In addition, three new genomes were recently sequenced by the Centres for Disease Control and annotated by PATRIC (Snyder *et al.*, 2007). Two of these genomes were identified as *Brucella*-like infections taken from human patients. One was isolated from a breast implant wound in a woman with clinical symptoms consistent with brucellosis (De *et al.*, 2008) and the other from a lung biopsy of a patient with chronic destructive pneumonia (Tiller *et al.*, 2010a). Another unusual isolate came from an Australian rodent (Tiller *et al.*, 2010b). All of these genomes, as well as those previously sequenced, were re-annotated using the RAST system (Aziz *et al.*, 2008). These annotations are available at PATRIC (http://patricbrc.vbi.vt.edu/portal/portal/patric/Home) and a summary of this data is provided (Table 2.1). Included in the summary are the number of genes, CDSs, hypothetical proteins, proteins with functional assignments (in the SEED system), proteins with EC numbers and proteins with GO assignments. PATRIC also takes the annotated EC numbers and maps them to the KEGG system (Ogata *et al.*, 1999) and provides a summary of the number of KEGG pathways that the genome has proteins mapped to, as well as the number of proteins found in the pathway. The pathway summary is also provided as are statistics for each category (Table 2.1). The average size of the 40 *Brucella* genomes is 3.30 MB, the range being 3.11 (*Brucella* sp. NF2653) to 3.40 (*B. pinnipedialis* B2/94), the average genes annotated per genome is 3524, ranging from 3325 (*Brucella* sp. 83/13) to 3709 (*B. ceti* Cudo).

These data have been summarized on a per-clade basis (Table 2.2). Looking at the averages within the different clades, it is clear that they are all strikingly similar. Small differences, as in the percentage of hypothetical protein assignments per clade, can be explained by specific genomes that are of potential interest. For example, the *B. abortus* and *B. suis* clades both have an average 23% of their CDSs annotated as hypothetical. In the case of *B. abortus*, this is due to the strain 86/8/59 (biovar 2) and strain Tulya (biovar 3) genomes. Each of these genomes has almost 200 more hypothetical proteins annotated than are found among other members of this clade (Table 2.1). The 86/8/59 genome is the quite fractionated with 21 contigs, but the *B. abortus* strain B3196 (biovar 5) genome has 48 contigs and many fewer hypothetical proteins annotated (Table 2.1). There appears to be something unique about these two genomes that merit further exploration. Likewise, the *B. suis* clade both strain 686 (biovar 3) and strain ATCC 23445 have similar excesses in hypothetical proteins. Surprisingly, strain ATCC 23445 is a complete genome. All four of these genomes have more CDSs annotated than their other clade members and it would be interesting to explore these extra proteins.

Summarizing data across all genomes and across individual clades can indicate areas of potential interest. PATRIC takes the EC assignments, maps them to KEGG pathways and then provides a summary of how many pathways in the KEGG database have EC numbers assigned to particular genomes. Most of the genomes have CDSs in 132 pathways (Table 2.1). However, two clades, the *B. ovis* and the 83/13 clades, have only 130 pathways that have CDSs in them (Table 2.2) indicating a potential area worth exploring.

As yet, there has been no published analysis covering all 40 genomes, the closest being the comparison of 37 genomes by Bohlin *et al.* (2010) done by newly developed, and unproven methodology that appears to be at odds with all of the traditional methods that have been used to examine the *Brucella* to date. A variety of methods, uniquely developed by these authors, were used to make comparisons at the genomic DNA level and at the proteome level, with each of these methods giving a different phylogeny. There were no comparisons between these methods and the more traditional methods that have been used not only to examine evolutionary relationships among the bacteria, as well as animals, plants and all other groups. Some of these novel methods developed by Bohlin split the traditional *Brucella* groups based on host preference, groupings that almost every other study has confirmed. Some methods even separate the same genome that has been sequenced by two different methods, implying that the same sequence is not its own closest relative, phylogenetically speaking. The authors claim that the proteome methods most closely resemble traditionally phylogeny, yet even here the traditional groups are separated, most notably in the *ceti/pinnipedialis* clade. Other phylogenies by other authors have all shown that the marine *Brucella* always cluster together, although there is divergence within the clade. All of the methods developed by Bohlin break this clade, putting *B. pinnipedialis* M163/99/10 near the root of the *Brucella* tree. If one had faith in the methodology, one could speculate on what this meant in terms of the *Brucella* colonizing the marine mammals (something the authors fail to do), but as there is

Table 2.1 Summary of data from 40 *Brucella* genomes available at the Pathosystems Resource Integration Centre (PATRIC)

Clades	Genome	Contigs	Size (Mb)	PATRIC CDS	PATRIC Genes	Hypothetical proteins	Functional assignments	EC numbers	GO assignments	Number of Pathways	CDSs in All Pathways
abortus	*Brucella abortus* NCTC 8038	16	3276866	3439	3538	747	2692	875	647	132	759
	Brucella abortus S19	2	3283936	3476	3591	756	2720	879	650	132	761
	Brucella abortus str. 2308 A	9	3277197	3473	3580	752	2721	875	647	132	758
	Brucella abortus 2308	2	3278307	3467	3578	755	2712	877	647	132	760
	Brucella abortus bv. 2 str. 86/8/59	21	3294912	3574	3670	945	2629	862	636	131	746
	Brucella abortus bv. 1 str. 9-941	2	3286445	3476	3587	759	2717	881	651	132	763
	Brucella abortus bv. 4 str. 292	12	3272945	3429	3567	736	2693	874	648	132	759
	Brucella abortus bv. 9 str. C68	13	3269576	3406	3501	728	2678	871	644	132	757
	Brucella abortus bv. 5 str. B3196	48	3343441	3473	3567	766	2707	877	644	132	761
	Brucella abortus bv. 6 str. 870	13	3270344	3435	3488	747	2688	873	644	132	759
	Brucella abortus bv. 3 str. Tulya	13	3282764	3521	3614	914	2607	847	629	132	736
melitensis	*Brucella melitensis* bv. 1 str. 16M	2	3294931	3499	3598	740	2759	886	646	131	762
	Brucella melitensis bv. 1 str. 16M [Broad]	15	3291248	3538	3624	895	2643	857	631	131	743
	Brucella melitensis bv. 1 str. Rev.1	26	3305822	3440	3528	716	2724	878	647	131	763
	Brucella melitensis ATCC 23457	2	3311219	3486	3590	738	2748	879	644	132	758
	Brucella melitensis bv. 2 str. 63/9	60	3284951	3447	3533	728	2719	878	643	132	759
	Brucella melitensis bv. 3 str. Ether	13	3308792	3457	3548	731	2726	879	648	132	760
ovis	*Brucella ovis* ATCC 25840	2	3275590	3543	3656	780	2763	884	653	130	765
	Brucella sp. NVSL 07-0026	17	3297137	3531	3645	750	2781	905	663	130	778
ceti	*Brucella ceti* M490/95/1	17	3345471	3386	3460	706	2680	878	643	132	759
	Brucella ceti B1/94	14	3337159	3386	3465	706	2680	882	643	132	765

	Brucella sp. F5/99	18	3336716	3359	3433	697	2662	871	640	131	755
	Brucella ceti str. Cudo	7	3389269	3610	3709	757	2853	908	660	131	786
	Brucella pinnipedialis B2/94	19	3398674	3425	3502	725	2700	882	645	131	764
	Brucella pinnipedialis M292/94/1	15	3373519	3399	3470	710	2689	877	643	131	759
	Brucella pinnipedialis M163/99/10	89	3404050	3453	3541	744	2709	885	649	131	768
	Brucella ceti M644/93/1	17	3334689	3414	3496	719	2695	880	647	132	762
	Brucella ceti M13/05/1	22	3337230	3406	3491	717	2689	877	645	132	759
suis	*Brucella canis* ATCC 23365	2	3312769	3435	3523	725	2710	879	647	132	757
	Brucella suis bv. 4 str. 40	12	3307260	3388	3470	708	2680	870	643	131	751
	Brucella suis bv. 3 str. 686	23	3298529	3515	3593	916	2599	858	625	131	744
	Brucella suis 1330	2	3315175	3432	3521	728	2704	884	651	131	761
	Brucella suis ATCC 23445	2	3324607	3591	3681	952	2639	858	626	131	741
	Brucella suis bv. 5 str. 513	19	3323676	3377	3430	721	2656	870	639	131	752
	Brucella neotomae 5K33	11	3329623	3434	3509	745	2689	882	655	132	761
	Brucella microti CCM 4915	2	3337369	3410	3490	726	2684	874	641	132	752
83/13	*Brucella rodentia* NF2653	113	3110281	3167	3239	684	2483	823	606	130	703
	Brucella sp. 83/13	20	3153851	3167	3235	686	2481	823	612	130	704
BO	*Brucella inopinata* BO1	55	3366774	3330	3390	737	2593	841	623	131	721
	Brucella inopinata BO2	174	3305941	3251	3311	739	2512	833	613	130	717
Statistics across all individual genomes											
	Minimum		3110281	3167	3235	684	2481	823	606	130	703
	Maximum		3404050	3610	3709	952	2853	908	663	132	786
	Average		3306226.375	3436.125	3524.05	755.775	2680.35	871.8	641.45	131.375	753.7
	Median		3306600.5	3437	3530.5	737.5	2690.5	877	644	131.5	759

Table 2.2 Summary of clade-associated data

Totals and averages	Size (Mb)	PATRIC CDS	PATRIC genes	Hypothetical	Functional assignments	EC numbers	GO assignments	Number of pathways	CDSs in all pathways
Average across *B. abortus* clade	3285158	3469.9	3571	782.3	2687.6	871.9	644.3	131.9	756.3
% of average CDS	–	–	–	23	77	25	19	–	22
Average across *B. melitensis* clade	3299494	3477.8	3570.2	758	2719.8	876.2	643.2	131.5	757.5
% of average CDS	–	–	–	22	78	25	18	–	22
Average across *B. ovis* clade	3286364	3537	3650.5	765	2772	894.5	658	130	771.5
% of average CDS	–	–	–	22	78	25	19	–	22
Average across *B. ceti* clade	3361864	3426.4	3507.4	720.1	2706.3	882.2	646.1	131.4	764.1
% of average CDS	–	–	–	21	79	26	19	–	22
Average across *B. suis* clade	3313669	3456.3	3536.3	791.7	2664.7	869.8	638.5	131.2	751
% of average CDS	–	–	–	23	77	25	18	–	22
Average across 83/13 clade	3132066	3167	3237	685	2482	823	609	130	703.5
% of average CDS	–	–	–	22	78	26	19	–	22
Average across *B. inopinata* clade	3336358	3290.5	3350.5	738	2552.5	837	618	130.5	719
% of average CDS	–	–	–	22	78	25	19	–	22

no congruence even among the phylogenies based on the methods developed by these authors, one is at a loss to determine which methodology is best, or even reliable. A more conventional analysis of all of these genomes is under way (Wattam *et al.*, unpublished).

Phylogenomics

No analysis of the comparative genomics of any group would be complete without a discussion of phylogeny. Throughout the history of *Brucella* analyses there have been a number of phylogenetic trees constructed, by a variety of methods. While it seems easy to cluster the *Brucella* into their species-specific clades (the exception being *B. canis*, a well-documented member of the *B. suis* clade), resolving the order of divergence, as well as subspecific relationships, is more difficult. Below we analyse those methods and the trees that were produced.

RNA analysis

Ribosomal RNA – 16S

Bacteria use a ribosomal operon that is cleaved into three molecules known as 16S, 23S and 5S. For the last twenty years, microbiologists have relied on these rRNA gene sequences not only to identify bacteria, but also to classify them into related groups (Rajendhran and Gunasekaran, 2010) on the assumption of a bifurcating tree. The 16S rRNA is universally distributed and has a slow rate of sequence evolution (Cole *et al.*, 2003; Ettema and Andersson, 2009; Kuo and Ochman, 2009), and was even proposed for a bacterial evolutionary clock (Woese, 1987). The 16S rRNA sequence has been used to place *Brucella* in the larger bacterial world. When the first rRNA sequence for *Brucella* was published it was noted that its sequence showed strong similarity to the rRNA sequence from *Agrobacterium tumefaciens* (Dorsch *et al.*, 1989). Based on this similarity it was suggested that this genus belonged to the α-2 subclass of Proteobacteria. A subsequent phylogenetic analysis of this sequence along with the known α-Proteobacteria rRNAs available at the time, *Rochalimaea quintana* (now *Bartonella quintana*) and *A. tumefaciens*, definitively placed *Brucella* in this subdivision of the Proteobacteria (Moreno *et al.*, 1990). Since that time, rRNA has been used to place *Brucella* and its closest relatives together into the *Brucellaceae*. Scholz *et al.* (2008b) analysed the 16S rRNA and *recA* sequences from 106 strains of *Ochrobactrum* and eight *Brucella*. The analysis confirmed that *Brucella* is most closely related to *Ochrobactrum* and moreover showed that the *Brucella* cluster together. This study also revealed that the *Brucella* cluster exists deep within an *Ochrobactrum* tree with *Ochrobactrum* clades on either side of it.

An additional study using 16S rRNA compared *Brucella, Ochrobactrum* and *Mycoplana,* placing *O. intermedium* closer to *Brucella* (Thacker *et al.*, 2007). This supports previous analyses that showed that *O. intermedium* is more closely related to *Brucella* than to the other *Ochrobactrum* species (Velasco *et al.*, 1998; Lebuhn *et al.*, 2000). Interestingly, the object of this study was to examine a new strain of chromate-resistant *Brucella* that had been isolated from an industrial waste site in India heavily contaminated with chromium. A neighbour-joining tree was constructed using the rRNA sequence isolated from this strain comparing it to the six classical *Brucella* species, five *Ochrobactrum* and three *Mycoplana* species. The tree placed the new isolate outside the other *Brucella* species examined, and in particular outside of the branch that unites the *Brucella* to their next closest neighbour, *O. intermedium*. This novel strain is on the same branch as another novel partial sequence from a Chinese isolate, *B. melitensis* strain T2.

Analysis of another new *Brucella* isolate produced a similar tree. De *et al.* (2008) reported on an unusual *Brucella* strain isolated from a breast implant wound from a woman symptomatic for brucellosis. A neighbour-joining tree based on the 16S rRNA sequence from this isolate, which they called *Brucella* sp. BO1 (now classified as *B. inopinata* BO1), and from one other *Brucella*, *B. ovis,* and from eight *Ochrobactrum* species. Rooted with *Afipia felis* as the outgroup, this tree shows grouped BO1 and *B. ovis* together, and showed that the *O. intermedium* was the closest of the *Ochrobactrum* sequences to that branch. This indicated that the 16S rRNA sequence from BO1 strain had greater homology to the *Brucella* genome than it did to any of the sequences from the eight *Ochrobactrum* species examined.

Although 16S is useful for definition of members of the *Brucellaceae*, it is limited in differentiating between the six original *Brucella* species. Gee *et al.* (2004) looked at 16S rRNA sequences from 65 *Brucella* isolates from the six classical species and compared them to sequences from 17 other bacterial strains that include *Agrobacterium* and *O. anthropi*. All of the 65 gene sequences from *Brucella* were determined to be identical and clearly different from the other strains that were examined.

Ribosomal RNA – 23S

Using a different ribosomal gene (23S rRNA) and conserved protein-coding genes, Halling and Jensen (2006) built a phylogenetic tree. The tree was based on concatenated gene sequences from 23S *rrn*, *rplV*, *tuf*-1 and *tuf*-2 and was rooted by including a number of similar sequences from genomes closely related to *Brucella*. *Brucella* formed a well–supported cluster with a 100% bootstrap value that separated it from *Agrobacterium*, *Mesorhizobium* and others. This tree placed *B. abortus*, *B. suis*/*B. canis* and *B. melitensis* on separate branches. This approach was the first comparative study of this sort that separated *B. abortus* from *B. melitensis*. The analysis showed that the *B. abortus* sequences were more closely related to *B. ovis*, and the *B. melitensis* sequences to *B. neotomae*. *Brucella* that had been isolated from marine mammals diverged close to the *B. neotomae* and *B. melitensis* branches. All branches had high bootstrap support except for the one that clustered *B. ovis* with *B. abortus*.

Fragment length polymorphism

Restriction fragment length polymorphism (RFLP)

Traditionally, RFLP analysis was used to detect rearrangements within genomes and mutations affecting recognition sites used for digestion of genomic DNA. It has also been used to discriminate between species or strains of a single genus, or as a way to look at biodiversity within a natural community (Busse *et al.*, 1996). The advantage of this approach is that it reveals high levels of variation throughout the genome but the disadvantage is that it requires a large amount of DNA, and that it is a slow and somewhat tedious process. This has been a common approach applied to genes known to be polymorphic in *Brucella* (Al Dahouk *et al.*, 2005).

One of the earliest phylogenies of *Brucella* was based on the RFLP approach. Macrorestriction fragment fingerprints generated by *Pac*I and *Spe*I were separated by pulsed-field gels and the number of shared versus different fragment sizes, as well as cross-hybridization of restriction fragments, were used to calculate a genetic distance. The six traditional *Brucella* species, with a total of 16 type strains were used in this study (Michaux-Charachon *et al.*, 1997) and the unrooted tree showed the *abortus* biovars clustering tightly together, with the *B. melitensis* biovars grouped near *B. abortus*. At the other extreme end of the tree were the *B. suis* biovars, with *B. canis* clustering with *B. suis* biovar 1. The *B. ovis* branch diverged near the *B. melitensis* branch, and *B. neotomae* was intermediately placed between the *suis* and *abortus* extremes. This dendogram reflected the traditional phenetic classification of the genus *Brucella*.

Bourg *et al.* (2007) looked at the genomic structure using *Spe*I or *Pac*I restriction patterns of 24 *Brucella* strains isolated from marine mammals and found that they divided into three clonal groups. In this study, each of these groups corresponded to a specific type of host, which naturally increased the likelihood of host preference as having some genetic basis behind it. Members that clustered phylogenetically in the first group were isolated from common, grey and hooded seals, as well as a single isolation from an otter. The second group was isolated from common and striped dolphins and the third predominately from harbour porpoises, with single isolations from a white-sided dolphin and another from a Minke whale. Members in the first group, with the notable exception of the otter isolate, were all isolated from pinnipeds, while the second and third groups are in toothed whales and the one baleen whale (Minke). This is a pattern of three clonal groups suggests a phylogenetic basis of host preference. Species designations were given, bases on the preferred hosts (Foster *et al.*, 2007). The name

B. pinnipedialis is given to the isolated from seals, and *B. ceti* to those from cetaceans, dividing two obviously distinct *B. ceti* clades under one name.

Amplified fragment length polymorphism (AFLP)

Amplified fragment length polymorphism (AFLP) is a technique that uses polymerase chain reaction (PCR) to amplify specific regions, and the variable fragment sizes of restriction endonucleases are used as markers of shared sequence features. It is most commonly used for multilocus genomic fingerprinting (Meudt and Clarke, 2007). It has been found to be useful phylogenetic reconstruction, particularly for groups of closely related organisms, such as recent species radiations (Koopman, 2005), making it especially pertinent for an examination of the *Brucella*. In most instances, AFLP has high reproducibility and the ability to differentiate between clones (Foley *et al.*, 2007). However, concerns about repeatability and inter laboratory consistency, as well as the emergence of better typing techniques, has resulted in the limited application of this technique (Bensch and Akesson, 2005).

Whatmore and others (2005) used this technique in their analysis of *Brucella*, looking at subsets of genomic restriction fragments in 78 different isolates *B. canis, B. neotomae, B. ovis, B. suis*, marine isolates and representatives from all the biovars in from *B. abortus* and *B. melitensis*. They calculated a correlation matrix and an unrooted dendogram and found that isolates within the species fell into separate clusters except for *B. canis*, which clustered with *B. suis*. All field strains clustered into their appropriate species groups but one reference strain, *B. suis* biovar 5, failed to cluster as expected, appearing to be more closely related to *B. abortus* than to any of the *B. suis* strains. When AFLP patterns from *Ochrobactrum* were included, the now rooted dendogram definitely separated the *Ochrobactrum* species from the *Brucella*.

Multilocus variable-number tandem repeat (VNTR) analysis (MLVA)

Many bacterial genes and intergenic regions contain regions with short motifs that may be variable among strains with respect to the number of repeat units present or their individual primary structure (Lindsted, 2005). These so-called 'variable number of tandem repeat regions' (VNTRs) have been identified in essentially all pro- and eukaryotic species and have been successfully used for identification purposes (van Belkum, 2007). Recent studies have shown that MLVA demonstrated higher genetic resolution when applied to *Brucella* species than assays targeting other, more monomorphic molecular markers, such as the outer membrane proteins, insertion sequences and the *rpoB* gene (Tiller *et al.*, 2009).

Bricker *et al.* (2003) published the first application of VNTR-based typing in *Brucella* using *B. melitensis* 16M, *B. suis* 1330 and a draught genome sequence of *B. abortus*. They identified eight variable loci. Le Flèche *et al.* (2006) were they first to use this type of analysis in a phylogenetic context with *Brucella*. Eighty tandem repeat sequences (TRs) were identified using *B. suis* strain 1330, *B. melitensis* strain 16M and *B. abortus* strain 9-941 and were used to test 21 *Brucella* strains that included 18 reference strains and three from marine isolates. A maximum parsimony tree was generated from MLVA data and while it showed some consistencies, there were some surprises. The *abortus* and *melitensis* strains clustered by species as expected, as did the *suis–canis* biovars, the exception being *B. suis* biovar 5 strain, which was well separated from the rest of *suis*, and actually closer to *B. pinnipedialis*. Interestingly, *B. neotomae* separated the marine isolates and was found between the two *ceti* (called *cetaceae* in this paper) isolates and the single *B. pinnipedialis* representative.

Two strains of *B. microti* were incorporated into the MLVA data analysed previously (Le Flèche *et al.*, 2006; Al Dahouk *et al.*, 2007; García-Yoldi *et al.*, 2007) and re-analysed as per Le Flèche with some unique results (Scholz *et al.*, 2008a). The *B. microti* strains clustered together, but the marine strains were separated with *B. pinnipedialis* B2/94 coming out closer to *B. neotomae* at the base of the tree than to the two *B. ceti* strains examined. Two strains of *B. suis* biovar 5 were found to be well separated from the *suis* clade, and there appeared to be a geographic pattern in the *melitensis* clade with isolated from similar localities clustering together. Clustering that reflected different geographical isolates was seen in an MLVA analysis of *B. abortus* (Valdezate *et al.*, 2009).

The techniques developed for *Brucella* by Le Flèche were termed the MLVA-15 (for the 15 loci). A new assay, termed the MLVA-16, was used to examine 294 marine isolates and to compare 500 genotypes from terrestrial and marine isolates (Maquart *et al.*, 2009). This assay used 16 loci grouped into three panels that included both mini- and microsatellite loci (Maquart *et al.*, 2009). The marine mammal isolates, which originated from 173 animals and in addition from one human patient that presumably came from a marine origin. All of these clustered into 117 representative genotypes that could be sorted into three major groups, one specific to *B. ceti*, one to *B. pinnipedialis*, and one composed of a single human isolate that was related by pattern, but clearly an outlier to the other two. In addition, a neighbour joining tree of 500 genotypes from both marine and terrestrial genomes was rooted with *B. microti* and *B. neotomae*. The sequence types from the different species all group together (for example, all *abortus* isolates cluster together), the exception being *B. suis* biovar 5, which comes out separate from the rest of the *suis* sequences at the base of the tree near the *microti–neotomae* branch.

Another interesting visual aspect of the tree shows the three distinct clusters among the marine isolates with one *ceti* branch, and then a *ceti* and a *pinnipedialis* cluster coming off from the same branch point. The human isolate is at the extreme basal point of the *pinnipedialis* cluster. In addition, these authors had *B. melitensis* data from three different locales that they define as 'west Mediterranean,' 'east Mediterranean' and 'Americas.' All cluster on their own branches, but the 'west Mediterranean' branch is closer to *B. abortus* than the other two are. The *B. abortus* STs segregate into two branches, one with biovars 1, 2 and 4, and the other with biovars 3, 5, 6, 7 and 9.

Although most of the MLVA data described above used the gel-based method described by Le Flèche *et al.* (2006) and now identified as MLVA-15_{IGM}. A different, automated MLVA method (MLVA-15_{NAU}) has also been developed (Huynh *et al.*, 2008). Tiller *et al.* (2009) did a direct comparison of the two methodologies and found that while both were able to genetically correlate some strains originating from the same hospital or region within a country, the methodology described by Huynh presented higher strain-to-strain diversity in their test population from sporadic human brucellosis cases in Egypt, Qatar and Libya.

Groussaud *et al.* (2007) analysed the diversity of marine mammal isolates by MLVA and also by using another selection of VNTRs that included all eight loci defining the HOOF-prints MLVA assay described by Bricker *et al.* (2003) as well as 13 additional loci characterized previously (Le Flèche *et al.*, 2006; Whatmore *et al.*, 2006). This panel of 21 VNTR loci corresponded to a 21-locus MLVA scheme sharing nine loci with MLVA-16 with a higher degree of diversity. Like other analyses they found that the marine isolates cluster into three groups that appear to be host based, and that one of the *ceti* clades is more closely related to the *B. pinnipedialis* clade than it is to the other *ceti* clade.

The MLVA method used to examine *B. inopinata* BO1 and BO2 strains (Tiller *et al.*, 2010a) was the MLVA-15_{NAU} method (Huynh *et al.*, 2008). The *B. inopinata* BO1 and BO2 strains had unique VNTR profiles when compared with each other and also to other *Brucella* strains. The dendogram placed both *B. inopinata* strains at the root of the tree. It also showed the *abortus* clade, separated from the *melitensis* clade, was actually closer to *B. neotomae* than to anything else. The *melitensis* clade came off the same branch than the *suis/canis* clade. This same topology was seen when seven new *Brucella* isolates from Australian rodents were examined (Tiller *et al.*, 2010b). The rodent strains all clustered to together on the same branch next to *B. inopinata* strains BO1 and BO2, but the rest of the topology was as described in the earlier study (Tiller *et al.*, 2010a).

Multilocus sequence typing (MLST)

The term multilocus sequence typing (MLST) was coined in 1998 for the typing of human bacterial pathogens (Maiden *et al.*, 1998), but more generally it now refers to the use of DNA sequences from multiple regions in the genome for discriminating strains in populations (Xu, 2006) while also retaining signatures of longer evolutionary relationships (Margos *et al.*,

2008). It has been used to provide insights into population structure and in evolution of several pathogens (Gupta and Maiden, 2001). This technique was deployed to do a comprehensive analysis of *Brucella* (Whatmore *et al.*, 2007). Nine distinct genomic fragments from 27 different strains were identified as unique sequence types (STs) and ranged in size from 422 to 589 bp. A total of 4396 bp were examined. Neighbour joining trees were constructed from the concatenated nucleotide sequences from these fragments and once again *Brucella* isolates fell into their classical species taxonomic divisions. Generally, these species-clusters were well-behaved, being isolated from the other species-clusters on a single branch with diversification within the cluster developing after some distance along their individual branch. The groups that show more fragmented diversification were the *abortus* and the *suis* clusters. *Brucella suis* biovar 5 was found to be totally distinct from the other *B. suis* isolates on a separate branch closer to the marine mammal strains than to the *B. suis* ones.

This same technique seems to be gaining in popularity to examine *Brucella*. Scholz *et al.* (2008a) sequenced the *B. microti* genome and did a phylogenetic analysis identical to the one that used by Whatmore *et al.* (2007) except for the addition of the new genome. Both of these trees the divergence of the *B. abortus*, *B. melitensis* and *B. neotomae* clades are clear, but there appears to be an explosive, almost simultaneous divergence that resulted in five separate clades (*B. microti*, *B. suis* biovar 5, *B. ovis*, the marine isolates and the *B. suis*/*B. canis* clade). These five clades have experienced individual divergence since that initial separation and some, like the *B. suis*/*B. canis* clade, have diverged in interesting patterns. Determining the order that these five clades diverged in may be a difficult problem to resolve.

An additional study using this same technique was used to examine the new *Brucella* strain, BO2, isolated from a patient with chronic destructive pneumonia (Tiller *et al.*, 2010a). *Brucella inopinata* BO1, also recently isolated from a human patient (De *et al.*, 2008) was also included as well as the same 27 sequence types used in other studies (Whatmore *et al.*, 2007). This study did not include *B. microti*. What is surprising about this analysis is the length of the branch that separates BO1 and BO2 from the rest of the *Brucella*, indicating that these two strains are very different from the other *Brucella*. The individual branch lengths for BO1 and BO2 are also long, indicating that they are also very different from each other. This same topology was present when an isolate from an Australian rodent (*Rattus assimilis*) was included (Tiller *et al.*, 2010b), but in this case the long branch that contains *B. inopinata* strains BO1 and BO2 also included the new rodent isolate. When these three new isolates are added, a structure showing a 'core' *Brucella* group, containing the six classical strains as well as the marine isolates and *B. microti*, is well separated and distinct from the *B. inopinata* and new Australian rodent isolates. The structure of the tree with regard to the branching points for previously examined clades remains the same in these studies.

Whole-genome analysis

SNP-based

Single nucleotide polymorphisms (SNP) may become the marker of choice for multilocus population analyses. They have simple patterns of variation and the potential for automated detection, low mutation rates and low levels of homoplasy (Brito and Edwards, 2009). The first time SNP analysis was used in *Brucella* was in the early days of whole genome sequencing of this organism. Halling *et al.* (2005) used SNP to examine three genomes, *B. abortus*, *B. suis* and *B. melitensis*. They found that the mean was one SNP for 463 nucleotides. A neighbour-joining dendogram generated from an SNP-based genetic distance data used the three *Brucella* genomes and the *Mesorhizobium loti* genome as an outgroup. Obviously, three species isn't a widespread comparison for *Brucella*, but this was the first time whole genomes were used. The tree showed that *B. abortus* and *B. melitensis* branched off together from *B. suis*, indicating that they were more closely related to each other than to *B. suis*. This was also the first time that a closely related genome was used as an outgroup, and as would be expected, *Mesorhizobium loti* was at the root of the tree and distantly placed from the next branch that contained *B. suis*.

A more comprehensive comparison of the SNPs from whole genomes of 13 *Brucella* genomes representing five species (*B. abortus*, *B. canis*, *B. melitensis*, *B. ovis* and *B. suis*) with *O. anthropi* as an outgroup was completed recently (Foster *et al.*, 2009). Using *B. melitensis* 16M as a reference, they generated a matrix of the SNP state for each genome that included an SNP position in that genome. They found 20,154 orthologous SNPs that were shared in all genomes. The SNP sequences were concatenated for each sample and a neighbour-joining tree was. This tree showed the separation of species into distinct clusters. Very little divergence was seen in the *B. abortus* and *B. melitensis* clades, but the *B. suis* clade, which contains *B. canis*, showed extensive intraspecific genetic diversity. Rooted with *O. anthropi*, their tree showed *B. ovis* to be the most basal lineage and the authors suggested that the *Brucella* genomes examined in this study diverged from a common *B. ovis* ancestor. They also estimated the rate of evolution at a synonymous mutation rate of 1.4×10^{-10} mutations per base pair per generation. Using a molecular clock they estimate that most *Brucella* species diverged from a *B. ovis* ancestor in the last 86,000 to 296,000 years, preceding the domestication of livestock hosts.

It should be noted that some authors caution against attempts to establish molecular clocks in bacteria. Kuo and Ochman (2009) found that there was a wide range of substitution rates across genes and bacterial taxa, leading them to conclude that this high level of variation cautions against the assumption of a universal molecular clock for inferring divergence times in bacteria.

Protein family based

Molecular phylogenies based on a single or few orthologous genes often yield contradictory results and using multiple genes in a large concatenation was proposed to end these incongruences (Rokas *et al.*, 2003; Jeffroy *et al.*, 2006). In looking at genome sequences from eight yeast species, it was demonstrated that data sets consisting of single or a small number of concatenated genes had a significant probability of trees with conflicting topologies (Rokas *et al.*, 2003). However, when entire data sets were used, in this case 106 widely distributed orthologous genes, a fully resolved species tree with maximum support was produced. These authors found that a minimum concatenation of 20 genes could produce similar results.

Use of large set of concatenated genes has been used to resolve bacterial phylogenies at many levels (Williams *et al.*, 2007; 2010). This technique was also used to examine the *Brucella* when four new genomic sequences, including the first sequence from a marine isolate (*B. ceti* Cudo), and a sequence from a second rough strain (*B. canis* ATCC 23365) became available and were compared with nine other *Brucella* genomes and four outgroups (Wattam *et al.*, 2009). A total of 2246 protein families were clustered with OrthoMCL (Li *et al.*, 2003) and a maximum-likelihood tree was constructed using RaxML (Stamatakis, 2006). The tree sorted the genomes into four clades; a *B. melitensis-B. abortus* clade, a *B. ovis* clade, a *B. suis-B. canis* clade and a *B. ceti* clade. The authors noted that caution should be taken in assigning a strict evolutionary branching order to these four clades as they appear to have radiated explosively. The branching order was clear and well supported but the depth of divergence was quite shallow. This study used four outgroups with sequences from *O. anthropi* ATCC 49188, *O. intermedium*, *Bartonella quintana* Toulouse and *Mesorhizobium loti* MAFF 303099. The two *Ochrobactrum* genomes were the closest to *Brucella* clade.

A recent, novel approach to the study of *Brucella* phylogenomics is the 'pan' genome (Bohlin *et al.*, 2010). Thirty-two sequenced genomes were compared by variety of nucleotide- or proteome-based methods. To examine the nucleotide differences between the genomes the authors used a new technique they named the zero'th order Markov chain (ZOM) model (Bohlin and Skjerve, 2009). Using this model, the *Brucella* separated into five different groups, but these groups were not species-specific. In fact, members of the same species were often widely separated, something not often seen in any other phylogenetic study. They also used first and second order Markov chain models, which they called FOM and SOM, and generated trees that were also not species-specific, and did not have similar branching patterns. The authors claim that these nucleotide-based trees grouped the *Brucella* according to a clear marine or terrestrial host preference. However, when one examines the trees, there is not a single instance, despite the three Markov chain approaches, where either the marine isolates or the terrestrial isolates are united into a single group.

To compare the proteins, Bohlin *et al.* (2010) constructed a BLAST matrix using pairwise gene comparisons for all genomes based on a previously described technique (Usssery *et al.*, 2009). The gene families computed were stored as a pan-matrix, and the presence or absence of these gene families was used in a high-resolution clustering of the genomes. Gene families were divided into the essential genes, or the cloud, and the accessory genes, or the shell. They constructed a pan-genome tree based on relative Manhattan distances between genomes (Bohlin *et al.*, 2010). Unfortunately, the description of the method used to construct the tree was vague and not supported by additional references. Care should be taken in interpreting these results and the level of confidence one should place in them. Briefly, *B. inopinata* BO1 and BO2 are at the root in both the clouds and shell tree, with *Ochrobactrum* as the nearest outgroup. *Brucella pinnipedialis* M163/99/10 is oddly placed near the root of both trees, and also is at the root in the trees that were developed using the ZOM, SOM and FOM approaches. This perplexing result places this genome outside of the grouping with its species, outside of the marine isolates, and in some of the trees even outside the outgroups. One can only conclude that these novel approaches require validation from other studies before they become standards of comparative genomic methodologies.

Phylogeny summary

Ever since DNA–DNA hybridization studies revealed more than 90% homology between the six classical species (Hoyer and McCullough, 1968a,b; Verger *et al.*, 1985) there has been division in the *Brucella* community with some believing that *Brucella* should be a monospecific genus (Verger *et al.*, 1985, 1987), and others thinking that there is continuing value dividing the genus by the clear host specificity that can be assigned to each of the classical species (Moreno *et al.*, 2002).

A variety of techniques, both in terms of data analysis and phylogenetic reconstruction, have been applied to *Brucella*. All techniques place isolates identified with the six classical species into their own groups with only one exception (Bohlin *et al.*, 2010). Addition of new isolates resulting in new species, specifically *B. pinnipedialis, B. ceti, B. microti* and *B. inopinata* does not disrupt these associations. The consistent species-specific clustering, by a variety of different phylogenetic analyses, points to a genetic basis behind these designations.

Outliers to the classical *Brucella*

While the separation of clades associated with individual species does not seem to be in question, the order of the branching patterns is not easily determined. Analysis that includes the new isolates from humans, *B. inopinata* isolate BO1 (De *et al.*, 2008), BO2 (Tiller *et al.*, 2010a) and the Australian rodent isolate (Tiller *et al.*, 2010b) shows that they diverged significantly from the classical *Brucella* strains, and in fact might represent an ancient and early branch of *Brucella*. Fig. 2.1, taken from Tiller *et al.* (2010b) and adapted to highlight species-based clades, clearly illustrates the length of the branch between these new isolates and the other 'core' *Brucella*. Determining the order in which the branching patterns within this 'core' may be difficult owing to a sudden, explosive radiation (Wattam *et al.*, 2009). *Brucella microti* appears to be part of this core, branching off simultaneously with *B. ovis*, the marine isolates, and the *B. suis–B. canis* clade (Fig. 2.2, adapted from Scholz *et al.*, 2008).

Brucella core members

B. melitensis and *B. abortus*

In almost every study, the eight biovars from *B. abortus* and the three from *B. melitensis* cluster within their designated species. In the analysis by Scholz *et al.* (2008b), the *B. abortus* and *B. melitensis* genomes seemed to branch of clearly from the rest of the core (Fig. 2.2). In many analyses it looks like *B. abortus* and *B. melitensis* had a common ancestor and have since diverged (Bourg *et al.*, 1997; Whatmore *et al.*, 2007; Foster *et al.*, 2009; Maquart *et al.*, 2009; Wattam *et al.*, 2009),

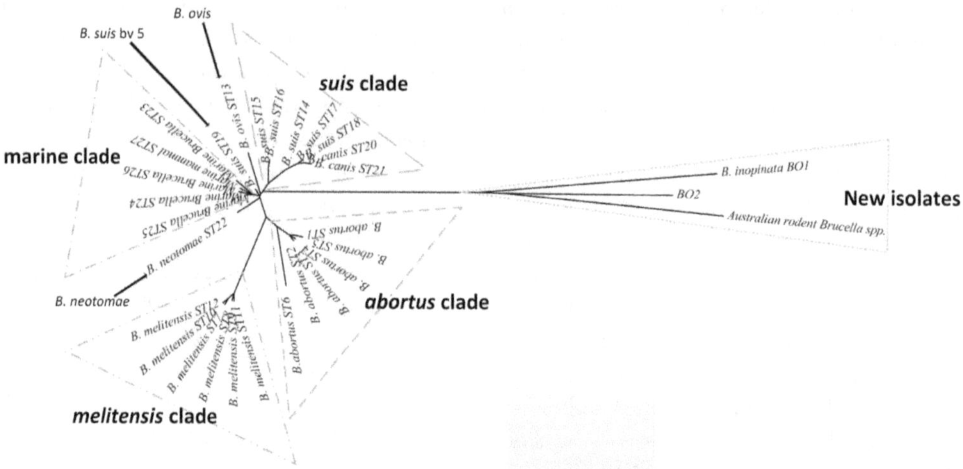

Figure 2.1 Adapted from figure presented by Tiller *et al.* (2010b). A neighbour-joining tree of concatenated sequence data (MLST) from nine loci that includes 4396 bp. Species-specific clades are highlighted. The isolates from marine mammals (marine clade) and *B. suis* biovar 5 appear to have the same point of divergence, with *B. ovis*, the *B. suis* clade and the new isolates branching off in very close proximity.

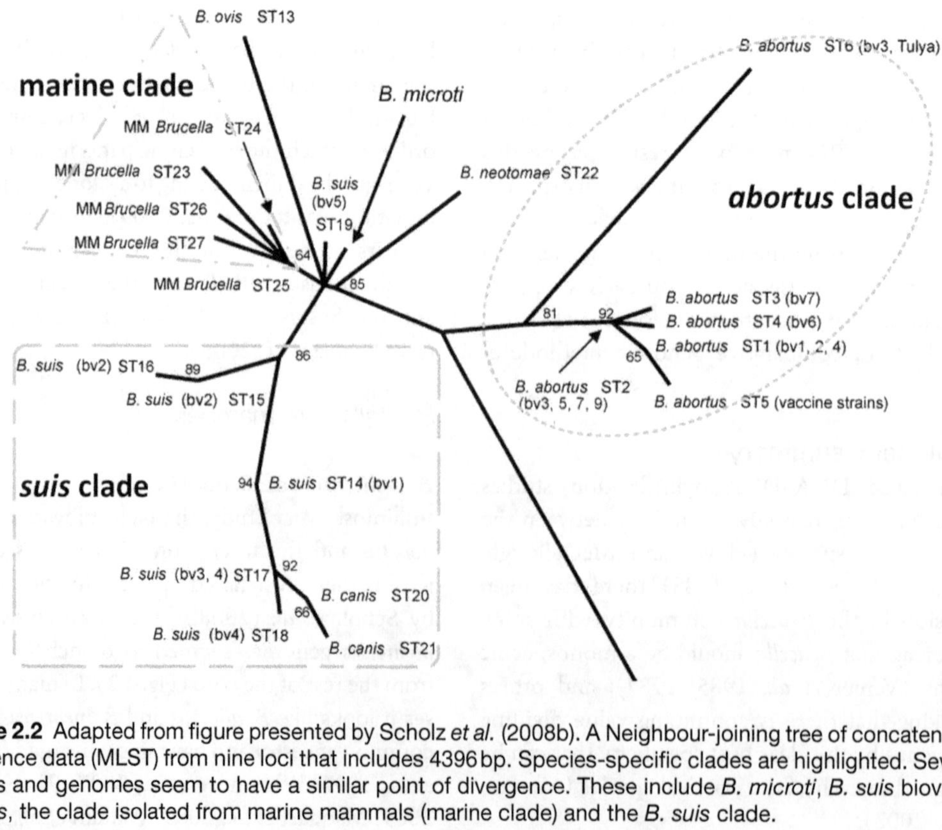

Figure 2.2 Adapted from figure presented by Scholz *et al.* (2008b). A Neighbour-joining tree of concatenated sequence data (MLST) from nine loci that includes 4396 bp. Species-specific clades are highlighted. Several clades and genomes seem to have a similar point of divergence. These include *B. microti*, *B. suis* biovar 5, *B. ovis*, the clade isolated from marine mammals (marine clade) and the *B. suis* clade.

although there are two studies that place *B. ovis* between them (Whatmore *et al.*, 2005; Le Flèche *et al.*, 2006).

There has been some question about biovars representing a true genetic grouping. In both the MLSA and MLVA analyses there appears to be scant evidence linking genotype to biovar, meaning that the same biovars do not necessarily group together (Le Flèche *et al.*, 2006; Whatmore *et al.*, 2006, 2007). Some MLVA studies (Al Dahouk *et al.*, 2007; Maquart *et al.*, 2009) showed that the clustering pattern of *B. melitensis* correlated with the geographic origin of the strains. This is an interesting result that should be further studied as, although it appears that there is a genetic basis to the species differentiation in *Brucella*, this might not be true where biovars are concerned. If this is true, patterns within the individual species probably reflect evolutionary change associated with geographic isolation.

B. ceti and *B. pinnipedialis*

There is some discrepancy in the divergence within the marine mammal clade and in the structuring among the different isolates. MLST analysis placed all isolates into one cluster, although bootstrap analysis for the group was rather low (Whatmore *et al.*, 2007). MLVA analysis showed three distinct clusters, two corresponding to *B. ceti* and one to *B. pinnipedialis* (Le Flèche *et al.*, 2006). Several different types of analyses including VNTR, genome fingerprints, VNTR analysis and multilocus sequencing (Bourg *et al.*, 2007; Groussaud *et al.*, 2007; Dawson *et al.*, 2008) have indicated that isolates currently described as *B. ceti* correspond to two distinct groups with apparently distinct preferred hosts of either porpoises or dolphins. These groups appear less closely related to each other than either is to *B. pinnipedialis*. This same relationship was also seen in a study using the VNTR approach outlined by Whatmore *et al.* (2006) identified three marine clusters that corresponded to isolates from three different types of hosts, seals, porpoises and dolphins (Groussaud *et al.*, 2007). Moreover, a more recent study (Maquart *et al.*, 2009), which included marine mammal isolates from European waters that extended the number of isolates, also grouped all isolates into two *ceti* clusters and one *pinnipedialis*. The Maquart study also noted a human isolate, presumably of marine origin that clustered grossly with all marine isolates, but was a definite outlier to the three subclusters.

One marine isolate of interest is *B. pinnipedialis* 163/99/10. Maquart *et al.* (2009) placed this strain deep within the *B. pinnipedialis* cluster. In a separate study, this strain was found to be near the root of a tree comparing all species of *Brucella* (Bohlin *et al.*, 2010). If the latter study is true, this particular isolate is very different from the rest of its species group.

B. suis

There are five *B. suis* biovars that have been isolated from a variety of mammalian hosts (Whatmore, 2009). Biovars 1 and 3 are associated with swine and biovar 2 with swine and hares. Biovar 4 was isolated from reindeer and caribou, and biovar 5 was found in rodents in the Soviet Union. *Brucella canis*, associated with dogs, is also included in the *suis* clade.

While most studies show the *B. suis* clade clustering together, albeit with long branch lengths suggesting significant divergence between strains, *B. suis* biovar 5 appears to be an outlier to the rest of the *B. suis* clade. It has been suggested that *B. suis* biovar 5 might be removed from *B. suis* as it appears to be consistently different from the other clade members (Whatmore, 2009). However, isolation of this particular biovar is rare and perhaps additional sequences from new strains will further define this problem.

Foster *et al.* (2009) looked at several different *B. suis* biovars. Biovar 2 was the most basal member of the clade, and biovars 3 and 4 were found to be closely related. Biovar 4 had only 253 SNP differences separating it from *B. canis*.

Single genomes: *B. ovis*, *B. neotomae* and *B. microti*

The exact point of divergence for each of these species is difficult to reconcile. One study shows *B. ovis* at the base of the *Brucella* tree and hypothesizes that if is the ancestral *Brucella* type (Foster *et al.*, 2009). Others show it near the base or at any number of positions. *B. neotomae* also moves around depending upon the study and analysis method, and while *B. microti* has not been included in many studies, there is some data that points to it being one of the earlier diverging *Brucella* (Whatmore, 2009). A detailed analysis of the 40 available genomes that might shed some light on branching patterns with respect to these, as well as all the other clades is currently under way (Wattam *et al.* unpublished).

Comparative genomics of virulence pathways

Several studies have looked across *Brucella* at genes that work together in some fashion, or work in a specific pathway. Some of these genes identified as virulence factors are intensely studied. We describe three of these virulence pathways below.

Lipopolysaccharide (LPS) genes

The lipopolysaccharide (LPS) has long been known as one of the key virulence factors in *Brucella* pathogenesis (Porte *et al.*, 2003). The presence of LPS has been shown to aid in avoiding an optimal inflammatory response in the host, and also can actively interfere with the host acquired immune response (Roop *et al.*, 2009). In *Brucella* the LPS is composed of three structural regions, a lipid core, a core oligosaccharide and a distal O-polysaccharide (O-PS) side chain (Godfroid *et al.*, 2000). There are two LPS phenotypes in *Brucella*; rough and smooth. In rough LPS, the core is the most distal part of the molecule, but in smooth LPS, the core links to lipid A, and then to the O-antigen. The O-antigen is a molecule with many repeated units that each consist of one to five sugar moieties (Godfroid *et al.*, 2000). Presence of an O-side chain of lipopolysaccharide (LPS) classifies a *Brucella* as having a smooth phenotype, and the lack of it gives a rough phenotype. There are only two *Brucella* species that are naturally rough, *B. ovis* and *B. canis*, and both are infective to their preferred hosts.

González *et al.* (2008) did an extensive study of the genes involved in the construction of LPS in *B. melitensis*. Using transposon mutagenesis, they found that disruption of 13 genes (*wboA*, *wboB*, *wa***, *wbkE*, *manB*, *wbkA*, *gmd*, *per*, *wzm*, *wbkF*, *wbkD*, *prm* and *manB*$_{core}$) resulted in a rough phenotype in *B. melitensis*, which is normally smooth. An additional six genes were also indicated as playing some role in LPS production that they were unable to fully determined. These 19 genes were specifically examined in the two rough strains, *B. ovis* and *B. canis* (Wattam *et al.*, 2009). Many of the 19 genes considered necessary for complete LPS synthesis were found to be in areas of the genome that probably entered by lateral transfer. With three exceptions, all 19 of these genes were found present and presumably active in all of the smooth *Brucella*. However, they found that *B. ovis* was missing two genes, *wboA* and *wboB*, which encode enzymes that polymerize *N*-formylperosamine (González *et al.*, 2008) and their loss in *B. ovis* was noted previously (Vizcaíno *et al.*, 2004). Both of these genes reside in areas that were identified as probably having entered *Brucella* via lateral transfer (Rajashekara *et al.*, 2008; Wattam *et al.*, 2009) The *B. ovis* genome also had a truncated *wzt*, encoding a protein that works as a part of an ABC transporter with *wzm*. Like *wboA* and *wboB*, this gene was found in an area that most probably entered *Brucella* by lateral transfer.

The second rough strain, *B. canis*, was found to have truncations in two of the 19 LPS synthesis genes, *wbkF*, an undecaprenyl-glycosyltransferase, and *wbkD*, an epimerase/dehydratase (Wattam *et al.*, 2009). A truncated enzyme could still be functional, but the fact that *B. canis* is rough, and that all other enzymes appear to be fully functional indicates that at least one of these genes is responsible for the production of the rough phenotype. As these are different genes than are noted for producing roughness in *B. ovis*, the rough phenotype appears to have developed twice independently (Wattam *et al.*, 2009).

The Type IV secretion system

Type IV secretion systems are one of seven major families of secretion systems that are capable of exporting virulence effectors proteins across the envelopes of pathogenic bacteria (Fronzes *et al.*, 2009). O'Callaghan *et al.* (1999) first identified an 11,860-bp region of the *B. suis* genome encoding a type IV secretion system which they found to be homologous to the VirB system of *A. tumefaciens*. *Brucella* is taken up by macrophages and the vacuole they are contained in becomes acidified. This seems to provide a signal for the virB gene expression (Boschiroli *et al.*, 2002). The acidified *Brucella*-containing vacuole (BCV) acquires markers like LAMP-1, CD63 and RAB7 while interacting with the late endosome/lysosomes (Starr *et al.*, 2008). Wild-type *Brucella* is able to exclude the late endosomal/lysosomal markers from the BCV and interact with endoplasmic reticulum, which in turn eventually allows for bacterial replication. Mutants that lack the *virB* genes, on the other hand, remain associated with the lysosomal compartment and fail to replicate (Delrue *et al.*, 2001; Celli *et al.*, 2003, 2005; den Hartigh *et al.*, 2008). The *virB* genes are also required for persistence in both the mouse, and in a natural host, the goat (Hong *et al.*, 2000; Sieira *et al.*, 2000; Kahl-McDonagh *et al.*, 2006; Zygmunt *et al.*, 2006). The role of the *Brucella* VirB T4SS is discussed in detail Chapter 11.

This region, found on Chr II, was detected by Southern blotting in the reference strains of *B. abortus, B. canis, B. melitensis, B. neotomae* and *B. ovis* (O'Callaghan *et al.*, 1999). In an examination of ten *Brucella* genomes that included three *B. abortus* strains (9–941, S19 and 2308), two *B. melitensis* strains (16M and 23457), *B. ceti* (Cudo), *B. canis, B. ovis*, and two *B. suis* strains (1330 and 23445), all 12 of the *vir*B genes were present (Wattam *et al.*, 2009). The region that contained these genes, identified as SAR 1–2, appear to have entered the genus by lateral transfer.

Flagellar genes

Among the *Brucellaceae*, both *Mycoplana* and *Ochrobactrum* produce functional flagella (Urakami *et al.*, 1990; Scholz *et al.*, 2008a). *Brucella* strains are uniformly described as being non-motile, yet genome sequence data suggests that they have the genetic capacity to produce flagella. Almost all of the *Brucella* genomes that have been sequenced have flagellar genes. The flagellar regulon of *Brucella* contains 31 genes clustered in three loci on the small chromosome (Léonard *et al.*, 2007). In some genomes, many of the genes necessary to produce a functioning flagellum appear to be disrupted. Under certain growth conditions *B. melitensis* 16M produces a polar organelle that resembles a flagellum (Fretin *et al.*, 2005). Despite the fact that *Brucella* is non-motile and does not appear to produce a complete flagellum, mutations of some of the *B. melitensis* flagellar genes (*flgI, fliF, fliC, flhA, motB* and *flgE*) have been found to exhibit attenuation in experimentally infected mice (Fretin *et al.*, 2005), indicating that their presence is necessary for virulence and may have some function other than motility. It has been hypothesized that the polar organelle produced by these genes could be secretion apparatus (DelVecchio *et al.*, 2002; Roop *et al.*, 2009) noting that the components of bacterial Type III secretion systems share many similarities with those involved in the transport and assembly of flagella.

Audic *et al.* (2009) carefully examined the flagellar genes in the *B. microti* and compared them with nine sequenced *Brucella* genomes as well as *O. anthropi*. *Ochrobactrum anthropi* had a complete complement of the 31 genes necessary to produce a flagellum, but of the *Brucella* strains examined, only *B. microti* had a similar, complete set. Despite this, *B. microti* has not been seen with a visible flagellum (Audic *et al.*, 2009). Every other *Brucella* genome had at least one disrupted gene, and most had more than one. Some of the disrupted genes appeared to be shared across clades of related genomes, an example being that the three *B. abortus* genomes examined (*B. abortus* S19, *B. abortus* 9–941, and *B. abortus* 2308) all had the same four genes disrupted (*fli*G, *fli*M, *flg*I and *flh*A), indicating that these mutations occurred in a common ancestor. The members of the *B. suis* clade examined (*B. canis, B. suis* 1330 and *B. suis* ATCC 23445) did not present as unified a picture. They all had one disrupted gene in common (*fli*G), but *B. suis* ATCC 23445 had three additional disrupted genes (*fli*L, *fli*R and *fli*F) implying that these two genes mutated after this strain diverged from the other suis clade members. Audic *et al.* (2009) noted that *fli*G, the motor switch gene, is the most frequently impaired gene among the *Brucella* strains they examined.

Transcriptional regulators also play a part in gene expression. A new master regulator of the flagellar system, FtcR (flagellar two-component regulator), has recently been identified in *B. melitensis* 16M and is conserved in other member of the α-Proteobacteria (Léonard *et al.*, 2007). This protein has been found to bind directly to the upstream region of *fliF* gene and specifically to the *pfliF* promoter region. Inactivation *ftcR* led to a decrease in flagellar gene expression and to impaired *Brucella* virulence, suggesting that FtcR could be considered a flagellar class I master regulator.

An intriguing exception is the *Brucella* sp. F5/99 genome that was recently sequenced. This genome belongs to the *B. ceti* clade and it has lost a 47-kb segment of the genome that contains all these genes, as well as many others (Wattam *et al.*, unpublished). When this strain was originally isolated, it was found to be avirulent when compared with known *Brucella* species (Ewalt *et al.*, 1994). Could the lack of the flagellar genes be the reason this *Brucella* strain is avirulent?

Future directions

With 40 sequenced genomes representing all species and biovars, the opportunity to examine detailed aspects across *Brucella* is now available. A complete phylogenetic analysis will point to species-specific clades that can be examined for unique regions or areas of loss. Clade-specific differences across protein families can also be assembled, as can differences in metabolic and biochemical pathway data. Differences among the individual genomes should also be identified. These types of data can be crossed with the known biochemical differences that exist between these species, perhaps revealing genetic differences that result in a particular phenotype. This type of analysis may provide clues to the evolution, virulence, intracellular survival, host preference and pathogenesis of this unique pathogen.

Acknowledgements

This project has been funded in whole or in part with federal funds from the National Institute of Allergy and Infectious Diseases, National Institutes of Health, Department of Health and Human Services, under Contract No. HHSN272200900040C, awarded to BWS Sobral.

References

Al Dahouk, S., Tomaso, H., Prenger-Berninghoff, E., Splettstoesser, W.D., Scholz, H.C., and Neubauer, H. (2005). Identification of *Brucella* species and biotypes using polymerase chain reaction-restriction fragment length polymorphism (PCR RFLP). Crit. Rev. Microbiol. *31*, 191–196.

Al Dahouk, S., Flèche, P.L., Nöckler, K., Jacques, I., Grayon, M., Scholz, H.C., Tomaso, H., Vergnaud, G., and Neubauer, H. (2007). Evaluation of *Brucella* MLVA typing for human brucellosis. J. Microbiol. Methods *69*, 37–45.

Audic, S., Lescot, M., Claverie, J.M., and Scholz, H.C. (2009). *Brucella microti*: the genome sequence of an emerging pathogen. BMC Genomics *4*, 10, 352.

Aziz, R.K., Bartels, D., Best, A.A., DeJongh, M., Disz, T., Edwards, R.A., Formsma, K., Gerdes, S., Glass, E.M., Kubal, M., *et al.* (2008). The RAST Server: rapid annotations using subsystems technology. BMC Genomics *8*, 9, 75.

Bandara, A.B., Contreras, A., Contreras-Rodríguez, A., Martins, A.M., Dobrean, V., Poff-Reichow, S., Rajasekaran, P., Sriranganathan, N., Schurig, G.G., and Boyle, S.M. (2007). *Brucella suis* urease encoded by ure1 but not ure2 is necessary for intestinal infection of BALB/c mice. BMC Microbiol. *19*, 57.

van Belkum, A. (2007). Tracing isolates of bacterial species by multilocus variable number of tandem repeat analysis (MLVA). FEMS Immunol. Med. Microbiol. *49*, 22–27.

Bensch, S., and Akesson, M. (2005). Ten years of AFLP in ecology and evolution: why so few animals? Mol. Ecol. *14*, 2899–2914.

Blasco, J.M. (1990). *Brucella ovis*. In Animal Brucellosis, Nielsen, K., and Duncan, J.R., eds. (Boca Raton: CRC Press), pp: 351–378.

Bohlin, J., Snipen, L., Cloeckaert, A., Lagesen, K., Ussery, D., Kristoffersen, A.B., and Godfroid, J. (2010). Genomic comparisons of *Brucella spp.* and closely related bacteria using base compositional and proteome based methods. BMC Evol. Biol. *13*, 249.

Bohlin, J., and Skjerve, E. (2009). Examination of genome homogeneity in prokaryotes using genomic signatures. PLoS One *4*, e8113.

Boschiroli, M.L., Ouahrani-Bettache, S., Foulongne, V., Michaux-Charachon, S., Bourg, G., Allardet-Servent, A., Cazevieille, C., Lavigne, J.P., Liautard, J.P., Ramuz, M., *et al.* (2002). Type IV secretion and *Brucella* virulence. Vet. Microbiol. *90*, 341–348.

Bourg, G., O'Callaghan, D., and Boschiroli, M.L. (2007). The genomic structure of *Brucella* strains isolated from marine mammals gives clues to evolutionary history within the genus. Vet. Microbiol. *125*, 375–380.

Bricker, B.J., Ewalt, D.R., and Halling, S.M. (2003). *Brucella* 'HOOF-Prints': strain typing by multilocus analysis of variable number tandem repeats (VNTRs). BMC Microbiol. *11*, 15.

Brito, P.H., and Edwards, S.V. (2009). Multilocus phylogeography and phylogenetics using sequence-based markers. Genetica *135*, 439–455.

Burns, D.L. (2003). Type IV transporters of pathogenic bacteria. Curr. Opin. Microbiol. *6*, 29–34.

Busse, H.J., Denner, E.B., and Lubitz, W. (1996). Classification and identification of bacteria: current approaches to an old problem. Overview of methods used in bacterial systematics. J. Biotechnol. 47, 3–38.

Buddle, M.B. (1956). Studies on *Brucella ovis* (*n. sp.*), a cause of genital disease of sheep in New Zealand and Australia. J. Hyg. (Lond). *54*, 351–364.

Celli, J., de Chastellier, C., Franchini, D.M., Pizarro-Cerdá, J., Moreno, E., and Gorvel, J.P. (2003). *Brucella* evades macrophage killing via VirB-dependent sustained interactions with the endoplasmic reticulum. J. Exp. Med. *198*, 545–556.

Chain, P.S., Comerci, D.J., Tolmasky, M.E., Larimer, F.W., Malfatti, S.A., Vergez, L.M., Aguero, F., Land, M.L., Ugalde, R.A., and García, E. (2005). Whole-genome analyses of speciation events in pathogenic *Brucella*. Infect. Immun. 73, 8353–8361.

Cole, J.R., Chai, B., Marsh, T.L., Farris, R.J., Wang, Q., Kulam, S.A., Chandra, S., McGarrell, D.M., Schmidt, T.M., Garrity, G.M., *et al.* (2003). The Ribosomal Database Project (RDP-II): previewing a new autoaligner that allows regular updates and the new prokaryotic taxonomy. Nucleic Acids Res. *31*, 442–443.

Crasta, O.R., Folkerts, O., Fei, Z., Mane, S.P., Evans, C., Martino-Catt, S., Bricker, B., Yu, G., Du, L., and Sobral, B.W. (2008). Genome sequence of *Brucella abortus* vaccine strain S19 compared to virulent strains yields candidate virulence genes. PLoS One *14*, e2193.

Crawford, R.P., Huber, J.D., and Adams, B.S. (1990). Epidemiology and surveillance. In Animal Brucellosis, Klaus, N., and Robert, D., eds. (CRC Press: Boca Raton, FL), pp. 131–151.

Dawson, C.E., Stubberfield, E.J., Perrett, L.L., King, A.C., Whatmore, A.M., Bashiruddin, J.B., Stack, J.A., and Macmillan, A.P. (2008). Phenotypic and molecular characterisation of *Brucella* isolates from marine mammals. BMC Microbiol. *8*, 224.

De, B.K., Stauffer, L., Koylass, M.S., Sharp, S.E., Gee, J.E., Helsel, L.O., Steigerwalt, A.G., Vega, R., Clark, T.A., Daneshvar, M.I., *et al.* (2008). Novel *Brucella* strain (BO1) associated with a prosthetic breast implant infection. J. Clin. Microbiol. *46*, 43–49.

Delrue, R.M., Martínez-Lorenzo, M., Lestrate, P., Danese, I., Bielarz, V., Mertens, P., De Bolle, X., Tibor, A., Gorvel, J.P., and Letesson, J.J. (2001). Identification of *Brucella spp.* genes involved in intracellular trafficking. Cell. Microbiol. 3, 487–497.

DelVecchio, V.G., Kapatral, V., Redkar, R.J., Patra, G., Mujer, C., Los, T., Ivanova, N., Anderson, I., Bhattacharyya, A., Lykidis, A., *et al.* (2002). The genome sequence of the facultative intracellular pathogen *Brucella melitensis*. Proc. Natl. Acad. Sci. U.S.A. *99*, 443–448.

Dorsch, M., Moreno, E., and Stackebrandt, E. (1989). Nucleotide sequence of the 16S rRNA from *Brucella abortus*. Nucleic Acids Res. *25*, 1765.

Ettema, T.J., and Andersson, S.G. (2009). The alpha-proteobacteria: the Darwin finches of the bacterial world. Biol. Lett. 23, 429–432.

Evguenieva-Hackenberg, E. (2005). Bacterial ribosomal RNA in pieces. Mol. Microbiol. *57*, 318–325.

Ewalt, D.R., Payeur, J.B., Martin, B.M., Cummins, D.R., and Miller, W.G. (1994). Characteristics of a *Brucella* species from a bottlenose dolphin (*Tursiops truncatus*). J. Vet. Diagn. Invest. *6*, 448–452.

Foley, S.L., Zhao, S., and Walker, R.D. (2007). Comparison of molecular typing methods for the differentiation of *Salmonella* foodborne pathogens. Foodborne Pathog. Dis. *4*, 253–276.

Foster, G., Osterman, B.S., Godfroid, J., Jacques, I., and Cloeckaert, A. (2007). *Brucella ceti* sp. nov. and *Brucella pinnipedialis* sp. nov. for *Brucella* strains with cetaceans and seals as their preferred hosts. Int. J. Syst. Evol. Microbiol. *57*, 2688–2693.

Foster, J.T., Beckstrom-Sternberg, S.M., Pearson, T., Beckstrom-Sternberg, J.S., Chain, P.S., Roberto, F.F., Hnath, J., Brettin, T., and Keim, P. (2009). Whole-genome-based phylogeny and divergence of the genus *Brucella*. J. Bacteriol. *191*, 2864–2870.

Fretin, D., Fauconnier, A., Köhler, S., Halling, S., Leonard, S., Nisjkens, C., Ferooz, J., Lestrate, P., Delrue, R.M., Danese, I., *et al.* (2005). The sheathed flagellum of *Brucella melitensis* is involved in persistence in a murine model of infection. Cell. Microbiol. *7*, 687–698.

Fronzes, R., Christie, P.J., and Waksman, G. (2009). The structural biology of type IV secretion systems. Nat. Rev. Microbiol. *7*, 703–714.

García-Yoldi, D., Le Fleche, P., De Miguel, M.J., Muñoz, P.M., Blasco, J.M., Cvetnic, Z., Marín, C.M,. Vergnaud, G., and López-Goñi, I. (2007). Comparison of multiple-locus variable-number tandem-repeat analysis with other PCR-based methods for typing *Brucella suis* isolates. J. Clin. Microbiol. *45*, 4070–4072.

Gee, J.E., De, B.K., Levett, P.N., Whitney, A.M., Novak, R.T., and Popovic, T. (2004). Use of 16S rRNA gene sequencing for rapid confirmatory identification of *Brucella* isolates. J. Clin. Microbiol. *42*, 3649–3654.

Godfroid, F., Cloeckart, A., Taminiau, B., Danese, I., Tibor, A., de Bolle, X., Mertens, P., and Letesson, J.J. (2000). Genetic organisation of the lipopolysaccharide O-antigen biosynthesis region of *Brucella melitensis* 16M (*wbk*). Res. Microbiol. *151*, 655–668.

González, D., Grilló, M.J., De Miguel, M.J., Ali, T., Arce-Gorvel, V., Delrue, R.M., Conde-Alvarez, R., Muñoz, P., López-Goñi, I., Iriarte, M., *et al.* (2008). Brucellosis vaccines: assessment of *Brucella melitensis* lipopolysaccharide rough mutants defective in core and O-polysaccharide synthesis and export. PLoS One *23*, 3(7):e2760.

Groussaud, P., Shankster, S.J., Koylass, M.S., and Whatmore, A.M. (2007). Molecular typing divides marine mammal strains of *Brucella* into at least three groups with distinct host preferences. J. Med. Microbiol. *56*, 1512–1518.

Gupta, S., and Maiden, M.C. (2001). Exploring the evolution of diversity in pathogen populations. Trends Microbiol. *9*, 181–185.

Halling, S.M., Peterson-Burch, B.D., Bricker, B.J., Zuerner, R.L., Qing, Z., Li, L.L., Kapur, V., Alt, D.P., and Olsen, S.C. (2005). Completion of the genome sequence of *Brucella abortus* and comparison to the highly similar genomes of *Brucella melitensis* and *Brucella suis*. J. Bacteriol. *187*, 2715–2726.

Halling, S.M., and Jensen, A.E. (2006). Intrinsic and selected resistance to antibiotics binding the ribosome: analyses of *Brucella* 23S *rrn*, L4, L22, EF-Tu1, EF-Tu2, efflux and phylogenetic implications. BMC Microbiol. *2*, 84.

den Hartigh, A.B., Rolán, H.G., de Jong, M.F., and Tsolis, R.M. (2008). VirB3 to VirB6 and VirB8 to VirB11, but not VirB7, are essential for mediating persistence of *Brucella* in the reticuloendothelial system. J. Bacteriol. *190*, 4427–4436.

Hong, P.C., Tsolis, R.M., and Ficht, T.A. (2000). Identification of genes required for chronic persistence of *Brucella abortus* in mice. Infect. Immun. *68*, 4102–4107.

Hoyer, B.H., and McCullough, N.B. (1968a). Polynucleotide homologies of *Brucella* deoxyribonucleic acids. J. Bacteriol. *95*, 444–448.

Hoyer, B.H., and McCullough, N.B. (1968b). Homologies of deoxyribonucleic acids from *Brucella ovis*, canine abortion organisms, and other *Brucella*. J. Bacteriol. *96*, 1783–1790.

Hubálek, Z., Scholz, H.C., Sedlácek, I., Melzer, F., Sanogo, Y.O., and Nesvadbová, J. (2007). Brucellosis of the common vole (*Microtus arvalis*). Vector Borne Zoonotic Dis. *7*, 679–688.

Huynh, L.Y., Van Ert, M.N., Hadfield, T., Probert, W.S., Bellaire, B.H., Dobson, M., Burgess, R.J., Weyant, R.S., Popovic, T., Zanecki, S., *et al.* (2008). Multiple locus variable number tandem repeat (VNTR) analysis (MLVA) of *Brucella spp.* identifies species specific markers and insights into phylogenetic relationships, In National Institute of Allergy and Infectious Disease, NIH, vol. 1. Frontiers in research. St. Georgiev, V., Western, K.A., and McGowan, J.J., eds. (Humana Press: Totowa, NJ), pp. 47–54.

Hsu, D., Zee, Y.C., Ingraham, J., and Shih, L.M. (1992). Diversity of cleavage patterns of *Salmonella* 23S rRNA. J. Gen. Microbiol. *138*, 199–203.

Jeffroy, O., Brinkmann, H., Delsuc, F., and Philippe, H. (2006). Phylogenomics: the beginning of incongruence? Trends Genet. *22*, 225–231.

Jiménez de Bagüés, M.P., Ouahrani-Bettache, S., Quintana, J.F., Mitjana, O., Hanna, N., Bessoles, S., Sánchez, F., Scholz, H.C., Lafont, V., Köhler, S., *et al.* (2010). The new species *Brucella microti* replicates in macrophages and causes death in murine models of infection. J. Infect. Dis. *202*, 3–10.

Kahl-McDonagh, M.M., Elzer, P.H., Hagius, S.D., Walker, J.V., Perry, Q.L., Seabury, C.M., den Hartigh, A.B., Tsolis, R.M., Adams, L.G., Davis, D.S., *et al.* (2006). Evaluation of novel *Brucella melitensis* unmarked deletion mutants for safety and efficacy in the goat model of brucellosis. Vaccine *24*, 5169–5177.

Kohler, S., Foulongne, V., Ouahrani-Bettache, S., Bourg, G., Teyssier, J., Ramuz, M., and Liautard, J.P. (2002). The analysis of the intramacrophagic virulome of *Brucella suis* deciphers the environment encountered by the pathogen inside the macrophage host cell. Proc. Natl. Acad. Sci. U.S.A. *99*, 15711–15716.

Koopman, W.J.M. (2005). Phylogenetic signal in AFLP data sets. Syst. Biol. *54*, 197–217.

Kuo, C.H., and Ochman, H. (2009). Inferring clocks when lacking rocks: the variable rates of molecular evolution in bacteria. Biol. Direct. *4*, 35.

Lebuhn, M., Achouak, W., Schloter, M., Berge, O., Meier, H., Barakat, M., Hartmann, A., and Heulin, T. (2000). Taxonomic characterization of *Ochrobactrum* sp. isolates from soil samples and wheat roots, and description of *Ochrobactrum tritici* sp. nov. and *Ochrobactrum grignonense* sp. nov. Int. J. Syst. Evol. Microbiol. *50*, 2207–2223.

Le Flèche, P., Jacques, I., Grayon, M., Al Dahouk, S., Bouchon, P., Denoeud, F., Nöckler, K., Neubauer, H., Guilloteau, L.A., and Vergnaud, G. (2006). Evaluation and selection of tandem repeat loci for a *Brucella* MLVA typing assay. BMC Microbiol. *9*, 6–9.

Léonard, S., Ferooz, J., Haine, V., Danese, I., Fretin, D., Tibor, A., de Walque, S., De Bolle, X., and Letesson, J.J. (2007). FtcR is a new master regulator of the flagellar system of *Brucella melitensis* 16M with homologs in Rhizobiaceae. J. Bacteriol. *189*, 131–141.

Li, L., Stoeckert, C.J. Jr, and Roos, D.S. (2003). OrthoMCL: identification of ortholog groups for eukaryotic genomes. Genome Res. *13*, 2178–2189.

Lindstedt, B.A. (2005). Multiple-locus variable-number tandem repeats analysis for genetic fingerprinting of pathogenic bacteria. Electrophoresis *26*, 2567–2582.

Maiden, M.C., Bygraves, J.A., Feil, E., Morelli, G., Russell, J.E., Urwin, R., Zhang, Q., Zhou, J., Zurth, K., Caugant, D.A., *et al.* (1998). Multilocus sequence typing: a portable approach to the identification of clones within populations of pathogenic microorganisms. Proc. Natl. Acad. Sci. U.S.A. *95*, 3140–3145.

Maquart, M., Le Flèche, P., Foster, G., Tryland, M., Ramisse, F., Djønne, B., Al Dahouk, S., Jacques, I., Neubauer, H., Walraven, K., *et al.* (2009). MLVA-16 typing of 295 marine mammal *Brucella* isolates from different animal and geographic origins identifies 7 major groups within *Brucella ceti* and *Brucella pinnipedialis*. BMC Microbiol. *20*, 145.

Margos, G., Gatewood, A.G., Aanensen, D.M., Hanincová, K., Terekhova, D., Vollmer, S.A., Cornet, M., Piesman, J., Donaghy, M., Bormane, A., *et al.* (2008). MLST of housekeeping genes captures geographic population structure and suggests a European origin of *Borrelia burgdorferi*. Proc. Natl. Acad. Sci. U.S.A. *105*, 8730–8735.

Meudt, H.M., and Clarke, A,C. (2007). Almost forgotten or latest practice? AFLP applications, analyses and advances. Trends Plant. Sci. *12*, 106–117.

Meyer, M.E. (1966). Metabolic characterization of the genus *Brucella*. V. Relationship of strain oxidation rate of i-erythritol to strain virulence for guinea pigs. J. Bacteriol. *92*, 996–1000.

Meyer, M.E. (1969). Phenotypic comparison of *Brucella ovis* to the DNA homologous *Brucella* species. Am. J. Vet. Res. *30*, 1757–1764.

Michaux-Charachon, S., Bourg, G., Jumas-Bilak, E., Guigue-Talet, P., Allardet-Servent, A., O'Callaghan, D., and Ramuz, M. (1997). Genome structure and phylogeny in the genus *Brucella*. J. Bacteriol. *179*, 3244–3249.

Moreno, E., Stackebrandt, E., Dorsch, M., Wolters, J., Busch, M., and Mayer, H. (1990). *Brucella abortus* 16S rRNA and lipid A reveal a phylogenetic relationship with members of the alpha-2 subdivision of the class *Proteobacteria*. J. Bacteriol. *172*, 3569–3576.

Moreno, E., Cloeckaert, A., and Moriyón, I. (2002). *Brucella* evolution and taxonomy. Vet. Microbiol. *90*, 209–227.

Morgan, W.J., and Corbel, M.J. (1976). Recommendations for the description of species and biotypes of the genus *Brucella*. Dev. Biol. Stand. *31*, 27–37.

O'Callaghan, D., Cazevieille, C., Allardet-Servent, A., Boschiroli, M.L., Bourg, G., Foulongne, V., Frutos, P., Kulakov, Y., and Ramuz, M. (1999). A homologue of the *Agrobacterium tumefaciens* VirB and *Bordetella pertussis* Ptl type IV secretion systems is essential for intracellular survival of *Brucella suis*. Mol. Microbiol. 33, 1210–1220.

Ogata, H., Goto, S., Sato, K., Fujibuchi, W., Bono, H., and Kanehisa, M. (1999). KEGG: Kyoto Encyclopedia of Genes and Genomes. Nucleic Acids Res. *27*, 29–34.

Paulsen, I.T., Seshadri, R., Nelson, K.E., Eisen, J.A., Heidelberg, J.F., Read, T.D., Dodson, R.J., Umayam, L., Brinkac, L.M., Beanan, M.J., *et al.* (2002). The *Brucella suis* genome reveals fundamental similarities between animal and plant pathogens and symbionts. Proc. Natl. Acad. Sci. U.S.A. *99*, 13148–13153.

Porte, F., Naroeni, A., Ouahrani-Bettache, S., and Liautard, J.P. (2003). Role of the *Brucella suis* lipopolysaccharide O antigen in phagosomal genesis and in inhibition of phagosome–lysosome fusion in murine macrophages. Infect. Immun. *71*, 1481–1490.

Rajashekara, G., Covert, J., Petersen, E., Eskra, L., and Splitter, G. (2008). Genomic island 2 of *Brucella melitensis* is a major virulence determinant: functional analyses of genomic islands. J. Bacteriol. *190*, 6243–6252.

Rajendhran, J., and Gunasekaran, P. (2010). Microbial phylogeny and diversity: Small subunit ribosomal RNA sequence analysis and beyond. Microbiol Res. *166*, 99–110.

Rokas, A., Williams, B.L., King, N., and Carroll, S.B. (2003). Genome-scale approaches to resolving incongruence in molecular phylogenies. Nature *425*, 798–804.

Roop, R.M. 2nd, Gaines, J.M., Anderson, E.S., Caswell, C.C., and Martin, D.W. (2009). Survival of the fittest: how *Brucella* strains adapt to their intracellular niche in the host. Med. Microbiol. Immunol. *198*, 221–238.

Sangari, F.J., Seoane, A., Rodríguez, M.C., Aguero, J., and García-Lobo, J.M. (2007). Characterization of the urease operon of *Brucella abortus* and assessment of its role in virulence of the bacterium. Infect. Immun. *75*, 774–780.

Scholz, H.C., Hubalek, Z., Sedlácek, I., Vergnaud, G., Tomaso, H., Al Dahouk, S., Melzer, F., Kämpfer, P., Neubauer, H., Cloeckaert, A., *et al.* (2008a). *Brucella microti* sp. nov., isolated from the common vole *Microtus arvalis*. Int. J. Syst. Evol. Microbiol. *58*, 375–382.

Scholz, H.C., Al Dahouk, S., Tomaso, H., Neubauer, H., Witte, A., Schloter, M., Kämpfer, P., Falsen, E., Pfeffer, M., and Engel, M. (2008b). Genetic diversity and phylogenetic relationships of bacteria belonging to the *Ochrobactrum-Brucella* group by recA and 16S rRNA gene-based comparative sequence analysis. Syst. Appl. Microbiol. *31*, 1–16

Scholz, H.C., Hubalek, Z., Nesvadbova, J., Tomaso, H., Vergnaud, G., Le Flèche, P., Whatmore, A.M., Al Dahouk, S., Krüger, M., Lodri, C., *et al.* (2008c). Isolation of *Brucella microti* from soil. Emerg. Infect. Dis. *14*, 1316–1317.

Sieira, R., Comerci, D.J., Sánchez, D.O., and Ugalde, R.A. (2000). A homologue of an operon required for DNA transfer in *Agrobacterium* is required in *Brucella abortus* for virulence and intracellular multiplication. J. Bacteriol. *182*, 4849–4855.

Smith, H., Williams, A.E., Pearce, J.H., Keppie, J., Harris-Smith, P.W., Fitz-George, R.B., and Witt, K. (1962). Foetal erythritol: a cause of the localization of *Brucella abortus* in bovine contagious abortion. Nature *193*, 47–49.

Snyder, E.E., Kampanya, N., Lu, J., Nordberg, E.K., Karur, H.R., Shukla, M., Soneja, J., Tian, Y., Xue, T., Yoo, H., *et al.* (2007). PATRIC: the VBI PathoSystems Resource Integration Center. Nucleic Acids Res. *35*, D401–406.

Stamatakis, A. (2006). RAxML-VI-HPC: maximum likelihood-based phylogenetic analyses with thousands of taxa and mixed models. Bioinformatics *22*, 2688–2690.

Starr, T., Ng, T.W., Wehrly, T.D., Knodler, L.A., and Celli, J. (2008). *Brucella* intracellular replication requires trafficking through the late endosomal/lysosomal compartment. Traffic *9*, 678–694.

Thacker, U., Parikh, R., Shouche, Y., and Madamwar, D. (2007). Reduction of chromate by cell-free extract of *Brucella* sp. isolated from Cr(VI) contaminated sites. Biores. Technol. *98*, 1541–1547.

Tiller, R.V., De, B.K., Boshra, M., Huynh, L.Y., Van Ert, M.N., Wagner, D.M., Klena, J., Mohsen, T.S., El-Shafie, S.S., Keim, P., *et al.* (2009). Comparison of two multiple-locus variable-number tandem-repeat analysis methods for molecular strain typing of human *Brucella melitensis* isolates from the Middle East. J. Clin. Microbiol. *47*, 2226–2231.

Tiller, R.V., Gee, J.E., Lonsway, D.R., Gribble, S., Bell, S.C., Jennison, A.V., Bates, J., Coulter, C., Hoffmaster, A.R., and De, B.K. (2010a). Identification of an unusual *Brucella* strain (BO2) from a lung biopsy in a 52 year-old patient with chronic destructive pneumonia. BMC Microbiol. *27*, 10, 23.

Tiller, R.V., Gee, J.E., Frace, M.A., Taylor, T.K., Setubal, J.C., Hoffmaster, A.R., and De, B.K. (2010b). Characterization of novel *Brucella* strains originating from wild native rodent species in North Queensland, Australia. Appl. Environ. Microbiol. *76*, 5837–5845.

Tsoktouridis, G., Merz, C.A., Manning, S.P., Giovagnoli-Kurtz. R., Williams, L.E., Mujer, C.V., Hagius, S., Elzer, P., Redkar, R.J., Patra, G., *et al.* (2003). Molecular characterization of *Brucella abortus* chromosome II recombination. J. Bacteriol. *185*, 6130–6236.

Tsolis, R.M., Seshadri, R., Santos, R.L., Sangari, F.J., Lobo, J.M., de Jong, M.F., Ren, Q., Myers, G., Brinkac, L.M., Nelson, W.C., *et al.* (2009). Genome degradation in *Brucella ovis* corresponds with narrowing of its host range and tissue tropism. PLoS One *4*, e5519.

Urakami, T., Oyanagi, H., Araki, H., Suzuki, K.I., and Komagata, K. (1990). Recharacterization and emended description Mycoplana and description of two new of the genus species, *Mycoplana ramosa* and *Mycoplana segnis*. Int. J. Sys. Bacteriol. *40*, 434–442.

Ussery, D.W., Kiil, K., Lagesen, K., Sicheritz-Pontén, T., Bohlin, J., and Wassenaar, T.M. (2009). The genus *Burkholderia*: analysis of 56 genomic sequences. Genome Dyn. *6*, 140–157.

Valdezate, S., Navarro, A., Rubio, V., Garin-Bastuji, B., Albert, D., Hernández, P., Alonso, P.M., and Saéz-Nieto, J.A. (2009). Emergence of a clonal lineage of *Brucella abortus* biovar 3 in clinical cases in Spain. J. Clin. Microbiol. *47*, 2687–2688.

Velasco, J., Romero, C., López-Goñi, I., Leiva, J., Díaz, R., and Moriyón, I. (1998). Evaluation of the relatedness of *Brucella* spp. and *Ochrobactrum anthropi* and description of *Ochrobactrum intermedium* sp. nov., a new species with a closer relationship to *Brucella* spp. Int. J. Syst. Bacteriol. *48*, 759–68.

Verger, J.M., Grimont, F., Grimont, P.A.D., and Grayon M. (1985). *Brucella*, a monospecific genus as shown by deoxyribonucleic acid hybridization. Int. J. Syst. Bacteriol. 35, 292–295.

Verger, J.M., Grimont, F., Grimont, P.A., and Grayon, M. (1987). Taxonomy of the genus *Brucella*. Ann. Inst. Pasteur Microbiol. *138*, 235–238.

Vizcaíno, N., Cloeckaert, A., Verger, J., Grayon, M., and Fernández-Lago, L. (2000). DNA polymorphism in the genus *Brucella*. Microbes Infect. 2, 1089–1100

Vizcaíno, N., Cloeckaert, A., Zygmunt, M.S., and Fernández-Lago, L. (2001). Characterization of a *Brucella* species 25-kilobase DNA fragment deleted from *Brucella abortus* reveals a large gene cluster related to the synthesis of a polysaccharide. Infect. Immun. *69*, 6738–6748.

Vizcaíno, N., Caro-Hernández, P., Cloeckaert, A., and Fernández-Lago, L. (2004). DNA polymorphism in the omp25/omp31 family of *Brucella* spp.: identification of a 1.7-kb inversion in *Brucella cetaceae* and of a 15.1-kb genomic island, absent from *Brucella ovis*, related to the synthesis of smooth lipopolysaccharide. Microbes Infect. *6*, 821–834.

Wattam, A.R., Williams, K.P., Snyder, E.E., Almeida, N.F., Jr., Shukla, M., Dickerman, A.W., Crasta, O.R., Kenyon, R., Lu, J., Shallom, J.M., *et al.* (2009). Analysis of ten *Brucella* genomes reveals evidence for horizontal gene transfer despite a preferred intracellular lifestyle. J. Bacteriol. *191*, 3569–3579.

Whatmore, A.M. (2009). Current understanding of the genetic diversity of *Brucella*, an expanding genus of zoonotic pathogens. Infect. Genet. Evol. *9*, 1168–1184.

Whatmore, A.M., Murphy, T.J., Shankster, S., Young, E., Cutler, S.J., and MacMillan, A.P. (2005). Use of amplified fragment length polymorphism to identify and type *Brucella* isolates of medical and veterinary interest. J. Clin. Microbiol. *43*, 761–769.

Whatmore, A.M., Shankster, S.J., Perrett, L.L., Murphy, T.J., Brew, S.D., Thirlwall, R.E., Cutler, S.J., and MacMillan, A.P. (2006). Identification and characterization of variable-number tandem-repeat markers for typing of *Brucella* spp. J. Clin. Microbiol. *44*, 1982–1993.

Whatmore, A.M., Perrett, L.L., and MacMillan, A.P. (2007). Characterisation of the genetic diversity of *Brucella* by multilocus sequencing. BMC Microbiol. *20*, 7, 34.

Williams, K.P., Sobral, B.W., and Dickerman, A.W. (2007). A robust species tree for the alpha-proteobacteria. J. Bacteriol. *189*, 4578–4586.

Williams, K.P., Gillespie, J.J., Sobral, B.W., Nordberg, E.K., Snyder, E.E., Shallom, J.M., and Dickerman, A.W. (2010). Phylogeny of gamma-proteobacteria. J. Bacteriol. *192*, 2305–2314.

Woese, C.R. (1987). Bacterial evolution. Microbiol. Rev. *51*, 221–271

Xu, J. (2006). Microbial ecology in the age of genomics and metagenomics: concepts, tools, and recent advances. Mol. Ecol. *15*, 1713–1731.

Zygmunt, M. S., S. D. Hagius, J. V. Walker, and P. H. Elzer. (2006). Identification of *Brucella melitensis* 16M genes required for bacterial survival in the caprine host. Microbes. Infect 8, 2849–2854.

The *Brucella* Genomic Islands

3

Marcos Mancilla

Abstract

The genomic islands (GIs) are DNA sequences of several kilobases (kb) that contain genes conferring adaptive advantages to the host bacteria. Despite the enclosed intracellular lifestyle, the sequence analysis of *Brucella* has revealed the presence of various GIs scattered through its genome, a fact that strongly suggests the acquisition of DNA by horizontal transfer events. In addition, the analysis of the genetic content of these regions indicates that they might represent a source of virulence factors. Recently, it has been demonstrated that some predicted GIs are unstable and can be excised from the chromosome by recombination. This experimental evidence points out the foreign origin of such loci and suggests an explanation for the polymorphism related to these large chromosomal regions exhibited by *Brucellae*.

This chapter summarizes the advances in the identification and characterization of *Brucella* GIs. A detailed analysis of their genetic content and its relation to pathogenicity is also included. Recent data obtained for GI instability is discussed in the context of genome plasticity and virulence attenuation observed in smooth *Brucellae*.

Introduction

According to recent reports of the World Health Organization, brucellosis is still one of the major zoonotic infections worldwide spread and every year there are more than half a million new human cases reported (Godfroid *et al.*, 2005; Ficht *et al.*, 2009). The disease is produced by Gram-negative, facultative intracellular microorganisms grouped into the genus *Brucella*. Currently, *Brucella* species are recognized based on the preferential host range and other phenotypic features: *B. abortus* affects primarily cattle and other *bovidae* as well as *cervidae*; *B. suis* is isolated from swine and several forms of wild life; *B. melitensis* infects mainly sheep and goats. *B. canis* is found in dogs; *B. neotomae* is found in desert woodrats; and *B. ovis* is responsible for epididymitis in rams (Moreno *et al.*, 2002). In addition, three new species have been proposed recently: *B. microti*, isolated from common voles (Scholz *et al.*, 2008) and *B. ceti* and *B. pinnipedialis*, both isolated from marine mammals (Foster *et al.*, 2007).

A major aspect of the pathogenicity of *Brucella* is its capacity to survive and replicate within host macrophages. However, since the *Brucella* genome does not seem to encode classical virulence factors found in other well-known bacterial pathogens, the pathogenic nature is yet a matter of discussion and intense research. Among the main virulence factors recognized so far are the poorly endotoxic lipopolysaccharide (LPS) (Goldstein *et al.*, 1992), the type IV secretion system VirB (O'Callaghan *et al.*, 1999), the β-cyclic periplasmic glucan (Arellano-Reynoso *et al.*, 2005) and the virulence control exerted by the BvrR/BvrS system (López-Goñi *et al.*, 2002). These molecular traits are key components of the stealthy strategy used by *Brucella* to avoid the early detection by the host innate immunity, which allows the parasite to reach the replicative intracellular niche and the establishment of a successful infection (Barquero-Calvo *et al.*, 2007).

An interesting research topic that remains only partially explored in *Brucella* is the characterization of gene clusters acquired by horizontal transfer. These loci are more generally referred to as genomic islands (GIs). They are large chromosomal regions which usually encode genes that confer new phenotypic traits upon transfer-acquisition, thus increasing bacterial fitness (Hacker and Kaper, 2000). Pathogenicity islands (PAIs) are an important subclass of GIs found in bacterial pathogens. The term was coined by Hacker and co-workers since virulence genes were found clustered in uropathogenic *E. coli* strains and these clusters exhibited a high frequency of deletions (Hacker *et al.*, 1990). The availability of thousands of genome sequences has permitted comparisons that unexpectedly revealed the presence of GIs among non-pathogenic strains. In recent years, it has become clear that not only virulence genes, but also genes for antibiotic resistance, metabolic fitness and symbiosis are carried on by GIs (Dobrindt *et al.*, 2004). Therefore, genes harboured by GIs could be important in host–pathogen interactions and bacterial survival under challenging environmental conditions. For that reason, GIs are primary targets to look for genetic polymorphisms, which accounts for differences in host or ecological niche specificity.

Modern bioinformatic techniques allow us to identify a GI prior to get experimental 'wet lab' evidence. Current software tools take advantage of the atypical nucleotide composition shown by most of the GIs compared with their host chromosome. Additionally, the search for flanking short direct repeat sequences, that are usually part of tRNA genes, is useful to delimit the borders of GIs. Such computational predictions are confirmed by experimental evidence which mainly involves the assessment of instability. This is carried out by searching for GIs in strain collections and by detecting *in vitro* deletions and transfer events (Gal-Mor and Finlay, 2006). Nevertheless, the most compelling data arise from studies using counter-selectable markers, approach called also island probing (Rajakumar *et al.*, 1997). By this means, strains lacking a complete GI can be obtained, however it is still a difficult and time-consuming task.

The concept of that *Brucella* genome carries GIs has recently gained attention, because *in silico* and experimental studies have not only shown the presence of GIs, but also of loci that can be considered PAIs. In this chapter, an in-depth review is presented along with a discussion of the latest advances in the area.

Bioinformatic prediction of GIs

Once a mobile element has been acquired, following chromosomal integration, positive selection triggers genetic changes such as mutation of replicative origin, inactivation of mobility genes, recombination or gene gain/loss that might result into a GI (Dobrindt *et al.*, 2004). However, GIs often retain their respective codon usages unless amelioration, i.e. the trend to adopt the host genome composition by adaptive mutations, takes place (Marri and Golding, 2008). Therefore, the major feature of a GI sequence is its atypical nucleotide composition (Karlin, 2001). This statement specially applies when the horizontal transfer event took place among non-related bacteria, because the difference in GC content is going to be clearly distinctive. Conventionally, this molecular trait can be rapidly detected at a whole-genome scale by GC content scanning using 5–10 kb sliding windows. Another way to get similar results is by assessment of genomic signature deviation. It is obtained by computing and comparing the dinucleotide abundances through genome segments. Other indexes like the codon usage bias and the amino acids usages can also be assessed with similar outcomes. Software tools developed over recent years have reduced dramatically the calculation time of the analyses and have also incorporated filters to remove highly expressed genes to avoid false-positive predictions (Merkl, 2004). The results of such calculations are visualized by genome viewers in which those DNA segments with a deviation exceeding the threshold can be easily detected (Rutherford *et al.*, 2000). The outcome of a compositional analysis focused on the detection of GIs is a putative 'alien' chromosomal region that can often be accompanied by other common features: presence of transposases, also known as insertion sequences (IS), phage-related integrases/recombinases (*int*) and flanking direct repeat sequences (DRs), all elements related to mobility of GIs. Despite the absence of biological data supporting a minimum size and, because genes displaying an unusual sequence composition have a low likelihood to be randomly together, conventionally clusters of >8 kb are indicative of a horizontal transfer event (Langille *et al.*, 2010), even though some authors named smaller segments between 1–10 kb 'islets' (Gal-Mor and Finlay, 2006). The most important characteristic of GI-like loci is that they usually encode genes whose expression confers adaptive advantages under specific environmental conditions: metabolic degradation pathways (Gaillard *et al.*, 2006), antibiotic resistance (Miriagou *et al.*, 2006), symbiosis (Ramsay *et al.*, 2006) and, of course, pathogenicity traits (Schmidt and Hensel, 2004). As to the latter, in different studies describing spontaneous mutants, a loss of virulence factors encoded by the GI candidates were found responsible for the attenuation (Dobrindt *et al.*, 2004; Middendorf *et al.*, 2004).

Although historical evidence of GIs in *Brucella* genomes were provided by studies on LPS genetics ten years ago (Cloeckaert *et al.*, 2000; Godfroid *et al.*, 2000; Vizcaíno *et al.*, 2001), computational studies including predictions of *Brucella* GIs have also been reported (Mantri and Williams, 2004; Hsiao *et al.*, 2005). The first genome sequence published was that of *B. melitensis*, but in this article GI-like loci were not described (DelVecchio *et al.*, 2002). When the complete genome sequence of *B. suis* was available, it was suggested that three regions shared features of GIs (Paulsen *et al.*, 2002). The article describes a specific island that contains 18 ORFs, termed later as IncP island by Lavigne and colleagues (Lavigne *et al.*, 2005b). In addition, two regions of >7 kb, in this review termed as GIBs (genomic island of *Brucella suis*), were reported. A comparative study performed with the whole genome sequences of the main zoonotic species *B. melitensis* 16M, *B. suis* 1330 and the *B. abortus* 9-941 draught sequence identified the IncP island as a *B. suis* element, a common transposon fragment of 3.9 kb termed Tn*2020* (Halling and Zuerner, 2002), two genetic islands of 21 kb (GI-3) and 25 kb (GI-8) (Vizcaíno *et al.*, 2001) which are absent in *B. suis* and *B. abortus*, respectively, and a phage remnant of 3.5 kb absent in *B. melitensis* previously reported by Paulsen and colleagues (2002) (Halling *et al.*, 2004) (see Table 3.1 for details). Later, a work based on comparative genomics identified nine chromosomal regions with GI common features across the genomes of the six main *Brucella* species (Rajashekara *et al.*, 2004). Interestingly, identical GI deletions were found in multiple strains of the same species, a fact that led the authors to suggest that the acquisition-deletion of GIs may be related to the host preference manifested by the corresponding species. In contrast with previous data, this report suggested that those regions, probably horizontally acquired, played an important role in shaping the *Brucella* genome. One year later, with the whole-genome sequence of *B. abortus* 9-941 was finished and a new comparison was made. In this study, DNA fragments unique to *B. abortus*, *B. suis* or *B. melitensis* were identified (Halling *et al.*, 2005). Several of these corresponded to GIs reported by Rajashekara and co-workers in 2004. A locus of interest is that encoding an autotransporter in *B. abortus*, but a pseudogene in *B. melitensis* and *B. suis*. This gene is part of 4 kb region located in chromosome II, a locus suggested to be involved in host specificity.

Sequence comparisons with other intracellular pathogens have revealed the presence of virulence factors in the *mgtC* region of chromosome II, which is absent in the genome of closely relatives (Blanc-Potard and Lafay, 2003). A prediction based on the identification of two relevant features of GIs, the integration site and the cognate integrase, was included in the paper of Mantri and Williams in 2004. In this work, only GI-2 and GI-3 were identified in *B. melitensis* and listed in the database along with GI candidates of other prokaryotic genomes (http://kementari.bioinformatics.vt.edu/cgi-bin/islander.cgi). The poor detection of GIs presented in this study can be attributed to the stringency of the analysis, because GIs carrying integrases that are also integrated in tRNA genes are less frequent. An extensive *in silico* identification of bacterial GIs was performed by Hsiao and colleagues in 2005. In this report, twenty one putative *Brucella* GIs were predicted taking into consideration the dinucleotide bias and the presence of mobility genes as predictors. Several of the GIs described in this study matched

Table 3.1 Relevant features and distribution of *Brucella* genomic islands

Locus	Size (kb)	GC%	Site of integration	DR length	Mobility genes	ORF content	Chr[a]	Presence in[b]					
								Ab	Me	Su	Ov	Ca	Mi
wbk	20.3[c]	50.0	–	–	IS	26 ORFs encoding enzymes for LPS transport and O-PS synthesis. Several transposases	I	+	+	+	+	+	+
virB	12.8	56.0	–	–	—	11 ORFs encoding T4SS required for virulence	I	+	+	+	+	+	+
IncP	18.3	55.3	–	52	*int*	18 plasmid-related ORFs	II	–	–	+[d]	–	+	+
mgtC	8.0	59.3	$tRNA_{Val}$	–	–	7 ORFs, Mg^{2+} uptake-related genes *mgtB-mgtC* and HPs	II	+	+	+	+	+	+
GIBs2	7.0	60.7	–	–	–	10 ORFs, mostly phage-related proteins	I	+	–[e]	+	+	+	+
GIBs3	7.7	56.1	–	–	–	6 ORFs, ABC binding proteins, amino acid dehydrogenase and conserved HP	II	+	–	+	+	+	+
4 kb[f]	10.1	56.2	–	–	–	6 ORFs, HPs and a autotransporter	II	+	+	+	+	+	+
13 kb	13.1	56.5	–	–	–	2 ORFs, a HP and a autotransporter	II	+	+	+	+	+	+
GI-1	8.1	53.3	$tRNA_{Leu}$	87	*int*	9 ORFs, mostly HPs and phage-related proteins	I	+	+	+	–	+	+
GI-2	15.1	51.3	$tRNA_{Gly}$	41	*int*, IS	20 ORFs, LPS glycosiltransferases *wboA-wboB*, *omp25*, HPs and phage-related	I	+	+	+	–	+	+
GI-3	20.9	52.3	$tRNA_{Phe}$	18	*int*	30 ORFs, mostly HPs. TIR domain protein	I	+	+	–	+	–	–
GI-5	44.1	57.0	Between $tRNA_{Ser/Met}$	61	IS	42 ORF, peptide ABC transporters, regulators, HPs, antibiotic activators.	II	+	+	+	–	+	+
GI-6[g]	7.5	54.7	$tRNA_{Ser}$	–	IS	10 ORFs, transposases, transcriptional regulator and a haemagglutinin	II	+	+	+	+	+	+
GI-8	25.1	58.0	–	4	–	25 ORFs, sugar metabolism and LPS synthesis	II	–	+	+	+	+	+
SAR1–2	8.1	53.7	$tRNA_{Thr}$	34[h]	*int*	11 ORFs, HPs and phage-related genes	I	+	+	+	+	+	+
SAR1–5	13.8	56.5	$tRNA_{Arg}$	–	*int*, IS	13 ORFs, exonuclease, acetyltransferase, phage regulator and HPs	I	+	+	+	+	+	+

SAR1–17	19.4	51.1	–	–	–	14 ORFs, putative Ser-protease and adhesion proteins, exonuclease, ribosomal protein L28, Mer-regulator, Cd^{2+} transporter	I	+	+	+	+	+	+
SAR2–10	14.3	56.4	$tRNA_{Thr}$	–	–	13 ORFs, Restriction-modification system and HPs	II	+	+	+	+	+	+
26.5 kb[i]	26.5	53.8	–	–	*int*[j], IS	32 ORFs, transposases, an ABC transporter, a putative haemagglutinin and HPs	II	–	–	–	+	–	–
12 kb	11.7	53.2	$tRNA_{Leu}$	7	*int*[j]	5 ORFs, phage-related proteins	II	–	–	–	–	–	+

[a]Chr, chromosome I or II. [b]Ab, *B. abortus* 2308; Me, *B. melitensis* 16M; Su, *B. suis* 1330; Ov, *B. ovis* ATCC 25840; Ca, *B. canis* ATCC 23365; Mi, *B. microti*; +, presence; –, absence. [c]Size calculated based on the report of Wattam *et al*. (2009). [d]Absent in *B. suis* biovar 5. [e]It is present in *B. melitensis* biovar 2 and partially absent in *B. melitensis* 16M. [f]Only *B. abortus* seems to encode a full-length autotransporter. [g]GI-6 is absent in *B. neotomae* (Rajashekara *et al*., 2004). [h]Bioinformatic predicted DR pair, confirmed by PCR (see Fig. 3.4). [i]This island is also present in *Brucella* strains isolated from bottlenose dolphins. [j]Pseudogenes. ORF-based co-ordinates: *wbk*, BMEI1393–1422; *virB*, BMEII0024–0034; IncP, BRA0361–0380; *mgtC*, BRA0036–0042; GIBs2, BR0584–0593; GIBs3, BRA0630–0635; 4 kb, BruAb2_0165–0171; 13 kb, BruAb2_1085–1086; GI-1, BMEI0899–0907; GI-2, BMEI0993–1012; GI-3, BMEI1674–1703; GI-5, BMEII0185–0226; GI-6, BMEII0710–0719; GI-8, BMEII0826–0850; SAR1–2, BMEI1656–1664; SAR1–5, BMEI1218–1231; SAR1–17, BMEI0051–0065; SAR2–10, BMEII0447–0457; 26.5 kb, BOV_A0482–0515 and 12 kb, BMI_II1048–1054.

with the formerly reported GIs, but new ones were also predicted. However, most of new loci were under the cut-off size and carried hypothetical genes without assigned function, a fact that was interpreted by the authors as a consequence of the GIs being a source of novel genes. Another sequence analysis-based report was recently published (Wattam *et al.*, 2009). In this work, seventeen shared anomalous regions (SARs) were described, a prediction at genus level based on a comparison of ten *Brucella* genomes from six different species. SARs were shared among all the species examined and they were found completely or partially absent from the phylogenetically close relatives like *Ochrobactrum*. It is to note that this prediction was made using the GI predictor Alien_Hunter, which displays the highest recall, but at the expense of precision, and therefore is highly recommended for an extensive capture at the first glance (Langille *et al.*, 2008). It is worth considering the similarity between the results of the compositional analyses performed by Hsiao and Wattam, and their contrast to the GIs identified by comparative genomics in Rajashekara's report. Such discrepancy is probably due to the methods chosen by the authors: while the compositional approach is mainly theoretical and might contain false-positive predictions, comparative genomics is based on biological evidence and related to the absence of gene clusters in genomes of closely related microorganisms. Regarding Hsiao and Wattam reports, the differences between their results mainly concern the number and size of the GIs. The latter depends on the identification of the integration site and flanking repeats sequences. Since the compositional approach does not incorporate these analyses, the outcome is a GI candidate with diffused boundaries. In order to test GI instability, a more refined analysis can be performed by flanking DR sequences prediction (Kurtz *et al.*, 2001). Recently, a genomic study performed on *Brucella* species affecting marine mammals revealed the presence of a GI not shared with other species. This island was shown to carry metabolic genes (Maquart *et al.*, 2008). At the time of writing, neither strains of marine species nor *B. neotomae* had fully annotated genomes, so they were not included in this review. Tsolis and co-workers reported a 26.5 kb region located on chromosome II that corresponds to a GI only found in *B. ovis* and *B. ceti* strains (Tsolis *et al.*, 2009). The structure of this locus suggests a composite transposon that encodes several transposases, ABC transporters, a gene encoding a haemagglutinin and hypothetical proteins (HPs). In the case of *B. microti*, an island of 12 kb located on chromosome II was identified (Audic *et al.*, 2009). This locus has been exclusively detected in *B. microti* and its organization resembles a phage remnant.

In the comparison presented in Table 3.1, we have compiled the data of genome sequences and bioinformatic analyses reported to date. There is a trend in previous reports to consider different kinds of horizontally acquired sequences as GIs, so the aim was to do a review in the context of the broadest definition of the term, but we think that some loci do not meet all the criteria accepted to be a 'canonical' GI. For simplicity and because several of these regions do not reach the conventional cut-off length, in our analysis we have considered those regions sized near or larger than 8 kb. Table 3.1 summarizes the characteristics of 20 GIs that show an atypical composition or have been identified by comparative genomics and meet one of the following criteria: size >7 kb, carrying mobility-related genes, and a tRNA gene or DR sequences as integration sites. We have maintained the names of *wbk*, *virB* and IncP for convenience and because they are well known in the literature. The GIs range in size from 7 to 44 kb. Remarkably, altogether they represent almost 10% of the genome, a similar fraction found in other microorganisms (Nakamura *et al.*, 2004). Most GIs correlate with a marked GC deviation, e.g. *wbk*, GI-2, GI-3 and SAR1–17. Many of them are integrated within a tRNA gene and encode mobility genes. On the contrary, few loci are found delimited by a well-defined pair of DRs. Interestingly, four GIs present all the hallmarks described for this kind of loci: IncP, GI-1, GI-2 and GI-3. There are GIs with known virulence genes e.g. *virB*, *wbk*, *mgtC*, GI-2 and GI-3. On the other hand, we found GIs carrying genes with putative metabolic functions (GI-5, GI-8, 26.5 kb), and some encoding outer membrane proteins (GIBs3, 4 kb, 13 kb) and phage-related proteins (GIBs2, GI-1, SAR1-2, 12 kb); but in general, these loci harbour genes with unknown functions. Indeed, comparative genomics studies indicate that this pool of novel genes is found preferentially in GI-like loci, which can not be an artefact generated by the computational methods (Hsiao *et al.*, 2005; Ho Sui *et al.*, 2009). It is observed a homogenous distribution of GIs between both chromosomes. Furthermore, there is an apparent selective pressure to maintain some loci, for example: *wbk*, *virB* and *mgtC* along with SAR1-2, 1-5, 1-17 and 2-10 are present in all the *Brucella* genomes available so far, whereas others specific islands such as those identified in *B. ovis* and *B. microti* are mainly restricted to the corresponding species, with the exception of the *B. ovis* 26.5 kb GI that is also found in some marine *Brucella* strains (Tsolis *et al.*, 2009). The GIs with distinctive genome signatures mentioned in this comparison represent those candidates which are not affected by amelioration, or which have been transferred by other bacteria with a distinctive genome signature.

To illustrate how GIs are predicted, a simple bioinformatic exercise can be done. If we take the genome sequence of *B. abortus* 2308 (Chain *et al.*, 2005), put it into a genome viewer (Artemis or similar) and use it to assess the GC content and genomic signature deviations, we get the plot displayed in Fig. 3.1. We observe several deviation zones, but 15 well-defined peaks displaying a difference in genomic signature, i.e. $\delta > 0.05$ (a threshold commonly used), mostly with a lower GC content (average 57%), which strongly suggests that these regions might be horizontally acquired sequences. Because highly expressed

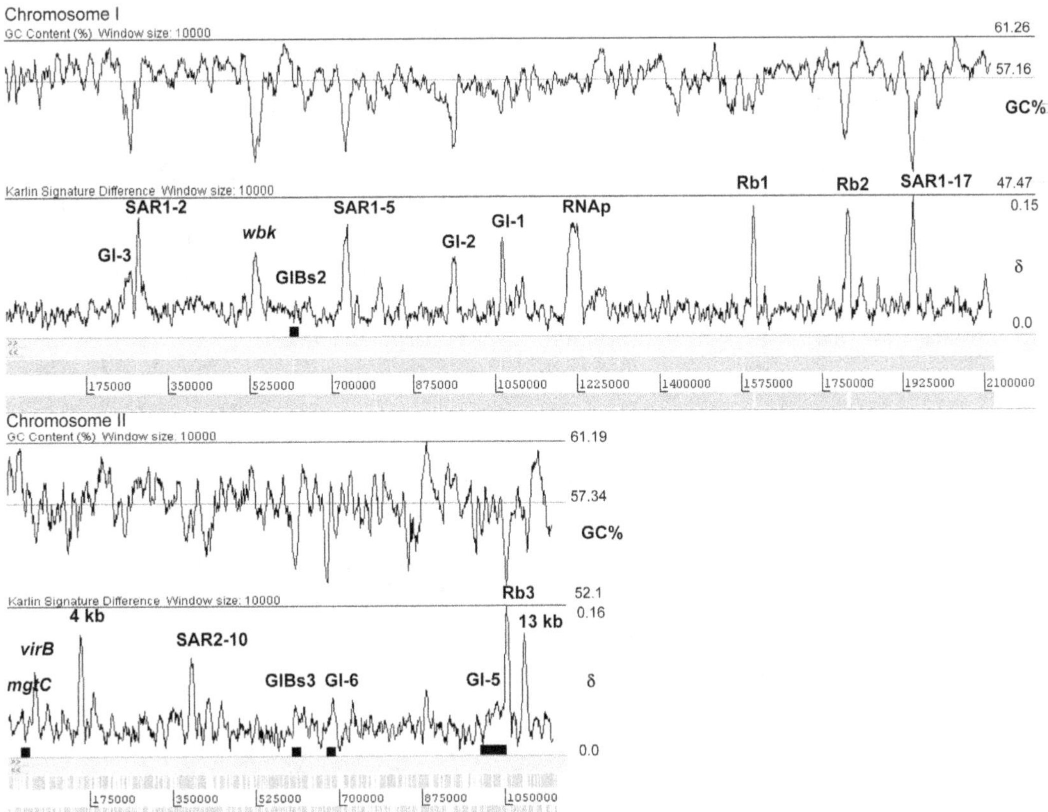

Figure 3.1 Whole genome scanning for horizontal transfer events in *B. abortus*. Artemis screenshots of *B. abortus* 2308 chromosomes AM040264 and AM0402645 showing the GC content and Karlin signature difference (δ) throughout the corresponding sequences (sliding window=10 kb). The average of GC content is shown by continuous line (nearly 57% in both chromosomes). Only those peaks with δ>0.05 are indicated. Black boxes are regions that do not display a marked genomic signature deviation, but they have been detected by comparative genomics. A full description of each GI is found in Table 3.1.

genes, such as those encoding the transcriptional machinery and ribosomal RNAs, were not filtered, so corresponding clusters also appear with larger deviations. This results in false-positive predictions and is therefore marked as RNAp and Rb 1, 2 and 3. Most of the *Brucella* GIs described in Table 3.1 are reflected in the ensuing graph. There are two clusters encoding autotransporters (4 kb and 13 kb) that show a strong Karlin signature variation. One of these signals, the 13 kb island, is found overlapping the locus I of flagellar genes required for full virulence in mice (Fretin *et al.*, 2005). Obviously, those GIs that are absent in the *B. abortus* genome do not appear: IncP, GI-8 and the specific 26.5 kb *B. ovis* and 12 kb *B. microti* islands. The comparison of the bioinformatic-predicted GI borders with those reported in comparative studies shows minor differences with the exception of *wbk*. The LPS-related *wbk* locus contains numerous transposases; this means that several sets of repeated sequences can be identified making it difficult to predict the boundaries with accuracy. In fact, this GI seems to have evolved from more than one acquisition event since a cluster of six genes is also found in *Yersinia enterocolitica* O:9, a pathogen with identical O-polysaccharide structure to that of *Brucella* (Skurnik *et al.*, 2007), but the nearby glycosyltransferase *wbkA* flanked by transposases IS*Bm1* is absent (Fig. 3.2). Fig. 3.1 shows the contribution of the horizontal transfer to the actual *Brucella* genome and it is consistent with the idea that *Brucella* might have evolved from an ancient free-living organism able to uptake and integrate foreign genetic material (Ficht, 2010).

Genetic repertoire of *Brucella* GIs

Mobile elements such as GIs might have evolved from a combination of functional gene modules originated from multiple origins, which makes difficult their categorization (Juhas *et al.*, 2009). Based on the functional classes of genes contained in *Brucella* GIs, we have divided them into two groups: GIs with well-known or putative virulence factors and GIs harbouring genes other than virulence factors, e.g. metabolic, plasmid or phage-related islands and hypothetical or unknown genes. Bioinformatic predictions and the cumulative experimental evidence are in agreement with the fact that *Brucella* GIs have clustered genes for virulence and metabolic flexibility, but which *Brucella* GIs are related to pathogenicity? In this section, we will try to address this question and we will discuss the information related to virulence in the context of mutational analysis performed on ORFs harboured by GIs.

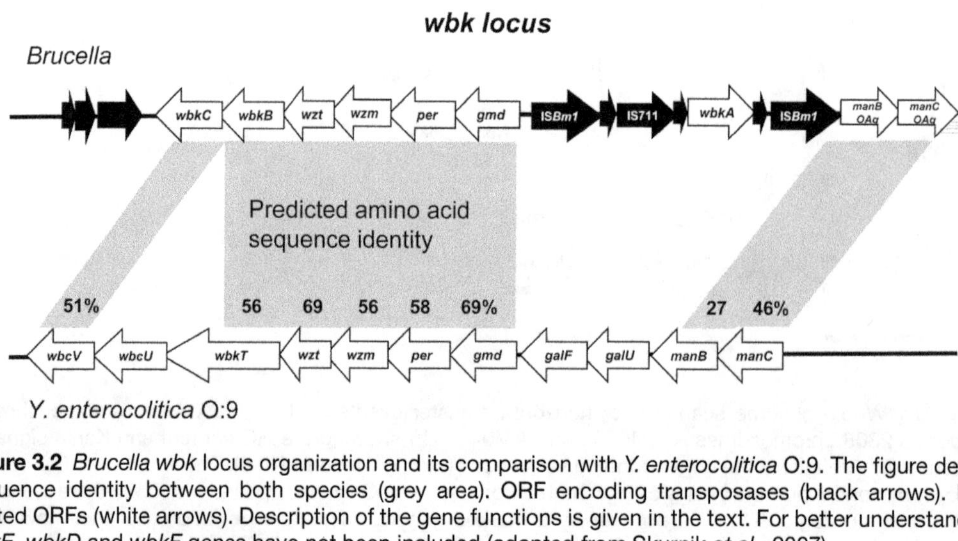

Figure 3.2 *Brucella wbk* locus organization and its comparison with *Y. enterocolitica* O:9. The figure depicts sequence identity between both species (grey area). ORF encoding transposases (black arrows). LPS-related ORFs (white arrows). Description of the gene functions is given in the text. For better understanding, *wbkE*, *wbkD* and *wbkF* genes have not been included (adapted from Skurnik *et al.*, 2007).

GIs carrying virulence factors

It has long been known that LPS is one of the major virulence factors of *Brucellae*. Genes that participate in LPS biosynthesis are clustered in two main regions of chromosome I: *wbk* and *wbo*. The former has been reported as a locus of ten LPS-related genes, including those encoding O-polysaccharide glycosiltransferases (*wbkA* and *wbkC, wbkE*), enzymes for LPS precursors (*gmd, per, wbkD, wbkF*) and proteins involved in O-polysaccharide transport across the inner membrane (*wzm, wzt*). Sequence analysis showed that several transposases split up the locus into two sections: an array of six genes (*gmd, per, wzm, wzt, wbkB* and *wbkC*) and inversely orientated the *wbkA* glycosyltransferase flanked by a pair of IS*Bm1* transposases (Godfroid *et al.*, 2000) (see Fig. 3.2). The glycosiltransferases tandem *wboA-wboB* is part of GI-2 and both are required to build up a complete LPS O-polysaccharide molecule (McQuiston *et al.*, 1999; Rajashekara *et al.*, 2008). GI-2 includes its own *int*, a copy of the IS*711* transposase, several HPs and a gene encoding the immunogenic protein Omp25, which is functional in *B. melitensis* (Vizcaíno *et al.*, 2004). *Brucellae* also contain LPS-related genes clustered in GI-8. It has been speculated that genes belonging to this island might be involved in the biosynthesis of a polysaccharide not present in *B. abortus* since the island is absent in this species (Vizcaíno *et al.*, 2001). Although the expression of several ORFs of this region has been recently reported (Rossetti *et al.*, 2009), there is not biochemical evidence of such unknown polysaccharide.

Among islands encoding virulence factors other than LPS, a cluster with a well-proven link to virulence is *virB*. It is known that the *virB* locus encodes a T4SS essential for full virulence (O'Callaghan *et al.*, 1999). Because it shows high similarity with the T4SS of *Agrobacterium tumefaciens* and *Bordetella pertussis*, it is expected to have analogous functions in *Brucella*. Thus, it has been hypothesized that this secretion system is a channel to export molecules that subvert host-defence mechanisms, but despite the discovery of protein candidates (de Jong *et al.*, 2008), the role is not yet well established.

As we mentioned above, an 8-kb locus encoding genes involved in Mg^{2+} uptake was detected by sequence similarity with well-known intracellular parasites (Blanc-Potard and Falay, 2003). Since Mg^{2+} is a cofactor required for the intracellular replication of *Salmonella enterica* and *Mycobacterium tuberculosis*, the authors suggested that *mgt* genes might have analogous roles in *Brucella*. Indeed, a *B. suis* insertional mutant of *mgtC* (BRA0040) was shown a lower intramacrophagic multiplication rate (Lavigne *et al.*, 2005a). The partial abolishment of such effect by the addition of Mg^{2+} to culture medium not only indicated that *Brucella*-containing phagosomes are Mg^{2+}-deprived, but also the ability to survive in a low-Mg^{2+} environment is a common strategy used by unrelated intracellular pathogens.

The 21 kb GI-3 island carries a gene encoding a TIR domain protein Btp1 that has been linked to virulence (Salcedo *et al.*, 2008). The protein is involved in the down-regulation of the NF-κB cascade and in the arrest of host dendritic cells maturation, thus it is interfering with TLR2 signalling. As a consequence, inflammatory cytokines secretion and antigen presentation to T lymphocytes is impaired, that contributes to avoid the pathogen recognition by the host immune system.

Concerning SAR1-2, a transposon mutational-based study in *B. suis* indicated that the ORF BMEI1658, which encodes a hypothetical protein of 94 amino acids, is involved in the multiplication within human macrophages (Kohler *et al.*, 2002). SAR1-5 encodes an exonuclease (BMEI1229), which participates in the acute phase of infection. This gene was identified during a screening of signature-tagged mutants tested in mice (Lestrate *et al.*, 2003; Delrue *et al.*, 2004).

Recently, a transcriptional regulator BasR/PmrA (BMEI0066) has been reported as essential for *B. melitensis* replication in murine macrophages (Zhang *et al.*, 2009). The protein is thought to be related to resistance against hyperosmotic growth conditions and the corresponding gene is harboured by SAR1-17. This island also contains a coding region for an autotransporter in *B. suis*, BR2013, that presents non-sense mutations in *B. melitensis* and *B. abortus* (Chain *et al.*, 2005). In *B. abortus* 9-941, both the 4 kb and 13 kb regions were shown to carry genes similar to BR2013, BruAb2_0168 and BruAb2_1085, respectively (Halling *et al.*, 2005). Autotransporters are large proteins possessing a unique structure that facilitate their own transport across the bacterial membrane. They are grouped into a big family of proteins found in Gram-negative pathogens (Nishimura *et al.*, 2010). These proteins are exported to the outer membrane interacting directly with host-cells membranes, therefore it has been speculated that they might be related to *Brucella* virulence and host-preference (Moreno and Moriyón, 2002). In fact, the autotransporter encoded by the *ompA* gene carried by SAR1-17 influences the survival of *B. suis* 1330, because the isogenic mutant was cleared from spleens of infected mice faster than the parental strain (Bandara *et al.*, 2005). SAR1-17 might carry more virulence factors as suggested by the presence of a gene encoding a putative adhesin (BMEI0063). Although the ORF is annotated as a pseudogene in *B. abortus* 2308 (BAB1_2009), the corresponding protein has been detected in a library of secreted/surface exposed proteins in this species (Marchesini *et al.*, 2004). Important functions other than virulence also seem to be encoded by the same island, because it contains a housekeeping gene (ribosomal protein L28) and efforts to obtain insertional mutants of *xth-2*, the homologue of exonuclease *xth-1*, have been unsuccessful (Hornback and Roop, 2006).

With regards to GI-6, Rajashekara and colleagues showed that a *B. melitensis* GI knock-out mutant had a similar behaviour to that of the parental strain in the murine model, even though a recent report indicated that the haemagglutinin gene carried by the GI may contribute to the abortion rate observed in caprine model of infection (Perry *et al.*, 2010).

GIs encoding genes other than virulence factors

Within the second group of GIs there are those carrying genes with a putative metabolic role. GIBs3 is predicted to encode two ABC transporters. One of them contains a putative glutamate binding protein encoded by BAB2_0612, that was detected in *B. abortus* 2308, but not in *B. melitensis* 16M (Eschenbrenner *et al.*, 2006). The *B. ovis* 26.5 kb island carries five ORFs encoding the same class of transporters found in GIBs3, but to date there is no evidence of their functionality. Whole-genome expression studies carried out on *B. melitensis* and *B. abortus* indicate that several ORFs of GI-5 encoding ABC transporters are transcriptionally active (Rossetti *et al.*, 2009; Viadas *et al.*, 2009). In addition, there is proteomic evidence for genes within the same region (BAB2_1055 and BAB2_1062) that are being differentially expressed in *B. abortus* 2308 (Eschenbrenner *et al.*, 2006; Lamontagne *et al.*, 2009). Surprisingly, several deletion mutants encompassing the entire 44.1 kb GI-5 island of *B. melitensis* were not attenuated in mice (Rajashekara *et al.*, 2008). While transcriptomic studies indicate gene expression in both *B. abortus* and *B. melitensis*, proteomic data is only available for *B. abortus* suggesting the functionality in this species. It is known that homologues of the transporters carried by GI-5 are involved in host–pathogen interactions in other bacteria, therefore its absence in *B. ovis* may be related to susceptibility of this species to cationic peptides (Rajashekara *et al.*, 2008). Noteworthy, since a clearly metabolic trait associated to these GIs is difficult to prove, the possibility cannot be discarded of individual genes contained in these loci having a subtle participation in fitness, but in a whole they may make a significant contribution.

Another category of GIs includes the ones encoding plasmid or phage-related genes. The IncP locus serves as an example. Insertional mutants constructed on three ORFs of that region were tested on cell lines (*int*, *traL* and BRA0375) and did not show any intracellular growth defect (Lavigne *et al.*, 2005b). Since this island encodes plasmid-derived genes and HPs and because it is absent in *B. melitensis* and *B. abortus*, it is probable that it does not have a role in virulence. A complete deletion mutant of GI-1 tested on RAW macrophages and mice showed a pattern similar to that of the parental *B. melitensis* 16M (Rajashekara *et al.*, 2008), indicating that this island is not essential for intracellular replication. GIBs2 and the specific island found in *B. microti* (12 kb) are also classified as phage-related islands. To date, there is no data available for any of the ORFs contained in both GIBs2 and 12 kb that relate with pathogenicity.

The final category to discuss consists of those GIs bearing hypothetical or unknown genes. Such is the case of SAR2-10 which also carries a putative restriction-modification system. GI-3 fits into this category owing to the fact its ORFs encode proteins of unknown function, with the exception of the *btp1* gene, the flagella-related gene *flgJ* and the *int*. Interestingly, it seems that almost all ORFs carried by the island are expressed under standard growth conditions (Ratushna *et al.*, 2006), which suggests that there are other unknown functions associated to this locus.

In summary, studies focused on demonstrating a role in pathogenicity for *Brucella* GIs suggest that GI-1, GI-5, GI-6 and the IncP island are not critical for establishing a successful infection at least in mice, but since they are absent in specific species, one possibility is that they are needed for full virulence in the corresponding animal host. On the other hand, genes encoding known and putative virulence factors clustered in *wbk*, *virB*, *mgtC*, GI-2, GI-3, SAR1-2, SAR1-5, SAR1-17, 4 kb and 13 kb islands indicate that the horizontal transfer has been important in the transition from a commensally free-living state to the enclosed pathogenic intracellular life-style.

Genome plasticity and evolution of *Brucellae*

Computational predictions should always be supported by experimental evidences in order to demonstrate the robustness of such *in silico* analyses. Along with sequence features mentioned above, GIs often display instability, which means that they have the tendency to excise spontaneously from the host chromosome. This trait entails the propagation of GIs among bacterial populations in a particular environmental condition (Dobrindt *et al.*, 2004). GI instability is part of a set of genomic rearrangements promoted by recombination known as genome plasticity. Spontaneous deletions affecting GIs is one of the outcomes of genome plasticity, resulting in the absence of one or several GIs in a specific species or strain. The GI-deletion-associated polymorphism might be related with the virulence manifested by different pathogenic strains (Lloyd *et al.*, 2009), whereas in environmental bacteria it would result in phenotypic variation (Lechner *et al.*, 2009). Additionally, it has been speculated that the excision and subsequent loss of GIs could be related with an advantageous phenotypic change that allowed a fast adaptation, and in the case of pathogens, eventually favoured the chronicity of the infection (Hacker *et al.*, 2003).

Several elements are necessary for GI instability to take place (Hochhut *et al.*, 2006). DNA motifs or *cis* elements represented by short identical or nearly identical direct repeats sequences with around or less than 100 bp should be present at the borders of GIs. Integrases (Int) are the main enzymes that recognize DRs as a target for site-specific recombination, thus they are catalysing the integration/excision of GIs. These enzymes are not only associated to the mobilization of GIs, but they also mediate recombination of the attachment sites (*att*) of bacteriophages and a class of self-transmissible islands called integrative-conjugative elements (ICEs) (Juhas *et al.*, 2009). Finally, excisive recombination is often stimulated by small proteins called excisionases (Xis). The *trans* elements mentioned above, Int and Xis, are frequently encoded by the cognate GI and their genes have been identified taking advantage of the similarity to components of the well-known model of lambda phage integration/excision, where these proteins are essential to accomplish the viral life cycle (Lewis and Hatfull, 2001). The assessment of instability considers four main aspects: (i) to identify any GI-linked deletion event by comparative genomics among microorganisms closely related. If this is not feasible, DR sequences prediction will help to define the theoretical GI borders and that makes subsequent steps easier; (ii) if there is an intact *int* within the GI, the detection of recombination products (by PCR) is mandatory. At this point, it is important to keep in mind that the detection of the chromosomal scar and the circular intermediate is not trivial, because the amount of both recombination products depends on the frequency of the GI excision event, which is generally quite low; (iii) in order to demonstrate the phenotype, the construction of a mutant of the complete GI is recommended; (iv) mutations on *int* or *xis* genes will help to get an insight into the control of the excision and transfer (if occurs) of the GI.

Table 3.2 Integrases/recombinases found in *Brucella* GIs

Locus	*B. abortus* 2308	*B. melitensis* 16M	*B. suis* 1330	*B. ovis* ATCC25840	*B. canis* ATCC23365	*B. microti*
IncP	–	–	BRA0362 (402)	–	BCAN_B0365 (402)	BMI_II357 (402)
GI-1	BAB1_1100 (414)	BMEI0907 (82)	BR1078 (414)	–	BCAN_A1095 (414)	BMI_I1085 (414)
GI-2	BAB1_0983 (341)	BMEI1012 (336)	BR0966 (341)	–	BCAN_A0980 (341)	BMI_I967 (341)
GI-3	BAB1_0250 (398)	BMEI1702 (398)	–	BOV_0239 (384)	–	–
SAR1–2	BAB1_0288 (588)	BMEI1664 (588)	BR0256 (584)	BOV_0277 (pseudo)	BCAN_A0261 (588)	BMI_I261 (584)
SAR1–5	BAB1_0747 (704)	BMEI1223 (653)	BR0729 (704)	BOV_0723 (pseudo)	BCAN_A0744 (578)	BMI_I727 (704)

The length in amino acids of each ORF is indicated in brackets.

Integrases/recombinases carried by *Brucella* GIs

The *Brucella* GIs contain at least six different full-length *int* genes that are listed in Table 3.2. All of them belong to the superfamily of tyrosine recombinases. The *int* carried by SAR1-5 presents an integrase and a Mu-transposase domain found in several phage-related *int*. Remarkably, the *int* of GI-2 is a XerC/XerD-like integrase. Members of this family have been linked to the excision of PAIs in the plant pathogen *Pseudomona syringae* pv. *phaseolicola* (Pitman *et al.*, 2005) and *Pseudomonas aeruginosa* (Qiu *et al.*, 2006). Genes encoding *int* of IncP and GI-3 islands show characteristics of P4-phage integrases that mediate the excision/transfer of well-known GIs, e.g. of the *Mesorhizobium loti* R7A symbiosis island (Ramsay *et al.*, 2006), the *clc* element of *Pseudomonas sp* B13 (Ravatn *et al.*, 1998), or PAIs from uropathogenic *E. coli* strains (Hochhut *et al.*, 2006), *Yersinia pseudotuberculosis* (Lesic *et al.*, 2004) and *Shigella flexneri* (Sakellaris *et al.*, 2004). We have studied the expression of *int* genes carried by GI-1, GI-2, GI-3 and SAR1-2 in the three main zoonotic species *B. melitensis*, *B. abortus* and *B. suis* (Fig. 3.3). The expression of GI-2 and GI-3 *int* was detected by conventional reverse transcription PCR (RT-PCR) (Fig. 3.3A), but using real-time RT-PCR (qRT-PCR), GI-1 and SAR1-2 *int* mRNA were also identified (Fig. 3.3B). In a previous work, all ORFs encoded by GI-3 shown to be active with the exception of *int* (Ratushna *et al.*, 2006), however, it is clear that the gene is being expressed (Fig. 3.3A). These results suggest that the expression of some *int* under standard growth conditions is constitutive, while others are poorly expressed or may be even down-regulated. As it is shown in Table 3.2, the *int* genes carried by the GIs of the three species are similar in length. Curiously, *B. melitensis* GI-1 apparently encodes an Int of 82 amino acids (BMEI0907); hence it would be smaller than the corresponding *B. abortus* and *B. suis* orthologues BAB1_1100 and BR1078. However, ORF prediction performed with Glimmer using the sequence of *B. melitensis* 16M revealed an ORF similar to its *B. abortus* orthologue (M. Mancilla, unpublished). The finding of an amplicon identical to the one obtained for *B. abortus* and *B. suis* (Fig. 3.3B, upper panel) confirms the discrepancy in the annotation of BMEI0907. The expression of *int* genes, along with the presence of flanking DRs strongly indicates that such GIs may be unstable elements. In fact, the expression of the IncP *int* has been linked to the appearance of site-specific recombination products derived from the GI excision (Lavigne *et al.*, 2005b). More recently, the functionality of GI-2 *int* has also been demonstrated by mutational analysis (Mancilla *et al.*, 2010). Additionally, the circular intermediates and chromosomal scars corresponding to the recombination products of excisions of GI-1, GI-3 and SAR1-2 islands have also been identified (Fig.

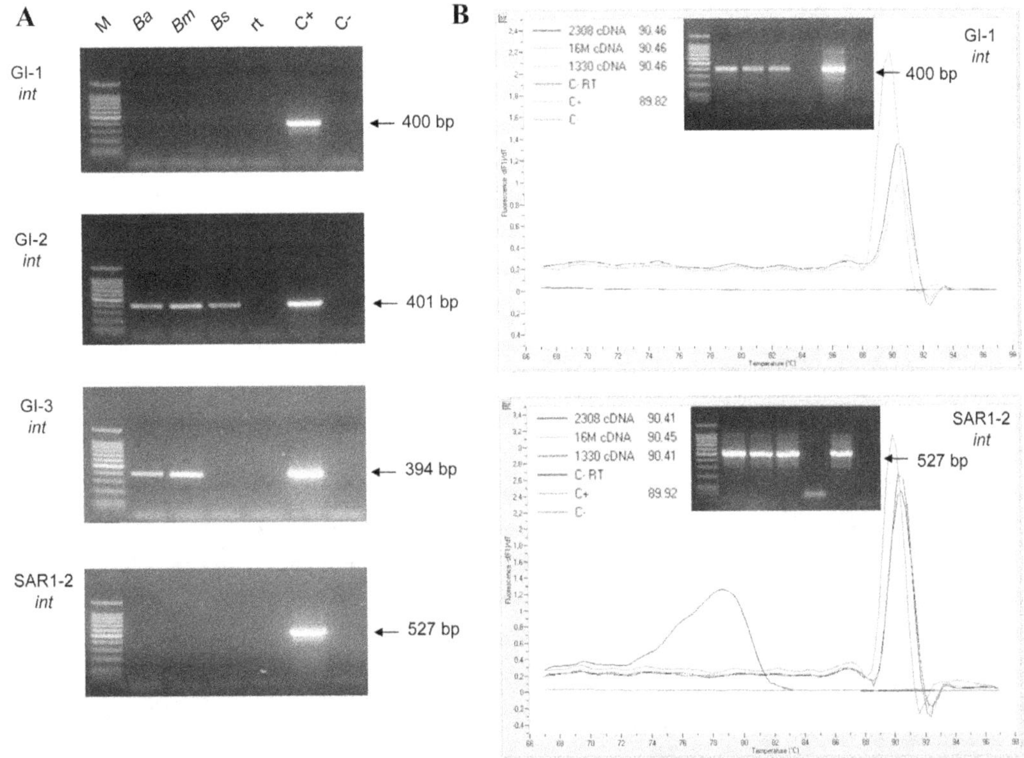

Figure 3.3 Difference in expression level of GI integrases/recombinases. (A) Conventional RT-PCR analysis. 100 ng of total RNA samples were treated with DNAse and were reversed transcribed. Amplicons for *B. abortus* 2308 (Ba), *B. melitensis* 16M (Bm) and *B. suis* 1330 (Bs) are shown. Note the result for GI-3 *int* in *B. suis*, which is due to the absence of GI-3 from its genome. Negative control corresponds to PCR performed on identical RNA samples without reverse transcription (rt). Positive control (C+), DNA of *B. melitensis* 16M. PCR negative control (C⁻), water. M, DNA ladder. (B) Real-time RT-PCR analysis of GI-1 and SAR1-2 *int*. Graphics represent SYBR green melting curves. At the top, it is shown the results of gel electrophoresis of qRT-PCR amplicons.

3.4). Altogether, it is noticeable that the activity of *int* might have played a role in the deletion of their cognate GIs in some *Brucella* species.

Regulation of excision/integration of GIs

The dynamics of GI excision catalysed by phage-related integrases relies on the expression of *int*, however accessory proteins may also participate. Similar to Xis protein of the lambda phage, some small proteins termed recombination directionality factors or RDFs promote excisive recombination by binding to the DNA near the recombination site, forming a complex with the Int (Lewis and Hatfull, 2001). The role of RDFs is architectural rather than catalytic; therefore, to accomplish the excision a functional Int is required. The genes encoding RDFs (*xis*) are small (coding sequence < 100 amino acids), usually include a DNA binding domain helix–turn–helix or HTH, and are often located in close proximity to the *int*, an array called integration–excision module or *int-xis*. The search for *int-xis* tandems in *Brucella* reveals that at least two GIs with full-length *int* contain this kind of modules: IncP and GI-3. Adjacent to the *int* of IncP, an ORF (BRA0363, 76 amino acids) is found with a DNA binding domain AlpA. Similarity searches using known RDFs sequences (http://www.pitt.edu/~gfh/rdf.html) show that GI-3 carries a putative *xis* gene encoding a similar binding domain of IncP *xis* (BAB1_0252, 72 amino acids). The specific GI of *B. microti* also carries a phage-related excisionase, the ORF BMI_II1049

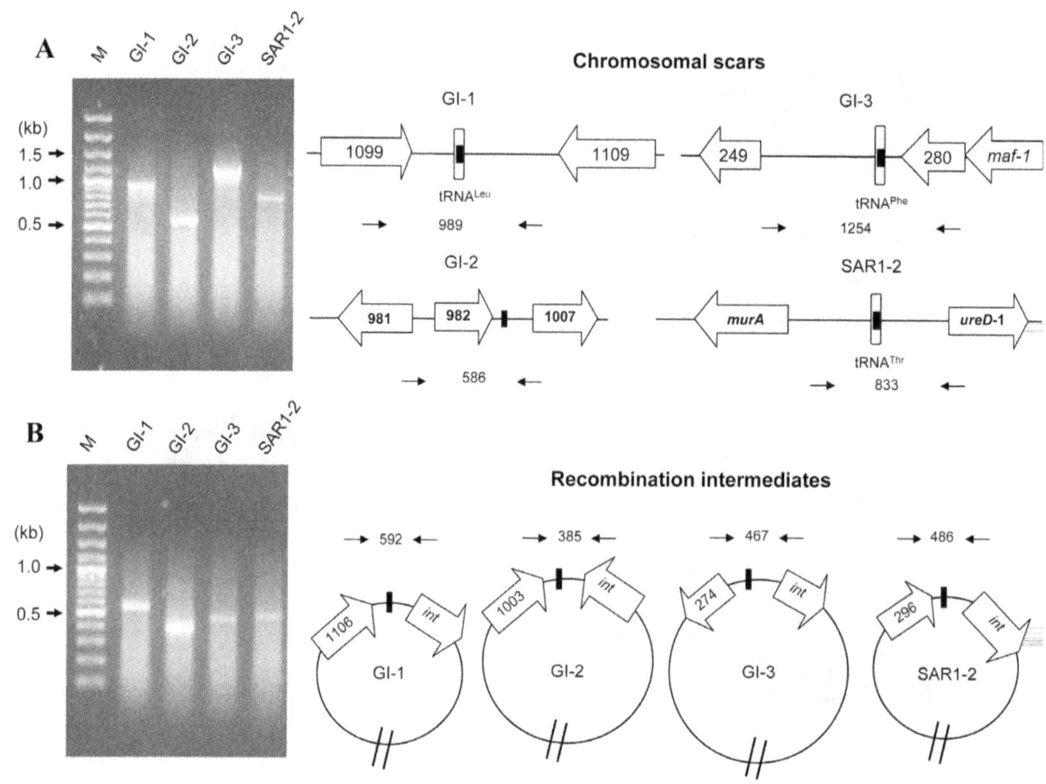

Figure 3.4 Detection of excisive recombination products linked to *int* activity. (A) PCR detection of chromosomal scars resulting from GI excisions. (B) PCR detection of recombination intermediates. 0.5 μg of genomic DNA purified from cultures of *B. abortus* 2308 were used as template. The schemes represent the results of the sequencing of PCR fragments based on *B. abortus* 2308 genome annotation (not in scale). Black boxes represent recombination sites. Black arrows indicate the position of primers. Numbers between primers show the molecular size of expected PCR products.

(Audic *et al.*, 2009), but it is probable that it is part of a functional module due to the *int* currently being a pseudogene. By contrast, neither GI-2, and GI-1 nor SAR1-2 seems to harbour *xis*. The detection of recombination intermediates and chromosomal scars suggest the possibility that RDFs might participate in the excision of *Brucella* GIs, a role that has already been demonstrated for GIs of the closely related genus *Mesorhizobium* (Ramsay *et al.*, 2006). The fact that GI-3 displays an apparently higher excision frequency than GI-1 and SAR1-2 (Fig. 3.3), and that IncP and GI-3 *int* belong to the same gene family argue for the functionality of the corresponding *xis* genes.

GI excision is mainly controlled by the expression of *int*, thus it might be affected by environmental stimuli. Regulation at promoter level as well as the unusual generation of alternative promoters by after the formation of the circular intermediates have been reported (Sentchilo *et al.*, 2003) It is known that some growth conditions like cell density and growth temperature influence the frequency of GI excision (Middendorf *et al.*, 2004). DNA damage and some antibiotics have shown to provoke not only increasing excision rates, but also transfer of GIs (Beaber *et al.*, 2004; Ubeda *et al.*, 2005; Soto *et al.*, 2006). Although the mechanisms behind these stimuli are poorly understood, they certainly cause profound changes in the structure of bacterial genomes. In smooth *Brucellae*, it has been shown that a mild acidic medium promotes the excision of 15.1 kb GI-2, and consequently, the appearance of rough mutants (Mancilla *et al.*, 2010). With regards to the maintenance of GIs, it has been suggested that mutation affecting mobility genes might be involved in the shifting from an excision-prone state to a more stable association to their host chromosomes (Dobrindt *et al.*, 2004). The stabilization of SAR1-2, SAR1-5 and 26.5 kb in *B. ovis* and the *B. microti* 12 kb specific island probably is an ongoing process in which the corresponding integrases underwent pseudogenization.

GI deletion-acquisitions and evolution of *Brucellae*

From an evolutionary point of view, the *Brucella* GIs, as well as the *Mycobacterium* regions of difference (RD), can be considered as markers of unidirectional deletions: once lost they cannot be regained (Ahmed *et al.*, 2008). Phylogenetic trees generated on the basis of comparisons of different markers (Audic *et al.*, 2009; Whatmore, 2009) are consistent with a hypothetical evolutionary scenario in which GI deletions, along with other speciation events, might have originated modern *Brucella* species (Fig. 3.5). This hypothesis is supported by the occurrence of genome deletions affecting GIs. In this context, traits to survive as a free living organism were acquired by the *Brucella* ancestor through a genomic expansion (Batut *et al.*, 2004). At this point of evolution, we can suggest that most of GIs were captured and integrated into the genome. In the evolving road towards an intracellular, more enclosed life-style, the selective pressure induced by the interactions with different hosts triggered those speciation events that lead to a loss of GIs. Strikingly, it seems that during the course of their evolutive histories, *B. ovis* and *B. microti* experienced genome reduction, but also GI acquisition, as well as the ancestor of *B. suis* and its related species *B. microti* and *B. canis*. Although it is difficult to set them up into a timescale, those deletion-acquisitions may be part of long-term evolution changes related to speciation: genomic data and the survey in strain collections indicate that GI-deletion polymorphism is accounting only at species level.

There are good examples in which genomic rearrangements, including GI-deletions, have also been observed in a shorter timescale. Current research focused on clarifying those genomic changes related them with rapid pathogen adaptation. A paradigm of short-term evolution instigated by host–pathogen interactions is given by uropathogenic strains of *E. coli*. During bladder colonization, along with other mutations, these strains show a marked tendency to suffer spontaneous deletion of PAIs. This form of virulence factor discarding is thought to contribute to their persistence into the bladder of asymptomatic patients (Zdziarski *et al.*, 2010). This effect has also been observed in the plant pathogen *P. syringae* pv. *phaseolicola* (Pitman *et al.*, 2005). When the bacteria are exposed to host-resistant mechanisms, restricted genome reduction affecting a particular GI is launched leading to the selection of mutants lacking the GI. However, the experimental evidence accumulated so far does not permit to state that *Brucella* GI-deletions, in a period as short as during host–pathogen interaction, actually occur.

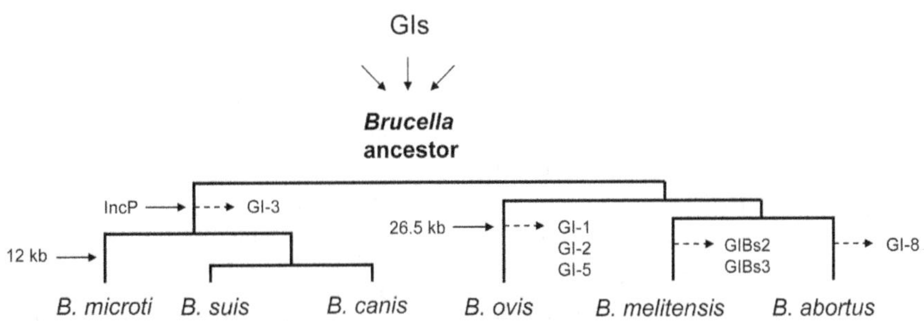

Figure 3.5 Hypothetical evolutive scenario concerning acquisition-deletions of GIs. Continuous arrows indicate acquisitions, while discontinuous ones deletions. Note that the tree is merely explicative of those speciation events involved in gaining/losing of GIs during the evolution of *Brucella* species, it is not a GI-based phylogenetic reconstruction.

Concluding remarks and open questions

Bacterial chromosomes are dynamic and constantly changing, and it seems that *Brucella* is not an exception of this rule. Major driving forces for prokaryotic evolution are point mutations, insertions, deletions and genome rearrangements, and also its natural selection. It is well recognized that horizontal transfer of GIs is the 'fast track' means for bacterial evolution, because point mutations introduce a low degree of variability in the population, but the acquisition of blocks of genes may result in genetic quantum leaps (Hacker and Carniel, 2001). The unusual *Brucella* LPS is an example of a major feature that has evolved from several horizontal gene acquisitions and modern smooth *Brucellae*, under not well-defined conditions, are able to change its O-polysaccharide structure by inverting the process, i.e. through GI deletion.

As discussed in Chapter 2, the *Brucella* genome, like most bacterial genomes, is assembled from two classes of genes: those encoding proteins with essential functions (core genome) and the ones encoding additional functions not essential for growth (flexible gene pool), which have been mainly horizontally acquired (Dobrindt and Hacker, 2001). The latter consists of horizontal acquisitions among which GIs, taking into account its fraction from total genome, are the most abundant category found in *Brucella*. Considering the evidence, it is possible to propose that the core genome of *Brucella* encompass nearly 3 Mb, while the flexible genome reaches approximately 0.3 Mb, the size of all GIs together. The fact that the unstable *Brucella* GIs are able to be excised argues in favour of that those islands are still undergoing an integrative process and suggests that they were recently acquired. A key aspect of the information presented above concerns the regulation of the GI instability: if we are able to maintain a very low Int activity or to switch it off, it is probable we will get strains with a reduced capacity of genome plasticity, therefore, enhanced stability. A fastidious and recurring problem in smooth vaccine production is the spontaneous appearance of rough phenotypes. Vaccine lots with mixed cultures of both smooth and rough cells that contain a fraction over 5% of rough types are not accepted and should be discarded (Alton *et al.*, 1988). This vaccine quality requirement has been established owing to the less protective effects showed by rough vaccines (Moriyón *et al.*, 2004). Since LPS-related GIs as GI-2 are targets for excisive recombination, the manipulation of their deletion rate by mutation of *cis* elements or genes involved in the excision emerges as a new strategy to improve the stability of current smooth vaccines.

Mutational analysis of uncharacterized *int* is currently in progress and will allow us to reconstruct part of the evolution of *Brucella* and, more interestingly, to elucidate if Int activity plays a role in the host-specific pathogen adaptation. In addition, future studies focused on the regulation of the instability of GIs will provide useful data to design strains in order to develop improved diagnostic antigens and more stable vaccines. Are *Brucella* GIs involved in the host specificity? Are there more virulence genes encoded by the GIs hidden as hypothetical proteins? If some GIs are found as circular intermediates, are they capable to be transferred too? Where? How? These are examples of several questions that arise from the most recent data and remain unanswered. On the other hand, genes with unknown functions contained in GIs might represent a source of novel genes that are waiting for characterization. It will not be a surprise if, soon, we find interesting phenotypes regarding virulence, metabolic features or even resistance mechanisms. With current methods, it is possible to obtain customized GI-cured mutants that can be used to prove their respective role in any of the traits mentioned above. We are sure that *Brucella* GIs will prompt further research into not only new pathogenicity features, but also fitness determinants that shed more light on the marked differences in host-specificity shown by this group of pathogens.

References

Ahmed, N., Dobrindt, U., Hacker, J., and Hasnain, S.E. (2008). Genomic fluidity and pathogenic bacteria: applications in diagnostics, epidemiology and intervention. Nat. Rev. Microbiol. *6*, 387–394.

Alton, G., Jones, L., Angus, R., and Verger, J.M. (1988). The production of *Brucella* vaccines. In Techniques for the Brucellosis Laboratory, (Paris, INRA), pp. 143–156.

Arellano-Reynoso, B., Lapaque, N., Salcedo, S., Briones, G., Ciocchini, A.E., Ugalde, R., Moreno, E., Moriyón, I., and Gorvel, J.P. (2005). Cyclic beta-1,2-glucan is a *Brucella* virulence factor required for intracellular survival. Nat. Immunol. *6*, 618–625.

Audic, S., Lescot, M., Claverie, J.M., and Scholz, H.C. (2009). *Brucella microti*: the genome sequence of an emerging pathogen. BMC Genomics *10*, 352.

Bandara, A.B., Sriranganathan, N., Schurig, G.G., and Boyle, S.M. (2005). Putative outer membrane autotransporter protein influences survival of *Brucella suis* in BALB/c mice. Vet. Microbiol. *109*, 95–104.

Barquero-Calvo, E., Chaves-Olarte, E., Weiss, D.S., Guzmán-Verri, C., Chacón-Diaz, C., Rucavado, A., Moriyón, I., and Moreno, E. (2007). *Brucella abortus* uses a stealthy strategy to avoid activation of the innate immune system during the onset of infection. PLoS One 2, e631.

Batut, J., Andersson, S., and O'Callaghan, D. (2004). The evolution of chronic infection strategies in the α-proteobacteria. Nat. Rev. Microbiol. *2*, 933–945.

Beaber, J.W., Hochhut, B., and Waldor, M.K. (2004). SOS response promotes horizontal dissemination of antibiotic resistance genes. Nature *427*, 72–74.

Blanc-Potard, A.B., and Lafay, B. (2003). MgtC as a horizontally acquired virulence factor of ontracellular bacterial pathogens: evidence from molecular phylogeny and comparative genomics. J. Mol. Evol. *57*, 479–486.

Cloeckaert, A., Grayon, M., Verger, J.M., Letesson, J.J., and Godfroid, F. (2000). Conservation of seven genes involved in the biosynthesis of the lipopolysaccharide O-side chain in *Brucella* spp. Res. Microbiol. *151*, 209–216.

Chain, P.S., Comerci, D.J., Tolmasky, M.E., Larimer, F.W., Malfatti, S.A., Vergez, L.M., Aguero, F., Land, M.L., Ugalde, R.A., and Garcia, E. (2005). Whole-genome analyses of speciation events in pathogenic Brucellae. Infect. Immun. *73*, 8353–8361.

Delrue, R.M., Lestrate, P., Tibor, A., Letesson, J.J., and De Bolle, X. (2004). *Brucella* pathogenesis, genes identified from random large-scale screens. FEMS Microbiol. Lett. *231*, 1–12.

DelVecchio, V.G., Kapatral, V., Redkar, R.J., Patra, G., Mujer, C., Los, T., Ivanova, N., Anderson, I., Bhattacharyya, A., Lykidis, A., *et al.* (2002). The genome sequence of the facultative intracellular pathogen *Brucella melitensis*. Proc. Natl. Acad. Sci. U.S.A. *99*, 443–448.

Dobrindt, U., and Hacker, J. (2001). Whole genome plasticity in pathogenic bacteria. Curr. Opin. Microbiol. *4*, 550–557.

Dobrindt, U., Hochhut, B., Hentschel, U., and Hacker, J. (2004). Genomic islands in pathogenic and environmental microorganisms. Nat. Rev. Microbiol. *2*, 414–424.

Eschenbrenner, M., Horn, T.A., Wagner, M.A., Mujer, C.V., Miller-Scandle, T.L., and DelVecchio, V.G. (2006). Comparative proteome analysis of laboratory grown *Brucella abortus* 2308 and *Brucella melitensis* 16M. J. Proteome Res. *5*, 1731–1740.

Ficht, T. (2010). *Brucella* taxonomy and evolution. Future Microbiol. *5*, 859–866.

Ficht, T.A., Kahl-McDonagh, M.M., Arenas-Gamboa, A.M., and Rice-Ficht, A.C. (2009). Brucellosis: the case for live, attenuated vaccines. Vaccine *27* (Suppl. 4), D40–43.

Foster, G., Osterman, B.S., Godfroid, J., Jacques, I., and Cloeckaert, A. (2007). *Brucella ceti* sp. nov. and *Brucella pinnipedialis* sp. nov. for *Brucella* strains with cetaceans and seals as their preferred hosts. Int. J. Syst. Evol. Microbiol. *57*, 2688–2693.

Fretin, D., Fauconnier, A., Kohler, S., Halling, S., Leonard, S., Nijskens, C., Ferooz, J., Lestrate, P., Delrue, R.M., Danese, I., *et al.* (2005). The sheathed flagellum of *Brucella melitensis* is involved in persistence in a murine model of infection. Cell. Microbiol. *7*, 687–698.

Gaillard, M., Vallaeys, T., Vorholter, F.J., Minoia, M., Werlen, C., Sentchilo, V., Puhler, A., and van der Meer, J.R. (2006). The clc element of *Pseudomonas* sp. strain B13, a genomic island with various catabolic properties. J. Bacteriol. *188*, 1999–2013.

Gal-Mor, O., and Finlay, B.B. (2006). Pathogenicity islands: a molecular toolbox for bacterial virulence. Cell. Microbiol. *8*, 1707–1719.

Godfroid, F., Cloeckaert, A., Taminiau, B., Danese, I., Tibor, A., de Bolle, X., Mertens, P., and Letesson, J.J. (2000). Genetic organisation of the lipopolysaccharide O-antigen biosynthesis region of *Brucella melitensis* 16M (*wbk*). Res. Microbiol. *151*, 655–668.

Godfroid, J., Cloeckaert, A., Liautard, J.P., Kohler, S., Fretin, D., Walravens, K., Garin-Bastuji, B., and Letesson, J.J. (2005). From the discovery of the Malta fever's agent to the discovery of a marine mammal reservoir, brucellosis has continuously been a re-emerging zoonosis. Vet. Res. *36*, 313–326.

Goldstein, J., Hoffman, T., Frasch, C., Lizzio, E.F., Beining, P.R., Hochstein, D., Lee, Y.L., Angus, R.D., and Golding, B. (1992). Lipopolysaccharide (LPS) from *Brucella abortus* is less toxic than that from *Escherichia coli*, suggesting the possible use of *B. abortus* or LPS from *B. abortus* as a carrier in vaccines. Infect. Immun. *60*, 1385–1389.

Hacker, J., and Kaper, J.B. (2000). Pathogenicity islands and the evolution of microbes. Annu. Rev. Microbiol. *54*, 641–679.

Hacker, J., and Carniel, E. (2001). Ecological fitness, genomic islands and bacterial pathogenicity. A Darwinian view of the evolution of microbes. EMBO Rep. *2*, 376–381.

Hacker, J., Bender, L., Ott, M., Wingender, J., Lund, B., Marre, R., and Goebel, W. (1990). Deletions of chromosomal regions coding for fimbriae and hemolysins occur *in vitro* and *in vivo* in various extraintestinal *Escherichia coli* isolates. Microb. Pathog. *8*, 213–225.

Hacker, J., Hentschel, U., and Dobrindt, U. (2003). Prokaryotic chromosomes and disease. Science *301*, 790–793.

Halling, S.M., and Zuerner, R.L. (2002). Evidence for lateral transfer to Brucellae: characterization of a locus with a Tn-like element (Tn*2020*). Biochim. Biophys. Acta *1574*, 109–116.

Halling, S.M., Gibas, C., and Boyle, S.M. (2004). Comparative Genomics of *Brucella melitensis*, *B. suis*, and *B. abortus*. In *Brucella*: Molecular and Cellular Biology, López-Goñi, I., and Moriyón, I., eds. (Horizons Bioscience: Pamplona), pp. 85–102.

Halling, S.M., Peterson-Burch, B.D., Bricker, B.J., Zuerner, R.L., Qing, Z., Li, L.L., Kapur, V., Alt, D.P., and Olsen, S.C. (2005). Completion of the genome sequence of *Brucella abortus* and comparison to the highly similar genomes of *Brucella melitensis* and *Brucella suis*. J. Bacteriol. *187*, 2715–2726.

Ho Sui, S.J., Fedynak, A., Hsiao, W.W., Langille, M.G., and Brinkman, F.S. (2009). The association of virulence factors with genomic islands. PLoS One *4*, e8094.

Hochhut, B., Wilde, C., Balling, G., Middendorf, B., Dobrindt, U., Brzuszkiewicz, E., Gottschalk, G., Carniel, E., and Hacker, J. (2006). Role of pathogenicity island-associated integrases in the genome plasticity of uropathogenic *Escherichia coli* strain 536. Mol. Microbiol. *61*, 584–595.

Hornback, M.L., and Roop, R.M., 2nd (2006). The *Brucella abortus* xthA-1 gene product participates in base excision repair and resistance to oxidative killing but is not required for wild-type virulence in the mouse model. J. Bacteriol. *188*, 1295–1300.

Hsiao, W.W., Ung, K., Aeschliman, D., Bryan, J., Finlay, B.B., and Brinkman, F.S. (2005). Evidence of a large novel gene pool associated with prokaryotic genomic islands. PLoS Genet. *1*, e62.

de Jong, M.F., Sun, Y.H., den Hartigh, A.B., van Dijl, J.M., and Tsolis, R.M. (2008). Identification of VceA and VceC, two members of the VjbR regulon that are translocated into macrophages by the *Brucella* type IV secretion system. Mol. Microbiol. *70*, 1378–1396.

Juhas, M., van der Meer, J.R., Gaillard, M., Harding, R.M., Hood, D.W., and Crook, D.W. (2009). Genomic islands: tools of bacterial horizontal gene transfer and evolution. FEMS Microbiol. Rev. *33*, 376–393.

Karlin, S. (2001). Detecting anomalous gene clusters and pathogenicity islands in diverse bacterial genomes. Trends Microbiol. *9*, 335–343.

Kohler, S., Foulongne, V., Ouahrani-Bettache, S., Bourg, G., Teyssier, J., Ramuz, M., and Liautard, J.P. (2002). The analysis of the intramacrophagic virulome of *Brucella suis* deciphers the environment encountered by the pathogen inside the macrophage host cell. Proc. Natl. Acad. Sci. U.S.A. *99*, 15711–15716.

Kurtz, S., Choudhuri, J.V., Ohlebusch, E., Schleiermacher, C., Stoye, J., and Giegerich, R. (2001). REPuter: the manifold applications of repeat analysis on a genomic scale. Nucleic Acids Res. *29*, 4633–4642.

Lamontagne, J., Forest, A., Marazzo, E., Denis, F., Butler, H., Michaud, J.F., Boucher, L., Pedro, I., Villeneuve, A., Sitnikov, D., *et al.* (2009). Intracellular adaptation of *Brucella abortus*. J. Proteome Res. *8*, 1594–1609.

Langille, M.G., Hsiao, W.W., and Brinkman, F.S. (2008). Evaluation of genomic island predictors using a comparative genomics approach. BMC Bioinformatics *9*, 329.

Langille, M.G., Hsiao, W.W., and Brinkman, F.S. (2010). Detecting genomic islands using bioinformatics approaches. Nat. Rev. Microbiol. *8*, 373–382.

Lavigne, J.P., O'Callaghan, D., and Blanc-Potard, A.B. (2005a) Requirement of MgtC for *Brucella suis* intramacrophage growth: a potential mechanism shared by *Salmonella enterica* and *Mycobacterium tuberculosis* for adaptation to a low-Mg^{2+} environment. Infect. Immun. *73*, 3160–3163.

Lavigne, J.P., Vergunst, A.C., Bourg, G., and O'Callaghan, D. (2005b). The IncP island in the genome of *Brucella suis* 1330 was acquired by site-specific integration. Infect. Immun. *73*, 7779–7783.

Lechner, M., Schmitt, K., Bauer, S., Hot, D., Hubans, C., Levillain, E., Locht, C., Lemoine, Y., and Gross, R. (2009). Genomic island excisions in *Bordetella petrii*. BMC Microbiol. *9*, 141.

Lesic, B., Bach, S., Ghigo, J.-M., Dobrindt, U., Hacker, J., and Carniel, E. (2004). Excision of the high-pathogenicity island of *Yersinia pseudotuberculosis* requires the combined actions of its cognate integrase and Hef, a new recombination directionality factor. Mol. Microbiol. *52*, 1337–1348.

Lestrate, P., Dricot, A., Delrue, R.M., Lambert, C., Martinelli, V., De Bolle, X., Letesson, J.J., and Tibor, A. (2003). Attenuated signature-tagged mutagenesis mutants of *Brucella melitensis* identified during the acute phase of infection in mice. Infect. Immun. *71*, 7053–7060.

Lewis, J.A., and Hatfull, G.F. (2001). Control of directionality in integrase-mediated recombination: examination of recombination directionality factors (RDFs) including Xis and Cox proteins. Nucleic Acids Res. *29*, 2205–2216.

López-Goñi, I., Guzmán-Verri, C., Manterola, L., Sola-Landa, A., Moriyón, I., and Moreno, E. (2002). Regulation of *Brucella* virulence by the two-component system BvrR/BvrS. Vet. Microbiol. *90*, 329–339.

Lloyd, A.L., Henderson, T.A., Vigil, P.D., and Mobley, H.L. (2009). Genomic islands of uropathogenic *Escherichia coli* contribute to virulence. J. Bacteriol. *191*, 3469–3481.

Mancilla, M., López-Goñi, I., Moriyón, I., and Zárraga, A.M. (2010). Genomic island-2 is an unstable genetic element contributing to *Brucella* lipopolysaccharide spontaneous smooth to rough dissociation. J. Bacteriol. *192*, 6346–6351.

Mantri, Y., and Williams, K.P. (2004). Islander: a database of integrative islands in prokaryotic genomes, the associated integrases and their DNA site specificities. Nucleic Acids Res. *32*, D55–58.

Maquart, M., Fardini, Y., Zygmunt, M.S., and Cloeckaert, A. (2008). Identification of novel DNA fragments and partial sequence of a genomic island specific of *Brucella pinnipedialis*. Vet. Microbiol. *132*, 181–189.

Marchesini, M.I., Ugalde, J.E., Czibener, C., Comerci, D.J., and Ugalde, R.A. (2004). N-terminal-capturing screening system for the isolation of *Brucella abortus* genes encoding surface exposed and secreted proteins. Microb. Pathog. *37*, 95–105.

Marri, P.R., and Golding, G.B. (2008). Gene amelioration demonstrated: the journey of nascent genes in bacteria. Genome *51*, 164–168.

McQuiston, J.R., Vemulapalli, R., Inzana, T.J., Schurig, G.G., Sriranganathan, N., Fritzinger, D., Hadfield, T.L., Warren, R.A., Lindler, L.E., Snellings, N., *et al.* (1999). Genetic characterization of a Tn5-disrupted glycosyltransferase gene homolog in *Brucella abortus* and its effect on lipopolysaccharide composition and virulence. Infect. Immun. *67*, 3830–3835.

Merkl, R. (2004). SIGI: score-based identification of genomic islands. BMC Bioinformatics *5*, 22.

Middendorf, B., Hochhut, B., Leipold, K., Dobrindt, U., Blum-Oehler, G., and Hacker, J. (2004). Instability of pathogenicity islands in uropathogenic *Escherichia coli* 536. J. Bacteriol. *186*, 3086–3096.

Miriagou, V., Carattoli, A., and Fanning, S. (2006). Antimicrobial resistance islands: resistance gene clusters in *Salmonella* chromosome and plasmids. Microbes Infect. *8*, 1923–1930.

Moreno, E., and Moriyón, I. (2002). *Brucella melitensis*: a nasty bug with hidden credentials for virulence. Proc. Natl. Acad. Sci. U.S.A. *99*, 1–3.

Moreno, E., Cloeckaert, A., and Moriyón, I. (2002). *Brucella* evolution and taxonomy. Vet. Microbiol. *90*, 209–227.

Moriyón, I., Grilló, M.J., Monreal, D., Gonzalez, D., C., M., López-Goñ, I., Mainar-Jaime, R.C., Moreno, E., and Blasco, J.M. (2004). Rough vaccines in animal brucellosis: structural and genetic basis and present status. Vet. Res. *35*, 1–38.

Nakamura, Y., Itoh, T., Matsuda, H., and Gojobori, T. (2004). Biased biological functions of horizontally transferred genes in prokaryotic genomes. Nat. Genet. *36*, 760–766.

Nishimura, K., Tajima, N., Yoon, Y.H., Park, S.Y., and Tame, J.R. (2010). Autotransporter passenger proteins: virulence factors with common structural themes. J. Mol. Med. *88*, 451–458.

O'Callaghan, D., Cazevieille, C., Allardet-Servent, A., Boschiroli, M.L., Bourg, G., Foulongne, V., Frutos, P., Kulakov, Y., and Ramuz, M. (1999). A homologue of the *Agrobacterium tumefaciens* VirB and *Bordetella pertussis* Ptl type IV secretion systems is essential for intracellular survival of *Brucella suis*. Mol. Microbiol. *33*, 1210–1220.

Paulsen, I.T., Seshadri, R., Nelson, K.E., Eisen, J.A., Heidelberg, J.F., Read, T.D., Dodson, R.J., Umayam, L., Brinkac, L.M., Beanan, M.J., *et al.* (2002). The *Brucella suis* genome reveals fundamental similarities between animal and plant pathogens and symbionts. Proc. Natl. Acad. Sci. U.S.A. *99*, 13148–13153.

Perry, Q.L., Hagius, S.D., Walker, J.V., and Elzer, P.H. (2010). Evaluating the virulence of a *Brucella melitensis* hemagglutinin gene in the caprine model. Vaccine *28 (Suppl. 5)*, F6–11.

Pitman, A.R., Jackson, R.W., Mansfield, J.W., Kaitell, V., Thwaites, R., and Arnold, D.L. (2005). Exposure to host resistance mechanisms drives evolution of bacterial virulence in plants. Curr. Biol. *15*, 2230–2235.

Qiu, X., Gurkar, A.U., and Lory, S. (2006). Interstrain transfer of the large pathogenicity island (PAPI-1) of *Pseudomonas aeruginosa*. Proc. Natl. Acad. Sci. U.S.A. *103*, 19830–19835.

Rajakumar, K., Sasakawa, C., and Adler, B. (1997). Use of a novel approach, termed island probing, identifies the *Shigella flexneri* she pathogenicity island which encodes a homolog of the immunoglobulin A protease-like family of proteins. Infect. Immun. *65*, 4606–4614.

Rajashekara, G., Glasner, J.D., Glover, D.A., and Splitter, G.A. (2004). Comparative whole-genome hybridization reveals genomic islands in *Brucella* species. J. Bacteriol. *186*, 5040–5051.

Rajashekara, G., Covert, J., Petersen, E., Eskra, L., and Splitter, G. (2008). Genomic island 2 of *Brucella melitensis* is a major virulence determinant: functional analyses of genomic islands. J. Bacteriol. *190*, 6243–6252.

Ramsay, J.P., Sullivan, J.T., Stuart, G.S., Lamont, I.L., and Ronson, C.W. (2006). Excision and transfer of the *Mesorhizobium loti* R7A symbiosis island requires an integrase IntS, a novel recombination directionality factor RdfS, and a putative relaxase RlxS. Mol. Microbiol. *62*, 723–734.

Ratushna, V.G., Sturgill, D.M., Ramamoorthy, S., Reichow, S.A., He, Y., Lathigra, R., Sriranganathan, N., Halling, S.M., Boyle, S.M., and Gibas, C.J. (2006). Molecular targets for rapid identification of *Brucella* spp. BMC Microbiol. *6*, 13.

Ravatn, R., Studer, S., Zehnder, A.J., and van der Meer, J.R. (1998). Int-B13, an unusual site-specific recombinase of the bacteriophage P4 integrase family, is responsible for chromosomal insertion of the 105-kilobase clc element of *Pseudomonas* sp. Strain B13. J. Bacteriol. *180*, 5505–5514.

Rossetti, C.A., Galindo, C.L., Lawhon, S.D., Garner, H.R., and Adams, L.G. (2009). *Brucella melitensis* global gene expression study provides novel information on growth phase-specific gene regulation with potential insights for understanding *Brucella*: host initial interactions. BMC Microbiol. *9*, 81.

Rutherford, K., Parkhill, J., Crook, J., Horsnell, T., Rice, P., Rajandream, M.A., and Barrell, B. (2000). Artemis: sequence visualization and annotation. Bioinformatics *16*, 944–945.

Sakellaris, H., Luck, S.N., Al-Hasani, K., Rajakumar, K., Turner, S.A., and Adler, B. (2004). Regulated site-specific recombination of the she pathogenicity island of *Shigella flexneri*. Mol. Microbiol. *52*, 1329–1336.

Salcedo, S.P., Marchesini, M.I., Lelouard, H., Fugier, E., Jolly, G., Balor, S., Muller, A., Lapaque, N., Demaria, O., Alexopoulou, L., *et al.* (2008). *Brucella* control of dendritic cell maturation is dependent on the TIR-containing protein Btp1. PLoS Pathog. *4*, e21.

Schmidt, H., and Hensel, M. (2004). Pathogenicity islands in bacterial pathogenesis. Clin. Microbiol. Rev. *17*, 14–56.

Scholz, H.C., Hubalek, Z., Sedlacek, I., Vergnaud, G., Tomaso, H., Al Dahouk, S., Melzer, F., Kampfer, P., Neubauer, H., Cloeckaert, A., *et al.* (2008). *Brucella microti* sp. nov., isolated from the common vole

Microtus arvalis. Int. J. Syst. Evol. Microbiol. *58*, 375–382.

Sentchilo, V., Ravatn, R., Werlen, C., Zehnder, A.J., and van der Meer, J.R. (2003). Unusual integrase gene expression on the clc genomic island in *Pseudomonas* sp. strain B13. J. Bacteriol. *185*, 4530–4538.

Skurnik, M., Biedzka-Sarek, M., Lubeck, P.S., Blom, T., Bengoechea, J.A., Perez-Gutierrez, C., Ahrens, P., and Hoorfar, J. (2007). Characterization and biological role of the O-polysaccharide gene cluster of *Yersinia enterocolitica* serotype O:9. J. Bacteriol. *189*, 7244–7253.

Soto, S.M., Jiménez de Anta, M.T., and Vila, J. (2006). Quinolones induce partial or total loss of pathogenicity islands in uropathogenic *Escherichia coli* by SOS-dependent or -independent pathways, respectively. Antimicrob. Agents. Chemother. *50*, 649–653.

Tsolis, R.M., Seshadri, R., Santos, R.L., Sangari, F.J., Lobo, J.M., de Jong, M.F., Ren, Q., Myers, G., Brinkac, L.M., Nelson, W.C., *et al.* (2009). Genome degradation in *Brucella ovis* corresponds with narrowing of its host range and tissue tropism. PLoS One *4*, e5519.

Ubeda, C., Maiques, E., Knecht, E., Lasa, I., Novick, R.P., and Penades, J.R. (2005). Antibiotic-induced SOS response promotes horizontal dissemination of pathogenicity island-encoded virulence factors in staphylococci. Mol. Microbiol. *56*, 836–844.

Viadas, C., Rodríguez, M.C., Garcia-Lobo, J.M., Sangari, F.J., and López-Goñi, I. (2009). Construction and evaluation of an ORFeome-based *Brucella* whole-genome DNA microarray. Microb. Pathog. *47*, 189–195.

Vizcaíno, N., Cloeckaert, A., Zygmunt, M.S., and Fernández-Lago, L. (2001). Characterization of a *Brucella* species 25-kilobase DNA fragment deleted from *Brucella abortus* reveals a large gene cluster related to the synthesis of a polysaccharide. Infect. Immun. *69*, 6738–6748.

Vizcaíno, N., Caro-Hernandez, P., Cloeckaert, A., and Fernández-Lago, L. (2004). DNA polymorphism in the omp25/omp31 family of *Brucella* spp.: identification of a 1.7-kb inversion in *Brucella cetaceae* and of a 15.1-kb genomic island, absent from *Brucella ovis*, related to the synthesis of smooth lipopolysaccharide. Microbes Infect. *6*, 821–834.

Wattam, A.R., Williams, K.P., Snyder, E.E., Almeida, N.F., Jr., Shukla, M., Dickerman, A.W., Crasta, O.R., Kenyon, R., Lu, J., Shallom, J.M., *et al.* (2009). Analysis of ten *Brucella* genomes reveals evidence for horizontal gene transfer despite a preferred intracellular lifestyle. J. Bacteriol. *191*, 3569–3579.

Whatmore, A.M. (2009). Current understanding of the genetic diversity of *Brucella*, an expanding genus of zoonotic pathogens. Infect. Genet. Evol. *9*, 1168–1184.

Zdziarski, J., Brzuszkiewicz, E., Wullt, B., Liesegang, H., Biran, D., Voigt, B., Gronberg-Hernandez, J., Ragnarsdottir, B., Hecker, M., Ron, E.Z., *et al.* (2010). Host imprints on bacterial genomes – rapid, divergent evolution in individual patients. PLoS Pathog. *6*, e1001078.

Zhang, X., Ren, J., Li, N., Liu, W., and Wu, Q. (2009). Disruption of the BMEI0066 gene attenuates the virulence of *Brucella melitensis* and decreases its stress tolerance. Int. J. Biol. Sci. *5*, 570–577.

Recent Advances in Molecular Approaches to *Brucella* Diagnostics and Epidemiology

4

Adrian M. Whatmore and Krishna K. Gopaul

Abstract

For many years, the diagnosis of brucellosis depended on tried and tested cultural and serological approaches. However, whilst these techniques have been instrumental in successful control and eradication schemes such as the one undertaken in Great Britain, these methods are not ideal. As a highly infectious zoonotic pathogen culture of *Brucella* is hazardous whilst serological tests targeting lipopolysaccharide can be compromised by non-specific cross-reactions. With rapid advances in molecular biology, and driven by the need for improved diagnostic and epidemiological approaches, novel techniques based on the genetic component of the pathogen have become more popular in recent years. In this chapter we will discuss the current status of development and implementation of such techniques applied to *Brucella* whilst also looking forward at potential avenues for further molecular advancement.

Introduction

In the last 25 years the advent of polymerase chain reaction (PCR) technology has revolutionized diagnostics with the *Brucella* field being no exception. The first description of what is now known as PCR came in 1985, although at that time the polymerase driving the reaction was *Escherichia coli* Klenow polymerase (Saiki *et al.*, 1985). In this landmark paper, a complicated set of temperature steps for denaturing of template, hybridization of primers and extension of product was described. Further, there was an enzyme addition step in every cycle because of enzyme inactivation at every denaturing step. Rapid improvement in the technique followed with the first work using *Taq* polymerase allowing for removal of the multiple enzyme addition steps soon reported (Saiki *et al.*, 1988).

While conventional PCR soon became an invaluable research tool, and of increasing diagnostic relevance, a major limitation was the need to visualize products by running them on agarose gels, limiting sensitivity, and increasing dangers of cross contamination. The next great leap forward in terms of molecular diagnostics was thus the development of real-time PCR chemistry following description of a technique to measure the intercalation of ethidium bromide into DNA using a video camera (Higuchi *et al.*, 1993). This observation of amplification in 'real-time' coupled with more sensitive detection without post-PCR processing enabled rapid generation of results. It should be noted that it is purely the methodology of detection that makes real-time PCR more sensitive than its conventional counterpart. The amplification kinetics are identical in both techniques.

A further advantage of the new real-time PCR platforms was that, as detection was now automated, the products generated could be much smaller that those that would be conventionally run on an agarose gel. This in turn allowed for the shortening in time of PCR cycles improving the rapidity of the technique. However, the downside of shorter product size was that 'primer dimers', reflecting oligonucleotide primers binding to each other, became an increasingly common artefact of the new technology. To overcome this problem and improve the overall specificity of real-time PCR, researchers started to use a number of probe technologies including fluorescence resonance emission transfer (FRET) and hydrolysis probes such as Taqman® (Roche Molecular Systems, CA) probes (see Espy *et al.*, 2006 for a complete review of real-time PCR). The incorporation of these probes gave real-time PCR a higher level of specificity, in so far as the technique is dependent on at least three regions within the target, and has become the method of choice.

These advances in PCR technology have facilitated a revolution in the ability to quickly detect and characterize isolates of *Brucella*, an organism that can be difficult and hazardous to grow conventionally, and form the basis of most of what will be described in this chapter.

Developments in molecular diagnostics

This section focuses largely on assays designed simply to detect the presence or absence of *Brucella* in clinical or field material while later parts of the chapter will examine tests that allow further characterization of the *Brucella* detected.

Targets for assay development

While there are numerous reports of PCR assays for *Brucella*, both conventional and real-time, there are only a relatively small number of targets that are commonly utilized. These are summarized in Table 4.1 and described in more detail below.

43 kDa protein

It did not take very long after the description of *Taq* based PCR for the first proposed implementation of the technique as a *Brucella* diagnostic test (Fekete *et al.*, 1990). Given that the use of PCR was still in its infancy it is unsurprising to see that the majority of the article describes optimization of assay conditions such as temperature and salt concentration. However, an interesting omission from this early paper is that of primer sequence or indeed any description of target other than the fact that it was a 635 bp fragment of a 43 kDa outer membrane protein. The authors do state that there were commercial reasons to withhold this information and that this would be disclosed in a later publication. However, a follow up article (Fekete *et al.*, 1992), applying the assay to detection of *Brucella* in different bovine tissues, still contained no mention of primer sequence.

In terms of sensitivity, the assay described was noted as detecting down to 100 fg of *B. abortus* S19 genomic DNA, calculated to be 20 organisms based on a 5 fg genome. This heightened detection brought about another problem that is still common to PCR testing today; namely the problem of contamination. To overcome this problem, the authors described the sterile technique employed, including the changing of tips between the addition of reagents and the use of a laminar hood. It should be pointed out that 20 years later, even though PCR technology has changed, we still employ the same means to reduce the possibility of contamination.

In specificity testing, amplification with the described assay was *Brucella* specific with no amplification occurring with organisms suspected to cause serological cross-reactions, such as *Yersinia enterocolitica* and *Serratia marcescens*. Nevertheless, a surprising omission was the failure to test *Ochrobactrum anthropi* (Holmes *et al.*, 1988) that had recently been shown to have a high level of 16S rRNA similarity with *Brucella* (De Ley *et al.*, 1987). It is unfortunate that there is insufficient information present in these papers to facilitate repetition of the 43 kDa work with *Ochrobactrum* as it is these organisms that have caused the most problems in the specificity testing of a number of molecular brucellosis diagnostics. Reflecting this lack of technical information, there have been no subsequent reports of use of this target despite its early lead.

Ribosomal RNA

The use of 16S rRNA as a target for the detection of various bacteria through molecular methods is a common approach. Although there can be multiple copies present within the genome, the sequence of potential duplications is generally conserved (Case *et al.*, 2007). Furthermore, there are regions of the target that are conserved between all bacteria and regions that can be used to specifically identify bacteria to genus and in some cases, even species level. Herman and De Ridder (1992) designed primers based on a variable

Table 4.1 Molecular targets commonly used in PCR assays to confirm infection with *Brucella*. This table does not include assays that have been used only in testing *in vitro* cultures

Target	Date of first use	Conventional/real-time assay	Publications utilizing target	Example references
43 kDa	1992	Conventional only	1	Fekete *et al.* (1992)
*omp*2	1995	Conventional only	9	Leal Klevezas *et al.* (1995b), Ohishi *et al.* (2004)
*omp*31	2006	Conventional only	4	el Razik *et al.* (2007), Kattar *et al.* (2007)
16S rRNA	1995	Conventional/real-time PCR assays	11	Romero *et al.* (1995b), El Kholy *et al.* (2009)
16S-23S rRNA spacer	2006	Conventional only	5	Kattar *et al.* (2007), Keid *et al.* (2010)
*bcsp*31	1996	Conventional/real-time PCR assays	29	Matar *et al.* (1996), Colmanero *et al.* (2005)
IS*711*	1994	Conventional/real-time PCR assays	15	Ewalt and Bricker (2000), Hinić *et al.* (2009)
*vir*B2	2006	Conventional only	2	Kim *et al.* (2006), Omata *et al.* (2006)
BMEI1162- IS*711*	1994	Conventional/real-time PCR assays	5	Amin *et al.* (2001), Navarro *et al.* (2006)
B. ovis specific islands	2010	Conventional only	1	Xavier *et al.* (2010)

region of the 16S rRNA sequence of *B. abortus* that had previously been sequenced (Dorsch *et al.*, 1989). Although 16S rRNA sequences of 'classical' *Brucella* were later all shown to be identical (Gee *et al.*, 2004) at this time it was unclear whether this region was conserved in all *Brucella* and the authors opted to confirm primer specificity by dot blot hybridization. Using this technique it was confirmed that the primers were *Brucella* specific although one of the primers did hybridize weakly with *O. anthropi*. Then, using the primers designed, PCR was undertaken and it was found again that whilst amplification was in the main, *Brucella* specific, there was weak amplification with *O. anthropi*.

Whilst Herman and De Ridder may have been the first to publish a *Brucella* PCR assay based around the 16S rRNA target, a later 16S rRNA assay (Romero *et al.*, 1995a) has been taken up more widely. However, false-positive results were also observed with this assay when testing *O. anthropi*. Despite this it has been used by a number of groups for both human and animal diagnosis (Nimri, 2003; O'Leary *et al.*, 2006; Mukherjee *et al.*, 2007) with varying degrees of success. Indeed both O'Leary *et al.* (2006) and Mukherjee *et al.* (2007) obtained better sensitivities using other targets (IS*711* and *omp*2/*bcsp*31, respectively). Nevertheless, the 16S rRNA target is still a relatively popular target as demonstrated by the number of published studies using assays based around it (Table 4.1).

In addition to the 16S rRNA gene itself, the 16S-23S rRNA spacer has also been employed as a molecular diagnostic target particularly for milk samples (Rijpens *et al.*, 1996) and as a target applied to detecting canine brucellosis (Keid *et al.*, 2007a,b, 2009, 2010).

*bcsp*31

A further popular target for molecular diagnosis of brucellosis is the gene encoding the 31-kDa *Brucella* cell surface protein (*bcsp*31) shown to be expressed in five of the six *Brucella* species known at the time of initial studies (Mayfield *et al.*, 1988). The only species that was found not to express this protein was *B. ovis* although the gene was shown to be present (Bricker *et al.*, 1988; Halling and Zehr, 1990). The first PCR assay developed around this target concentrated on only two *Brucella* species, *B. abortus* and *B. melitensis* (Baily *et al.*, 1992). Whilst some specificity testing was done in this study, the test panel was small and did not include relevant *Ochrobactrum*. Subsequently in other studies it was shown that there was indeed some cross-reaction with both *O. intermedium* and *O. anthropi*, using primers directed to *bcsp*31 (Da Costa *et al.*, 1996; Casañas *et al.*, 2001). Nevertheless, this target has become one of the most popular used in molecular assays. In addition to its use in conventional PCR, the same primers were also used successfully in a real-time PCR assay based on the intercalating dye SYBR Green with associated melt curve (Queipo-Ortuño *et al.*, 2005). Two later real-time assays, a probe based FRET PCR assay (Al Dahouk *et al.*, 2007b) and a 5′- nuclease probe (Taqman®) based assay (Probert *et al.*, 2004) showed no cross-reactions with *Ochrobactrum* demonstrating that the addition of probe to the PCR reaction adds a further level of specificity to the molecular technique. These finding were in line with earlier *Brucella* work noting that real time assays that used probes were more specific than those that used intercalating dyes (Newby *et al.*, 2003). From Table 4.1 it can be seen that of all the targets that have been used in the development of molecular assays for *Brucella* detection, *bcsp*31 remains extremely popular.

IS*711* (IS*6501*)

The insertion sequence *IS711* (also described as IS*6501*) is another frequent target for molecular diagnosis of brucellosis, in part down to its presence in all published *Brucella* species in multiple copies. First described as causing an insertion at the 5′ end of *bcsp*31 in *B. ovis* (Halling and Zehr, 1990) it may have been the reason this protein was not detected in *B. ovis* by Western blotting (Bricker *et al.*, 1988). This element was subsequently named and sequenced independently by two groups (Halling *et al.*, 1993; Ouahrani *et al.*, 1993), both noting that there were multiple copies of this insertion sequence although the number varied from species to species. However, aside from two studies on an insertion hot-spot (Halling and Bricker, 1994) and active IS*711* transposition in two *Brucella* species (Ocampo-Sosa and García-Lobo, 2008) there has been little work done to ascertain any other role for this element. However, IS*711* is considered specific to *Brucella*, and its presence has been used to support the placement of novel isolates within the group as new species (Clavereau *et al.*, 1998; De *et al.*, 2008; Tiller *et al.*, 2010a).

Early conventional PCR assays using this target were designed to identify *Brucella* to the species and biovar level, rather than just the genus level, based on specific IS*711* insertion locations (Bricker and Halling, 1994; Ouahrani Bettache *et al.*, 1996). In later work, both Redkar *et al.* (2001) and Probert *et al.* (2004) used these species specific insertions as the basis of real time species defining assays that have since been used as direct diagnostics for *B. melitensis* infection (Amin *et al.*, 2001; Mutnal *et al.*, 2007; Navarro *et al.*, 2006; Castanõ and Solera, 2009; Navarro-Martínez *et al.*, 2008).

More recently a study was published examining three targets: *per*, *bcsp*31 and IS*711* using both conventional and real-time PCR chemistry with hydrolysis probes (Bounaadja *et al.*, 2009) showing that in all cases a conventional PCR assay was less sensitive than a real time equivalent. Furthermore a conventional IS*711* PCR was ten times more sensitive than the equivalent *bcsp*31 or *per* assays in detecting *B. abortus*, *B. ovis* and *B. canis* (minimum detection was 100 fg in comparison with 1 pg for the other two assays) although there was no difference in any of the three conventional assays when testing *B. melitensis*. In the case of real-time PCR assays the IS*711* assay was ten times more sensitive than the other targets for *B. melitensis*, *B. ovis* and *B. canis* (minimum detection in this case being 2 fg in comparison with 20 fg) although in the case of *B. abortus* all three assays detected down to 2 fg. With these credentials, it would seem that the IS*711* element would be the ideal marker for a PCR assay although there are some reasons for caution. While IS*711* appears to be present in all *Brucella* examined to date, in other bacteria insertion sequence based diagnostics can be compromised by the lack of elements in some isolates, such as *Mycobacterium tuberculosis* isolates that lack any copies of IS*6110* (Yuen *et al.*, 1993). A further area for caution is minor sequence variation between elements in different *Brucella* species. We have found that single base changes within regions of IS*711* used as targets for Taqman® probe based assays can cause very major changes in the assay sensitivity (K.K. Gopaul, unpublished).

Other *Brucella* targets

In addition to the targets mentioned above, other molecular markers have been investigated as diagnostic targets. In addition to examining *bcsp*31 and 16S rRNA, Da Costa *et al.* (1996) studied four heat shock protein-encoding genes but found that in all cases there was cross reaction with *O. anthropi*. The *omp*2 locus is composed of two genes, *omp*2a and *omp*2b which are roughly the same size in most *Brucella* except in *B. abortus* biovar 1, which has a 115bp deletion in *omp2a* (Ficht *et al.*, 1989). This was exploited to develop an assay designed as a rapid diagnostic for *B. abortus* biovar 1 given the obvious difference in pattern generated by PCR (Leal Klevezas *et al.*, 1995a). In addition to this work, *omp*2 has been used as a confirmatory PCR in a number of situations including in human blood (Sifuentes-Ricon *et al.*, 1997), cattle blood and milk (Mukherjee *et al.*, 2007) and whale testes (Ohishi *et al.*, 2004).

Omp2 is not the only outer membrane protein target that can be used as a molecular diagnostic. Some publications have used the *omp*31 gene (not to be confused with *bcsp*31), a target first described and sequenced by Vizcaíno *et al* (1996). However, one problem in using this target as a *Brucella* genus marker is that this gene is deleted in *B. abortus* meaning that false negative results could be generated. Other targets that merit mention in this section include the *vir*B2 gene, a gene involved in a secretion pathway required for virulence (den Hartigh *et al.*, 2004). This has been used in at least two published studies; one testing for *B. canis* (Kim *et al.*, 2006) with the other being part of a conventional multiplex PCR assay for *Neospora caninum*, *Toxoplasma gondii* and *Brucella* in killer whales (Omata *et al.*, 2006). Yet another outer membrane protein target that has been investigated is *omp*28 (also referred to as bp26) which was used by Mitka *et al.* (2007) in a study of *Brucella* detection in human blood.

The perosamine synthase (*per*) gene is involved in the generation of the lipopolysaccaride O-chain and is conserved in all *Brucella* (Cloeckaert *et al.*, 2000). However, there are homologues of this gene in other bacteria, most notably the serological cross-reacting *Y. enterocolitica* (Bogdanovich *et al.*, 2004) and a conventional multiplex PCR to discriminate between *Y. enterocolitica* and *Brucella* has been reported (Lubeck *et al.*, 2003). Primers from this study have been used in conjunction with a number of hydrolysis probes to create a real-time PCR assay to *per* (Bogdanovich *et al.*, 2004). The optimum assay developed was shown to have a minimum sensitivity of 200fg when testing DNA obtained from *B. neotomae* and *B. ovis*. A lower sensitivity (2pg) was observed for the remaining species tested although the authors mentioned that this may have been attributable to poor extraction.

In summary, it is clear that a large number of targets have been examined as potential diagnostic markers for *Brucella*. Although there have been problems with specificity with some targets, in the main, these assays do allow the selective detection of *Brucella*. With more sequencing data becoming available, not only for *Brucella* type strains, but also *Brucella* clinical isolates and non-*Brucella* such as *Ochrobactrum*, increasingly specific targets may be located in the future that will enable the development of alternative molecular assays.

Molecular assays for the identification of vaccines

Although this section is concerned with the identification of *Brucella* organisms, it should be noted that there are also diagnostic PCR assays available for the identification of the approved live vaccine strains, *B. abortus* S19, *B. abortus* RB51 and *B. melitensis* Rev1. These are particularly valuable as distinction of vaccine strains from field isolates can be useful during control and eradication programs.

The erythritol-sensitive phenotype displayed by the USDA strain of the *B. abortus* S19 vaccine has been attributed to a 700bp deletion in the *ery*CD operon which facilitated development of a PCR assay for specific S19 detection (Sangari and Agüero, 1994). This assay has been incorporated into other multiplex assays such as the AMOS PCR (Bricker and Halling, 1994) and 'Bruceladder' (García-Yoldi *et al.*, 2006) which are described in more detail later. However, work undertaken on a variant of S19 pre-dating the USDA strain identified a possible problem with the *ery*CD target in as far as it did not exist in older derivations of the vaccine (Mukherjee *et al.*, 2005). Using data obtained by the complete genome sequencing of the S19 vaccine strain (Crasta *et al.*, 2008), we have subsequently developed an assay based on single nucleotide polymorphism (SNPs) that are specific to a variety of S19 strains that either possess or lack the *ery*CD deletion (Gopaul *et al.*, 2010).

B. abortus RB51 has a disrupted *wbo*A gene reflecting an IS*711* insertion event (Vemulapalli *et al.*, 1999) which is the basis of the diagnostic PCR recommend by the OIE. Whilst subsequent work established that this insertion was not the cause of the rough phenotype, it nonetheless provided a good marker for this vaccine and was incorporated into the 'Bruceladder' multiplex PCR assay described later in this chapter (García-Yoldi *et al.*, 2006). The most commonly used molecular assay for distinction of *B. melitensis* Rev1 is PCR-RFLP. There are currently two published assays. One approach exploits a SNP located within *rps*L attributed to the streptomycin resistant phenotype displayed by the vaccine strain (Cloeckaert *et al.*, 2002). The other assay, developed by an Israeli group, utilizes a discriminating SNP in *omp*2 (Bardenstein *et al.*, 2002).

Commercial PCR kits

Whilst a number of publications in the field of *Brucella* molecular diagnostics mention the use commercial kits for the extraction of DNA from sample, the actual DNA amplification in most cases is performed using in house assays. However, there are also an increasing number of commercially available *Brucella* PCR kits that have been optimized for either conventional or real-time PCR application. Examples of some of these kits are outlined in Table 4.2 for information although the authors have no personal experience in the use of any commercial kit based PCR approaches and there are no detailed peer-reviewed comparisons of kits.

Sample extraction

A crucial step in any assay that is to be used for detection of *Brucella* in field material is the sample extraction step. However, whilst there many publications assessing *Brucella* PCRs, only a proportion take a further step and include the use of clinical or field material as opposed to testing directly on laboratory cultured material. Such studies allow determination of the analytical specificity and sensitivity of an assay. However, the major problem with this type of validation is that it is not a true representation of real field samples. In the case of infected material, the molecular method has to be able to specifically detect pathogen DNA within a mixture of host and other microbial DNA. Further, it is widely known that PCR can be inhibited by components of field samples such as haem, fats and metal ions. To overcome these problems, extraction protocols have to be devised to obtain good quality DNA. Having a good extraction protocol in place can be an important determinant of the sensitivity of a molecular diagnostic (Schuurman *et al.*, 2005). Many early studies used classical phenol-based extraction approaches. However, as interest in molecular tests has expanded, so too have the number of kit-based protocols that have become available to the scientist, many of which have been applied to *Brucella*. Whilst there are a plethora of trade names, each promising good recovery methods, the kits can be categorized by function.

One very popular approach for manual kits is to employ the classical technique of a combination of guanidine thiocyanate to lyse cells and silica particles to bind to the free DNA in solution (Boom *et al.*, 1990). These particles then undergo a number of washes before being re-suspended in Tris-EDTA buffer. For subsequent testing it is the supernatant and not the particle that is tested as silica inhibits PCR (Shoffner *et al.*, 1996). This technology (referred to as 'Boom technology') is used by a number of kits, particularly those using silica gel matrices in columns. Another extraction procedure that makes use of guanidine thiocyanate is that described by Chomczynski *et al.* (1997). Essentially after addition of a solution containing the chaotropic salt, the DNA is recovered by ethanol precipitation. Kits have also been designed around the principle of 'salting out' (Miller *et al.*, 1988). For this technique, which is commonly applied to blood samples, after the addition of a lysis buffer, SDS and proteinase K, a concentrated salt solution is added to denature proteins omitting the need for phenol chloroform extraction. The DNA can then be recovered by ethanol precipitation. Yet another process of extraction utilized by some kits is that of DNA

Table 4.2 Examples of commercial PCR kits available for the diagnosis of brucellosis

Kit name	Kit manufacturer	Conventional/real-time PCR	Target	Notes
Vita-Tech *Brucella* PCR kit	Idexx (Canada)	Conventional	16S rRNA	Used by Nimri (2003)
Amplisens *Brucella* PCR kit	*E. coli* (Russia)	Real time	Not given	–
Lightmix kit *Brucella* genus	TIBMolbiol (Germany)	Real time	IS*711*	FRET assay
Foodproof *Brucella* detection kit	Biotecon diagnostic (Germany)	Real time	Not given	FRET and Taqman® assays available
RAPID *Brucella* test	Idaho Technologies (USA)	Real time	Not given	To be used with hardware provided by kit manufacturer
Powercheck *Brucella* kit	Korgen Biotech (Korea)	Real time	Not given	–
Liferiver *Brucella* Kit	Shanghai ZJBiotech (China)	Conventional and real-time versions available	Not given	FRET and Taqman® assays available

chelation using the resin compound Chelex-100 (de Lamballerie *et al.*, 1992). In extractions using this process, the sample undergoes lysis, typically by boiling in the presence of Chelex resin. The sample is then cooled and spun with the supernatant being used for the molecular assay. The Chelex resin binds the polar cellular components after breaking open cells whilst the non-polar nuleic acids remain in the supernatant.

In addition to the above mentioned techniques there are also automated extraction methodologies based around the use of magnetic beads. Basically a lysis buffer and magnetic beads are added to the clinical sample. Once lysed, depending on the technology used, either an interaction between a chaotropic salt, DNA and magnetic silica bead (for example in the case of the Roche MagnaPure system) or a detergent, DNA, and magnetic bead (as in the case of Dynabead technology) followed by exposure to a magnetic field allows separation of DNA from possible inhibitors. Whilst these procedures are more costly in terms of hardware required, they do allow for automation and high throughput processing.

A number of papers have compared the performance of some of the extraction methodologies and kits. Romero and López-Goñi (1999), using their 16S rRNA conventional PCR with *B. abortus* spiked bovine milk samples, found that using kits based on the Chelex and Boom technology approaches gave poorer results than those obtained using an in house lysis solution and phenol chloroform extraction, based on the band intensity of PCR product generated. Marianelli *et al.* (2008a) made similar observations when examining milk samples from water buffalo using conventional and real-time PCR assays. Using a conventional *bcsp*31 PCR assay, they observed a 100-fold difference in sensitivity between extractions using the in house approach described above (Romeo and López-Goñi, 1999) (4×10^3 cfu/ml) and a commercial kit based on Boom technology (4×10^5 cfu/ml). Other authors have reported PCR failure using a commercial extraction kit, while a modified salting out procedure applied to the same samples subsequently allowed amplification by conventional PCR assays targeted to *omp*2 or *bcsp31* (El Kholly *et al.*, 2009). However, even though these papers imply that generic extraction methods may perform better than some commercial kits, in more recent publications, there has been a shift towards comparing commercial kits only. To date there have been three major publications in this area examining sera (Queipo-Ortuño *et al.*, 2008b), spiked swabs (Dauphin *et al.*, 2009), and infected tissue (Tomaso *et al.*, 2010).

Considering spiked swabs, and using an IS*711* assay, it was found that, regardless of the material the swab was made of, a kit based on salting out

(Masterpure, Epicentre, Madison, USA) was the best method for extraction of DNA (Dauphin *et al.*, 2009). The minimum reproducible detection level for the *B. melitensis* and *B. suis* assays was claimed to be 10 cfu/ml whilst in the case of *B. abortus*, a lower detection limit of 1 cfu/ml was noted, at least 10-fold more sensitive than the next best method in this study (MagNAPure, Roche, Basel, Switzerland). The infected tissue comparison used a variety of tissue samples taken *post mortem* from two serologically positive camels (Tomaso *et al.*, 2010) from which subsequent culturing indicated that both animals had disseminated *B. melitensis* infection. This material was processed for PCR using five commercially available DNA extraction kits; four of which were column based using Boom technology, with the remaining one based on salting out. Using a real time assay based on IS*711* the best performing kit overall was the Qiagen Qiamp kit (QIAGEN, Hilden, Germany) based on improved sensitivity of the IS*711* assay using its extractions. Interestingly, although the same real-time PCR assay was used to detect the DNA it was found that there were differences in recovery (manifesting itself in the sensitivity of the real-time PCR using the extracted material) between the four column based kits tested. This may be explained by intrinsic differences in the compositions of the reagents and protocol used in the different kits. This reasoning may also account for the results seen in a study comparing serum extraction (Queipo-Ortuño *et al.*, 2008b) examining, amongst others, two column based kits that both used Boom technology. Using a SYBR green assay based around *bcsp*31 it was found that two column based kits, QIAamp (Qiagen, Hilden, Germany) and Ultraclean DNA blood spin kit (Mobio, Carlsbad, USA) gave minimum sensitivities of 10 fg although, interestingly, the authors noted false-positive results with the QIAamp but not with the Ultraclean kit (Queipo-Ortuño *et al.*, 2008b). As the same procedure and samples were used for both kits it appears that a factor intrinsic to kits or buffers may be leading to false positive reactions in this study.

In summary, when implementing a molecular assay, in addition to using a sensitive, specific and well-optimized PCR it is important to take into consideration the method by which clinical material is processed to obtain target DNA. It may be worthwhile testing a number of extraction procedures to determine optimum sensitivity.

Application of molecular assays for diagnosis of brucellosis in animals

There are a variety of sample types that may be intrinsically suitable for molecular testing including infected tissue (such as fetal tissues or lymph nodes), vaginal swabs, blood, milk and semen and all have been examined to a greater or lesser extent in recent years.

Tissues

The first recorded use of PCR to test fetal tissue was use of the 43 kDa protein targeted assay to identify *Brucella* from both an aborted calf and its mother (Fekete *et al.*, 1992). As well as looking at fetal tissues (placenta and liver), this paper was the first to describe a critical evaluation of sensitivity through the spiking of rat tissue with varying amounts of *B. abortus*. A later study examined a conventional multiplex PCR for the detection of *Brucella* and *Leptospira* testing pooled tissue samples (lung, spleen, liver and kidney from individual animals homogenised to give one sample per animal) taken from aborted bovine fetuses (Richtzenhain *et al.*, 2002). Comparing culture results obtained with results generated by the *bcsp*31 component of the multiplex, it was found that whilst 6/11 samples were culture positive, 10/11 were positive by PCR. One explanation for this discrepancy was that material was not fresh but had been shipped into the laboratory. In support of this a repeat of the exercise examining infected hamster tissue gave comparable cultural and PCR results.

A number of studies have also examined the use of PCR to confirm the presence of *Brucella* in lymph nodes post-mortem with varying degrees of success. In work on cattle using a real-time PCR assay (Redkar *et al.*, 2001) for direct detection of *B. abortus* it was possible to detect three out of four culture positive animals when testing supramammary lymph node tissue by the molecular method (O'Leary *et al.*, 2006). Interestingly, in the testing of retropharyngeal lymph nodes with the same assay, although real-time PCR only detected one out of three culture positive samples, it did detect *Brucella* DNA in a sample that was culture negative, although a biopsy from another lymph node in the same animal was culture positive (O'Leary *et al.*, 2006). Negative PCRs could reflect low bacterial loads or inhibition problems. The reasons given for disparity between culture and PCR from tissue could equally be applied to other workers using a *B. melitensis* specific assay based on a specific IS*711* insertion (Bricker and Halling, 1994) to test seropositive sheep (Razik *et al.*, 2007). Although the authors stated that the PCR assay used in their study was more sensitive than culture, detecting 27/33 animals compared with 26/33 for culture, when the data are examined closely two culture positive animals are seen to be negative by PCR. In comparison a further study examining seropositive sheep using the same assay found that all culture-positive animals (28 in total) were also positive by PCR, although PCR managed to detect a further 19 positive cases including two serologically negative animals (Ilhan *et al.*, 2008). More recently a study of canine brucellosis found comparable results when comparing PCR and culture of lymph node tissue (Aras and Uçan, 2010).

Other studies tested reproductive tissue (testicles and uterine tissue) and spleens of 53 wild boar by PCR (Hinić *et al.*, 2009). Within this population four animals that were culture positive were also PCR positive. However, PCR detected *Brucella* in a further nine culture-negative animals. Most of the PCR positive animals gave positive results from all the various tissue samples tested illustrating how the infection had dissipated through the animal. All the work described above involved the use of homogenates of fresh or frozen material. However, for long-term storage, histological analysis and safe transportation, it is common for material to be fixed in paraffin and embedded in wax. The possibility of testing this type of material using a PCR multiplex assay for a number of pathogens including *B. melitensis* has been examined (Navarro *et al.*, 2009). A multiplex PCR specifically identified *B. melitensis* in all the archived material that was used for validation of the technique opening up the possibility of retrospective work to determine the presence of *Brucella*.

In summary, tissue samples have shown substantial potential as a target for PCR diagnostics. However, given that, with the obvious exception of fetal material, such material usually has to be taken *post mortem* it could only serve as a confirmatory tool following slaughter rather as an active screen for on-going infection.

Stomach contents

In the case of abortion, fetal stomach contents are frequently examined for bacterial content and have also been examined as a target for PCR diagnostics. Indeed, the first application of PCR as a diagnostic tool for brucellosis described earlier used such material (Fekete *et al.*, 1992). In later studies a conventional PCR based on 16S rRNA was compared with culture to test the stomach contents of 16 aborted ovine fetuses (Cetinkaya *et al.*, 1999). Whilst both culture and PCR identified *Brucella* in five individuals, one animal was culture positive but PCR negative with the converse situation in another animal, leading to a suggestion that there was some mismanagement in sample handling. Other studies noted that PCR from stomach contents generated marginally more positive results than culture (Richtzenhain *et al.*, 2002). In contrast, a further study, using a conventional *B. melitensis*-specific IS*711*-based assay, found that detection of *B. melitensis* by PCR and culture were roughly equivalent with 36/37 culture positive samples being positive by the molecular method while no culture-negative samples were PCR positive (Leyla *et al.*, 2003). In summary, stomach contents, where abortions or stillbirths have occurred, provide a good sample for confirmation of the presence of *Brucella* that, from the studies highlighted above, appears to generate molecular results that are on par with conventional culture.

Milk

Zoonotic transmission of *Brucella* from livestock to humans can occur in a number of ways, one of the most common being the ingestion of unpasteurized dairy products. Furthermore milk has also been reported as a possible transmission route from mother to offspring (Grilló *et al.*, 1997). Studies carried out over three years revealed that *B. melitensis* could be repeatedly isolated from infected sheep (Tittarelli *et al.*, 2005), whilst in cattle, it has been shown that an infected cow shed *B. abortus* in its milk 9 years after primary infection (Lapraik and Moffat, 1982). With this persistence, and the ease and non-invasive nature of sample collection, milk is a potentially attractive target for molecular diagnostic approaches.

Work examining the potential of this approach has examined milk from naturally and artificially infected goats (Leal-Klevezas *et al.*, 1995b). DNA was isolated by centrifugation of the cream layer of milk and resuspension of the pellet in a simple lysis buffer prior to a phenol chloroform extraction. PCR using the *omp*2 target was positive in 65% of the animals tested and in testing of spiked milk samples the approach appeared highly sensitive with the technique claiming to detect 10 organisms in 1 ml of milk. A 16S rRNA-based PCR assay has also been used to test milk for the presence of *Brucella* in cattle and compared with serological status as determined by ELISA (Romero *et al.*, 1995b). In this case whole milk was extracted and after digestion (which included RNase treatment) and centrifugation, the supernatant was taken for phenol–chloroform extraction. However, even using this protocol, it was noted that there was PCR inhibition in two culture-positive samples that the authors attributed to components of the extraction procedure (Romero *et al.*, 1995b). Specificities of the serological and molecular methods were identical in the test panel (both 100% in the negative animals). However, sensitivity of the ELISA (55/56 culture-positive animals) was slightly better than that of PCR (49/56 culture-positive animals). Even so, an animal that was ELISA negative was found to be culture and PCR positive, indicating that serological approaches can be misleading in certain circumstances.

In a different approach an assay based on the16S–23S rRNA spacer region and a line probe assay (LiPA) has been examined (Rijpens *et al.*, 1996). The LiPA assay involved a nested PCR of the 16S–23S rRNA spacer region with biotinylated primers, with the resulting products being incubated with a membrane containing specific probes thus producing bands where amplification was successful. The technique itself was shown to be both specific and more sensitive than conventional PCR. Indeed whilst the conventional PCR could detect 2.5×10^5 *Brucella* cfu/ml, using the LiPA gave a tenfold improvement in detection. Advances in probe chemistry probably account for the lack of subsequent publications using this technique. However, the idea of probe hybridization has been used in a multiplex assay based on *bcsp*31 to detect *B. abortus* and *Mycobacterium bovis* in milk (Sreevatsan *et al.*, 2000) with probe bound to microtitre plates rather than membranes. The multiplex included a host directed internal amplification control that could be used to assess for false negative reactions caused by PCR inhibition. Examination of spiked samples suggested sensitivity of 40 genome equivalents though specificity was not tested. The authors also examined 20 milk samples from field infection, 15 of which were positive for brucellosis by the MRT (milk ring test) and found 13/15 MRT positive samples to be positive on molecular testing as well as 4/5 MRT negative samples. In testing of bulk milks, where samples of infected and uninfected milks were mixed, the molecular test identified four out of six tainted batches, with the authors suggesting that the low bacterial load may have accounted for the false negative results.

In addition to caprine and bovine milk, PCR has been used to identify *Brucella* from sheep, camel and water buffalo milk (Hamdy and Amin, 2002; Marianelli *et al.*, 2008a). Using an IS*711*-based conventional PCR assay for the detection of *B. abortus* and *B. melitensis* it was found that testing milk from serologically positive cattle, goats and camels produced more PCR-positives than direct culture of the sample (Hamdy and Amin, 2002). In testing of cattle, PCR detected *B. abortus* in all 24 culture-positive, as well as five culture-negative, samples, whilst in goats, PCR detected *B. melitensis* in all 11 culture-positive, as well as two culture-negative, samples. All PCR-negative samples in cattle, goats and camels were also negative by culture. However, while all ten PCR-positive sheep milk samples were also culture-positive, there was a discrepancy in the ovine study where 2 of 11 milk samples found to be negative for *B. melitensis* by PCR were actually culture-positive. One possible reason for this discrepancy could lie in the sensitivity of the *B. melitensis* assay used. Using spiked milk, Hamdy and Amin (2002) established a minimum detection limit of 1000 cfu/ml of *B. melitensis* by PCR. If the bacterial load was lower, although this might test negative by the molecular method, detection might still be possible by culture. The authors also suggested that inhibition of PCR could account for the two culture-positive samples testing negative by the molecular method. However, to counter these suggestions, PCR also detected *B. melitensis* in milk from a serologically positive camel that was culture-negative. The authors did not give any explanation of this observation but one could speculate that this could be attributable to a sample where the organism was no longer viable.

Marianelli *et al.* (2008a), using a real-time PCR assay based on *bcsp31*, saw similar discrepancies in the testing of milk taken from serologically positive buffalo. Although 26 of the culture-positive samples were also positive by real-time PCR, nine culture-positive samples were negative by the molecular method. As with the work of Hamdy and Amin (2002), this could easily be explained as an issue of bacterial load and the assay detection limit. However, 9 of 16 culture-negative samples were positive by real-time PCR, again with a possible explanation being the detection of non-viable material by PCR. Despite these observations, PCR has also been used as a confirmatory tool demonstrating experimentally infected cattle shedding *B. abortus* RB51 in milk (Longo *et al.*, 2009), and real-time PCR, based on intercalating dye technology, has been used to identify infected buffalo shedding *Brucella* in milk with comparable results to concomitant culture (Capparelli *et al.*, 2009).

The main advantage of considering milk as material for testing is ease of availability although obviously this is confined to lactating animals. Nevertheless, a diagnostic that could reliably be applied to milk could also be adapted to other dairy produce and thus may have useful applications in food safety studies.

Blood

A further attractive and accessible matrix for *Brucella* testing is blood although there are obvious potential difficulties with this approach, not least the length, extent and potential variability of a bacteraemic phase. Leal-Klevezas *et al* (1995b) managed to detect *Brucella* in the blood of all of a cohort of artificially infected and culture confirmed goats. Furthermore, PCR positive results were obtained from 21 (52.5% of the sample population) in contrast with serology (42.5%) and culture (10%) suggesting that the molecular method was the most sensitive in this scenario. In contrast, no control animal tested positive by PCR, showing the specificity of the molecular method. A more recent study of seropositive sheep noted that numbers of PCR-positive results from lymph node tissue and blood were virtually identical and significantly greater than the number of culture-positives obtained from the same samples (Ilhan *et al.*, 2008).

However, there are difficulties in using blood and the outcome of other studies has been far less clear-cut. A small study of bovines found no PCR positive results using either a conventional or real time *IS711* PCR assay in bloods taken from serologically positive animals, although by the same token they could not isolate the organism (O'Leary *et al.*, 2006). Studies in dogs examining cohorts of blood culture positive and blood culture and serologically negative animals showed an exact correlation with the PCR testing (Keid *et al.*, 2007b). However, a third group of animals that were suspected of being infected gave PCR-positive results for only a proportion of animals. Other animals, that were both serologically positive and demonstrating clinical signs of infection, appeared PCR negative. The authors' explanation was that in these cases the infection might have been abacteraemic. A similarly confusing picture was seen in a study of wild boar where roughly equal numbers tested positive by serological and molecular methods (Hinić *et al.*, 2009). However, when the data were combined, it was found that only two out of nearly 200 tested animals were positive by both methods. Similar testing in water buffalo found that while PCR alone identified nearly 30% of the animals as infected, only around half this number were both serologically and PCR positive (Guarino *et al.*, 2000). Unfortunately, it was not possible in either study to obtain cultures from bloods tested (Guarino *et al.*, 2000; Hinić *et al.*, 2009). However, an interesting finding in the Guarino *et al* (2000) study was that lymph node tested from a PCR positive but serologically negative animal was found to be culture positive.

Whilst the testing of serum by PCR is common in the diagnosis of human brucellosis, there conversely is little published on the use of animal sera for molecular diagnostics. One of the few such articles examined use of a conventional PCR based on the 16S-23S rRNA region (Keid *et al.*, 2010). Testing of whole blood and sera from 35 culture positive dogs resulted in 34/35 PCR positives using whole blood but only 9/35 using serum (Keid *et al.*, 2010). This observation and the relatively long bacteraemia associated with *B. canis*, suggest that molecular testing of serum for diagnosis of brucellosis in animals may be difficult.

Blood, like milk, is an attractive, accessible sample type with the additional advantage that it is not limited to a subset of animals and thus could readily be applied in a screening situation. However, a major problem with using this type of material is the uncertainties surrounding the bacteraemic phase which can vary between species and individuals within a species. In the case of dogs literature suggests that the period of *Brucella* bacteraemia is relatively long (Wanke, 2004). In the case of pigs, a time period of 8 weeks has been cited for *Brucella* bacteraemia although it can persist for longer periods of time in certain individuals (Deyoe and Manthei, 1964). In unpublished work done at the Veterinary Laboratories Agency, where four cows were experimentally infected with *B. abortus* and periodically bled, there was brief and intermittent detection of the organism by culture and no detection at any time point by PCR.

Reproductive fluids and urine

Molecular testing of vaginal swabs or uterine discharge seems an obvious approach to confirming *Brucella* at least immediately post abortion; however, there is remarkably little literature in this area. There is a mention of uterine/vagina discharge molecular testing being useful in bovines although there is no breakdown of the samples tested in this publication (Fekete *et al.*, 1992). In the only other study of note including screening of vaginal swabs studies in canines found that whilst the results of culturing vaginal swabs was poor (6/144 positive), the results of PCR on the same material (52/144 positive) were similar to those obtained by blood culture (49/144 positive) and PCR from whole blood (54/144 positive) (Keid *et al.*, 2007a).

In the case of testing male animals for brucellosis there has been some work examining molecular detection in semen. Detection of *B. melitensis* directly from bovine and ovine sperm using a non-kit based method for DNA extraction and a previously described PCR assay (Bricker and Halling, 1994) has been described (Amin *et al.*, 2001) with a limit of detection of 100 cfu/ml. Using fractionated sperm samples all the fractions tested were PCR positive and the molecular method was more sensitive than culture, detecting more positives among seropositive animals. In contrast later work using an IS*711* based PCR assay found that molecular methods were comparable, rather than superior to culture, for the detection of *B. ovis* infection in rams (Manterola *et al.*, 2003). Using the same assay as part of a multiplex assay for the detection of three pathogens causing reproductive problems in sheep, *Brucella* was demonstrated in three samples by molecular assay in agreement with cultural findings (Saunders *et al.*, 2007). Recent work described a specific assay for the diagnosis of *B. ovis* in semen with equivalent sensitivity to the earlier studies (Xavier *et al.*, 2010). In this case, the targets selected were two regions defined in a previous study as being specific to *B. ovis* (Tsolis *et al.*, 2009). These assays were also used to detect *Brucella* in the urine of infected rams with the assay being more sensitive in urine than when applied to semen. Hinić *et al* (2009), in a study of wild boar, also managed to detect *Brucella* in urine by PCR, although this was confined to a single animal with no associated cultural data. Semen has also been used in the detection of *Brucella* in dogs with *vir*B2 being the *Brucella* target in a multiplex nested PCR assay to detect *B. canis* and *Leptospira interrogans* and able to detect down to 100 cfu *Brucella* (Kim *et al.*, 2006). A subsequent study, also in dogs, showed PCR to be more sensitive than culture of semen (Keid *et al.*, 2007a).

Given that brucellosis in animals is associated primarily with reproductive problems it seems that this would be a productive avenue for further research. From the limited information that is available it can be seen that molecular methods are as efficient as direct culture from such material and thus might effectively be used as a replacement to give safer and more rapid diagnosis of infection.

Application of molecular assays for diagnosis of human brucellosis

Given that brucellosis remains a major zoonotic disease issue in many parts of the world there has been substantial effort to investigate the potential of molecular diagnostics for human brucellosis. The earliest such study tested culture positive blood from three individuals and found that the PCR assay based around *omp2* was positive for all (Leal-Klevezas *et al.,* 1995b). A subsequent, much more substantial, study examined 50 whole blood samples that were brucellosis positive by either culture or serology and compared these with 60 control blood samples (Queipo-Ortuño *et al.,* 1997), Using a non-kit-based extraction methodology and a PCR assay based on the *bcsp*31 target, all 50 of the serologically or culturally positive samples were also PCR positive. Furthermore the assay identified an individual who had clinical brucellosis two months prior to the study but whose blood sample was both negative serologically and by culture.

The value of PCR in monitoring drug therapy in brucellosis patients has also been investigated (Morata *et al.,* 1999). This study monitored 30 individuals diagnosed with brucellosis all of whom were PCR positive pretreatment. After completing a full course of treatment it was found that only two individuals remained positive by PCR and it was later shown that the two positive results came from patients who had relapsed. In a similar study monitoring 20 patients pre and post treatment, whilst residual antibody titres remained, PCR was negative in all but two relapse patients (Elfaki *et al.,* 2005). However, in stark contrast other studies have noted that although bacterial load (as determined by standard curve of the real-time PCR assay) decreased during treatment, there was still detection of DNA after successful treatment and patients becoming asymptomatic (Navarro *et al.,* 2006; Vrioni *et al.,* 2008). Navarro *et al* (2006) noted that after a course of treatment at the 60th day, 40% of the test groups were still PCR-positive. Vrioni *et al* (2008) looked at whole-blood samples from a group of 39 individuals who were either culture confirmed or serologically positive for brucellosis prior to a course of treatment. At the start of the study 39/39 patients were PCR-positive using a real-time PCR assay based on *bcsp*31. By the end of the treatment phase (6–12 weeks), 34/39 patients were still positive by PCR. As time progressed the number of positive PCR samples dwindled but it was noted that even at 3 years after treatment, 7/10 patients were still PCR-positive.

Whilst there are many studies applying PCR to human blood different researchers have applied their approaches to different blood components. Many authors have used whole peripheral blood as their sample of choice and regardless of the molecular target used, results from all these studies were either equivalent or superior to matched blood culture (Nimri, 2003; Kattar *et al.,* 2007; El Kholly *et al.,* 2009; Kantartzi *et al.,* 2010). Other researchers have cited serum as a useful fraction from which to test for the presence of *Brucella*. Indeed whilst the action of the complement system in serum can make culture difficult, PCR from serum has been effectively undertaken for a variety of pathogens (van Standen *et al.,* 1998; Zhang *et al.,* 1998). One advantage of testing sera over whole blood is that the same sample can be used for serological studies and thus confirmation of infection from the same material can be determined through two means. In addition potential inhibitors of PCR amplification in blood, such as haem, are not present in serum provided there is no haemolysis. The inhibitory nature of haem to *Brucella* PCR was previously aptly demonstrated by studies showing that increased washing of the leucocyte pellet 'until all reddish colour was gone', improved detection (Morata *et al.,* 1998). However, it should be noted that IgG present within serum has been shown to inhibit PCR amplification, although this can be overcome by simply diluting the sera and testing directly (Queipo-Ortuño *et al.,* 2008a). Elsewhere testing of serum derived from culture positive blood by PCR detected 12/14 culture-positive samples, with no positives in the control samples (Debeaumont *et al.,* 2005).

Given the preferred intracellular habitat of *Brucella* there has been significant interest in using white blood cells (also referred to as the 'buffy coat' fraction). Matar *et al* (1996) used buffy coat derived by splitting whole blood with ficol hypaque and testing with a conventional *bcsp*31 PCR and obtained 20 PCR-positive results from 20 seropositive patients. Successful PCR has also been reported from pus taken from the abscesses of patients serologically positive for brucellosis (Colmenero *et al.*, 2002).

Although there have been studies that were instigated to determine if whole blood or blood fractions gave the better results in terms of sensitivity when tested by PCR, the findings published seem to be contradictory. Zerva *et al* (2001) examined serum and whole blood samples taken from a study group of 76 individuals containing 31 patients with clinical brucellosis and 45 healthy controls. Using a conventional *bcsp*31 targeted assay it was found that 58% (18/31) of the known brucellosis patients were PCR positive examining both whole blood and sera. None of the healthy controls were PCR positive from either whole blood or sera. However, of the remaining 13 patients known to have brucellosis, 11/13 were PCR-positive only when testing sera with 1/13 being PCR-positive only when testing blood. Interestingly blood culture of 24 of the 31 infected patients revealed only 13 positives. Based on these observations the authors suggested improved sensitivity when testing sera rather than matching whole blood.

A further study used three different PCR targets (*omp*2, *bcsp*31 and *omp*28) to examine differences in PCR detection rates in matched samples of whole blood, buffy coat and sera from 200 patients beginning antibiotic therapy for brucellosis (Mitka *et al.*, 2007) The individuals tested were all previously diagnosed with brucellosis on the basis of a clinical assessment. Blood culturing prior to treatment detected 148 culture-positive cases whilst 183 patients were serologically positive. From the molecular testing done prior to treatment, it was found that, using the buffy-coat, all 200 patients were PCR-positive to all targets. The *bcsp*31 assay also managed to detect 100% of the whole blood samples as PCR-positive. However, for the *omp*2 and *omp*28 assays there was a slight decrease in sensitivity to between 98.5% and 99% in whole blood. Further there was also a drop in sensitivity registered for all the molecular tests when serum was used as the test sample, with the best value coming from *bcsp*31 at 97% and the worst being the *omp*2 assay at 95.5%.

Looking at both these studies, there was very little difference in the execution of the work with both examining similar matrices, using the same *bcsp*31 based assay and indeed some authors in common. However, one key difference was the extraction kit used, which as mentioned in an earlier section, is a critical factor in assay sensitivity. One study (Zerva *et al.*, 2001) used an Isoquick Nucleic acid extraction kit (Orca Research, Bothell, USA) whilst the other (Mitka *et al.*, 2007) used the Qiamp Kit (Qiagen, Hilden, Germany). In an unrelated study comparing recovery of viral DNA from spiked samples (Fahle and Fisher, 2000) it was demonstrated that the Qiamp Kit recovered more DNA and gave rise to more reproducible results (6/6 tested) in whole blood that the Isoquick kit (4/6 tested). A similar phenomenon may account for the earlier observations of serum samples producing more positive results than whole blood (Zerva *et al.*, 2001).

Matrices other than blood components have also been used as templates for PCR to diagnose complications of *Brucella* infection. For example testing of cerebral spinal fluid has been used successfully to diagnose neurobrucellosis (Morata *et al.*, 2001; Colmanero *et al.*, 2005; Queipo-Ortuño *et al.*, 2009), direct testing of synovial fluid has been used to diagnose *Brucella* related osteoarthritis (Debeaumont *et al.*, 2001; Morata *et al.*, 2001; Queipo-Ortuño *et al.*, 2009) and bone tissues have been used to detect *Brucella* related spondylitis and osteomyelitis (Navarro-Martínez *et al.*, 2008, Queipo-Ortuño *et al.*, 2009). As in animals urine has also been a target for PCR based diagnostics. Human epididymo-orchitis caused by *Brucella* has been successfully identified by PCR direct from urine (Morata *et al.*, 2001; Queipo-Ortuño *et al.*, 2006). In the latter study nine individuals who were blood culture positive also gave positive PCR results from their urine although culture was only successful from one of these samples. Recently the presence of *Brucella* was demonstrated by PCR of the semen from an individual who was not displaying overt clinical symptoms (Meltzer *et al.*, 2010).

Molecular diagnosis of infection has become integral in human medicine, allowing rapid and specific results that can allow for earlier treatment and improved patient outcome. In the case of brucellosis, PCR has provided a useful tool in monitoring disease and gauging treatment. Although evidence is limited studies to date demonstrate that, unlike the situation with animals, serum may be a useful sample type for the molecular diagnosis of human infection.

Alternative molecular approaches for *Brucella* detection

An alternative molecular approach for the detection of *Brucella* DNA is direct use of hybridization probes. As long ago as 1989 biotin labelled *B. abortus* S19 genomic DNA was used as a probe to detect *B. abortus* in mouse liver tissue (Hopper *et al.*, 1989). Extracted DNA from these livers was dot blotted onto nitrocellulose paper and then hybridized with the biotin labelled *Brucella* DNA. The assay developed was shown to be specific when tested against some non-*Brucella* although there were weak cross-reactions with *Y. enterocolitica* and *Serratia marcescens*. More recently the use of a hybridization probe based on the 16S rRNA sequence was reported for the detection of *Brucella* using flow cytometry although cross-reaction was noted with phylogenetically related organisms (Fernández-Lago *et al.*, 2000). To overcome this problem Wellinghausen *et al* (2006) used an unlabelled competitive probe with one base difference. This base difference facilitates preferential binding of the unlabelled probe to non-*Brucella* organisms and application of the *in situ* hybridization assay to whole blood allowed specific identification of *Brucella* even when challenged with the phylogenetic near neighbour *O. anthropi*. While such *in situ* hybridization approaches offer an alternative to PCR as a molecular diagnostic for brucellosis they have yet to find widespread use.

Future strategies for molecular diagnostics

There are a number of emerging techniques that utilize isothermal amplification but the most popular by far is loop-mediated amplification (LAMP). In these assays amplification occurs at a constant temperature eliminating the need for expensive thermocyclers. First described for *Escherichia coli* (Notomi *et al.*, 2000) these assays require six primers; two internal primers to run the reaction, two external primers to dislodge the product off the template without the need for temperature denaturation and two 'loop' primers to speed up the LAMP reaction (Nagamine *et al.*, 2002).

In addition to its utilization of isothermal amplification, another benefit in using this technique is that it is apparently less sensitive to agents within samples (such as haem in whole blood) that can inhibit PCR (Poon *et al.*, 2006). However, this needs further investigation as there are also cases where it has been shown that LAMP itself can be inhibited, such as in testing of milk spiked with *Salmonella* (Li *et al.*, 2009). Another potential benefit of using LAMP is the variety of methodologies available to detect amplification products. In addition to the use of agarose gels and intercalating dyes such as EvaGreen, precipitation of insoluble magnesium pyrophosphate, a by-product of the reaction can be detected by eye (Mori *et al.*, 2001). The creation of this precipitate can also be detected colormetrically through the addition of agents such as hydroxynapthol blue which changes from violet to blue in a positive reaction (Goto *et al.*, 2009).

Currently there is only one published LAMP assay for the detection of *Brucella* organisms (Ohtsuki *et al.*, 2008). This work describes the testing of two LAMP assays based around the *bcsp*31 target with both assays specific for *Brucella* when tested against a panel which included *O. anthropi*. This is an important observation given that the there are specificity issues with the conventional PCR assay based on this target (Baily *et al.*, 1992). One reason for the improved specificity in LAMP could be that, whilst PCR uses two regions to direct amplification, the unique composition of the internal primers in LAMP means that four regions are required.

In terms of sensitivity, the LAMP assay could detect down to 10fg of purified genomic *B. melitensis* DNA, which was equivalent to a real-time PCR assay based on the same target (Ohtsuki *et al.*, 2008). However, LAMP amplification was complete in 30 min, significantly less time than an equivalent real-time PCR assay. The assay also successfully detected *Brucella* from both spiked milks and tissues, although in both cases, this was after an extraction procedure using the DNAzol commercial DNA extraction kit (Ohtsuki *et al.*, 2008).

Although PCR diagnostics have facilitated rapid diagnosis of brucellosis one factor that could inevitably hamper their universal application, particularly in areas of the developing world where brucellosis is most problematic, is the cost of set up and use. Further, as mentioned earlier, there are potential inhibitors of PCR in many clinically relevant samples such as milk and blood that means an extraction procedure is required. Extraction adds time, manipulations and cost to the test that may make it unattractive in situations when a cheap and quick diagnosis is desirable. Furthermore, reducing the number of steps required between obtaining samples and reporting results could give vital time in the control of an outbreak. Using LAMP based assays could circumvent a number of problems highlighted and at the same time provide cheaper diagnostic testing that may be more accessible to the developing world.

Developments in molecular epidemiology

Early epidemiological tools

For many years, six 'classical' *Brucella* species were recognized: *B. abortus* (bovine), *B. melitensis* (caprine, ovine), *B. suis* (predominantly porcine), *B. canis* (canine), *B. ovis* (ovine) and *B. neotomae* (rodent) although in recent years a number of confirmed or potential new species have been identified (Whatmore, 2009). The divisions into classical species were based almost entirely on phenotypic differences and host specificity. Subtle differences in phage sensitivity, serological specificity, dye sensitivity and growth characteristics allowed further differentiation into a number of biovars within the three economically significant species (*B. abortus, B. suis* and *B. melitensis*). Initial attempts to establish molecular relationships between *Brucella* isolates and use this understanding to develop effective epidemiological tools were frustrated by the apparent genetic homogeneity of the genus. Reflecting this for many years the only tool available for intraspecific subdivision of isolates was biotyping.

A number of early molecular tools proved invaluable in helping establish the relationships between species. These include approaches such as genome mapping, repetitive element based PCRs, genomic fingerprinting based on IS*711* elements, AFLP, and RFLP of various polymorphic genes. The reader is referred elsewhere for a fuller discussion of these earlier approaches (Whatmore, 2009) some of which, notably IS*711* fingerprinting and certain RFLPs, remain valuable for comparison of novel isolates with extant databases. However, none of these tools provide significant and reliable subspecies resolution, a problem that persisted until the advent of the tandem-repeat based approaches described later.

IS*711*-based PCR typing tools

The differential location of IS*711* within *Brucella* chromosomes has been extensively exploited as the basis of a number of useful PCR tools for typing *Brucella* primarily to the species level. A series of multiplex PCRs have been developed where use of one common primer anchored in the IS*711* element and a species-specific primer that binds to unique sequence flanking that insertion site allows species discrimination based on different amplicon sizes. The prototype assay was known as the AMOS-PCR named after the *Brucella* species it could initially identify (*B. abortus* biovars 1, 2, and 4, *B. melitensis, B. ovis* and *B. suis* biovar 1) (Bricker and Halling, 1994). The assay was subsequently modified to include the vaccine strains S19 and RB51 (Bricker and Halling, 1995) allowing discrimination of field and vaccine isolates in eradication programmes (Ewait and Bricker, 2000). Some later modifications purported to allow differentiation of some additional biovars of *B. abortus* such as biovars 3b (an unofficial designation reflecting the identification of genetic variation within biovar 3 isolates), 5, 6 and 9 (Ocampo-Sosa *et al.*, 2005). However, this assay still has the disadvantage that it fails to detect all species or all biovars of some species.

In recent years the availability of genome sequences facilitated the development of a new generation of multiplex PCRs with a much

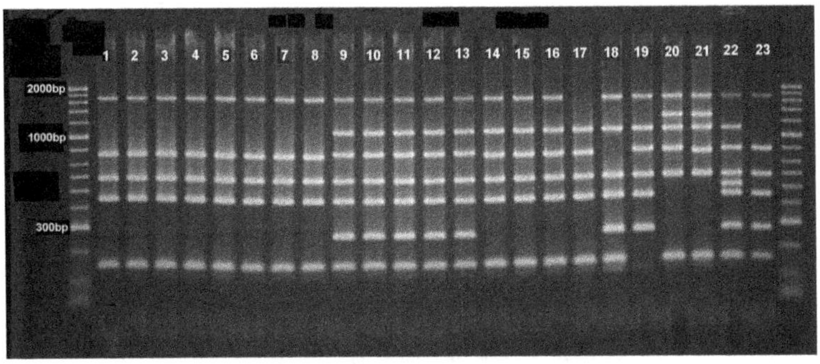

Figure 4.1 Example of 'Bruceladder' multiplex PCR for typing to species level carried out using the modifeid approach described by Mayer-Scholl *et al.* (2010). Lanes 1–8 *B. abortus* biotypes; Lanes 9–13, *B. suis* biotypes; lanes 14–16, *B. melitensis* biotypes; lane 17, *B. ovis*; lane 18, *B. canis*; lane 19, *B. neotomae*; lane 20, *B.pinnipedialis*; lane 21, *B. ceti*; lane 22, *B. microti*, lane 23, *B. inopinata*. The assay can also differentiate the three approved live vaccine strains (not shown). Reproduced with permission from Mayer-Scholl *et al.* (2010).

broader scope. The most established of these is an assay known as 'Bruceladder', a single tube PCR assay that can differentiate all six classical species, more recently described *Brucella* isolates from marine mammals and the vaccine strains *B. abortus* RB51 and S19 and *B. melitensis* Rev 1 (García-Yoldi *et al.*, 2006). The test uses eight primer pairs designed based on species-specific differences identified historically or by genomic approaches and results in a unique profile for each species following agarose gel electrophoresis (see Fig. 4.1). The assay has been extensively validated and found to be generally very effective with the rare exception of *B. canis* where some isolates can be erroneously identified as *B. suis* (López-Goñi *et al.*, 2008; Koylass *et al.*, 2010). More recent developments include the incorporation of primers allowing discrimination of two newly described species *Brucella inopinata* and *Brucella microti* (Mayer-Scholl *et al.*, 2010). This new assay was also reported to separate the marine mammal species *B. pinnipedialis* and *B. ceti* on the basis of a 19bp partial duplication. This finding appears surprising when considering the observed genetic relationships within the marine mammal *Brucella* and thus needs to be confirmed using a much more extensive isolate collection. A number of similar PCR-based assays have been described (Hinić *et al.*, 2008; Ratushna *et al.*, 2006) but they are not as comprehensive, nor as widely tested, as the 'Bruceladder' based assays.

Emerging epidemiological tools

In the last few years new classes of tools have been developed, notably multilocus sequencing and variable number of tandem repeat (VNTR) typing, which have become widely used tools in bacterial typing. These approaches, which generally have contrasting and complementary powers of resolution, have been applied to *Brucella* and have greatly increased understanding of genetic population structure as well as faciltating the identification of useful diagnostic markers.

Multilocus sequencing

Multilocus sequencing has become the accepted tool for understanding the global epidemiology of bacterial populations as well as providing data that can be invaluable in establishing phylogenetic relationships. Multilocus sequencing involves the sequencing of fragments of a number of genes, classically genes encoding 'housekeeping' functions, which are subject to purifying selection and slow evolution and within which variation is nearly neutral (Enright and Spratt, 1999). The use of multiple gene fragments can give high discriminatory power while retaining signatures of longer-term evolutionary relationships (Margos *et al.*, 2008).

Multilocus sequencing was first applied to *Brucella* in 2007 with the examination of nine discrete genetic loci, equating to 4396bp of sequence, from 160 *Brucella* isolates of all species and biovars

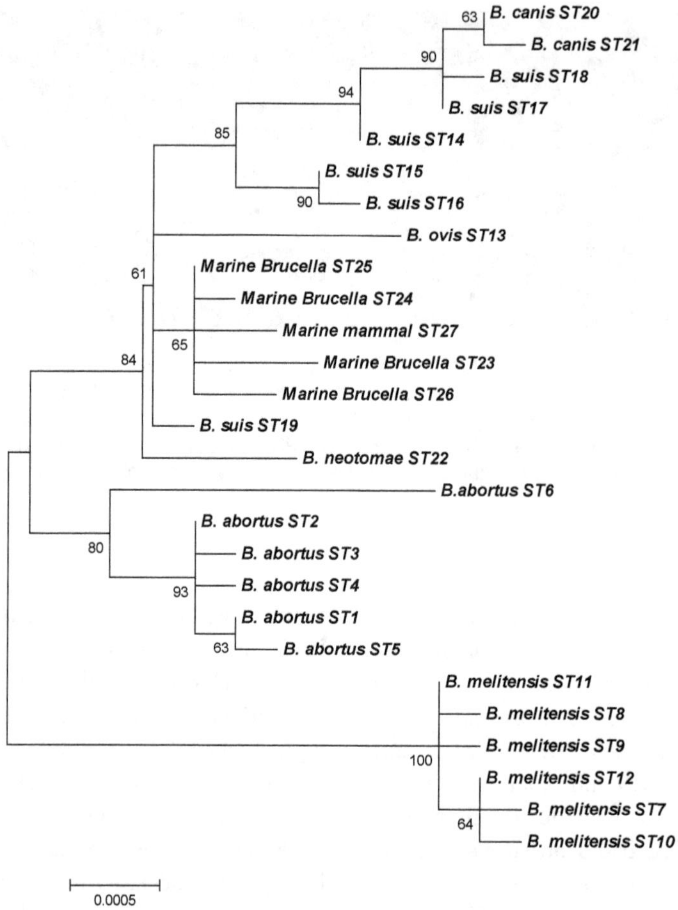

Figure 4.2 Phylogenetic reconstruction of the relationships between sequence types defined by multilocus sequencing. This tree was constructed with the concatenated sequence data of nine loci (4396 bp) using the Neighbour joining approach. The percentage bootstrap confidence levels of internal branches were calculated from 100 resamplings of the original data. Bar = 0.0005 substitutions per nucleotide position.

known at the time (Whatmore *et al.*, 2007). With only 1.5% of sites found to be polymorphic these studies confirmed long standing observations on the genetic homogeneity of the group and allowed the division of the 160 strains into 27 sequence types (STs) as shown in Fig. 4.2. This homogeneity means that, in contrast to the situation with some other bacteria, multilocus sequencing is of limited value as a short- to medium-term epidemiological tool for *Brucella*. However, it has proven valuable in establishing relationships at higher phylogenetic and taxonomic levels and in understanding global population structure. The preliminary analysis showed *B. abortus*, *B. melitensis*, *B. ovis* and *B. neotomae* corresponding to well-separated clusters. With the exception of biovar 5 (ST19), the relationship of which to *B. suis* had long been debated, *B. suis* isolates cluster together. *B. canis* isolates are located in a distinct subcluster alongside *B. suis* isolates, very closely related to, but distinguishable from, *B. suis* biovar 3 and 4 isolates (STs 17 and 18). Again this confirmed longstanding observations of a very close relationship between *B. canis* and *B. suis* despite the notable host specificity of the former. The recently described marine mammal strains were found to represent a distinct group of isolates (ST23–ST27) confirming that these isolates represented a distinct taxonomic cluster within the genera clearly separate from the 'classical species'.

Thus the classically described *Brucella* species, despite being identified on the basis of phenotype and host specificity, were found to represent remarkably accurately the observed genetic relationships within the genus.

This initial work has now been extended to examine 21 distinct genetic loci (c. 10.5 kbp) from over 600 strains resulting in the identification of some 100 STs (A. M. Whatmore, unpublished data). Analyses of these data maintain the same basic structure but give much more detail at the subspecies level and are particularly informative in establishing genetic relationships between biovars of *B. melitensis*, *B. abortus* and *B. suis*. While *B. suis*, and some *B. abortus* biovars do appear to correspond to true genetic entities there is no obvious relationship between *B. melitensis* biovars and genetic clustering.

One of the real advantages of multilocus sequencing is that, in contrast to many earlier approaches, it indexes known genetic diversity. Reflecting this, rather than just providing a new 'profile', the relationship of which to existing 'profiles' may not be obvious, it provides a framework allowing the accurate placement of any newly emerging or newly discovered genotypes relative to extant knowledge. Reflecting this multilocus sequencing has been instrumental in helping describe a number of novel groups within the genus. The genetic diversity within the genus was markedly widened recently by the publication of a novel species, *B. inopinata*, isolated from an infected prosthetic breast implant (Scholz *et al.*, 2010). Multilocus sequencing, including comparative analysis with members of the *Ochrobactrum*, the genus most closely related to *Brucella*, helped establish that, although rather distinct from previously described *Brucella*, classification of this organism as a new *Brucella* was appropriate (De *et al.*, 2008). Multilocus sequencing was also used in the characterization of *B. microti* (Scholz *et al.*, 2008) originally isolated from diseased voles (Hubálek *et al.*, 2007) although in this case the isolates, while representing a distinct novel ST, were very closely related to existing groups notably *B. suis* biovar 5. Unfortunately the marine mammal isolates were finally formally published as two species, *B. pinnipedialis* (from seals) and *B. ceti* (from porpoises and dolphins), just prior to the availability of multilocus sequencing data (Foster *et al.*, 2007). However, it has rapidly become clear that this taxonomic placement does not reflect genetic relationships. Both multilocus sequencing and tandem repeat-based typing have clearly shown that *B. pinnipedialis* consists of two genetic clusters which are less closely related to each other than either is to *B. ceti* (Groussaud *et al.*, 2007; Maquart *et al.*, 2009). These observations are in agreement with the genome organization shown by *Pac*I and *Spe*I physical maps (Bourg *et al.*, 2007) and have also been confirmed by a number of other molecular approaches such as IS*711* fingerprinting (Dawson *et al.*, 2008), location of IS*711* (Zygmunt *et al.*, 2010), *omp* fingerprinting (Dawson *et al.*, 2008) and unpublished genome sequence data all of which support a taxonomical reassessment of this group.

Finally multilocus sequencing has been used in characterization of a number of other novel *Brucella*, which though not yet formally described, will probably ultimately represent additional new species. These include novel isolates from baboons, which cluster within the classical *Brucella* but are clearly distinct from any known species (Schlabritz-Lousevitch *et al.*, 2009) and two groups that, like *B. inopinata*, extend the known diversity of the genus. These represent a single isolate, BO2, from human disease which has been described as '*B. inopinata*-like' although it does appear rather genetically distinct (Tiller *et al.*, 2010b) and a group of isolates from Australian rodents (Tiller *et al.*, 2010a).

Other sequencing

Sequence originating from a single locus should be treated with caution when used to place a bacterial isolate because of the dangers of incorrect conclusions being drawn from single loci whose evolution may not necessarily reflect that of the genome. Single genes are also probably of limited value because of a lack of discriminatory capacity when applied to a group such as *Brucella*. Despite this caveat sequences of single loci have been widely used in characterization of *Brucella* notably targeting outer membrane protein (*omp*) encoding genes and *rpoB*. Particularly well characterized is the *omp2* locus consisting of two genes, designated *omp2a* and *omp2b*, displaying around

85% sequence identity, and arranged in opposite orientations (Ficht *et al.*, 1989, 1996). DNA polymorphism at the *omp2* locus was also instrumental in the suggestion of the division of the new marine mammal *Brucella* isolates into two species (Cloeckaert *et al.*, 2001) as well as in descriptions of *B. microti* and *B. inopinata* (Scholz *et al.*, 2008, 2010). Over the years a number of PCR-RFLP tests based on *omp2* have been described as tools to type *Brucella* at the species/biovar level (Al Dahouk *et al.*, 2005b). An additional popular gene for describing *Brucella* diversity in recent years has been *rpoB* originally examined in reference strains where 17 alleles were described (Marianelli *et al.*, 2006) and which was later examined in field isolates, usually being used in conjunction with other molecular typing approaches (Marianelli *et al.*, 2007; Sayan *et al.*, 2009; Valdezate *et al.*, 2010).

SNP based approaches

In the last five years a number of single nucleotide polymorphism (SNP)-based approaches to typing *Brucella* have been described. Many of these are based on SNPs identified as a result of multilocus sequencing studies decribed above which have been utilized to devlop tools that, in contrast to multilocus sequencing, are simple and rapid enough for routine diagnostic use. These developments have been enabled by the availability of a population genetic framework which has allowed identification of 'canonical' SNPs (Keim *et al.*, 2004) that define groups of interest. Further population genetic analyses suggest a general lack of recombination within the genus (Whatmore *et al.*, 2007) which gives confidence in the robustness and stability of such diagnostic markers. A number of assays using such SNPs to characterize isolates to the species level have been described using approaches such as primer extension (Scott *et al.*, 2007), MGB real-time PCR probes (Foster *et al.*, 2008; Gopaul *et al.*, 2008) and high resolution melt curve analysis (Winchell *et al.*, 2010). Although all of these published assays include only the classical species and the marine mammal group these assays can easily be extended as new species are identified. For example, the real-time PCR assay described by Gopaul *et al* (2008) has already been updated to include markers for recently described species (K.K. Gopaul, unpublished data). Where useful SNPs have been identified assays have also been developed that can type at the subspecies level – these include SNP based assays to identify *B. suis* biovars (Fretin *et al.*, 2008) and to differentiate live vaccine strains as described earlier (Gopaul *et al.*, 2010).

Development of tandem repeat based approaches

Although useful for addressing certain questions none of the above approaches currently represent highly discrminatory approaches capable of substantial differentiation of isolates at the subspecies level. However, in recent years a new generation of typing approaches based on tandemly repeated sequences has emerged. These repeats display allelic hypervariability related to variation in the number of repeats caused by slipped strand mispairing or replication slippage (Levinson and Gutman, 1987) or by recombination between homologous repeat sequences (van Belkum *et al.*, 1998). These markers variously known as variable number of tandem repeats (VNTRs), microsatellites (shorter repeats) or mini-satellites (longer repeats generally of 10–100 bp repeats) are considered high-speed molecular clocks (van Belkum, 1999). Schemes based on such markers, sometimes referred to as MLVA (multilocus VNTR analysis), have now been widely applied to many bacterial species and can be particularly useful for discriminating within very conserved species and genera (Vergnaud and Pourcel, 2006). They rely on variation in the sizes of PCR products (that include the repetitive loci) that can be related to repeat number at a number of distinct repetitive loci to give a typing system that is unambiguous and portable allowing ready comparison between laboratories. The application of this approach has huge potential in facilitating the rapid detection of related disease clusters, in the detection of transmission chains and to trace the source of outbreaks. Such data may also prove invaluable in identifying risk factors, as strains can be more accurately linked to epidemiological data, and also in helping understand the basis of virulence by allowing the identification of clonal groups with variable pathogenic properties.

VNTR typing was first applied to *Brucella* in 2003 with the publication of a scheme designated

'Hoofprints' (hypervariable octomeric oligonucleotide fingerprints) (Bricker *et al.*, 2003) that was facilitated by analysis of the first *Brucella* whole genome sequences to become available. These analyses identified an eight base pair tandem repeat present at nine genetic loci of which eight displayed variation in repeat number. Analyses of these markers successfully differentiated all type strains for all species and biovars and among unrelated *B. abortus* biovar 1 field isolates. In contrast isolates from the same herd or from short term *in vitro* passage showed little or no variation. Later analyses using the same approach on a larger number of field isolates highlighted the remarkable discriminatory capabilities of this approach (Bricker and Ewalt, 2005).

Subsequent to the publication of the 'Hoofprints' approach two additional VNTR schemes were soon published. One included the original 'Hoofprints' loci, as well as an additional 13 newly described VNTR loci, to give a 21-locus scheme (Whatmore *et al.*, 2006) while the other examined variation at 15 repetitive loci (Le Flèche *et al.*, 2006). Although there is much overlap between the three schemes, the last two schemes differ from 'Hoofprints' in that they include loci with a much broader range of evolutionary speeds. Reflecting this, and in contrast to the 'Hoofprints' scheme, both these schemes provide some resolution at the 'species level' with species-specific alleles being identified at loci with slower evolutionary speeds. Both schemes therefore add value by retaining the discriminatory capabilities of 'Hoofprints' but allowing placement of isolates within higher taxonomic groups such that clustering into species is possible and groupings are congruent with other typing approaches such as multilocus sequencing (e.g. Groussaud *et al.*, 2007; Maquart *et al.*, 2009). Thus, for example while the VNTR-21 scheme (Whatmore *et al.*, 2006) retained the huge discriminatory power of 'Hoofprints' identifying 119 distinct genotypes when applied to a worldwide collection of 121 *Brucella* isolates clustering analysis showed that, with minor exceptions, groups defined by this approach correspond to conventional species designations. Thus clusters corresponding to *B. abortus*, *B. ovis*, *B. melitensis*, and *B. neotomae* were identified as well as a *B. suis* cluster including subclusters corresponding to biovars 2 and biovars 1, 3 and 4. Consistent with observations from other approaches *B. canis* appeared closely related to *B. suis*, particularly biovars 3 and 4 while *B. suis* biovar 5 appeared somewhat distinct from both other *B. suis* biovars and other *Brucella* species. Le Fleche *et al.* (2006) carried out the most extensive analysis of tandem repeats in *Brucella* to date examining eighty tandem repeat loci in 21 reference strains and again finding the same clusters that largely correspond to classical species. A subset of 15 loci that preserved this clustering was selected to establish a routine typing scheme comprising eight markers with good species identification capability ('mini-satellites') and seven with higher discriminatory power ('microsatellites'). The application of this 15 locus scheme to 236 clinical and reference isolates generated clustering consistent with other molecular and phenotypic characteristics and, similar to VNTR-21, with some relationship to biovar designations of *B. suis* and, to a lesser extent, *B. abortus*. This scheme has now been extensively used particularly within the European Union and a recent publication demonstrated clustering of more than 500 genotypes observed by typing over 750 strains (Maquart *et al.*, 2009). This clustering remained absolutely consistent with previous observations as illustrated in Fig. 4.3 with generally good separation of the six classical species, separation of the new species *B. microti*, and again the identification of three distinct clusters of isolates comprising the marine mammal *Brucella*.

An additional two MLVA schemes applicable to *Brucella* have since been described although again with much overlap with markers previously described. A 15 locus scheme described by Huynh *et al* (2008) was later used in a direct comparison with the European 15 locus MLVA approach (Le Flèche *et al.*, 2006). Comparing a local population of 101 *B. melitensis* isolates from Egypt the performance of both schemes was found to be similar with most advantages or disadvantages described reflecting technical aspects related to the design of one assay for automated fragment size analysis on a sequencer versus sizing by agarose gel electrophoresis (Tiller *et al.*, 2009). A 'single tube' eight locus MLVA scheme has also been described with the argument that its technical and economic

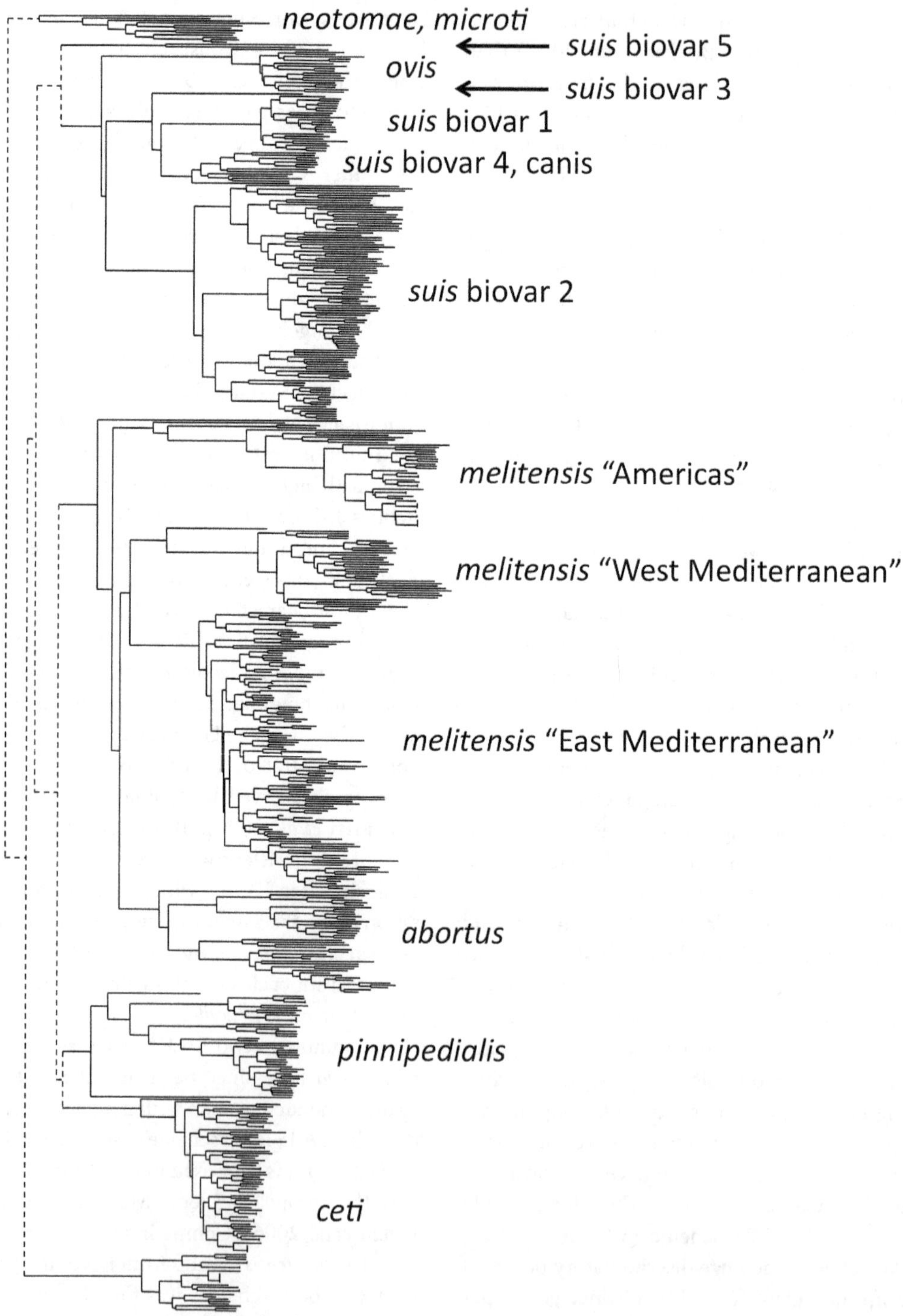

Figure 4.3 Current view of the global population structure of the *Brucella* genus based on 552 VNTR profiles obtained using the MLVA-16 approach (Al Dahouk *et al.*, 2007a). Clustering analysis was performed using the categorical coefficient (character state unordered) and the Neighbour joining algorithm with the *B. microti*/*B. neotomae* cluster used to root the tree. All loci were weighted equally. Courtesy of Gilles Vergnaud, Université Paris-Sud. Further details at http://mlva.u-psud.fr/brucella/.

feasibility might make it particularly applicable in resource limited settings (Rees *et al.*, 2009). The description of so many variant schemes in a short period highlights the value of promotion of a globalized MLVA scheme. Using a common set of markers would facilitate international comparison of profiles and avoid the situation where much of the value of inter-laboratory comparison has been lost because non-complementary VNTR schemes, based on different markers, are used (Whatmore, 2009). As a move in this direction a website has been set up (http://mlva.u-psud.fr/brucella/) designed to facilitate submission and international comparison of profiles based on an agreed European scheme (Al Dahouk *et al.*, 2007a).

Practical applications of tandem repeat based approaches

Reflecting the 'mini-satellite' elements of some of the second generation VNTR based assays, that facilitate discrimination at the species level, they have also found use as tools in taxonomy. Thus for example they have been instrumental in confirming the presence of at least three distinct groups within the marine mammal *Brucella* that are inconsistent with the currently accepted species (Groussaud *et al.*, 2007; Maquart *et al.*, 2009). Likewise MLVA-15 (Le Flèche *et al.*, 2006), when used in conjunction with an extensive comparative database, added to the body of evidence for identification of the recently described *Brucella* species such as *B. microti* and *B. inopinata* (Scholz *et al.*, 2008, 2010).

While tandem repeat based approaches have been useful in the types of approaches described above their unique value is in application as a highly discriminatory epidemiological tool, exploiting particularly the 'microsatellite' elements, of a kind not previously available for this group. A significant number of papers are now beginning to be published applying one of the schemes described above to address practical epidemiological issues. To date, these mostly relate to human infection, although clearly there are also potentially valuable applications in the veterinary arena. Common uses of VNTR data are to recognize human outbreaks that relate to a common source (Valdezate *et al.*, 2007; Kattar *et al.*, 2008; Rees *et al.*, 2009; Tiller *et al.*, 2009; Lucero *et al.*, 2010) to confirm the source of a laboratory infection (Marianelli *et al.*, 2008b; Rees *et al.*, 2009; Valdezate *et al.*, 2010) or to confirm relapse (Al Dahouk *et al.*, 2005a). There have also been a number of studies beginning to characterize diversity among isolates from restricted areas of endemicity such as Korea (Kang *et al.*, 2009; Her *et al.*, 2009), Egypt (Tiller *et al.*, 2009), Sicily (Marianelli *et al.*, 2007), Peru (Smits *et al.*, 2009) and Lebanon (Kattar *et al.*, 2008). Veterinary applications published to date include assessing the stability of a live vaccine (García-Yoldi *et al.*, 2007a), showing that wild boar and domestic pigs sharing localities can have identical *B. suis* genotypes (García-Yoldi *et al.*, 2007b), and demonstrating that isolates of *B. abortus* from cattle in Yellowstone Park, USA, are closely related to elk and not to bison isolates (Beja-Pereira *et al.*, 2009).

All these studies demonstrate the potential of tandem-based approaches to address epidemiological issues. The use of tandem repeat based data to link genotype to virulence properties is in its infancy but the potential of this approach has already been illustrated in a few studies. Most notably a significant association of different genotypes of *B. melitensis* in Peru with different pathological outcomes, as well differential age associations of genotypes, was recently reported (Nöckler *et al.*, 2009). VNTR data also formed part of the characterization of a clone of marine mammal *Brucella* apparently associated with zoonotic human disease (Groussaud *et al.*, 2007; Whatmore *et al.*, 2008).

Future perspectives

While molecular approaches for detection and characterization of *Brucella* have made huge advances over recent years the gold standard for diagnosis of brucellosis is still isolation of the bacterium from clinical or field material. As can be seen much effort has gone into development of various PCR tests but most have not made the step from research and development tools to more practical application. As a result PCR has still to find widespread use as a diagnostic tool in either the animal or human sectors. A number of issues still need to be resolved to allow this to happen. As pointed out elsewhere recently, methods still need to be extensively validated on clinical samples before they are acceptable for routine laboratory testing with issues such as thorough assessment of specificity and sensitivity, establishment of standard positive and negative controls, implementation of internal and inhibition controls, and quality assurance issues needing thorough investigation (Yu and Nielsen, 2010). With respect to human diagnostics, while the nature of clinical specimen, sample preparation procedure, and duration and stage of illness probably have a significant impact on results (Franco *et al.*, 2007) there are some promising reports with regard to the value of PCR applied to blood components. However, the observations of persistance of PCR signals many months or years after treatment and resolution of clinical symptoms, and with consistently negative blood culture, raise fundamental questions about the value and interpretation of positive PCR results. For animal brucellosis, although there are many inconsistencies and much work to be done, in general PCR appears to have similar or slightly enhanced sensitivity to blood culture despite the limits in the amount of material than can be physically included in a PCR relative to that that can be used for culture. Given these issues, questions regarding when in the course of infection particular sample types are appropriate, and probable animal to animal variablity, the most sensisible use of PCR currently would appear to be as a component of a battery of tests that cover serological, cultural and molecular approaches. The results of these should be considered in tandem to give an indication of the likelihood of *Brucella* infection ideally on a herd, rather than an individual animal, basis. PCR, and particularly real-time PCR, remains a relatively expensive technology and emerging alternative approaches, such as LAMP, may ultimately offer cheaper, more rapid and more robust diagnostic alternatives.

As discussed in this chapter new molecular approaches, based on the ever increasing genomic data available for the genus, are revolutionizing our ability to understand relationhips between strains within a group of organisms traditionally described as homogeneous. Tandem based approaches in particular, offering an unprecedented ability to resolve isolates, will continue to have significant value in areas such as epidemiological traceback and linkage as well as potentially revealing varying biological properties of different genotypes. The next major advances in this area probably reflect the current revolution in sequencing technologies. Already at the time of writing around 40 *Brucella* genome sequences are available and continuing advances in next generation sequencing approaches promise to provide the next major wave of molecular data for bacteria. The new ease in generating such data will allow the construction of extensive SNP based phylogenies and the identification of further 'canonical' SNPs that define lineages of importance. Relecting this SNP based typing approaches could ultimately offer high-throughput, highly sensitive and highly discriminatory diagnostic assays even at subspecies level.

Acknowledgements

Brucella research at the AHVLA is funded by the UK Department of Environment, Food and Rural Affairs (Defra). We thank Gilles Vergnaud and Anne Mayer-Scholl for contributing figures to this chapter.

References

Al Dahouk, S., Hagen, R.M., Nockler, K., Tomaso, H., Wittig, M., Scholz, H.C., Vergnaud, G., and Neubauer, H. (2005a). Failure of a short-term antibiotic therapy for human brucellosis using ciprofloxacin. A study on *in vitro* susceptibility of *Brucella* strains. Chemotherapy *51*, 352–356.

Al Dahouk, S., Tomaso H., Prenger-Berninghoff, E., Splettstoesser, W.D., Scholz, H.C., and Neubauer, H. (2005b). Identification of *Brucella* species and biotypes using polymerase chain reaction-restriction fragment length polymorphism (PCR-RFLP). Crit. Rev. Microbiol. *31*, 191–196.

Al Dahouk, S., Flèche, P.L., Nöckler, K., Jacques, I., Grayon, M., Scholz, H.C., Tomaso, H., Vergnaud, G., and Neubauer, H. (2007a). Evaluation of *Brucella* MLVA typing for human brucellosis. J. Microbiol. Methods *69*, 137–145.

Al Dahouk, S., Nöckler, K., Scholz, H.C., Pfeffer, M., Neubauer, H., and Tomaso, H. (2007b). Evaluation of genus-specific and species-specific real-time PCR assays for the identification of *Brucella* spp. Clin. Chem. Lab. Med. *45*, 1464–1470.

Amin, A.S., Hamdy, M.E., and Ibrahim, A.K. (2001). Detection of *Brucella melitensis* in semen using the polymerase chain reaction assay. Vet. Microbiol. *83*, 37–44.

Aras, Z., and Uçan, U.S. (2010). Detection of *Brucella canis* from inguinal lymph nodes of naturally infected dogs by PCR. Theriogenology *74*, 658–662.

Baily, G.G., Krahn, J.B., Drasar, B.S., and Stoker, N.G. (1992). Detection of *Brucella melitensis* and *Brucella abortus* by DNA amplification. J. Trop. Med. Hyg. *95*, 271–275.

Bardenstein, S., Mandelboim, M., Ficht, T.A., Baum, M., and Banai, M. (2002). Identification of the *Brucella melitensis* vaccine strain Rev.1 in animals and humans in Israel by PCR analysis of the PstI site polymorphism of its *omp2* gene. J. Clin. Microbiol. *40*, 1475–1480.

Beja-Pereira, A., Bricker, B., Chen, S., Almendra, C., White, P.J., and Luikart, G. (2009). DNA genotyping suggests that recent brucellosis outbreaks in the Greater Yellowstone Area originated from elk. J. Wildl. Dis. *45*, 1174–1177.

van Belkum, A. (1999). The role of short sequence repeats in epidemiologic typing. Curr. Opin. Microbiol. *2*, 306–311.

van Belkum, A., Schere, S., van Alphen, L., and Verbrugh, H. (1998). Short sequence DNA repeats in prokaryotic genomes. Microbiol. Mol. Biol. Rev. *62*, 275–293.

Bogdanovich, T., Skurnik, M., Lübeck, P.S., Ahrens, P., and Hoorfar, J. (2004). Validated 5′ nuclease PCR assay for rapid identification of the genus *Brucella*. J. Clin. Microbiol. *42*, 2261–2263.

Boom, R., Sol, C.J., Salimans, M.M., Jansen, C.L., Wertheim-van Dillen, P.M., and van der Noordaa, J. (1990). Rapid and simple method for purification of nucleic acids. J. Clin. Microbiol. *28*, 495–503.

Bounaadja, L., Albert, D., Chénais, B., Hénault, S., Zygmunt, M.S., Poliak, S., and Garin-Bastuji, B. (2009). Real-time PCR for identification of *Brucella* spp.: a comparative study of *IS711*, *bcsp31* and *per* target genes. Vet. Microbiol. *137*, 156–164.

Bourg, G., O'Callaghan, D., and Boschiroli, M.L. (2007). The genomic structure of *Brucella* strains isolated from marine mammals gives clues to evolutionary history within the genus. Vet Microbiol. *125*, 375–380.

Bricker, B.J. (2002). PCR as a diagnostic tool for brucellosis. Vet. Microbiol. *90*, 435–446.

Bricker, B.J., and Halling, S.M. (1994). Differentiation of *Brucella abortus* bv. 1, 2, and 4, *Brucella melitensis*, *Brucella ovis*, and *Brucella suis* bv. 1 by PCR. J. Clin. Microbiol. *32*, 2660–2666.

Bricker, B.J., and Halling, S.M. (1995). Enhancement of the *Brucella* AMOS PCR assay for differentiation of *Brucella abortus* vaccine strains S19 and RB51. J. Clin. Microbiol. *33*, 1640–1642.

Bricker B.J., and Ewalt, D.R. (2005). Evaluation of the HOOF-Print assay for typing *Brucella abortus* strains isolated from cattle in the United States: results with four performance criteria. BMC Microbiol. *5*, 37.

Bricker, B.J., Tabatabai, L.B., Deyoe, B.L., and Mayfield, J.E. (1988). Conservation of antigenicity in a 31-kDa *Brucella* protein. Vet. Microbiol. *18*, 313–325.

Bricker, B.J., Ewalt, D.R., and Halling S.M. (2003). *Brucella* 'HOOF-Prints': strain typing by multilocus analysis of variable number tandem repeats (VNTRs). BMC Microbiol. *3*, 15.

Capparelli, R., Parlato, M., Iannaccone, M., Roperto, S., Marabelli, R., Roperto, F., and Iannelli, D. (2009). Heterogeneous shedding of *Brucella abortus* in milk and its effect on the control of animal brucellosis. J. Appl. Microbiol. *106*, 2041–2047.

Casañas, M.C., Queipo-Ortuño, M.I., Rodríguez-Torres, A., Orduña, A., Colmenero, J.D., and Morata, P. (2001). Specificity of a polymerase chain reaction assay of a target sequence on the 31-kilodalton *Brucella antigen* DNA used to diagnose human brucellosis. Eur. J. Clin. Microbiol. Infect. Dis. *20*, 127–131.

Case, R.J., Boucher, Y., Dahllöf, I., Holmström, C., Doolittle, W.F., and Kjelleberg, S. (2007). Use of 16S rRNA and *rpoB* genes as molecular markers for microbial ecology studies. Appl. Environ. Microbiol. *73*, 278–288.

Castaño, M.J., and Solera, J. (2009). Chronic brucellosis and persistence of *Brucella melitensis* DNA. J. Clin. Microbiol. *47*, 2084–2089.

Cetinkaya, B., Ongör, H., Muz, A., Ertas, H.B., Kalender, H. and Erdogan, H.M. (1999). Detection of *Brucella* species DNA in the stomach content of aborted sheep fetuses by PCR. Vet. Rec. *144*, 239–240.

Chomczynski, P., Mackey, K., Drews, R., and Wilfinger, W. (1997). DNAzol: a reagent for the rapid isolation of genomic DNA. Biotechniques *22*, 550–553.

Clavareau, C., Wellemans, V., Walravens, K,. Tryland, M., Verger, J.M, Grayon, M., Cloeckaert, A., Letesson, J.J., and Godfroid, J. (1998). Phenotypic and molecular characterization of a *Brucella* strain isolated from a minke whale (*Balaenoptera acutorostrata*). Microbiology *144*, 3267–3273.

Cloeckaert, A., Grayon, M., Verger, J.M., Letesson, J.J., and Godfroid, F. (2000). Conservation of seven genes involved in the biosynthesis of the lipopolysaccharide O-side chain in *Brucella* spp. Res. Microbiol. *151*, 209–216.

Cloeckaert A., Verger J.M., Grayon M., Paquet J.Y., Garin-Bastuji B., Foster G., and Godfroid, J. (2001). Classification of *Brucella* spp. isolated from marine mammals by DNA polymorphism at the *omp2* locus. Microbes Infect. *3*, 729–738.

Cloeckaert, A., Grayon, M., and Grépinet, O. (2002). Identification of *Brucella melitensis* vaccine strain Rev.1 by PCR-RFLP based on a mutation in the *rpsL* gene. Vaccine *20*, 2546–2550.

Colmenero, J.D., Queipo-Ortuño, M.I., Reguera, JM., Suárez-Muñoz, A.M., Martín-Carballino, S., and Morata, P. (2002). Chronic hepatosplenic abscesses in Brucellosis. Clinico-therapeutic features and molecular diagnostic approach. Diagn. Microbiol. Infect. Dis. *42*, 159–167.

Colmenero, J.D., Queipo-Ortuño, M.I., Reguera, J.M., Baeza, G., Salazar, J.A., and Morata, P. (2005). Real time polymerase chain reaction: a new powerful tool for the diagnosis of neurobrucellosis. J. Neurol. Neurosurg. Psychiatry *76*, 1025–1027.

Crasta, O.R., Folkerts, O., Fei, Z., Mane, S.P., Evans, C., Martino-Catt, S., Bricker, B., Yu, G., Du, L., and Sobral, B.W. (2008). Genome sequence of *Brucella abortus* vaccine strain S19 compared to virulent strains yields candidate virulence genes. PLoS One *3*, e2193.

Da Costa, M., Guillou, J.P., Garin-Bastuji, B., Thiébaud, M., and Dubray, G. (1996). Specificity of six gene sequences for the detection of the genus *Brucella* by DNA amplification. J. Appl. Bacteriol. *81*, 267–275.

Dawson C.E., Stubberfield E.J., Perrett L.L., King A.C., Whatmore A.M., Bashiruddin J.B., Stack J.A., and Macmillan, A.P. (2008). Phenotypic and molecular characterisation of *Brucella* isolates from marine mammals. BMC Microbiol. *8*, 224.

Dauphin, L.A., Hutchins, R.J., Bost, L.A., and Bowen, M.D. (2009). Evaluation of automated and manual commercial DNA extraction methods for recovery of *Brucella* DNA from suspensions and spiked swabs. J. Clin. Microbiol. *47*, 3920–3926.

De, B.K., Stauffer, L., Koylass, M.S., Sharp, S.E., Gee, J.E., Helsel, L.O., Steigerwalt, A.G., Vega, R., Clark, T.A., Daneshvar, M.I., Wilkins, P.P., and Whatmore, A.M. (2008). Novel *Brucella* strain (BO1) associated with a prosthetic breast implant infection. J. Clin. Microbiol. *46*, 43–49.

Debeaumont, C., Falconnet, P.A., and Maurin, M. (2005). Real-time PCR for detection of *Brucella* spp. DNA in human serum samples. Eur. J. Clin. Microbiol. Infect. Dis. *24*, 842–845.

De Ley, J., Mannheim, W., Segers, P., Lievens, A., Denijn, M., Vanhoucke, M., and Gillis, M. (1987). Ribosomal ribonucleic acid cistron similarities and taxonomic neighborhood of *Brucella* and CDC Group Vd. Int. J. Syst. Bacteriol. *37*, 35–42.

Deyoe, B.L. and Manthei, C. A. (1964). Chapter 23, Brucellosis. In Diseases of Swine. Dunne, H.W., ed. (Iowa State University Press: Ames, USA), pp. 492–515.

Dorsch, M., Moreno, E., and Stackebrandt, E. (1989). Nucleotide sequence of the 16S rRNA from *Brucella abortus*. Nucleic Acids Res. *17*, 1765.

Elfaki, M.G., Al-Hokail, A.A., Nakeeb, S.M., and Al-Rabiah, F.A. (2005). Evaluation of culture, tube agglutination, and PCR methods for the diagnosis of brucellosis in humans. Med. Sci. Monit. *11*, MT69–74.

El Kholy, A.A., Gomaa, H.E., El Anany, M.G., and Abd El Rasheed, E. (2009). Diagnosis of human brucellosis in Egypt by polymerase chain reaction. East Mediterr. Health J. *15*, 1068–1074.

el-Razik, K.A, Ghazi, Y.A., and Salama, E.M. (2007). Monitoring of *Brucella* reactor does following milk examination using different techniques. Pak. J. Biol. Sci. *10*, 240–244.

Enright, M.C., and Spratt, B.G. (1999). Multilocus sequence typing. Trends Microbiol. *7*, 482–487.

Espy, M.J., Uhl, J.R., Sloan, L.M., Buckwalter, S.P., Jones, M.F., Vetter, E.A., Yao, J.D., Wengenack, N.L., Rosenblatt, J.E., Cockerill, F.R., 3rd, *et al.* (2006). Real-time PCR in clinical microbiology: applications for routine laboratory testing. Clin. Microbiol. Rev. *19*, 165–256.

Ewalt, D.R., and Bricker, B. J. (2000). Validation of the abbreviated *Brucella* AMOS PCR as a rapid screening method for differentiation of *Brucella abortus* field strain isolates and the vaccine strains, S19 and RB51. J. Clin. Microbiol. *38*, 3085–3086.

Fahle, G.A., and Fischer, S.H. (2000). Comparison of six commercial DNA extraction kits for recovery of cytomegalovirus DNA from spiked human specimens. J. Clin. Microbiol. *38*, 3860–3863.

Fekete, A., Bantle, J.A., Halling, S.M., and Sanborn, M.R. (1990). Preliminary development of a diagnostic test for *Brucella* using polymerase chain reaction. J. Appl. Bacteriol. *69*, 216–227.

Fekete, A., Bantle, J.A., and Halling, S.M. (1992). Detection of *Brucella* by polymerase chain reaction in bovine fetal and maternal tissues. J. Vet. Diagn. Invest. *4*, 79–83.

Fernández-Lago, L., Vallejo, F.J., Trujillano, I., and Vizcaíno, N. (2000). Fluorescent whole-cell hybridization with 16S rRNA-targeted oligonucleotide probes to identify *Brucella* spp. by flow cytometry. J. Clin. Microbiol. *38*, 2768–2771.

Ficht T.A., Bearden S.W., Sowa B.A., and Adams, L.G. (1989). DNA sequence and expression of the 36-kilodalton outer membrane protein gene of *Brucella abortus*. Infect. Immun. *57*, 3281–3291.

Ficht, T.A., Husseinen, H.S., Derr, J., and Bearden, S.W. (1996). Species-specific sequences at the *omp2* locus of *Brucella* type strains Int. J. Syst. Bacteriol. *46*, 329–331.

Foster, G., Osterman, B.S., Godfroid, J., Jacques, I., and Cloeckaert, A. (2007). *Brucella* ceti sp. nov. and *Brucella pinnipedialis* sp. nov. for *Brucella* strains with cetaceans and seals as their preferred hosts. Int. J. Syst. Evol. Microbiol. *57*, 2688–2693.

Foster, J.T., Okinaka, R.T., Svensson, R., Shaw, K., De, B.K., Robison, R.A., Probert, W.S., Kenefic, L.J., Brown, W.D., and Keim, P. (2008). Real-time PCR assays of single-nucleotide polymorphisms defining the major *Brucella* clades. J. Clin. Microbiol. *46*, 296–301.

Franco, M.P., Mulder, M., Gilman, R.H., and Smits, H.L. (2007). Human brucellosis. Lancet Infect. Dis. *7*, 775–786.

Fretin, D., Whatmore, A.M., Al Dahouk, S., Neubauer, H., Garin-Bastuji, B., Albert, D., Van Hessche, M., Ménart, M., Godfroid, J., Walravens, K., *et al.* (2008). *Brucella suis* identification and biovar typing by real-time PCR. Vet. Microbiol. *131*, 376–385.

García-Yoldi, D., Marín, C.M., de Miguel, M.J., Muñoz, P.M, Vizmanos, J.L., and López-Goñi, I. (2006). Multiplex PCR assay for the identification and differentiation of all *Brucella* species and the vaccine strains *Brucella abortus* S19 and RB51 and *Brucella melitensis* Rev1. Clin. Chem. *52*, 779–781.

García-Yoldi, D., Le Fleche, P., Marín, C.M., De Miguel, M.J., Muñoz, P.M., Vergnaud, G., and López-Goñi, I. (2007a). Assessment of genetic stability of *Brucella melitensis* Rev 1 vaccine strain by multiple-locus variable-number tandem repeat analysis. Vaccine *25*, 2858–2862.

García-Yoldi, D., Le Fleche, P., De Miguel, M.J., Muñoz, P.M., Blasco, J.M., Cvetnic, Z., Marín, C.M., Vergnaud, G., and López-Goñi, I. (2007b). Comparison of multiple-locus variable-number tandem-repeat analysis with other PCR-based methods for typing *Brucella suis* isolates. J. Clin. Microbiol. *45*, 4070–4072.

Gee, J.E., De, B.K., Levett, P.N., Whitney, A.M., Novak, R.T., and Popovic, T. (2004). Use of 16S rRNA gene sequencing for rapid confirmatory identification of *Brucella* isolates. J. Clin. Microbiol. *42*, 3649–3654.

Gopaul, K.K., Koylass, M.S., Smith, C.J., and Whatmore, A.M. (2008). Rapid identification of *Brucella* isolates to the species level by real-time PCR based single nucleotide polymorphism (SNP) analysis. BMC Microbiol. *8*, 86.

Gopaul, K.K., Sells, J., Bricker, B.J., Crasta, O.R., and Whatmore, A.M. (2010). Rapid and reliable single nucleotide polymorphism-based differentiation of *Brucella* live vaccine strains from field strains. J. Clin. Microbiol. *48*, 1461–1464.

Goto, M., Honda, E., Ogura, A., Nomoto, A., and Hanaki K. (2009). Colorimetric detection of loop-mediated isothermal amplification reaction by using hydroxy naphthol blue. Biotechniques *46*, 167–172.

Grilló, M.J., Barberán, M., and Blasco, J.M. (1997). Transmission of *Brucella melitensis* from sheep to lambs. Vet. Rec. *140*, 602–605.

Groussaud, P., Shankster, S.J., Koylass, M.S., and Whatmore, A.M. (2007). Molecular typing divides marine mammal strains of *Brucella* into at least three groups with distinct host preferences. J. Med. Microbiol. *56*, 1512–1518.

Guarino, A., Serpe, L., Fusco, G., Scaramuzzo, A., and Gallo P. (2000). Detection of *Brucella* species in buffalo whole blood by gene-specific PCR. Vet. Rec. *147*, 634–636.

Halling, S.M, and Bricker, B.J. (1994). Characterization and occurrence of two repeated palindromic DNA elements of *Brucella* spp.: Bru-RS1 and Bru-RS2. Mol. Microbiol. *14*, 681–689.

Halling, S.M., and Zehr, E.S. (1990). Polymorphism in *Brucella* spp. due to highly repeated DNA. J. Bacteriol. *172*, 6637–6640.

Halling, S.M., Tatum, F.M., and Bricker, B.J. (1993). Sequence and characterization of an insertion sequence, IS*711*, from *Brucella ovis*. Gene *133*, 123–127.

Hamdy, M.E., and Amin, A.S. (2002). Detection of *Brucella* species in the milk of infected cattle, sheep, goats and camels by PCR. Vet. J. *163*, 299–305.

den Hartigh, A.B., Sun, Y.H., Sondervan, D., Heuvelmans, N., Reinders, M.O., Ficht, T.A, and Tsolis, R.M. (2004). Differential requirements for VirB1 and VirB2 during *Brucella abortus* infection. Infect. Immun. *72*, 5143–5149.

Her, M., Kang, S.I., Cho, D.H., Cho, Y.S., Hwang, I.Y., Heo, Y.R., Jung, S.C., and Yoo, H.S. (2009). Application and evaluation of the MLVA typing assay for the *Brucella abortus* strains isolated in Korea. BMC Microbiol. *9*, 230.

Herman, L., and De Ridder, H. (1992). Identification of *Brucella* spp. by using the polymerase chain reaction. Appl. Environ. Microbiol. *58*, 2099–2101.

Higuchi, R., Fockler, C., Dollinger, G., and Watson, R. (1993). Kinetic PCR analysis: real-time monitoring of DNA amplification reactions. Biotechnology (N. Y.) *11*, 1026–1030.

Hinić, V., Brodard, I., Thomann, A., Cvetnić, Z., Makaya, P.V., Frey, J., and Abril, C. (2008). Novel identification and differentiation of *Brucella melitensis, B. abortus, B. suis, B. ovis, B. canis*, and *B. neotomae* suitable for both conventional and real-time PCR systems. J. Microbiol. Methods 75, 375–378.

Hinić, V., Brodard, I., Thomann, A., Holub, M., Miserez, R., and Abril, C. (2009) IS*711*-based real-time PCR assay as a tool for detection of *Brucella* spp. in wild boars and comparison with bacterial isolation and serology. BMC Vet. Res. *5*, 22.

Holmes, B., Popoff, M., Kiredjian, M. and Kersters, K. (1988) *Ochrobactrum anthropi* gen. nov., sp. nov. from human clinical specimens and previously known as Group Vd. Int. J. Syst. Bacteriol. *38*, 406–416.

Hopper, B.R., Sanborn, M.R., and Bantle, J.A. (1989). Detection of *Brucella abortus* in mammalian tissue, using biotinylated, whole genomic DNA as a molecular probe. Am. J. Vet. Res. *50*, 2064–2068.

Hubálek, Z., Scholz, H.C., Sedlácek, I., Melzer, F., Sanogo, Y.O., and Nesvadbová, J. (2007) Brucellosis of the common vole (*Microtus arvalis*). Vector Borne Zoonotic Dis. *7*, 679–687.

Huynh, L.Y., Van Ert, M.N., Hadfield, T., Probert, W.S., Bellaire, B.H., Dobson, M., Burgess, R.J., Weyant, R.S., Popovic, T., Zanecki, S., *et al.* (2008). Multiple locus variable number tandem repeat (VNTR) analysis (MLVA) of *Brucella* spp. identifies species specific markers and insights into phylogenetic relationships. In National Institute of Allergy and Infectious Disease, NIH, Vol. 1. Frontiers in Research, St. Georgiev, V., Western, K.A., and McGowan, J.J., eds. (Humana Press: Totowa, NJ), pp. 47–54

Ilhan, Z., Aksakal, A., Ekin, I.H., Gülhan, T., Solmaz, H., and Erdenlig, S. (2008). Comparison of culture and PCR for the detection of *Brucella melitensis* in blood and lymphoid tissues of serologically positive and negative slaughtered sheep. Lett. Appl. Microbiol. *46*, 301–306.

Kang, S.I, Her, M., Heo, E.J, Nam, H.M, Jung, S.C, and Cho, D. (2009). Molecular typing for epidemiological evaluation of *Brucella abortus* and *Brucella canis* isolated in Korea. J. Microbiol. Methods. *78*, 144–149.

Kantartzi, K., Panagoutsos, S., Kokkinou, V., Alepopoulou, E., Mourvati, E., Passadakis, P., and Vargemezis, V. (2010). Brucellosis in dialysis patients. Does it exist? Clin. Nephrol. *73*, 309–313.

Kattar, M.M., Zalloua, P.A., Araj, G.F., Samaha-Kfoury, J., Shbaklo, H., Kanj, S.S., Khalife, S., and Deeb, M. (2007). Development and evaluation of real-time polymerase chain reaction assays on whole blood and paraffin-embedded tissues for rapid diagnosis of human brucellosis. Diagn. Microbiol. Infect. Dis. *59*, 23–32.

Kattar, M.M., Jaafar, R.F., Araj, G.F., Le Flèche, P., Matar, G.M., Abi Rached, R., Khalife, S., and Vergnaud, G. (2008). Evaluation of a multilocus variable-number tandem-repeat analysis scheme for typing human

Brucella isolates in a region of brucellosis endemicity. J. Clin. Microbiol. *46*, 3935–3940.

Keid, L.B., Soares, R.M., Vasconcellos, S.A., Chiebao, D.P., Megid, J., Salgado, V.R., and Richtzenhain, L.J. (2007a). A polymerase chain reaction for the detection of *Brucella canis* in semen of naturally infected dogs. Theriogenology. *67*, 1203–1210.

Keid, L.B., Soares, R.M., Vieira, N.R., Megid, J., Salgado, V.R., Vasconcellos, S.A., da Costa, M., Gregori, F., and Richtzenhain, L.J. (2007b). Diagnosis of canine brucellosis: comparison between serological and microbiological tests and a PCR based on primers to 16S-23S rDNA interspacer. Vet. Res. Commun. *31*, 951–965.

Keid, L.B., Soares, R.M., Vasconcellos, S.A., Megid, J., Salgado, V.R., and Richtzenhain, L.J. (2009). Comparison of agar gel immunodiffusion test, rapid slide agglutination test, microbiological culture and PCR for the diagnosis of canine brucellosis. Res. Vet. Sci. *86*, 22–26.

Keid, L.B, Soares, R.M., Vasconcellos, S.A., Salgado, V.R., Megid, J., and Richtzenhain, L.J. (2010). Comparison of a PCR assay in whole blood and serum specimens for canine brucellosis diagnosis. Vet. Rec. *167*, 96–99.

Keim, P., Van Ert, M.N., Pearson, T., Vogler, A.J., Huynh, L.Y., and Wagner, D.M. (2004). Anthrax molecular epidemiology and forensics: using the appropriate marker for different evolutionary scales. Infect. Genet. Evol. *4*, 205–213.

Kim S., Lee, D.S., Suzuki, H., and Watarai, M. (2006). Detection of *Brucella canis* and *Leptospira interrogans* in canine semen by multiplex nested PCR. J. Vet. Med. Sci. *68*, 615–618.

Koylass, M.S, King, A.C, Edwards-Smallbone, J, Gopaul, K.K., Perrett, L.L. and Whatmore, A.M. (2010). Comparative performance of SNP typing and 'Bruce-ladder' in the discrimination of *Brucella suis* and *Brucella canis*. Vet. Microbiol. *142*, 450–454.

de Lamballerie, X., Zandotti, C., Vignoli, C., Bollet, C., and de Micco, P. (1992). A one-step microbial DNA extraction method using 'Chelex 100' suitable for gene amplification. Res. Microbiol. *143*, 785–790.

Lapraik, R.D., and Moffat, R. (1982). Latent bovine brucellosis Vet. Rec. *111*, 578–579.

Leal-Klevezas, D.S., López-Merino, A. and, Martínez-Soriano, J.P. (1995a). Molecular detection of *Brucella* spp.: rapid identification of *B. abortus* biovar I using PCR. Arch. Med. Res. *26*, 263–267.

Leal-Klevezas, D.S., Martínez-Vázquez, I.O., López-Merino, A., and Martínez-Soriano, J.P. (1995b). Single-step PCR for detection of *Brucella* spp. from blood and milk of infected animals. J. Clin. Microbiol. *33*, 3087–3090.

Le Flèche, P., Jacques, I., Grayon, M., Al Dahouk, S., Bouchon, P., Denoeud, F., Nöckler, K., Neubauer, H., Guilloteau, L.A, and Vergnaud, G. (2006). Evaluation and selection of tandem repeat loci for a *Brucella* MLVA typing assay. BMC Microbiol. *6*, 9.

Levinson G., and Gutman, G.A. (1987). Slipped-strand mispairing: a major mechanism for DNA sequence evolution. Mol. Biol. Evol. *4*, 203–221.

Leyla, G., Kadri, G., and Umran, O. (2003). Comparison of polymerase chain reaction and bacteriological culture for the diagnosis of sheep brucellosis using aborted fetus samples. Vet. Microbiol. *93*, 53–61.

Li, X., Zhang, S., Zhang, H., Zhang, L., Tao, H., Yu, J., Zheng, W., Liu, C., Lü, D., Xiang, R., and Liu, Y. (2009). A loop-mediated isothermal amplification method targets the *pho*P gene for the detection of *Salmonella* in food samples. Int. J. Food Microbiol. *133*, 252–258.

Longo, M., Mallardo, K., Montagnaro, S., De Martino, L., Gallo, S., Fusco, G., Galiero, G., Guarino, A., Pagnini, U., and Iovane, G. (2009). Shedding of *Brucella abortus* rough mutant strain RB51 in milk of water buffalo (*Bubalus bubalis*). Prev. Vet. Med. *90*, 113–118.

López-Goñi, I., García-Yoldi, D., Marín, C.M., de Miguel, M.J., Muñoz, P.M., Blasco, J.M., Jacques, I., Grayon, M., Cloeckaert, A., Ferreira, A.C., *et al.* (2008). Evaluation of a multiplex PCR assay (Bruce-ladder) for molecular typing of all *Brucella* species, including the vaccine strains. J. Clin. Microbiol. *46*, 3484–3487.

Lübeck, P.S., Skurnik, M., Ahrens, P., and Hoorfar, J. (2003). A multiplex PCR-detection assay for *Yersinia enterocolitica* serotype O:9 and *Brucella* spp. based on the perosamine synthetase gene. Application to *Brucella* diagnostics. Adv. Exp. Med. Biol. *529*, 451–453.

Lucero, N.E., Tenenbaum, M., Jacob, N.R., Escobar, G.I., Groussaud, P., and Whatmore A.M. (2010). Application of variable number of tandem repeats typing to describe familial outbreaks of brucellosis in Argentina. J. Med. Microbiol. *59*, 648–652.

Manterola, L., Tejero-Garcés, A., Ficapal, A., Shopayeva, G., Blasco, J.M, Marin, C.M, and López-Goñi, I. (2003). Evaluation of a PCR test for the diagnosis of Brucella ovis infection in semen samples from rams. Vet. Microbiol. *92*, 65–72.

Margos, G., Gatewood, A.G., Aanensen, D.M., Hanincová, K., Terekhova, D., Vollmer, S.A., Cornet, M., Piesman, J., Donaghy, M., Bormane, A., *et al.* (2008). MLST of housekeeping genes captures geographic population structure and suggests a European origin of *Borrelia burgdorferi*. Proc. Natl. Acad. Sci. U.S.A. *105*, 8730–8735.

Marianelli, C., Ciuchini, F., Tarantino, M., Pasquali, P., and Adone, R. (2006). Molecular characterization of the *rpoB* gene in *Brucella* species: new potential molecular markers for genotyping. Microbes Infect. *8*, 860–865.

Marianelli, C., Graziani, C., Santangelo, C., Xibilia, M.T., Imbriani, A., Amato, R., Neri, D., Cuccia, M., Rinnone, S., Di Marco, V., and Ciuchini, F. (2007). Molecular epidemiological and antibiotic susceptibility characterization of *Brucella* isolates from humans in Sicily, Italy. J Clin Microbiol. *45*, 2923–2928.

Marianelli, C., Martucciello, A., Tarantino, M., Vecchio, R., Iovane, G., and Galiero, G. (2008a). Evaluation of molecular methods for the detection of *Brucella* species in water buffalo milk. J. Dairy Sci. *91*, 3779–3786.

Marianelli, C., Petrucca, A., Pasquali, P., Ciuchini, F., Papadopoulou, S., and Cipriani, P. (2008b). Use of MLVA-16 typing to trace the source of a laboratory-acquired *Brucella* infection. J. Hosp. Infect. *68*, 274–276.

Maquart, M., Le Flèche, P., Foster, G., Tryland, M., Ramisse, F., Djønne, B., Al Dahouk, S., Jacques, I., Neubauer, H., Walravens, K., *et al.* (2009). MLVA-16 typing of 295 marine mammal *Brucella* isolates from different animal and geographic origins identifies 7 major groups within *Brucella ceti* and *Brucella pinnipedialis*. BMC Microbiol. *9*, 145.

Matar, G.M., Khneisser, I.A., and Abdelnoor, A.M. (1996). Rapid laboratory confirmation of human brucellosis by PCR analysis of a target sequence on the 31-kilodalton *Brucella* antigen DNA. J. Clin. Microbiol. *34*, 477–478.

Mayer-Scholl, A., Draeger, A., Göllner, C., Scholz, H.C., and Nöckler, K. (2010). Advancement of a multiplex PCR for the differentiation of all currently described *Brucella* species. J. Microbiol. Methods *80*, 112–114.

Mayfield, J.E., Bricker, B.J., Godfrey, H., Crosby, R.M., Knight, D.J., Halling, S.M., Balinsky, D., and Tabatabai, L.B. (1988). The cloning, expression, and nucleotide sequence of a gene coding for an immunogenic *Brucella abortus* protein. Gene *63*, 1–9.

Meltzer, E., Sidi, Y., Smolen, G., Banai, M., Bardenstein, S., and Schwartz, E. (2010). Sexually transmitted brucellosis in humans. Clin. Infect. Dis. *51*, e12–15.

Miller, S. A., Dykes, D. D. and Polesky, H. F. (1988). A simple salting out procedure for extracting DNA from human nucleated cells. Nucleic Acids Res. *16*, 1215.

Mitka, S., Anetakis, C., Souliou, E., Diza, E., and Kansouzidou, A. (2007). Evaluation of different PCR assays for early detection of acute and relapsing brucellosis in humans in comparison with conventional methods. J. Clin. Microbiol. *45*, 1211–1218.

Morata, P., Queipo-Ortuño, M.I, and Colmenero, J.D. (1998) Strategy for optimizing DNA amplification in a peripheral blood PCR assay used for diagnosis of human brucellosis. J. Clin. Microbiol. *36*, 2443–2446.

Morata, P., Queipo-Ortuño, M.I., Reguera, J.M., García-Ordoñez, M.A., Pichardo, C., and Colmenero, J.D. (1999). Posttreatment follow-up of brucellosis by PCR assay. J. Clin. Microbiol. 37, 4163–4166.

Morata, P., Queipo-Ortuño, M.I., Reguera, J.M., Miralles, F., Lopez-Gonzalez, J.J., and Colmenero, J.D. (2001). Diagnostic yield of a PCR assay in focal complications of brucellosis. J. Clin. Microbiol. *39*, 3743–3746.

Mori, Y., Nagamine, K., Tomita, N., and Notomi, T. (2001). Detection of loop-mediated isothermal amplification reaction by turbidity derived from magnesium pyrophosphate formation. Biochem. Biophys. Res. Commun. *289*, 150–154.

Mukherjee, F., Jain, J., Grilló, M.J., Blasco, J.M., and Nair, M. (2005). Evaluation of *Brucella abortus* S19 vaccine strains by bacteriological tests, molecular analysis of *ery* loci and virulence in BALB/c mice. Biologicals 33, 153–160.

Mukherjee, F., Jain, J., Patel, V., and Nair, M. (2007). Multiple genus-specific markers in PCR assays improve the specificity and sensitivity of diagnosis of brucellosis in field animals. J. Med. Microbiol. *56*, 1309–1316.

Mutnal, M.B., Purwar, S., Metgud, S. C., Nagmoti, M.B., and Patil, C. S. (2007). PCR confirmation of cutaneous manifestation due to *Brucella melitensis*. J. Med. Microbiol. *56*, 283–285.

Nagamine, K., Hase, T., and Notomi, T. (2002). Accelerated reaction by loop-mediated isothermal amplification using loop primers. Mol. Cell Probes *16*, 223–229.

Navarro, E., Segura, J.C, Castaño, M.J., and Solera, J. (2006). Use of real-time quantitative polymerase chain reaction to monitor the evolution of *Brucella melitensis* DNA load during therapy and post-therapy follow-up in patients with brucellosis. Clin. Infect. Dis. *42*, 1266–1273.

Navarro, J.A., Ortega, N., Buendia, A.J., Gallego, M.C, Martínez, C.M., Caro, M.R., Sánchez, J., and Salinas, J. (2009). Diagnosis of placental pathogens in small ruminants by immunohistochemistry and PCR on paraffin-embedded samples. Vet. Rec. *165*, 175–178.

Navarro-Martínez, A., Navarro, E., Castaño, M.J., and Solera, J. (2008). Rapid diagnosis of human brucellosis by quantitative real-time PCR: a case report of brucellar spondylitis. J. Clin. Microbiol. *46*, 385–387.

Newby, D.T., Hadfield, T.L., and Roberto, F.F. (2003). Real-time PCR detection of *Brucella abortus*: a comparative study of SYBR green I, 5′-exonuclease, and hybridization probe assays. Appl. Environ. Microbiol. *69*, 4753–4759.

Nimri, L.F. (2003). Diagnosis of recent and relapsed cases of human brucellosis by PCR assay. BMC Infect. Dis. *28*, 5.

Nöckler, K., Maves, R., Cepeda, D., Draeger, A., Mayer-Scholl, A., Chacaltana, J., Castañeda, M., Espinosa, B., Castillo, R., Hall, E., *et al.* (2009). Molecular epidemiology of *Brucella* genotypes in patients at a major hospital in central Peru. J. Clin. Microbiol. *47*, 3147–3155.

Notomi, T., Okayama, H., Masubuchi, H., Yonekawa, T., Watanabe, K., Amino, N., and Hase, T. (2000). Loop-mediated isothermal amplification of DNA. Nucleic Acids Res. *28*, E63.

Ohishi, K., Takishita, K,, Kawato, M., Zenitani, R., Bando, T., Fujise, Y., Goto, Y., Yamamoto, S., and Maruyama, T. (2004). Molecular evidence of new variant *Brucella* in North Pacific common minke whales. Microbes Infect. *6*, 1199–1204.

Ohtsuki, R., Kawamoto, K., Kato, Y., Shah, M.M., Ezaki, T., and Makino, S.I. (2008). Rapid detection of *Brucella* spp. by the loop-mediated isothermal amplification method. J. Appl. Microbiol. *104*, 1815–1823.

O'Leary, S., Sheahan, M., and Sweeney, T. (2006). *Brucella abortus* detection by PCR assay in blood, milk and lymph tissue of serologically positive cows. Res. Vet. Sci. *81*, 170–176.

Ocampo-Sosa, A.A., Agüero-Balbín J., and García-Lobo, J.M. (2005). Development of a new PCR assay to identify Brucella abortus biovars 5, 6 and 9 and the new subgroup 3b of biovar 3. Vet. Microbiol. *110*, 41–51.

Ocampo-Sosa, A.A, and García-Lobo, J.M. (2008). Demonstration of IS*711* transposition in *Brucella ovis* and *Brucella pinnipedialis*. BMC Microbiol. *8*, 17.

Omata, Y., Umeshita, Y., Watarai, M., Tachibana, M., Sasaki, M., Murata, K., and Yamada, T.K. (2006). Investigation for presence of *Neospora caninum*, *Toxoplasma gondii* and *Brucella*-species infection in killer whales (*Orcinus orca*) mass-stranded on the coast of Shiretoko, Hokkaido, Japan. J. Vet. Med. Sci. *68*, 523–526.

Ouahrani, S., Michaux, S., Sri Widada, J., Bourg, G., Tournebize, R., Ramuz, M., and Liautard, J.P. (1993). Identification and sequence analysis of IS*6501*, an insertion sequence in *Brucella* spp.: relationship between genomic structure and the number of IS*6501* copies. J. Gen. Microbiol. *139*, 3265–3273.

Ouahrani-Bettache, S., Soubrier, M.P., and Liautard, J.P. (1996). IS*6501*-anchored PCR for the detection and identification of *Brucella* species and strains. J.Appl. Bacteriol. *81*, 154–160.

Poon, L.L., Wong, B.W., Ma, E.H., Chan, K.H., Chow, L.M., Abeyewickreme, W., Tangpukdee, N., Yuen, K.Y., Guan, Y., Looareesuwan, S., *et al.* (2006). Sensitive and inexpensive molecular test for falciparum malaria: detecting *Plasmodium falciparum* DNA directly from heat-treated blood by loop-mediated isothermal amplification. Clin. Chem. *52*, 303–306.

Probert, W.S., Schrader, K.N., Khuong, N.Y., Bystrom, S.L., and Graves, M.H. (2004). Real-time multiplex PCR assay for detection of *Brucella* spp., *B. abortus*, and *B. melitensis*. J. Clin. Microbiol. *42*, 1290–1293.

Queipo-Ortuño, M.I., Morata, P., Ocón, P., Manchado, P., and Colmenero, J.D. (1997). Rapid diagnosis of human brucellosis by peripheral-blood PCR assay. J. Clin. Microbiol. *35*, 2927–2930.

Queipo-Ortuño, M.I., Colmenero, J.D., Reguera, J.M., García-Ordoñez, M.A., Pachón, M.E., Gonzalez, M., and Morata, P. (2005). Rapid diagnosis of human brucellosis by SYBR Green I-based real-time PCR assay and melting curve analysis in serum samples. Clin. Microbiol. Infect. *11*, 713–718.

Queipo-Ortuño, M.I., Colmenero, J.D., Muñoz, N., Baeza, G., Clavijo, E., and Morata, P. (2006). Rapid diagnosis of *Brucella* epididymo-orchitis by real-time polymerase chain reaction assay in urine samples. J. Urol. *176*, 2290–2293.

Queipo-Ortuño, M.I., De Dios Colmenero, J., Macias, M., Bravo, M.J., and Morata, P. (2008a). Preparation of bacterial DNA template by boiling and effect of immunoglobulin G as an inhibitor in real-time PCR for serum samples from patients with brucellosis. Clin. Vaccine Immunol. *15*, 293–296.

Queipo-Ortuño, M.I., Tena, F., Colmenero, J.D., and Morata, P. (2008b). Comparison of seven commercial DNA extraction kits for the recovery of *Brucella* DNA from spiked human serum samples using real-time PCR. Eur. J. Clin. Microbiol. Infect. Dis. *27*, 109–114.

Queipo-Ortuño, M.I., Colmenero, J.D., Bermúdez, P., Bravo, M.J., and Morata, P. (2009). Rapid differential diagnosis between extrapulmonary tuberculosis and focal complications of brucellosis using a multiplex real-time PCR assay. PLoS One *4*, e4526.

Ratushna, V.G., Sturgill, D.M., Ramamoorthy, S., Reichow, S.A., He. Y., Lathigra, R., Sriranganathan, N., Halling, S.M., Boyle, S.M., and Gibas, C.J. (2006). Molecular targets for rapid identification of *Brucella* spp. BMC Microbiol. *6*, 13.

Redkar, R., Rose, S., Bricker, B., and DelVecchio V. (2001). Real-time detection of *Brucella abortus*, *Brucella melitensis* and *Brucella suis*. Mol. Cell Probes *15*. 43–52.

Rees, R.K, Graves, M., Caton, N., Ely, J.M., and Probert, W.S. (2009). Single tube identification and strain typing of *Brucella melitensis* by multiplex PCR. J. Microbiol. Methods. *78*, 66–70.

Richtzenhain, L.J., Cortez, A., Heinemann, M.B., Soares, R.M., Sakamoto, S.M., Vasconcellos, S.A., Higa, Z.M., Scarcelli, E., and Genovez, M.E. (2002). A multiplex PCR for the detection of *Brucella* spp. and *Leptospira* spp. DNA from aborted bovine fetuses.Vet. Microbiol. 87, 139–147.

Rijpens, N.P., Jannes, G., Van Asbroeck, M., Rossau, R., and Herman, L.M. (1996). Direct detection of *Brucella* spp. in raw milk by PCR and reverse hybridization with 16S-23S rRNA spacer probes. Appl. Environ. Microbiol. *62*, 1683–1688.

Romero, C., and López-Goñi, I. (1999). Improved method for purification of bacterial DNA from bovine milk for detection of *Brucella* spp. by PCR. Appl. Environ. Microbiol. *65*, 3735–3737.

Romero, C., Gamazo, C., Pardo, M., and López-Goñi, I. (1995a). Specific detection of *Brucella* DNA by PCR. J. Clin. Microbiol. 33, 615–617.

Romero, C., Pardo, M., Grilló, M.J., Díaz, R., Blasco, J.M., and López-Goñi, I. (1995b). Evaluation of PCR and indirect enzyme-linked immunosorbent assay on milk samples for diagnosis of brucellosis in dairy cattle. J. Clin. Microbiol. *33*, 3198–3200.

Saiki, R.K., Scharf, S., Faloona, F., Mullis, K.B., Horn, G.T., Erlich, H.A., and Arnheim, N. (1985). Enzymatic amplification of beta-globin genomic sequences and restriction site analysis for diagnosis of sickle cell anemia. Science *230*, 1350–1354.

Saiki, R.K., Gelfand, D.H., Stoffel, S., Scharf, S.J., Higuchi, R., Horn, G.T., Mullis, K.B., and Erlich, H.A. (1988). Primer-directed enzymatic amplification of DNA with a thermostable DNA polymerase. Science *239*, 487–491.

Sangari, F.J., and Agüero, J. (1994). Identification of *Brucella abortus* B19 vaccine strain by the detection of DNA polymorphism at the *ery* locus. Vaccine *12*, 435–438.

Saunders, V.F., Reddacliff, L.A., Berg, T., and Hornitzky, M. (2007). Multiplex PCR for the detection of *Brucella ovis*, *Actinobacillus seminis* and *Histophilus somni* in ram semen. Aust. Vet. J. *85*, 72–77.

Sayan M, Yumuk Z, Bilenoglu O, Erdenlig S, and Willke, A. (2009). Genotyping of *Brucella melitensis* by *rpoB* gene analysis and re-evaluation of conventional serotyping method. Jpn. J. Infect. Dis. *62*, 160–163.

Schlabritz-Loutsevitch, N., Whatmore, A.M., Quance, C.R., Koylass, M.S., Cummins, L.B., Dick, E.J., Snider,

C.L., Capelli, D., Ebersole, J.L., Nathanielsz, P.W., *et al.* (2009). A novel *Brucella* isolate in association with two cases of still-births in non-human primates – first report. J. Med. Primatol. *38*, 70–73.

Scholz, H.C., Hubalek, Z., Sedlácek, I., Vergnaud, G., Tomaso, H., Al Dahouk, S., Melzer, F., Kämpfer, P., Neubauer, H., Cloeckaert, A., *et al.* (2008). *Brucella microti* sp. nov., isolated from the common vole *Microtus arvalis*. Int. J. Syst. Evol. Microbiol. *58*, 375–382.

Scholz, H.C., Nöckler, K., Göllner, C., Bahn, P., Vergnaud, G., Tomaso, H., Al Dahouk, S., Kämpfer, P., Cloeckaert, A., Maquart, M., *et al.* (2010). *Brucella inopinata* sp. nov., isolated from a breast implant infection. Int. J. Syst. Evol. Microbiol. *60*, 801–808.

Schuurman, T, van Breda, A., de Boer, R., Kooistra-Smid, M., Beld, M., Savelkoul, P., and Boom, R. (2005). Reduced PCR sensitivity due to impaired DNA recovery with the MagNA Pure LC total nucleic acid isolation kit. J. Clin. Microbiol. *43*, 4616–4622.

Scott J.C., Koylass M.S., Stubberfield M.R., and Whatmore, A.M. (2007). Multiplex assay based on single-nucleotide polymorphisms for rapid identification of *Brucella* isolates at the species level. Appl. Environ. Microbiol. *73*, 7331–7337.

Shoffner, M.A., Cheng, J., Hvichia, G.E., Kricka, L.J., and Wilding, P. (1996). Chip PCR. I. Surface passivation of microfabricated silicon-glass chips for PCR. Nucleic Acids Res. *24*, 375–379.

Sifuentes-Rincón, A.M, Revol, A., and Barrera-Saldaña, H.A. (1997). Detection and differentiation of the six *Brucella* species by polymerase chain reaction. Mol. Med. *3*, 734–739.

Smits, H.L., Espinosa, B., Castillo, R., Hall, E., Guillen, A., Zevaleta, M., Gilman, R.H., Melendez, P., Guerra, C., Draeger, A., *et al.* (2009). MLVA genotyping of human *Brucella* isolates from Peru. Trans. R. Soc. Trop. Med. Hyg. *103*, 399–402.

Sreevatsan, S., Bookout, J.B., Ringpis, F., Perumaalla, V.S., Ficht, T.A., Adams, L.G., Hagius, S.D., Elzer, P.H., Bricker, B.J., Kumar, G.K., *et al.* (2000). A multiplex approach to molecular detection of *Brucella abortus* and/or *Mycobacterium bovis* infection in cattle. J. Clin. Microbiol. *38*, 2602–2610.

van Staden, M., van der Ryst, E., Attwood, E.M., Hendricks, M.L., Joubert, G., and Weich, D.J. (1998). Detection of *Mycobacterium tuberculosis* in serum samples using the polymerase chain reaction. J. Infect. *36*, 273–277.

Tiller, R.V, De, B.K, Boshra, M., Huynh, L.Y., Van Ert, M.N., Wagner, D.M., Klena, J, Mohsen, T.S., El-Shafie, S.S., Keim, P., *et al.* (2009). Comparison of two multiple-locus variable-number tandem-repeat analysis methods for molecular strain typing of human *Brucella melitensis* isolates from the Middle East. J. Clin. Microbiol. *47*, 2226–2231.

Tiller, R.V., Gee, J.E., Frace, M.A., Taylor, T.K., Setubal, J.C., Hoffmaster, A.R, and De, B.K. (2010a). Characterization of novel *Brucella* strains originating from wild native rodent species in North Queensland, Australia. Appl. Environ. Microbiol. *76*, 5837–5845.

Tiller, R.V., Gee, J.E., Lonsway, D.R., Gribble, S., Bell, S.C., Jennison, A.V, Bates, J., Coulter, C., Hoffmaster, A.R., and De, B.K. (2010b). Identification of an unusual *Brucella* strain (BO2) from a lung biopsy in a 52 year-old patient with chronic destructive pneumonia. BMC Microbiol. *10*, 23.

Tittarelli, M., Di Ventura, M., De Massis, F., Scacchia, M., Giovannini, A., Nannini, D., and Caporale, V. (2005). The persistence of *Brucella melitensis* in experimentally infected ewes through three reproductive cycles. J. Vet. Med. B Infect. Dis. Vet. Public Health *52*, 403–409.

Tomaso, H., Kattar, M., Eickhoff, M., Wernery, U., Al Dahouk, S., Straube, E., Neubauer, H., and Scholz, H.C. (2010). Comparison of commercial DNA preparation kits for the detection of *Brucellae* in tissue using quantitative real-time PCR. BMC Infect. Dis. *10*, 100.

Tsolis, R.M., Seshadri, R., Santos, R.L., Sangari, F.J., Lobo, J.M., de Jong, M.F., Ren, Q., Myers, G., Brinkac, L.M., Nelson, W.C., *et al.* (2009). Genome degradation in *Brucella ovis* corresponds with narrowing of its host range and tissue tropism. PLoS One *4*, e5519.

Valdezate, S., Cervera, I., Hernandez, P., Navarro, A., and Saéz Nieto, J.A. (2007). Characterisation of human outbreaks of brucellosis and sporadic cases by the use of hyper-variable octameric oligonucleotide fingerprint (HOOF) variable number tandem repeats. Clin. Microbiol. Infect. *13*, 887–892.

Valdezate, S., Navarro, A., Villalón, P., Carrasco, G., and Saéz-Nieto, J.A. (2010). Epidemiological and phylogenetic analysis of Spanish human *Brucella melitensis* strains by multiple-locus variable-number tandem-repeat typing, hypervariable octameric oligonucleotide fingerprinting, and *rpoB* typing. J. Clin. Microbiol. *48*, 2734–2740.

Vemulapalli, R., McQuiston, J.R., Schurig, G.G., Sriranganathan, N, Halling, S.M, and Boyle, S.M. (1999). Identification of an IS*711* element interrupting the wboA gene of *Brucella abortus* vaccine strain RB51 and a PCR assay to distinguish strain RB51 from other *Brucella* species and strains. Clin. Diagn. Lab. Immunol. *6*, 760–764.

Vergnaud, G, and Pourcel, C. (2006). Multiple Locus VNTR (Variable Number of Tandem Repeat) Analysis. In Molecular Identification, Systematics, and Population Structure of Prokaryotes, Stackebrandt, E., ed. (Springer-Verlag: Berlin Heidelberg), pp. 83–104.

Vizcaíno, N., Cloeckaert, A., Zygmunt, M.S., and Dubray, G. (1996). Cloning, nucleotide sequence, and expression of the *Brucella melitensis omp*31 gene coding for an immunogenic major outer membrane protein. Infect. Immun. *64*, 3744–3751.

Vrioni, G., Pappas, G., Priavali, E., Gartzonika, C., and Levidiotou, S. (2008). An eternal microbe: *Brucella* DNA load persists for years after clinical cure. Clin. Infect. Dis. *46*, e131–136.

Wanke, M.M. (2004). Canine brucellosis. Anim. Reprod. Sci. *82–83*, 195–207.

Wellinghausen, N., Nöckler, K., Sigge, A., Bartel, M., Essig, A., and Poppert, S. (2006). Rapid detection of

Brucella spp. in blood cultures by fluorescence *in situ* hybridization. J. Clin. Microbiol. *44*, 1828–1830.

Whatmore, A.M. (2009). Current understanding of the genetic diversity of *Brucella*, an expanding genus of zoonotic pathogens. Infect. Genet. Evol. 9, 1168–1184.

Whatmore, A.M., Shankster, S.J., Perrett, L.L., Murphy, T.J., Brew, S.D., Thirlwall, R.E., Cutler, S.J., and Macmillan, A.P. (2006). Identification and characterization of variable-number tandem-repeat markers for typing of *Brucella spp.* J. Clin. Microbiol. *44*, 1982–1993.

Whatmore, A.M., Perrett, L.L., and Macmillan, A.P. (2007). Characterisation of the genetic diversity of *Brucella* by multilocus sequencing. BMC Microbiol. *7*, 34.

Whatmore, A.M., Dawson, C.E., Groussaud, P., Koylass, M.S., King, A.C., Shankster, S.J., Sohn, A.H., Probert, W.S., and McDonald, W.L. (2008). Marine mammal *Brucella* genotype associated with zoonotic infection. Emerg. Infect. Dis. *14*, 517–518.

Winchell, J.M, Wolff, B.J, Tiller, R., Bowen, M.D., and Hoffmaster, A.R. (2010). Rapid identification and discrimination of *Brucella* isolates by use of real-time PCR and high-resolution melt analysis. J. Clin. Microbiol. 48, 697–702.

Xavier, M.N., Silva, T.M., Costa, E.A., Paixão, T.A., Moustacas, V.S., Carvalho, C.A., Jr., Sant'Anna, F.M., Robles, C.A., Gouveia, A.M., Lage, A.P., *et al.* (2010). Development and evaluation of a species-specific PCR assay for the detection of *Brucella ovis* infection in rams. Vet. Microbiol. *145*, 158–164.

Yu, W.L., and Nielsen, K. (2010). Review of detection of *Brucella* spp. by polymerase chain reaction. Croat. Med. J. *51*, 306–313.

Yuen, L.K., Ross, B.C., Jackson, K.M., and Dwyer, B. (1993). Characterization of *Mycobacterium tuberculosis* strains from Vietnamese patients by Southern blot hybridization. J. Clin. Microbiol. *31*, 1615–1618.

Zerva, L., Bourantas, K., Mitka, S., Kansouzidou, A., and Legakis, N.J. (2001). Serum is the preferred clinical specimen for diagnosis of human brucellosis by PCR. J. Clin. Microbiol. *39*, 1661–1664.

Zhang, G.Q., Nguyen, S.V., To, H., Ogawa, M., Hotta, A., Yamaguchi, T., Kim, H.J., Fukushi, H., and Hirai, K. (1998). Clinical evaluation of a new PCR assay for detection of *Coxiella burnetii* in human serum samples. J. Clin. Microbiol. *36*, 77–80.

Zygmunt, M.S, Maquart, M., Bernardet, N, Doublet, B., and Cloeckaert, A. (2010). Novel IS*711*-specific chromosomal locations useful for identification and classification of marine mammal *Brucella* strains. J. Clin. Microbiol. 48, 3765–3769.

The Exploration of *Brucella* Transcriptome, from the ORFeome to RNAseq

5

Juan M. García-Lobo, María C. Rodríguez, Asunción Seoane, Félix J. Sangari and Ignacio López-Goñi

Abstract

In this chapter we will analyse the results available on the characterization of the *Brucella* transcriptome. After a summary of earlier work on transcription, two technical approaches will be mainly described. On one side the use of microarrays, specially that derived from the *Brucella* ORFeome that allows hybridization with mRNA derived cDNA to determine the relative abundance of transcripts from each *B. melitensis* ORF. On the other, RNAseq, consisting in the massive sequencing of cDNA libraries derived from mRNA obtained in this case from *B. abortus* grown in culture medium. Sequencing with the Illumina Genome-Analyser II platform (Illumina Inc., San Diego, CA, USA) produced 3 millions of 35-nt-long reads that annealed with single copy coding regions of the genome. This allowed a good coverage for every CDS and produced a new data set on the transcription of *Brucella*. We obtained a good correlation for the set of highly expressed genes from the microarrays and confirmed the previous observations obtained on the transcriptional asymmetry between chromosome transcription. Preliminary conclusions on intracellular transcription have been drawn from real-time polymerase chain reaction (RT-PCR) on selected candidate genes and from microarray data sets obtained from virulence related conditions. The RNAseq derived data allowed more versatile data mining, giving some new details on transcription from pseudogenes or intergenic regions.

Introduction

During the last 50 years, to understand the mechanisms working in bacterial cells, we have applied a reductionist approach consisting in the analysis of isolated parts of the organism at different complexity levels ranging from the genetics to the three-dimensional determination of complex molecular structures. This approach has been really fruitful and undoubtedly it will be our principal source of information in the near future. Newer approaches to the study of biological systems aim to analyse living organisms as a complex unit by looking at all the components of the organism as well as the interactions among them that result in an observable behaviour at the same time. This methodology is the hallmark of the field of Systems Biology. These procedures are being applied already into microbiology. Perhaps, the most appealing case has been the analysis of *Mycoplasma*, the free-living bacteria with the smallest genome (Ochman and Raghavan, 2009). After a lot of genomic data on *Mycoplasma* were available, a multinational team has analysed in depth the transcriptome of the bacteria using both, RNA sequencing and tiling microarrays (Güell *et al.*, 2009). They have also studied the proteome of the bacteria, including protein interactions (Kühner *et al.*, 2009) and finally they have reconstructed the metabolism by analysing any metabolic pathway integrating transcriptional, proteomic and metabolic flux data sets (Yus *et al.*, 2009). This piece of work signals the path for the study of more complex organisms using the same methodologies. Perhaps, the most important conclusion to take into account is that some classic paradigms of Molecular Microbiology are changing. This is mainly the case for bacterial transcription. The analysis of the transcriptome of *Mycoplasma* has revealed extensive transcription from both DNA strands of the same DNA, alternative transcripts from operons including only some of the operon genes, and existence of many short RNA's usually non-detected by automatic annotation pipelines.

As a desirable objective we should consider performing a similar study on *Brucella*. Taking into account genome size only, we expect a much more complex problem to solve. While this effort is organized, the progressive application of high throughput technologies to *Brucella* may build a set of data that could be later used to sketch the blueprint of our beloved bacteria.

Early transcriptional analysis

Application of molecular analysis to genetic information flow in *Brucella* started near 30 years ago. To our knowledge the first paper published on this field (that somehow inspired work of our group) was a description of mutagenesis produced by transposon Tn5 in *B. abortus* published in 1987 (Smith and Heffron, 1987). The first nucleotide sequences of *Brucella* rRNA were deposited in GenBank in 1989.

Focus on transcription started soon after, targeting first surface antigen genes with a clear interest in the improvement of brucellosis vaccines and diagnostic tests. For instance, the analysis of the 36-kDa outer membrane protein gene of *B. abortus*, included the detection of promoter activity by fusion to LacZ, identification of Shine-Dalgarno sequences, and rho-independent terminators (Ficht *et al.*, 1989).

Besides the transcriptional analysis of a few dozens of individual genes performed in the following years, a first experiment addressing global analysis of transcription was performed by R.C. Essemberg, who made a library of *Sau*3AI generated DNA fragments fused to a promoterless luciferase reporter and studied transcription arising from the cloned fragments and the effect that different sugars produced on the transcription levels. In this way fragments containing promoters with activity dependent on erythritol, glucose, galactose and succinate were detected. This work was not published, but was communicated at the Chicago Brucellosis meeting and the sequence of the cloned fragments was deposited in GenBank (accession numbers AF075168, AF072121, AF072119, AF074323, AF073884, AF072580, AF072569, AF072120, AF072570, AF072571, AF072572, AF072573, AF072574, AF072575, AF072576, AF072577, AF072578, AF072579).

Methodological overview of massive transcriptomic analysis

The *Brucella* ORFeome

The ORFeome of *Brucella* is an ordered collection of plasmids devised to contain all the ORF's from the sequenced genome of *B. melitensis* 16M. The available annotation was refined to correct the start or the end points of the ORF's according to the best homology with known database products and proximity of Shine-Dalgarno sequences near ORF 5′ end. 908 ORF's limits were corrected out of 3198 ORF's examined. Amplification primers designed for each ORF contained a specific sequence from the ORF and a constant part containing *attB1* and *attB2* sequences at the 5′ and 3′ end of the ORF respectively. This strategy produced ds DNA fragments containing the complete ORF flanked by constant sequences that served as substrate for site-specific recombination reactions in a process known as recombinational cloning. The 3198 amplification products were introduced into the Gateway compatible vector pDONR201 and transformed into *E. coli* DH5α. 3091 clones (96.7%) were successfully demonstrated to contain the appropriate ORF by sequencing (Dricot *et al.*, 2004). The genomic data available on *Brucella* have shown than more than 90% of the genes are common (over 99% identity) between different species indicating that the *Brucella* ORFeome can be used as a general tool for high throughput studies in the genus *Brucella*.

Brucella microarrays

One of the uses of the *Brucella* ORFeome was the construction of a microarray containing all the *Brucella* ORF's printed on a glass slide. The ORFeome was used to isolate plasmid DNA from each clone and every insert was PCR amplified using a single pair of primers containing the *attB1* and *attB2* sequences flanking each ORF in the ORFeome. Each DNA was spotted by duplicate and appropriate positive and negative controls were spiked over the whole microarray. As a positive control we used a gene from *Brucella* that was considered to be constitutively expressed, namely the IF-1 gene (Eskra *et al.*, 2001; Hernández-Castro *et al.*, 2003). As negative controls we included 128 spots containing an *Arabidopsis thaliana* gene (*porB*, encoding NADPH-protochlorophyllide oxidoreductase), known not to hybridize with *Brucella* DNA, and another 128 spots containing spotting buffer but no DNA. Microarrays were validated by ethidium bromide staining showing that they were uniform and could be used to analyse transcription in *Brucella*. Total *B. abortus* 2308 RNA was prepared from 10 ml of cells growing to mid stationary phase in *Brucella* Broth (Pronadisa, Madrid, Spain) with the RNeasy mini System (Quiagen, Cortaboeuf, France). RNA was depleted of DNA by repeated treatment with DNAse and enriched in messenger RNA by subtraction of rRNA with the MicrobExpress kit (Ambion, Life Technologies, Foster City, CA, USA). This enriched mRNA was amplified with the Message AmpII-bacteria amplification kit (Ambion) and labelled with Cy3 fluorescent dye. 10 µg of this labelled aRNA was used to hybridize the microarrays (Viadas *et al.*, 2009).

The ORFeome derived microarray was used first to analyse global transcription in *B.* abortus 2308 grown in laboratory conditions and to evaluate what genes were being expressed or remained silent in these conditions. The same microarray was later used to analyse the function of the two component regulatory system *bvrR/S* (Viadas *et al.*, 2010) and the effect of erythritol on transcription in *Brucella*.

In addition to the ORFeome microarray, at least two more oligonucleotide based microarrays have been used to analyse the *Brucella* transcriptome. A microarray was built from 8768 70-mer unique synthetic oligonuleotides derived from *B. melitensis* ORFs and additional ORFs from *B. suis*. This microarray was hybridized with cDNA derived from total RNA and with genomic DNA as an internal control (Rossetti *et al.*, 2009). Another oligonucleotide based microarray constructed by NimbleGen Systems (Madison, WI, USA) contained 20 24-mer oligonucleotides per CDS, and has also been used to investigate transcriptional regulation (Uzureau *et al.*, 2010, discussed below).

RNA seq: preparation of samples

The more recent approach to study the transcriptome of *Brucella* takes advantage of new generation sequencing systems. RNAseq is a method consisting in the massive sequencing of RNA from a cell, usually after conversion into the appropriate cDNA library. The Illumina platform is especially well suited to this aim since it is capable to produce 10^9 reads per run. Size of reads is not as crucial in RNAseq as is in genome sequencing, and short 35 nt reads are usually satisfactory. Preparation of RNA samples for *Brucella* RNAseq was essentially as described for microarrays with several modifications to improve DNA and rRNA removal as well as mRNA integrity. All samples were processed using the Qiagen RNeasy Protect bacteria mini kit with on-column RNase free DNase I digestion for effective removal of genomic DNA. Samples were assayed for RNA integrity using an Agilent 2100 Bioanalyzer and were quantified on a NanoDrop 1000 spectrophotometer (Thermo, Waltham, MA, USA). 5 µg of pure RNA was depleted of rRNA with the Ambion MicrobExpress kit. The kit is devised to use 10 µg RNA per reaction; using only half of the kit capacity we found a significant improvement in rRNA removal. This step results in a better proportion of non-rRNA reads in the final result. 100 ng of this enriched mRNA were used to construct the library for subsequent sequencing.

Libraries were generated using the Illumina® sample preparation protocols. Following purification, the mRNA was fragmented into small pieces using divalent cations at 94°C during 5 min. The cleaved RNA fragments were copied to first strand cDNA using SuperScript II reverse transcriptase

(Invitrogen, Life Technologies) and random primers. This was followed by second strand cDNA synthesis using DNA polymerase I and RNase H. These cDNA fragments went through an end repair process and then were ligated to the adapters. The products of the ligation reaction, of sizes around 250 bp, were further amplified by PCR using Phusion DNA polymerase (Finnzymes, Thermo, Vantaa, Finland). These cDNA libraries were diluted to 6 pM. cDNA was hybridized to the flow-cell using a Single-Read Cluster Generation Kit v2 (Illumina Inc.). Further amplification of each DNA single molecule was performed by 35 cycles of isothermal amplification in the flow-cell on the Cluster Station. After amplification, one of the strands is removed and the genomic sequencing primer hybridized. The flow-cell was transferred to the Genome Analyser II (Illumina, Inc.) and sequencing was performed for 36 cycles using SBS Sequencing Kit v3 (Illumina Inc.)

RNAseq.: data analysis

The raw data of the sequencing run consisted of 13,835,287 sequences 35 nt long. These sequences were aligned to the reference *B. abortus* 2308 genome by two separate methods. First we used the Illumina proprietary software ELAND (Illumina Corp). This procedure uses a filter file containing the rRNA coding sequences. Those reads aligning with sequences in the filter file are not reported in the output. A total 2,979,705 sequences were found to align with the coding region of the genome fulfilling at the same time the quality criteria of the software.

On the other hand, the reads were aligned using the open software package MAQ (Li *et al.*, 2008). This program allows a better user control on the aligning process. 8,762,781 (63.3% of the total reads) were found to align with the reference genome with two or less mismatches. 5,731,599 (65% of the aligned reads) corresponded to rRNA sequences. The remaining 3,031,182 reads were considered to align to 'single copy' DNA. The summary data of the RNAseq experiment is shown in Table 5.1.

To count how many times each base of the reference genome was contained in a single read we used the Maq pileup utility. From the output of this program we parsed a file containing the complete reads per base of the genome. This file contains a single column of numbers indicating how many times this base was read and a row per base of the genome (3,278,307 rows). Such a file may be used by the visualizing program Artemis (Rutherford *et al.*, 2000) as a user plot to indicate graphically the transcription level along the complete genome. Fig. 5.1 shows a screenshot of the Artemis program containing the erythritol operon region of the *B. abortus* 2308 genome (chromosome II). The results of the RNAseq experiment are shown in the top of the figure as user loaded plot containing the reads per base of the complete genome. The plot shows the data for 100 bp sliding window centred in each base. The region shown contains the four genes of the *ery* operon (*eryABCD*: BAB2_0372, BAB2_0371, BAB2_0370, BAB2_0369) encoded in the bottom chain of the DNA. The location of the *ery* deletion in strain S19 is also shown. The transcription plot for the region ranges from very low values (0 counts upstream the *eryA* gene) to a maximum of 41.6 counts, indicating how many reads containing that base have been sequenced. The characteristic saw-shaped profile of the plot is usually seen in RNAseq experiments (Güell *et al.*, 2009). The observed transcription level seems to be higher for *eryB* and *eryC* than for *eryA* and *eryD*, which is unexpected from the paradigm of operon behaviour. In the case shown in the screen, there is a sharp increase in transcription upstream

Table 5.1 Summary of the RNAseq sequencing data

Total reads	1,383,5287
Aligned reads (maq, *n*=35, mis=2)	8,762,781 (63.3%)
rRNA	5,731,599 (65%)
Unique	3,031,182
Aligned ELAN	2,979,705

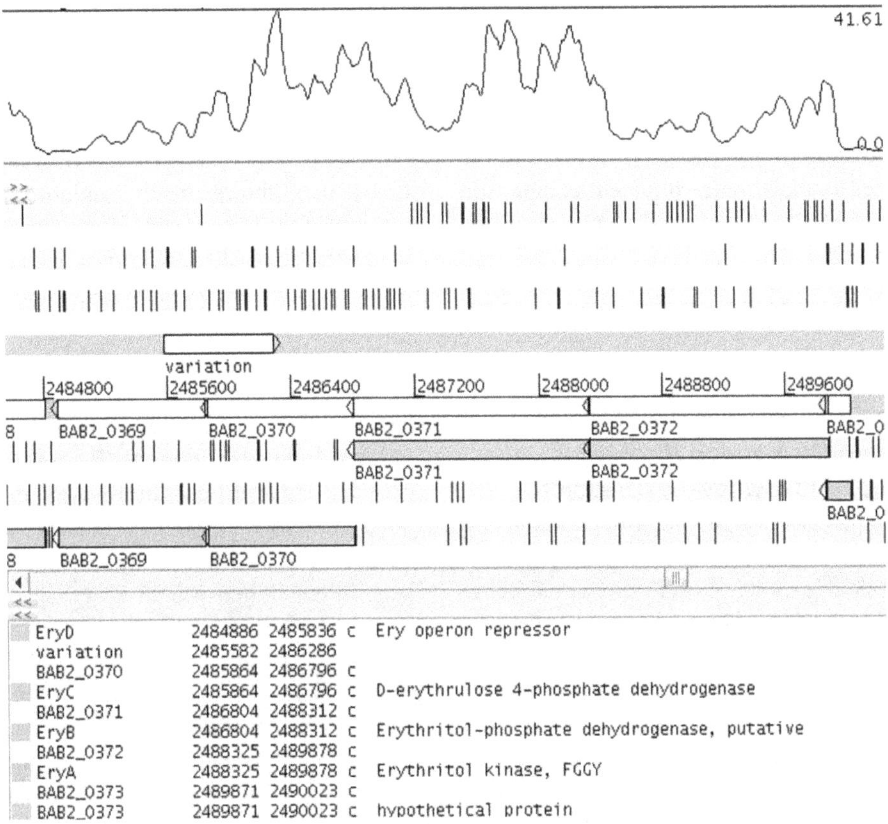

Figure 5.1 Artemis screenshot showing the Chr II region containing the *ery* operon of *B. abortus* 2308. In the top part of the figure appears the corresponding RNA transcription plot read from the reads per base file. Base numbering corresponds to the two fused chromosomes (2,121,359 bp is the length of *B. abortus* 2308 Chr I).

the *eryA* gene in a position neatly coincident with the experimentally determined transcriptional start point for the *ery* operon (Sangari *et al.*, 2000). From these data, the existence of the gene BAB2_0373, annotated in this genome as a hypothetical protein could be questioned. Fig. 5.1 summarizes the informative content of the data obtained. We may visualize the relative level of transcription of any gene. In some cases, such as the one presented in Fig. 5.1, the absence of background allows detection of the approximate site of the translation initiation point, which may be useful to identify and characterize *Brucella* promoters. Nevertheless specific libraries are needed in order to selectively keep the 5′ ends of every transcript, as has been applied for the global detection of transcriptional start sites of the gastric pathogen *Helicobacter pylori* (Sharma *et al.*, 2010). Other applications of the RNAseq method as new gene discovery, or identification of sRNAs from antisense regions, may obtain some information from this approach, but they will need to ascertain the actual DNA chain that is being transcribed in each position which in turn will require a different method of library preparation keeping directionality of mRNA (Vivancos *et al.*, 2010), which is lost in the process of library construction followed in this experiment.

The reads per base obtained as described were integrated using the GenBank annotation of *B. abortus* 2308 selecting both the CDSs and genes annotated as pseudogenes which totalized 3350 annotations (3034 CDSs and 316 pseudogenes). The result of this operation provided for each annotation (either gene or pseudogene) an expression value that represents the sum of the reads for all the bases of the CDS (or pseudogene), and accordingly gives a relative value to the extent of transcription of the element. This annotation could be improved specially in the part concerning to pseudogenes; however, we decided to use it in order to allow more consistent comparison between the RNAseq and microarray data.

Brucella transcriptional snapshots

Identification of gene sets in exponentially growing *B. abortus* with high and low level expression

Two comparable data sets were obtained in our laboratories on the global transcription of *B. abortus* 2308 growing at mid-exponential phase on *Brucella* Broth medium, one using the *Brucella* microarray (Viadas *et al.*, 2009) and the other by RNAseq. We used these data to analyse with them the genes more expressed and those with the poorest expression in the mentioned conditions. In both sets of data, the expression indexes were normalized to gene size and Llog$_2$ transformed to compact the data. Then, the Gene Expression Index (GEI) was sorted and the top and bottom 10% of each list were selected as the more and less expressed genes in *B. abortus* 2308 growing in exponential conditions. The dynamic range of the quantitative data obtained by each method was as different as could be expected by the different technologies. Microarray results were obtained from the fluorescence intensity that needs to be normalized and background corrected before analysis. In our microarray experiment the log$_2$ of the fluorescence intensity range from the more to the less expressed genes was 14.23–2.73. The cut points for the high and low expression categories were 9.19 and 6.50. In the RNAseq experiment we count how many reads were contained in the interval defining a gene, then to avoid selecting for large genes, counts were normalized by gene size and expressed as counts per kb. Twenty-six genes obtained 0 counts in the RNAseq experiment showing that they are indeed not expressed. On the other hand we have a null background due to contaminant genomic DNA. Perhaps this is one of greatest differences that favour RNAseq versus microarrays for estimation of low expression genes (see below). The highest expressed gene (after filtration of rRNAs) in the RNAseq experiment was the porin Omp26 BAB1_0660. This was also the most expressed gene in microarray experiment. BAB1_0660 showed nearly 5 million counts, this means that the 1128 nucleotides that the gene contains were read 4,955,906 times or that 141,600 reads contained BAB1_0660 sequences. The GEI calculated for BAB1_0660 was 4397.43 (counts/size in bp). The last gene in the high expression set in RNAseq showed a GEI of 73.5. These values were similar to those observed in the microarrays (12.2–6.2 log$_2$ transformed). The low expression gene set deduced from the RNAseq experiment ranged in GEI value between and 1.18 and 0.001 (log$_2$ transformed from 1.18 to −10). There is a large difference between the lower values detected in the two methods. In the microarray experiment the lower log$_2$ observed was 2.73. Detection of lower values, as those obtained in the RNAseq method are not reliable owing to the high background. The range of values observed spanned 22 logs for RNAseq but only 12 in the microarrays.

We then analysed the coincidence among the gene sets obtained by both methodologies. Results were rather consistent in the high expression data set, 185 genes (near 60%) were common in the two data sets, while only 59 genes (less than 20%) were common in the low-expression sets. The Venn diagram in Fig. 5.2 was produced with BioVenn (Hulsen *et al.*, 2008) accessed at the web server (http://www.cmbi.ru.nl/cdd/biovenn/index.php). The level of coincidence was as expected, taken into account the inherent divergence found in global assays and the differences in methodology. As discussed above, probably the low expression data set obtained from RNAseq should reflect best the reality than the microarray-derived data.

A more functional way to compare the results was to analyse the distribution of the selected genes in the functional COG categories (Tatusov *et al.*, 2003). The COG category of the selected genes was taken from the precomputed refseq ptt files corresponding to the *B. abortus* 2308 genome obtained from Gene Bank. The distribution of the more and less expressed genes in COG categories is shown in Table 5.2. The table contains the COG

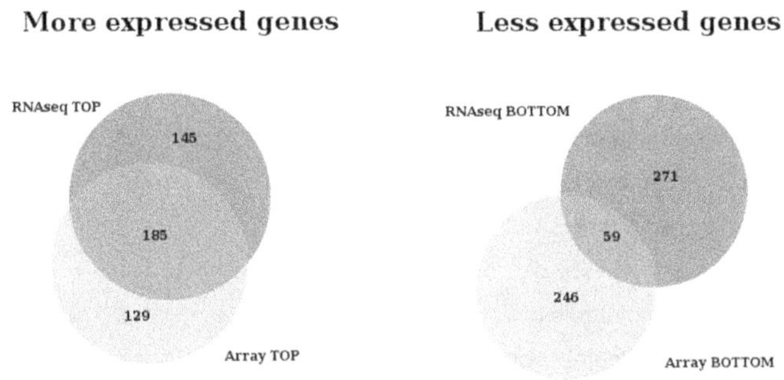

Figure 5.2 Coincidence degree among the high and low expression gene sets determined by the microarray and the RNAseq technologies.

distribution for the whole genome as a reference. This allowed to detect functional differences between the two distributions. Looking only to RNAseq results, in the group of seq results, in the group of more expressed genes the more over-represented COG categories were 'translation', 'post-translational modification, protein turnover, chaperones' and 'energy production and conversion'. As expected, the categories overrepresented in the high expression gene-set were almost absent in the low expression data set. The low expression group of genes was enriched in the 'cell motility' category owing to the very low expression of many of the genes related to biosynthesis of the flagellum. The possibility of the presence of a flagellum in *Brucella* was deduced from the existence of flagellar genes in the *Brucella* genome (Halling, 1998). Later it was shown that *Brucella* may produce a flagellar structure involved in persistence in mice and that it was produced only at the beginning of the exponential phase (Fretin *et al.*, 2005). Another category overrepresented in this set was 'replication, recombination and repair'. This result seems to be a result of the presence in the set of a large number of transposases, and site-specific recombinases, rather to the lack of expression of DNA polymerase or general recombination genes. It was also noteworthy that more than half of the genes in this set could not be assigned to any of the COG categories. This proportion is much more lower (28.8%) in the whole *Brucella* genome. This enrichment could be explained by the existence in the *B. abortus* 2308 annotation of some genes, many of them small, predicted by automatic annotation systems but not corresponding to real genes. Most of these will not be assigned to any COG. A summary of the analysis of distribution among COG categories was obtained by plotting in an histogram the frequency of the genes in each category relative to frequency in *B. abortus* 2308 genome, piling the high and low expression data sets in the same column. This plot contained in Fig. 5.3, shows the extreme enrichment of the J category in the high expression and N and L categories in the low expression data set.

Another way to look at the functional classification of the above data sets was by means of the Database for Annotation, Visualization and Integrated Discovery (DAVID v6.7) accessible via web at http://david.abcc.ncifcrf.gov/ (Huang *et al.*, 2007). The gene functional classification tool of DAVID computes a list of genes to detect whether some descriptive terms taken from the more used databases are enriched in the list with respect to the reference genome. This enrichment is calculated with a variation of the Exact Fisher *P*-value. DAVID also clusters the list in groups of genes associated with some biological function allowing a functional analysis of the gene list. Using this tool we found that the more represented functional groups among the more expressed genes in exponentially growing *Brucella* were those involved in ribosomal biogenesis (55 and 36 genes with this function were present in the sequencing and microarray gene sets). Another cluster of genes in this set was related with respiratory electron transport mainly implicated in reduction of NADH. Enriched

Table 5.2 COG category distribution of the genes in the low and high expression data sets

Code	Description	Genome	Top_arr	Top_seq	Bot_arr	Bot_seq
A	RNA processing and modification	0	0	0	0	0
B	Chromatin structure and dynamics	0	0	0	0	0
C	Energy production and conversion	155	37	45	17	12
D	Cell cycle control, mitosis and meiosis	27	3	5	5	0
E	Amino acid transport and metabolism	294	12	11	38	26
F	Nucleotide transport and metabolism	58	3	3	3	0
G	Carbohydrate transport and metabolism	131	6	13	22	12
H	Coenzyme transport and metabolism	105	10	6	16	5
I	Lipid transport and metabolism	100	7	7	15	5
J	Translation	131	61	63	5	0
K	Transcription	172	15	12	22	6
L	Replication, recombination and repair	114	4	1	20	22
M	Cell wall/membrane biogenesis	143	21	19	18	8
N	Cell motility	24	0	0	9	16
O	Post-translational modification, protein turnover, chaperones	103	14	24	6	4
P	Inorganic ion transport and metabolism	136	10	12	28	13
Q	Secondary metabolites biosynthesis, transport and catabolism	61	4	2	12	4
R	General function prediction only	283	19	13	33	14
S	Function unknown	164	25	24	17	10
T	Signal transduction mechanisms	74	3	6	13	2
U	Intracellular trafficking and secretion	46	7	8	10	9
V	Defence mechanisms	27	0	0	5	2
W	Extracellular structures	1	0	0	0	0
Z	Cytoskeleton	0	0	0	0	0
–	Not in COGs	950	79	69	29	179

Code: One letter code of COG categories. Genome: indicates the number of genes in each COG category in the complete *B. abortus* 2308 genome. Top_arr, Top_seq, Bot_arr, Bot_seq: indicate the number of genes in each COG category in the 10% more expressed genes from the microarray, 10% more expressed genes from RNAseq, 10% less expressed genes from the microarray and 10% less expressed genes from RNAseq, respectively.

terms in this cluster were 'electron transport', 'iron–sulphur', and 'NADH-quinone oxyreductase'. Oxidative phosphorylation, ATP synthase and tricarboxylic acid cycle related terms were also enriched. The more significant category identified by DAVID in the low expression gene sets were again the flagellum related genes, also detected by COG distribution comparison. Another functional group of genes identified by the DAVID engine was a large and heterogeneous group of genes encoding nucleotide binding proteins many of them transport related. This group included many ATP-binding proteins from small molecule transport systems but also components of the type IV secretion system. A group of eight genes encoding transposases and related proteins also was found significant enriched only in the RNAseq low expression gene set.

Transcription during cell invasion

While no direct analysis of global intracellular transcription has been performed, there have been several lateral approaches that are worth a comment in this section.

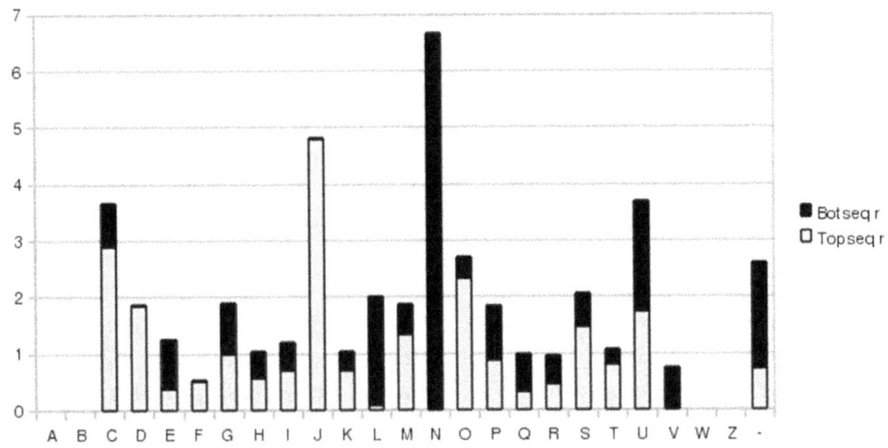

Figure 5.3 Distribution in COG categories of the more and less expressed genes of *B. abortus* 2308. The histogram shows the per cent on each class (black or light grey bars) of genes in each COG category as indicated. Similar height of the colours in each column indicates similar distribution in the two classes. Colour asymmetries indicate different distribution. This is extreme in some classes (D, F and J categories are poorly represented in the low expression set and L and N are almost absent in the high expression gene set). One letter code to COG categories are described in Table 5.2.

From the observation that *B. melitensis* 16M at late-log phase of growth were more invasive to epithelial cells that were bacteria at mid-log or stationary growth phases, it has been assumed that genes preferentially expressed at this phase could be implicated at least in the initial phases of the cell invasion process (Rossetti *et al.*, 2009). Then, analysis of transcription at late-log phase of growth may be considered a surrogate of the invasion process. Comparative microarray analysis revealed a greater number of genes up-regulated in late-log phase cultures than in stationary phase cultures: of the 454 genes significantly altered in *B. melitensis* during late-log phase (the most invasive), 414 (91%) were up- and 40 (9%) were down-regulated compared with the bacteria in stationary phase (the less invasive). According to the conditions compared, the majority of gene expression changes were associated with growth and metabolism. Among the up-regulated genes were those associated with DNA replication, transcription and translation (57 genes), nucleotide, amino acid, lipid and carbohydrate metabolism (65 genes), energy production and conversion (24 genes), membrane transport (56 genes) and cell envelope, biogenesis and outer membrane (26 genes), while the 40 down-regulated genes were more heterogeneous in nature, demonstrating no predominant functional category. Several genes whose products are known to be associated with *Brucella* virulence were differentially expressed between the most and the less invasive cultures. These included: SP41 gene, a surface protein that enables *B. suis* to attach and penetrate non-phagocytic cells (Castaneda-Roldán *et al.*, 2006); three genes components of the *virB* locus (*virB1*, *virB3* and *virB10*), that plays a critical role in *Brucella* virulence and intracellular multiplication (see also Chapter 11); five genes (*fliC*, *fliF*, *fliN*, *flhA* and *flgD*) which encode parts of the flagellar apparatus; several transcriptional regulators; and cell envelope and outer membrane biogenesis genes (LPS and peptidoglycan biosynthesis and outer membrane proteins and lipoproteins). Interestingly 22% of the differentially expressed genes identified in this study had an uncharacterized function, and may contain unknown essential information to completely understand the virulence factors utilized for *Brucella* to invade and infect the host.

Other transcriptional problem targeted with microarray experiments was the effect of the transcriptional regulator VjbR. Mutants missing this regulator are attenuated in all models of infection and suggest that quorum sensing plays an essential role in the regulation of virulence, specially of those genes required for adaptation to growth in the host. Nimblegen oligonucleotide microarrays were used to compare transcriptional profiles of wild-type and *vjbR* mutants (Uzureau *et al.*, 2010). This analysis showed near a 10% of *B. melitensis* genes differentially expressed and henceforth probably involved in the process of host invasion. These differentially expressed genes, corroborated by other experimental methods, were involved in central metabolic pathways, virulence and stress responses which has been interpreted as proof of the central role of *vjbR* (and quorum sensing) in the adaptation of *Brucella* to the oxidative, pH, and nutritional stresses encountered within the host.

Intracellular transcription of *Brucella* genes

The description of the bacterial genes being expressed in the intracellular compartment would be extremely informative to understand the mechanisms the *Brucellae* use to thrive inside cells. A lot of effort has been dedicated to the characterization of *Brucella* genes required for *in vivo* infection, either in animals or in cultured cells. In some way a correlation may be established between requirement of the genes for *in vivo* growth and gene expression. In other words, the set of genes required for intracellular growth must be equivalent to the genes expressed in intracellular conditions. Kohler *et al.* used expression of Green Fluorescent Protein (GFP) as a means to identify *Brucella* promoters induced specifically in macrophages (Kohler *et al.*, 1999). The gene *nikA* was found to be induced in *B. suis* growing into J774 macrophages, which prompted an extensive analysis of the *nik* operon involved in nickel uptake and urease activity (Jubier-Maurin *et al.*, 2001). The analysis of intracellular expression with the use of gene-fusions with GFP to measure intracellular fluorescence either by microscopy or flow cytometry was also applied to study the behaviour of the *virB* promoter of *B. suis* growing inside J774 macrophages (Boschiroli *et al.*, 2022).

Later, a collection of 10,272 mini-Tn5 transposon mutants of a constitutively fluorescent *B. suis* were screened by fluorescence microscopy for lack of intracellular multiplication in human macrophages, and 131 attenuated mutants affected in 59 different genes were detected. This study identified genes involved in global adaptation to intracellular environment, amino acid and nucleotide synthesis, sugar and nitrogen metabolism, redox reactions, regulation of transcription, disulphide bond formation, and LPS biosynthesis. The analysis allowed also to define the environment that the bacteria encountered in the macrophage and to better understand the nature of the phagosome: it is poor in nutrients, as shown by the requirement of functional genes of various biosynthesis pathways; it most probably has a neutral pH; it is also characterized by low oxygen tension and, alternatively, nitrate ions are available for anaerobic respiration.

The limitation in the amount and quality of *Brucella* RNA recovered from infected cells has prevented by now the global analysis of intracellular transcription either by microarray or RNAseq experiments. For this reason direct analysis of intracellular transcription has been performed only on selected genes by RT-PCR reactions using as template RNA obtained from intracellular bacteria. This analysis was done on a set of 32 genes selected because of their known relation with virulence (Viadas *et al.*, 2010). Previously, a whole-genome microarray analysis determined the genes regulated by the two-component system BvrR/BvrS, related to *Brucella* virulence (see also Chapter 10). Most of the genes candidate to be regulated by BvrR/BvrS identified in the microarray experiments can be involved with the changes needed for intracellular survival of *Brucella*. In order to investigate if these genes were expressed intracellularly, bacterial RNA was obtained from *B. abortus* wild type recovered from BHK21 infected cells and from the same strain (*B. abortus* 2308) grown in laboratory conditions (Viadas *et al.*, 2010) and RT-PCR performed with specific primers for the selected genes. The results showed significant differences in the expression of at least fifteen genes: (i) those highly expressed intracellularly included genes related to virulence (*virB8*), carbon and nitrogen metabolism (*malF, pckA, fumB, norC*), fatty acid and LPS biosynthesis and a transcriptional regulator (*vjbR*), and (ii) genes with lower levels of intracellular expression were related to cell envelope proteins (*omp25d*, lipoproteins) and denitrification (*nosZ, nirK* and glutaminase). Interestingly, *virB8* was the gene with the highest level of expression intracellularly, which is in concordance with previous reports (Boschiroli *et al.*, 2002; Sieira *et al.*, 2004). In relation to nitrogen metabolism, it is significant that where the nitrite reductase gene (*nirK*) was poorly expressed intracellularly, the nitric oxide reductase (*norC*) was highly expressed. This regulation of the denitrification route could be use by *Brucella* to survive using nitrogen oxides as terminal electron receptors and limiting the production of reactive nitrogen intermediates by the host.

Highly transcribed intergenic regions

One of the advantages of the RNAseq approach is that data are obtained for all the positions of the genome (counts per base) in an annotation independent way. This raw result may be integrated for any set of subsequences the analyst could imagine by only defining start and endpoints of the features of interest. Taking advantage of this versatility, we have also analysed transcription from the intergenic regions of the *Brucella* 2308 genome. We added to the CDS and pseudogene annotation the data for tRNA and mRNA. To avoid short intercistronic regions of operons, intergenic spaces smaller than 20 bp were eliminated. The total counts per base were integrated for these intergenic spaces as was made for CDSs, and a relative 'intergene' expression index was also calculated. GEI for the more expressed CDS varied from 73 to 4400. Surprisingly 156 intergenic spaces showed GEI's higher than 73, reaching a maximal value of 1800, very close to the more expressed genes. This finding indicates that in addition to annotated CDS and RNAs there are many loci in the genome that are actively transcribed and presumably play functions in the biology of the *Brucella*. We have analysed the top members of the transcribed intergenic regions and detected that some of them corresponded to well known structural or functionally relevant non-translated RNAs. While most of the highly expressed intergenic regions are between highly expressed genes (i.e. ribosomal proteins), some of them encoded for elements related to mRNA processing or stability such as RNAse P (Scott *et al.*, 2009) and tmRNA (Liang and Deutscher, 2010).

Transcription of pseudogenes

Twenty-four of the genes selected from the RNAseq data were annotated as pseudogenes in the *B. abortus* 2308 genome, which was considered a rather unexpected finding. By comparison with other *Brucella* genomes we can reduce the list of highly expressed pseudogenes to 16 (often, truncated parts of a gene are annotated as different pseudogenes especially in *B. abortus* 2308). This seems contradictory since high transcription of these genes, which should be not able to translate into functional proteins, will be contrary to biological economy. The high levels of transcription observed for these genes strongly suggest that they could be active genes and their products may perform functions unreported in metabolic reconstructions. High pseudogene expression may also indicate that these are very recently produced pseudogenes that did not turned down transcription yet by accumulation of mutations in their promoter or control regions. It is also possible that these pseudogenes may contain sequencing errors and they are indeed active genes. On the other hand eighty two pseudogenes were found in the low expression class. This represents a significant enrichment of pseudogenes in the low expression set. This subset of pseudogenes showed a level of transcription consistent with their identification as inactivated, non-working genes, which also turned out transcription to avoid the energy burden of expressing partial or altered gene products.

A similar result showing pseudogene translation has been recently reported in a compilation of proteomic analysis, that described five peptide spots in a *B. abortus* 2308 proteome analysis corresponding to genes annotated as pseudogenes (Lamontagne *et al.*, 2010).

The absence of pseudogenes in the microarray data set was due to the method used to convert *B. melitensis* annotation to *B. abortus* 2308. This conversion was done by BlastP and Tblastn comparisons against *B. abortus* 2308 protein database, in which pseudogenes are not included.

Transcriptional differences between the two chromosomes

The structure of the *Brucella* genome in two separate replicons has been interpreted as a solution to organize the more used genes in the large (or principal) chromosome while the small chromosome contains genes useful only for particular situations. An analysis of COG category distribution among the two chromosomes confirms this view. Genes related to replication and recombination and translation, post-translational modification, protein turnover, chaperones, etc. are overrepresented in Chr I, while Chr II contains more genes related to transport. According to this view a higher level of transcription would be expected from genes in Chr I than from genes in Chr II in fast growth favourable conditions as those used in our experiments. The asymmetric distribution of highly expressed genes in Chr I had been already detected in the microarray analysis of transcription. The RNA seq data confirmed this observation as demonstrates the plot shown in Fig. 5.4. The plot of density frequencies of genes by expression level shows a global difference of near one log (two times) between expression of the genes in both chromosomes. In the RNAseq gene lists, 273 genes of the high expression class were in Chr I while only 146 of the low expressed genes were from Chr I.

Some differences in gene expression have been attributed to gene dosage effects. Genes located near the replication origin may have copy numbers higher than those close to the terminus, specially at fast growth rates (Cooper *et al.*, 2010), as could be the case considered in our data sets. To analyse this possibility, we decided to study the correlation between the level of expression and the distance to the origin for the two chromosomes for each gene. To locate the origin of replication of *Brucella* chromosomes we used the DoriC program, which is a systematic method comprising: (i) the Z-curve analysis for nucleotide distribution asymmetry; (ii) DnaA box distribution; (iii) the presence of genes adjacent to candidate *oriC*s in other species; and (iv) phylogenetic relationships (http://tubic.tju.edu.cn/doric/). The *B. abortus* 2308 Chr I origin was located at 2007638..2008089. This *oriC* region was defined as a 452 bp containing three putative DnaA boxes (TGTGGAAAA) and is adjacent to *haem*, a gene found close to the replication origin in other α-Proteobacteria (Brassinga *et al.*, 2001). For Chr II the origin of replication was predicted as a 368 nt sequence (1156735..154) containing two putative DNA boxes and close to a *repC* gene. Considering that replication of the two chromosomes was bidirectional from the origin we computed the distance of each gene to the origin but failed to detect any effect of the distance to the origin on the observed levels of transcription.

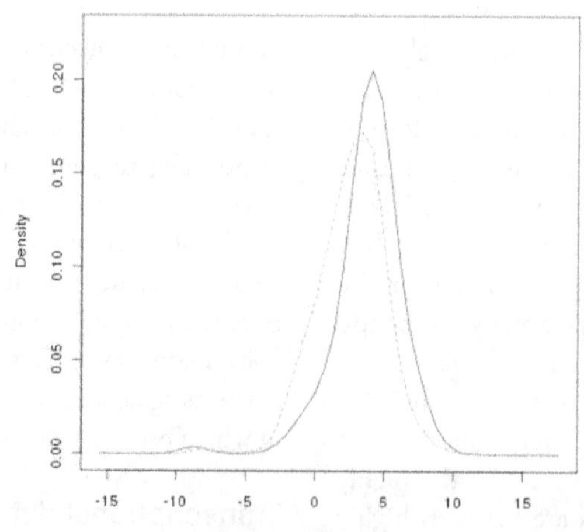

Figure 5.4 The figure shows the plot of the density of frequencies of genes showing a given transcription level expressed as the Log2 of the gene expression index (GEI) from either the Chr I (continuous line) or the Chr II (dashed line).

Future perspectives and conclusions

The transcriptome of *B. abortus* 2308 has been analysed in a comprehensive manner using two powerful technologies, a DNA microarray derived from the *Brucella* ORFeome and RNAseq. The data sets were used to identify the more and less expressed genes in bacteria grown in laboratory conditions. The two methods used performed well to identify the more expressed genes. The low expression sets were more divergent but still allowed the identification some gene families in this class, particularly those related to flagellar production in *Brucella*.

RNAseq data allow more possibilities to obtain information on the transcriptional landscape of the bacteria than the ORFeome derived microarray. Some interesting results on pseudogene expression, transcription from non-annotated regions, identification of clusters of genes and location of promoters may be obtained from the results. However, a different methodology, conserving directionality of mRNA in the sequencing libraries will be needed for a full definition of the transcriptome. Another objective that has to be pursued is the analysis of transcription in intracellular conditions. This has done already in other intracellular bacteria as *Yersinia pestis* (Fukuto *et al.*, 2010) and new methods available (Azhikina *et al.*, 2010) to improve purification of RNA from intracellular bacteria combined to the high power of RNAseq, should allow soon the detailed analysis of the transcriptome of intracellular *Brucella*.

Several proteomic analysis of *Brucella* have been performed using different species and experimental conditions. The results from many of them have been compiled in a recent publication (Lamontagne *et al.*, 2010) and are also discussed in Chapter 6. The combination of proteomic data with those of the transcriptome should reveal in the near future details of *Brucella* biology that we should transform in new tools for the practical control of brucellosis on earth.

References

Azhikina, T., Skvortsov, T., Radaeva, T., Mardanov, A., Ravin, N., Apt, A., and Sverdlov, E. (2010). A new technique for obtaining whole pathogen transcriptomes from infected host tissues. BioTechniques *48*, 139–144.

Boschiroli, M.L., Ouahrani-Bettache, S., Foulongne, V., Michaux-Charachon, S., Bourg, G., Allardet-Servent, A., Cazevieille, C., Liautard, J.P., Ramuz, M., and O'Callaghan, D. (2002). The *Brucella suis virB* operon is induced intracellularly in macrophages. Proc. Natl. Acad. Sci. U.S.A. *99*, 1544–1549.

Brassinga, A.K.C., Siam, R., and Marczynski, G.T. (2001). Conserved gene cluster at replication origins of the α-proteobacteria *Caulobacter crescentus* and *Rickettsia prowazekii*. J. Bacteriol. *183*, 1824–1829.

Castañeda-Roldán, E.I., Ouahrani-Bettache, S., Saldaña, Z., Avelino, F., Rendón, M.A., Dornand, J., and Girón, J.A. (2006). Characterization of SP41, a surface protein of *Brucella* associated with adherence and invasion of host epithelial cells. Cell. Microbiol. *8*, 1877–1887.

Cooper, V.S., Vohr, S. H., Wrocklage, S.C., and Hatcher, P.J. (2010). Why genes evolve faster on secondary chromosomes in bacteria. PLoS Comput. Biol. *6*, e1000732.

Dricot, A., Rual, J., Lamesch, P., Bertin, N., Dupuy, D., Hao, T., Lambert, C., Hallez, R., Delroisse, J., Vandenhaute, J., *et al.* (2004). Generation of the *Brucella melitensis* ORFeome version 1.1. Genome Res. *14*, 2201–2206.

Eskra, L., Canavessi, A., Carey, M., and Splitter, G. (2001). *Brucella abortus* genes identified following constitutive growth and macrophage infection. Infect. Immun *69*, 7736–7742.

Ficht, T.A., Bearden, S.W., Sowa, B.A., and Adams, L.G. (1989). DNA sequence and expression of the 36-kilodalton outer membrane protein gene of *Brucella abortus*. Infect. Immun. *57*, 3281–3291.

Fretin, D., Fauconnier, A., Köhler, S., Halling, S., Léonard, S., Nijskens, C., Ferooz, J., Lestrate, P., Delrue, R., Danese, I., *et al.* (2005). The sheathed flagellum of *Brucella melitensis* is involved in persistence in a murine model of infection. Cell. Microbiol. *7*, 687–698.

Fukuto, H.S., Svetlanov, A., Palmer, L.E., Karzai, A.W., and Bliska, J.B. (2010). Global gene expression profiling of *Yersinia pestis* replicating inside macrophages reveals the roles of a putative stress-induced operon in regulating type III secretion and intracellular cell division. Infect. Immun. *78*, 3700–3715.

Güell, M., van Noort, V., Yus, E., Chen, W., Leigh-Bell, J., Michalodimitrakis, K., Yamada, T., Arumugam, M., Doerks, T., Kühner, S., *et al.* (2009). Transcriptome complexity in a genome-reduced bacterium. Science *326*, 1268–1271.

Halling, S.M. (1998). On the presence and organization of open reading frames of the nonmotile pathogen *Brucella abortus* similar to class II, III, and IV flagellar genes and to LcrD virulence superfamily. Microb. Comp. Genomics 3, 21–29.

Hernández-Castro, R., Rodríguez, M.C., Seoane, A., and García Lobo, J.M. (2003). The aquaporin gene *aqpX* of *Brucella abortus* is induced in hyperosmotic conditions. Microbiology (Reading, Engl.) *149*, 3185–3192.

Huang, D.W., Sherman, B.T., Tan, Q., Kir, J., Liu, D., Bryant, D., Guo, Y., Stephens, R., Baseler, M. W., Lane, H.C., *et al.* (2007). DAVID Bioinformatics Resources: expanded annotation database and novel algorithms to better extract biology from large gene lists. Nucleic Acids Res. 35, W169–W175.

Hulsen, T., de Vlieg, J., and Alkema, W. (2008). BioVenn – a web application for the comparison and visualization of biological lists using area-proportional Venn diagrams. BMC Genomics *9*, 488–488.

Jubier-Maurin, V., Rodrigue, A., Ouahrani-Bettache, S., Layssac, M., Mandrand-Berthelot, M.A., Kohler, S., and Liautard, J.P. (2001). Identification of the *nik* gene cluster of *Brucella suis*: regulation and contribution to urease activity. *183*, 426–434.

Kohler, S., Ouahrani-Bettache, S., Layssac, M., Teyssier J., and Liautard, J.P. (1999). Constitutive and inducible expression of green fluorescent protein in *Brucella suis*. *67*, 6695–6697.

Kohler, S., Foulongne, V., Ouahrani-Bettache, S., Bourg, G., Teyssier, J., Ramuz, M., and Liautard, J.P. (2002). The analysis of the intramacrophagic virulome of *Brucella suis* deciphers the environment encountered by the pathogen inside the macrophage host cell. Proc. Natl. Acad. Sci. U.S.A. *99*, 15711–15716.

Kühner, S., van Noort, V., Betts, M.J., Leo-Macias, A., Batisse, C., Rode, M., Yamada, T., Maier, T., Bader, S., Beltran-Alvarez, P., *et al.* (2009). Proteome organization in a genome-reduced bacterium. Science *326*, 1235–1240.

Lamontagne, J., Béland, M., Forest, A., Côté-Martin, A., Nassif, N., Tomaki, F., Moriyón, I., Moreno, E., and Paramithiotis, E. (2010). Proteomics-based confirmation of protein expression and correction of annotation errors in the *Brucella abortus* genome. BMC Genomics *11*, 300.

Li, H., Ruan, J., and Durbin, R. (2008). Mapping short DNA sequencing reads and calling variants using mapping quality scores. Genome Res. *18*, 1851–1858.

Liang, W., and Deutscher, M.P. (2010). A novel mechanism for ribonuclease regulation: transfer-messenger RNA (tmRNA) and its associated protein SmpB regulate the stability of RNase R. J. Biol. Chem. *285*, 29054–29058.

Ochman, H., and Raghavan, R. (2009). Systems biology. Excavating the functional landscape of bacterial cells. Science *326*, 1200–1201.

Rossetti, C.A., Galindo, C.L., Lawhon, S.D., Garner, H.R., and Adams, L.G. (2009). *Brucella melitensis* global gene expression study provides novel information on growth phase-specific gene regulation with potential insights for understanding *Brucella*:host initial interactions. BMC Microbiol. *9*, 81.

Rutherford, K., Parkhill, J., Crook, J., Horsnell, T., Rice, P., Rajandream, M.A., and Barrell, B. (2000). Artemis: sequence visualization and annotation. Bioinformatics *16*, 944–945.

Sangari, F.J., Agüero, J., and García-Lobo, J.M. (2000). The genes for erythritol catabolism are organized as an inducible operon in *Brucella abortus*. Microbiology *146*, 487–495.

Scott, W.G., Martick, M., and Chi, Y. (2009). Structure and function of regulatory RNA elements: ribozymes that regulate gene expression. Biochim. Biophys. Acta *1789*, 634–641.

Sharma, C.M., Hoffmann, S., Darfeuille, F., Reignier, J., Findeiss, S., Sittka, A., Chabas, S., Reiche, K., Hackermüller, J., Reinhardt, R., *et al.* (2010). The primary transcriptome of the major human pathogen *Helicobacter pylori*. Nature *464*, 250–255.

Smith, L.D., and Heffron, F. (1987). Transposon Tn5 mutagenesis of *Brucella abortus*. Infect. Immun *55*, 2774–2776.

Sieira, R., Comerci, D.J., Pietrasanta, L.I., and Ugalde, R.A. (2004). Integration host factor is involved in transcriptional regulation of the *Brucella abortus virB* operon. Mol. Microbiol. *54*, 808–822.

Tatusov, R.L., Fedorova, N.D., Jackson, J.D., Jacobs, A.R., Kiryutin, B., Koonin, E.V., Krylov, D.M., Mazumder, R., Mekhedov, S.L., Nikolskaya, A.N., *et al.* (2003). The COG database: an updated version includes eukaryotes. BMC Bioinformatics *4*, 41.

Uzureau, S., Lemaire, J., Delaive, E., Dieu, M., Gaigneaux, A., Raes, M., De Bolle, X., and Letesson, J. (2010). Global analysis of quorum sensing targets in the intracellular pathogen *Brucella melitensis* 16 M. J. Proteome Res. *9*, 3200–3217.

Viadas, C., Rodríguez, M.C., García-Lobo, J.M., Sangari, F.J., and López-Goñi, I. (2009). Construction and evaluation of an ORFeome-based *Brucella* whole-genome DNA microarray. Microb. Pathog. *47*, 189–195.

Viadas, C., Rodríguez, M.C., Sangari, F.J., Gorvel, J., García-Lobo, J.M., and López-Goñi, I. (2010). Transcriptome analysis of the *Brucella abortus* BvrR/BvrS two-component regulatory system. PLoS ONE *5*, e10216.

Vivancos, A.P., Güell, M., Dohm, J.C., Serrano, L., and Himmelbauer, H. (2010). Strand-specific deep sequencing of the transcriptome. Genome Res. *20*, 989–999.

Yus, E., Maier, T., Michalodimitrakis, K., van Noort, V., Yamada, T., Chen, W., Wodke, J. A. H., Güell, M., Martínez, S., Bourgeois, R., *et al.* (2009). Impact of genome reduction on bacterial metabolism and its regulation. Science *326*, 1263–1268.

What Have we Learned from *Brucella* Proteomics?

6

Esteban Chaves-Olarte, Caterina Guzmán-Verri, Eustache Paramithiotis and Edgardo Moreno

Abstract

Members of the genus *Brucella* infect and cause disease in a wide variety of mammals, including humans. Despite this host diversity, the clinical and pathological manifestations of brucellosis seem to be conserved within certain range, being the reproductive tract the main target of infection. The pathogenesis of brucellosis is dependent on the ability of the bacterium to invade and replicate within the endoplasmic reticulum of epithelial and phagocytic cells. From this perspective, the use of comprehensive system biology approaches, like proteomics, has contributed to dissect and unravel some aspects of the life cycle of *Brucella* organisms. In this review, we describe the advantages and limitations of the proteomic approaches employed in the field of brucellosis to discern, at the cellular and the molecular level, the pathogenesis of brucellosis. The emerging picture is that of a pathogen that does not rely on discrete virulence determinants to establish successful infections, but rather of a microbe that utilizes synchronized complex metabolic and signalling systems to reach its replicating niche at the endoplasmic reticulum of host cells. These systems include secretion machineries like the type IV secretion VirB system and signalling complexes like the two component regulatory BvrR/BvrS system and the quorum sensing transcriptional regulator VjbR, all of which have a profound impact on the metabolic state of the bacterium and particularly on the homeostasis of the cell envelope. In addition, we also discuss how the systematic comparison at the proteomic level of different strains, species and mutants within the *Brucella* genus has improved the experimental approaches used for diagnosis, taxonomy and phylogeny.

Introduction

The life cycles followed by the facultative extracellular intracellular *Brucella* pathogens are highly complex, since these organisms must adapt to two very different environments (Moreno and Moriyón, 2002). At any given moment of the *Brucella* life cycle, numerous bacterial molecules are interacting with each other, either as singly entities or as part of multimolecular complexes that may vary according to their growth conditions. The interactions ultimately form networks, often with temporal and spatial elements to them. The arrangements themselves may functionally overlap, resulting in redundancy and co-regulatory feedback control. Although brucellologists have long assumed the existence of such complex relationships, the tools to directly interrogate many of their components simultaneously were not available until recently. The great potential of high-throughput technologies such as genomics, transcriptomics and proteomics, is that they enable insights at scales of the biological systems that otherwise would not be possible.

For instance, *in silico* analysis of the *Brucella* genomes has been highly valuable in predicting evolutionary pathways and functionality of the genus (Moreno and Moriyón, 2002; Tsolis *et al.*, 2002). The genomes of *Brucella* species are composed of two chromosomes (with the exception of *B. suis* biovar 3) estimated to code close to 3200 proteins with an average size of 49 kDa

(DelVecchio *et al.*, 2001; Paulsen *et al.*, 2002; Halling *et al.*, 2005). These numbers contrast to the 6275, 4444 and 1488 putative proteins coded in the genomes of *Sinorhizobium meliloti, Ochrobactrum anthropi* and *Bartonella henselae*, respectively all α-Proteobacteria that are the closest relatives of *Brucella* organisms. Despite these differences the comparison of the predicted *Brucella* proteins is in close agreement with the current taxonomy, phylogeny and biology of the genus (Lavín *et al.*, 2010). In addition, some functional correspondence among the orthologous proteins in the different α-Proteobacteria has been found (LeVier *et al.*, 2000; Barquero-Calvo *et al.*, 2009) and the putative codon expression analysis of the *Brucella* genome has predicted a higher synthesis of proteins coded in chromosome I (Chr I) than in chromosome II (Chr II) (Karlin *et al.*, 2003; see also Chapter 5). Indeed, proteins involved in the tricarboxylic acid cycle, aerobic respiration, chaperones and outer membrane composition (coded in Chr I) are predicted to be expressed at higher levels than those devoted to anaerobic pathways of glycolysis, fermentation and or flagellum synthesis (coded in Chr II) (Karlin *et al.*, 2003). As expected for intracellular bacteria that do not harbour plasmids, lysogenic phages or conspicuous recombinant machinery (Moreno 1998), the analysis of the putative *Brucella* open reading frames shows that the uptake of DNA is a rare event that has not occurred for eons (Wattam *et al.*, 2009; Bohlin *et al.*, 2010). Last but not least, computational analysis of the putative proteins coded in the *Brucella* genome have suggested functional clusters related to adaptation to different environments and virulent functions (DelVecchio *et al.*, 2002; Paulsen *et al.*, 2002; Halling *et al.*, 2005; Tsolis *et al.*, 2009). As expected, all these predictions have been highly valuable, particularly when they correlate with phenotypic characteristics obtained through observation and experimentation, ultimately explaining the biology of this intracellular pathogen.

Within this context, high-throughput analytical methods such as transcriptomics and proteomics are some of the avant-garde experimental strategies used to understand the overall molecular phenotype of bacterial organisms. The results acquired by transcriptomics do not necessarily correspond with those obtained by proteomics (Lamontagne *et al.*, 2007; Viadas *et al.*, 2010) a fact that is commensurate with the various levels of translation, different regulatory activities and homeostasis of an organism (De Groote *et al.*, 2007). Since the proteome corresponds to the framework over a large part of the structural, synthetic and metabolic properties of the bacterium function, its elucidation more readily represents and relates to the overall bacterial phenotype.

Considering that high-throughput proteomics is a rapidly evolving field, this review will concentrate in the most significant findings obtained in *Brucella* proteomic research during the last decade, and will attempt to describe some relevant models on the biology of this important intracellular pathogen. For other reviews on *Brucella* proteomics, the works of DelVecchio *et al.* (2002) and DelVecchio and Mujer (2004) are available.

On the verge of various proteomic strategies

Biological samples tend to be too complex to be meaningfully analysed without some form of sample fractionation and simplification. Multiple approaches exist to accomplish this. One effective method is to prepare subcellular compartments such as outer membrane fractions (OMF) or cell envelopes prior to proteomic analysis. This approach not only reduces the sample complexity, but the data generated from such samples retain protein localization information that may become particularly useful for interpretation of the results. Even with this approach, additional sample fractionation and simplification is often necessary. In the second half of the twentieth century, electrophoretic or chromatographic separations of proteins in combination with immunodetection of the separated proteins were the prime methods used to characterize complex protein mixtures. However, the ultimate identification of the individual protein was only possible by the relatively slow direct sequencing by the Edman degradation. In the last decade, the analysis of complex samples of proteins was sped up by arrival of large-scale study techniques now called proteomics.

Presently, the two most common sample fractionation and separation strategies employ either polyacrylamide gel-based or liquid chromatography (LC)-based methods to separate the sample into fractions for subsequent analysis. Both fractionation approaches use widely available, established technologies, and have been extensively discussed in the literature, though sometimes with contradictory conclusions. Our laboratories have extensive experience with one-dimensional (SDS-PAGE) and two-dimensional gel (2DGE) electrophoresis-based protein separation, as well as LC-based fractionation of either intact proteins or their proteolytically derived peptides.

The choice of sample fractionation method should not be made in isolation, but after due consideration of several key factors. The objective of the experiment is one key factor. Some types of protein post-translational modifications can be more readily visualized using 2DGE. For example, if protein splice variant analysis is the experimental objective, a 2DGE based method may be the most effective approach to obtain that data. On the other hand, if sample quantities are limiting, or the proteins of interest are membrane-bound, a liquid chromatography-based method might be the optimal approach, since it will allow a greater yield of data per microgram of sample. In addition to the experimental objectives, the amounts of sample required and the effort necessary to generate it are also key parameters. Familiarity in using the technology involved is another criterion, as is the cost of sample preparation. A conceptually similar argument was presented by Patel *et al.* (2009) with respect to using labelled or label-free mass spectrometry data acquisition methods.

Before we further explore the key distinctions between gel-based and LC-based sample fractionation methods, it should be noted that they are merely one part of a multistep process. Proteomics is a multidisciplinary field. Successful execution of proteomics experiments is not thus dependent on any one technology selected, but rather on the effective integration of multiple, often unrelated, technologies. Sample preparation and fractionation are indeed key considerations, but no less so than executing accurate and reproducible mass spectrometry analyses, executing proper integration and evaluation of the raw mass spectrometry data collected or to achieving the correct annotation of protein functions. Effective data visualization must also be considered here as well, especially if the results generated by the proteomics experiments are to be meaningful to a broad range of scientists. Thus, if proteomics is considered in this light, as a multistep process meant to generate reproducible and interpretable results at the end, then more systemic considerations arise. An example is quality control. Good quality control at each step of the entire process, from the initial sample preparation to the final statistical analyses, will have far higher impact on the quality of the final data than the selection of any one individual analytical method. Furthermore, the more steps a process has, especially if those include manual interventions or subjective judgement decisions, the greater variability and non-reproducibility the final product will have. This consideration is not applicable only to proteomics, but to any multistep process, whether manufacturing, analytical, or scientific. Thus, processes with the fewest most automatable steps will generate the most reproducible results.

Specifically with respect to sample fractionation methods for proteomics, all gel-based methods have lower sample loading capacity, comparatively lower resolution, and lower dynamic range than LC-based methods. Two-dimensional gel separation methods additionally suffer from poor resolution of hydrophobic proteins, including those with transmembrane domains, as well as more gel-to-gel variation than the other approaches, clearly a consequence of the extent of manual interventions required for this method. The analytical limitations of 2DGE have improved with the introduction of two-dimensional differential gel electrophoresis, which utilizes fluorescent dyes to label the analytical samples prior to electrophoresis (Lai, 2010). Although this approach has improved 2DGE resolution, sensitivity, and reproducibility, these improvements do not match the performance of one dimensional-gel based methods, much less the performance of LC-based methods.

LC-based methods may utilize either intact proteins, or peptides derived from enzymatic digestion of the proteins in the sample. Of these two methods, peptide-based fractionation typically using strong cation-exchange chromatography, has far better performance than intact protein fractionation that typically utilizes size exclusion chromatography or reverse-phase high-performance liquid chromatography. Strong cation exchange-based peptide fractionation is thus the preferred approach, with the consequence that proteomics processes that use this approach enzymatically digest their samples prior to fractionation. In contrast, gel-based proteomics processes separate intact, denatured, proteins and either extract them from gel plugs for digestion, or perform in-gel digests and then extract the resulting peptides for analysis. An advantage of in-gel digestion is that it allows the removal of detergents and salts that could hinder mass spectrometry analysis (Alvarado *et al.*, 2010). However, there are also many disadvantages, chief of which is that in-gel digestion is susceptible to large and variable losses. These arise mainly from the destaining process prior to the in-gel digestion, which may include a gel de-hydration step to aid the uptake of the proteolysis solution, and the incomplete peptide extraction from the gel after proteolysis (Granvogl *et al.*, 2007; Duan *et al.*, 2009). Losses ranging from 15% to 50% have been reported (Stewart, 2001), consistent with our own unpublished observations of 35% loss. The losses clearly increase sample quantity requirements. However, variability in the amount lost from gel to gel is a bigger issue because this can more seriously affect a study's statistical power and interpretation.

Most studies utilizing one dimensional-gel-based fractionation cut the gel lane into multiple slices, typically 20–40, and perform in-gel digestions of these slices in parallel. Proteins that span the borders of the gel slices will therefore be found in multiple gel fractions. Quantitation of these cross-slice proteins is difficult, and requires additional calculations and normalizations to reduce the effect of these proteins on the study's overall variability (Cañas *et al.*, 2007; Cutillas *et al.*, 2007). The effect of the cross-slice proteins appears to increase in direct proportion with the number of slices generated. Thus depending on the experimental design, the cross-slice proteins may affect a substantial portion of the data set generated. Liquid chromatography methods that generate multiple sample fractions are also susceptible to this cross-fraction effect. The optimal solution for comparative proteomics analyses appears to be to generate as few fractions as possible. This might be more easily attainable with liquid chromatography-based methods which have greater loading capacity per fraction than gel based methods.

Studies that have analysed the same samples using gel-based and liquid chromatography-based methods have directly demonstrated that the LC-based approaches were more effective (Mastroleo *et al.*, 2009; Patel *et al.*, 2009). The LC methods used in these studies yielded more protein identifications than the gel-based methods per microgram sample input, had higher confidence protein identifications, and took less time to complete. Although other investigators reported results that indicated fewer comparative advantages for LC-based methods (Gauci *et al.*, 2009), the observations made by Mastroleo and Patel match our own experience with *Brucella* proteome analysis (Lamontagne *et al.*, 2007, 2009, 2010).

Proto-proteomics of *Brucella* extracts

In addition to the lipopolysaccharide (LPS) and structurally related native hapten polysaccharide (NH) that have been described as the main *Brucella* antigens (Moreno and Moriyón, 2006), the species of this bacterial genus possess a large collection of proteins that have been demonstrated to induce antibody responses in infected and immunized animals (Olitzki, 1960; Raybould, 1982; Cherwonogrodzky *et al.*, 1990; Cloeckaert *et al.*, 1996; Moriyón and López-Goñi, 1998). In the early days of proto-proteomic analysis, up to 20 precipitin lines against proteins were demonstrated by immunoelectrophoresis (IEP), being most if not all of these proteins, shared by the different *Brucella* species (Díaz and Chordi, 1966; Díaz *et al.*, 1967, 1968). This fact clearly reinforced the taxonomy based in pure bacteriological and metabolic markers (Moreno and Moriyón, 2006). Moreover, one of the relevant arguments used to include the naturally occurring 'rough' *B. ovis* and *B. canis* (that do not possess the classical A and M determinants) within the genus, was the extensive cross reaction displayed by the proteins of these two species with those of the traditional smooth strains (Díaz *et al.*, 1967, 1968). Therefore, this early proto-proteomics helped to resolve some significant controversies regarding the taxonomical position of *Brucella* species.

The IEP analysis also contributed to understand some significant differences in the antibody responses against proteins in infected versus vaccinated animals, leading to the immunochemical characterization and purification of protein antigens (Schurig *et al.*, 1978, 1981; Raybould 1982). One major finding resolved by traditional immunochemical procedures was the dissociation of hypersensitivity reactions, unequivocally assigned to proteins and not to *Brucella* sugar containing molecules (Jones *et al.*, 1973; Jones and Berman 1975; Velasco *et al.*, 1997). In addition, IEP and other immunoprecipitation assays with protein fractions were useful in the diagnosis of chronic cases of human brucellosis (Díaz *et al.*, 1976, 1982), as well as canine and ovine brucellosis caused by *B. canis* and *B. ovis*, respectively (Myers *et al.*, 1972, 1974). An important contribution performed by these precipitation tests, was the finding that cross reaction of smooth *Brucella* with *Yersinia enterocolitica* O:9 (infecting cattle and causing diagnostic uncertainties), was due to the O-polysaccharide antigenic determinants of the LPS and NH, and not to membrane or cytosolic proteins (Díaz and Dorronsoro, 1971; Díaz and Bosseray, 1974).

A significant number of experimental works on the use of *Brucella* protein antigens for the diagnosis of cattle brucellosis were published (Olitzki, 1960; Raybould, 1982). However, none of these assays have demonstrated to be practical. The attempts to find a 'magic' protein antigen for differentiating smooth *Brucella*-infected animals from vaccinated individuals have been finally relegated as experimental relics, mainly because the sensitivity and specificity of protein based assays did not rival those developed with LPS or NH (Stemshorn, 1984; Moreno and Moriyón, 2006; Greiner *et al.*, 2009). As stated before, conspicuous exceptions are the protein antigens used for the diagnosis of canine and ovine brucellosis caused by natural rough *Brucella*, and the use of protein antigens in the diagnosis of chronic human brucellosis (Díaz *et al.*, 1976, 1982).

The introduction of other separation techniques in brucellosis research such as SDS-PAGE, isoelectric focusing and LC-based technologies, supported the findings revealed by the traditional immunochemical analysis of proteins and reinforced the classical taxonomical relatedness among the different *Brucella* species and the close phylogenetic relationship with other α-Proteobacteria (Morris, 1973; Tabatabai and Deyoe, 1984; Verstreate and Winter, 1984; Hill and Cook, 1994; Ewalt *et al.*, 1994; Velasco *et al.*, 1997, 1998; Vizcaíno *et al.*, 1997; Cloeckaert *et al.*, 1999). SDS-PAGE and isoelectric focusing not only enlarged the number of proteins demonstrated by chromatographic and immunoprecipitation procedures, but it was essential to describe for the first time the different groups of outer membrane proteins (Omps), which owing to their insolubility were precluded from traditional immunochemical analysis. It should be noticed that the conclusions generated by these proto-proteomic analysis were further confirmed and expanded by high-throughput analytical methods (Gupta and Monk, 2007). With the arrival of western blotting, monoclonal antibodies and immunoelectron microscopy to the field of brucellosis, combined with the use of more defined bacterial fractions, the number of Omps was more accurately described. Moreover, the topology and immunogenicity of these proteins was placed in their right context, revealing also the complexity of the *Brucella* cell envelope (Douglas *et al.*, 1984; Cloeckaert *et al.*, 1990, 1996, 1999; Kittelberger *et al.*, 1995).

Amino terminal sequencing of purified released proteins (Boigegrain *et al.*, 2004), cloning of different genes coding for specific proteins involved in function such as heat shock (Lin and Fitch, 1992) or oxidative metabolism (Bricker *et al.*, 1990), and those identified as immunogenic proteins by antibodies (Cloeckaert *et al.*, 1990, 1996, 1999) or cells (Brooks-Worrell *et al.*, 1992a,b), were the first efforts for relating protein function with specific signatures. Complementary indirect approaches to identify protein functions were carried out by mutagenesis of putative protein coding genes (Halling *et al.*, 1991; Latimer *et al.*, 1992; Tatum *et al.*, 1993), a strategy that has been expanded during the last two decades (Ficht *et al.*, 2010). Although valuable in many respects, mutagenesis is not straight forward, mainly owing to the pleiotropic or silence mutations. These phenomena make it difficult to correlate the absence of one protein with direct functionality, as evidenced by high-throughput analytical methods, in which the absence of one protein may cause overall changes in the proteomics and phenotype of the bacterium (Lamontagne *et al.*, 2007; Uzureau *et al.*, 2010; Wang *et al.*, 2010).

Significant efforts to detect *Brucella* putative 'virulence' factors released under acidic conditions (expected to be found within the intracellular vacuolar milieu), were undertaken by incubating *B. suis* and *B. melitensis* in acetate buffer at pH 4.0 medium (Boigegrain *et al.*, 2004). Under cultured conditions, *Brucella* seems to release soluble periplasmic proteins and OMF. The proteins present in these fluids were identified by traditional SDS-PAGE, Western blotting with monoclonal antibodies and Edman degradation. From the periplasmic subset, 13 proteins were identified while the OMF were enriched in Omps type III family, mainly Omp25 (also known as Omp3a) and Omp31. Interestingly, under these conditions the release of the identified periplasmic proteins was independent of the VirB type IV secretion system. However, it is important to point out that *Brucella* cells barely replicate in acidic media bellow pH 5.5 (Moreno and Moriyón, 2006). The analysis of *Brucella* OMF and secreted proteins by high-throughput proteomics have expanded some of these findings (Lamontagne *et al.*, 2007; Delpino *et al.*, 2009).

The early work in proto-proteomics has then contributed to the taxonomical classification of the different species within the genus as well as to the finding of protein antigens suitable for differentiation between infected and vaccinated animals. In contrast to other bacterial pathogens, this work did not allow the discovery of discrete virulence determinants as protein toxins for instance, an early indication of the particular pathogenic strategy used by *Brucella*. However, it generated relevant information for the understanding of the composition of the *Brucella* outer membrane.

The era of *Brucella* proteomics

Unexpectedly, the first 2DGE analysis in brucellosis research was not for the identification of proteins but for the characterization of *Brucella* LPS heterogeneity (Sowa *et al.*, 1986). Some of the pioneering investigations for analysing proteins by 2DGE were carried out to compare *B. abortus* strains (Sowa *et al.*, 1992), and for the identification of antigenic *Brucella* proteins recognized by lymphocytes from vaccinated cattle (Brooks-Worrell and Splitter, 1992). The proteomic era in brucellosis research in which matrix-assisted laser desorption ionization mass spectrometry (MALDI-MS) was used for determination of protein sequences isolated from 2DGE (Guzmán-Verri *et al.*, 2002; Mujer *et al.*, 2002; Wagner *et al.*, 2002), converged with the elucidation of the fist *Brucella* genomic sequences (Sánchez *et al.*, 2001; DelVecchio *et al.*, 2002). As expected, the availability of these and other gene sequences from related bacteria greatly facilitated the powerful search engine that uses mass spectrometry data to identify proteins and algorithms to assign the corresponding putative functions.

The first broad proteomic analysis of *Brucella* proteins using MALDI-MS strategies was carried out for the examination of soluble extracts of *B. melitensis* 16M grown for three days in solid media (Wagner *et al.*, 2002). In this study, close to 1160 protein spots detected, from which 440 spots were unambiguously identified by MALDI-MS. However, more than half of these spots matched with protein repetitions, reducing the number of identified proteins to 187, corresponding just to 6% of the predicted *Brucella* open reading frames. In this analysis, one inner membrane and ten Omps proteins were positively identified, demonstrating the improvement of the separation and the supremacy of MALDI-MS sequencing over other techniques. However, it was also evident that the redundancy of proteins distributed in different spots in the 2DGE would be part of the complications confronted during the proteomic analysis achieved by this method. It is relevant to highlight that what biologists need are proteins, and not spots. Indeed, the number of spots in a 2DGE seldom equals the number of proteins and frequently many spots may represent just one or a few proteins (Guzmán-Verri *et al.*, 2002).

Immunoproteomics of *Brucella* extracts

Noticeably, the first immunoproteomic approach for identifying proteins separated in 2DGE was for the discovery of nitrocellulose blotted *B. abortus* antigenic proteins recognized by lymphocytes from S19 vaccinated bovines (Brooks-Worrell and Splitter 1992a,b). The isolated sections containing the individual proteins were then assayed for lymphocyte activation, identifying 38 spots that induced cell proliferation. One of the proteins identified by N-terminal microsequencing was a ribosomal protein while a second one did not match with any of the known *Brucella* putative protein sequences.

The first significant proteomic trial for determination of large set antigenic proteins detected by antibodies from naturally infected animals was performed with proteins from *B. ovis* (Texeira-Gomes *et al.*, 1997b). By combination of 2DGE and Western blotting, close to 82 protein spots were recognized with sera from *B. ovis* infected rams, expanding the spectrum of antigenic proteins detected previously by other techniques. The more significant immunoreactivity was observed with proteins ranging between 20 and 90 kDa on the acidic side (pI 4.6–5.5). Some of these proteins were identified by monoclonal antibodies while others by N-terminal microsequencing. Among the proteins with the strongest immunoreactivity were the DnaK, GroEL, NikA, dihydrolipoamide succinyltransferase, and BP26. It is significant that most of these proteins are foreseen to be highly abundant in *Brucella,* according to the prediction of gene expression levels from codon biases in α-Proteobacteria (Karlin *et al.*, 2003).

In a similar immunoproteomic analysis performed with enriched cell envelope fractions from virulent *B. abortus* 2308, a total of 19 proteins out of the 163 identified, reacted against sera from *Brucella* infected human or bovine individuals (Connolly *et al.*, 2006). From these, six were Omps. The most conspicuous protein was the porin Omp2b, while four (Omp25, Omp25c, Omp22, Omp32b) belonged to the classical group III proteins defined by SDS-PAGE (Moriyón and López-Goñi, 1998). As expected, all Omps except one (68 kD) were identified in several spots, revealing the heterogeneity in the migration pattern of these hydrophobic proteins. In a comparable work, six proteins out of 383 spots were recognized with hyperimmune rabbit serum in several *B. abortus* S19 extracts (Al Dahouk *et al.*, 2006). Again, most of the antigenic proteins identified were distributed in more than one spot, suggesting that many of the protein spots recognized in previous works by combination of 2DGE and Western blotting with sera from infected animals (Texeira-Gomes *et al.*, 1997a,b) also corresponded to repetitions. The distribution of the same protein in several distinct spots is regularly observed with membrane proteins and chaperones, probably owing to the aggregation tendencies of these proteins, and/or interaction with other molecules such LPS and membrane lipids, and other cell envelope components. Some non-exclusive options are incomplete unfolding, partial degradation and secondary protein modification such as phosphorylations.

Comparison of proteomes between *Brucella* strains

The work of Sowa *et al.*, (1992) compared the proteomes of *B. abortus* virulent 2308 and attenuated vaccine S19 strains and their corresponding rough counterparts RB51 and S19M3 strains. In this study, the 2DGE analysis was performed with metabolically radioactive labelled proteins with ^{35}S-methionine, a procedure that ensured higher sensitivity. Indeed, close to one thousand spots from *B. abortus* extracts were detected by autoradiography. Assuming that each spot corresponded to a single different protein (which is incorrect), then, about one third of the total predicted *Brucella* proteome could be counted. At that time the interpretation was that significant quantitative (>10 times) or qualitative differences between the proteomes of the virulent and attenuated brucellae could identify those proteins related to virulence. By this procedure 14 spots were tentatively identified as 'putative virulence proteins'. In addition, 28 spots recognized by paired comparisons in both smooth and rough strains, were tentatively assigned as relevant for the biosynthesis of *Brucella* LPS. This strategy was valuable in describing for the first time significant proteome variations that could be related to differential protein synthesis which in turn would be related to different metabolic adaptations between the *Brucella* strains of high and low virulence, pointing out some significant number of candidate targets. Unfortunately, the small amounts of protein available in this radioactive labelling system precluded the isolation and characterization of these differentially expressed proteins.

The first significant identification of stained proteins by 2DGE using a combination of immunoblotting with amino terminal sequence of isolated spots was performed with rough *B. melitensis* B115 and *B. ovis* strains (Texeira-Gomes *et al.*, 1997a,b). The use of protein extracts obtained from rough strains has the advantage over those obtained from smooth strains that the rough LPS migrates to the bottom of the gel, generating less distortion of the electrophoretic protein patterns. Indeed, owing to its polymeric nature and the abundance of the *Brucella* smooth LPS complexes, these molecules migrate as smears in the 2DGE. Since LPS strongly interacts with some membrane proteins (Moriyón and Berman, 1983; Moriyón and López-Goñi, 1998), it frequently generates some distortion and lower resolution of the protein spots (Gamazo *et al.*, 1989). The 2DGE protein map of *B. melitensis* B115 and *B. ovis* revealed close to 600 protein spots, from which 25 and 21 were positively identified, respectively. In addition, seven proteins with different expression levels were readily recognized between both strains. The differences observed between both brucellae were assumed to be related to species differences. The protein spots recognized by immunoblotting with monoclonal antibodies were 89 kDa Omp, DnaK, bacterioferritin, CP24, and BP26, while the rest were identified by N-terminal microsequencing. Among this last group, the most relevant were the DnaK, GroEL GroES, bacterioferritin, malate dehydrogenase, Cu-Zn SOD, ribosomal L7/L12, succinyl-CoA synthetase, NikA, BP26, ClpP, Fe-Mn SOD, the nucleoside diphosphate kinase, amino acid ABC transporter and a subunit of electron transfer flavoprotein. Seven proteins corresponded to unknown proteins. Again, according to codon biases analysis, most of these proteins are predicted to be among the most abundant in α-Proteobacteria, including *Brucella* (Karlin *et al.*, 2003).

The first comparison between *Brucella* strains by high-throughput MALDI-MS proteomics was for evaluating the differences between the virulent *B. melitensis* 16M and the attenuated counterpart vaccine *B. melitensis* Rev1 (Mujer *et al.*, 2002). In this study, 15 proteins were differentially expressed between the strains (five increased in 16M and ten increased in Rev1), demonstrating again the various metabolic alternatives between *Brucella* of high and low virulence. More recently, the comparison between *B. abortus* 2308 and *B. melitensis* 16M grown to late stationary phase was carried out by MALDI-MS (Eschenbrenner *et al.*, 2006). In this analysis, 11 spots (corresponding to ten identified proteins) were overexpressed and 40 spots (34 proteins) were underexpressed in *B. abortus* 2308 as compared with the same proteins in *B. melitensis* 16M. Similarly, 45 spots (42 proteins) and 28 spots (27 proteins) were electrophoretically unique protein spots in *B. abortus* 2308 and *B. melitensis* 16M, respectively. With the exception of two conspicuous Omps, all the other differentially expressed spots corresponded to cytoplasmic and periplasmic proteins. Again, this fact stresses the limitations in resolving hydrophobic membrane proteins by 2DGE. The discrepancies observed between both bacterial species were attributed to proteins involved in membrane transport as well as secretion systems. However, the analysis did not reveal significant variations in group III Omps (Omp25/Omp31) between these two strains (Martin-Martin *et al.*, 2009). Moreover, the results did not bring light to the traditional differences observed between these two *Brucella* species in the oxidation of amino acids and carbohydrates (Meyer and Cameron, 1961). In any case, the proteomic differences observed between these two closely related bacteria reinforced the traditional concept that they correspond to two distinct species.

The comparative studies described in this section have contributed with valuable information regarding the identity of proteins differentially expressed by the various *Brucella* strains and have reinforced the biochemical and biological differences observed among strains of high and low virulence. The proteomes reported have been also useful for the correct annotation of the *Brucella* genomes. However, owing to the complexity of the survival strategies used by *Brucella*, it was difficult to gain insight into the mechanisms related with virulence.

Brucella proteomes and virulence systems

Three molecular systems crucial for the intracellular lifestyle of *Brucella* have been characterized by proteomic strategies: the two component regulatory system BvrR/BvrS, the type IV secretion system VirB and the quorum sensing transcriptional regulator VjbR. It is worth nothing that the involvement of these three systems in the virulence of *Brucella* was discovered through the generation of mini-Tn5 mutants and subsequent screening of smooth bacteria with attenuated phenotypes in cells and mice. The mini-Tn5 *B. abortus* containing independent insertions in the *bvrS* (coding for the sensor protein) or *bvrR* (coding for the regulatory protein) contiguous genes were initially selected for two characteristics: smooth phenotype and increased sensitivity to cationic peptides (Sola-Landa *et al.*, 1998). The complete biological characterization of these mutants determined that they were highly attenuated, presenting defects in cell invasion and multiplication in animal cells and unable to replicate in mice. The *bvrS* and *bvrR* mutants do not display polar defects, and reconstitution with either the regulator or sensor coding plasmids in trans, re-establishes the virulent phenotype and cell envelope characteristics of the mutants (Sola-Landa *et al.*, 1998; Guzmán-Verri *et al.*, 2002; Chaves-Olarte *et al.*, 2003; Manterola *et al.*, 2005, 2007; Lamontagne *et al.*, 2007; Martínez-Nuñez *et al.*, 2010; see also Chapter 10). The second genetic arrangement involved in virulence identified by mini-Tn5 mutagenesis was the type IV secretion VirB system (Foulongne *et al.*, 2000; Delrue *et al.*, 2001; Kim *et al.*, 2003). This secretion machinery has been repeatedly reported to be crucial for the trafficking of *Brucella* to the endoplasmic reticulum (O'Callaghan *et al.*, 1999; Sieira, 2000; Comerci *et al.*, 2001; see also Chapter 11). Finally, one of the screenings that identified mutants in the VirB operon, also picked up a mutant with an insertion in the gene encoding for the quorum sensing transcriptional regulator VjbR (Delrue *et al.*, 2001). This protein has been shown to control the transcription of the *virB* operon and other genes related to the intracellular survival of *Brucella* (Delrue *et al.*, 2005; De Jong *et al.*, 2008).

A significant proteomic assessment between *B. abortus* virulent and non-virulent *bvrR/bvrS* mutants was performed by comparing their purified OMF (Guzmán-Verri *et al.*, 2002). The BvrR/BvrS system, built of a histidine kinase sensor (BvrS) and regulatory (BvrR) protein, is homologous to other arrangements found in *Ochrobactrum, Agrobacterium, Sinorhizobium* and *Bartonella* (López-Goñi *et al.*, 2004). In *Brucella* this system is necessary for the invasion and intracellular survival of host cells (Sola-Landa *et al.*, 1998). It was determined by combination of 2DGE, MALDI-MS peptide mapping and monoclonal antibody probing that the phenotype of the *bvrR/bvrS* mutants correlated with the reduced number of group III Omps, specifically Omp3a and Omp3b (also known as Omp25 and Omp22). Later, it was determined that the *bvrS* and *bvrR* mutants also displayed LPS acylation differences in the lipid A. These modifications significantly influenced the cell envelope properties of the *Brucella* mutants, such as the hydrophobicity, permeability and sensitivity to bactericidal peptides (Sola-Landa *et al.*, 1998; Manterola *et al.*, 2005). Based on this, it was proposed that the BvrR/BvrS system was devoted to the homeostasis of the *Brucella* cell envelope. This proposal was further sustained by the overall proteomic analysis of OMF obtained from these two *B. abortus* mutants, the reconstituted plasmid-*bvrR* mutant (BvrRp) and the parental 2308 strains. However, an important distinction in the proteomic strategy was carried out: instead of analysing the OMF protein samples separated in 2DGE by MALDI-MS, the proteins in the purified OMF were first enzymatically digested, the resulting peptides separated by LC-based methods and then identified by extensive quantitative MS/MS analysis (Lamontagne *et al.*, 2007). In contrast to 2DGE separation, the LC-MS/MS based-method was able to identify many more proteins from the digested preparations and to estimate the amounts of proteins based on peptide intensities. Only the proteins differentially expressed among the various *Brucella* strains were considered.

Unsupervised principal component analysis of the peptides clearly separated the OMF proteomes of the various *B. abortus* strains (Fig. 6.1). Dendrogram analysis showed that the OMF proteins from the 2308-derived strains (BvrRp, BvrS, BvrR and RB51) were more closely related to each other than to the OMF of the vaccine *B. abortus* S19. Moreover, the OMF of the two avirulent *bvrS* and *bvrR* mutants, were closely related to each other. Similarly, the peptide profiles of the virulent 2308 and BvrRp were also closely related (Fig. 6.1), indicating reconstitution of the phenotype in the BvrRp revertant. Although the *B. abortus* RB51 rough strain (not included in the original analysis) clustered together with the 2308/BvrR/BvrS/BvrRp group, it was located farther apart. This revealed that the lack of O-polysaccharide and NH moieties on the bacterial surface significantly influences the composition of the outer membrane in relation to smooth phenotype, a fact that correlates with the atypical biological behaviour of this rough strain (Freer at al., 1996; Sola-Landa *et al.*, 1998; Moreno and Moriyón, 2006). Comparisons of peptide intensities across strains were able to identify those that were differentially expressed (Fig. 6.1). As noted, the disruption of *bvrS* or *bvrR* genes resulted in large scale but non-identical OMF protein changes. The reintroduction of wild type BvrR into the *bvrR*-mutant (BvrRp) reversed most of these changes (Fig. 6.1).

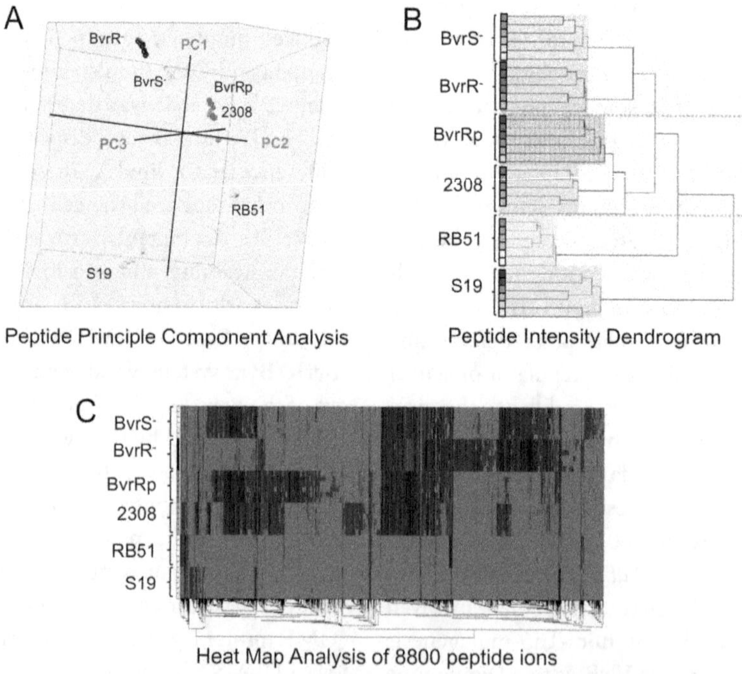

Figure 6.1 Comparative proteomic analysis. Peptides derived from the OMF of *B. abortus* strains were analysed by mass spectrometry (*n*=5–6 replicates per strain). (A) Principal component analysis was performed using the intensity values of all the peptide ions detected. Unsupervized sorting of blinded replicates along three axes of variance (PC1–PC3) was conducted. (B) Dendrograms were generated by hierarchical clustering using Pearson correlation and average linkage on the intensities of all the peptides detected. (C) Heat map analysis. Replicates are indicated along the x-axis and peptides are indicated along the y-axis. Log_2 values of the intensities of 8800 peptides that were differentially expressed in any two-strain combinations are displayed. The *B. abortus* 2308 wild type; the BvrS, and BvrR avirulent mutant strains, the reconstituted BvrRp strain and the attenuated vaccine strains smooth S19 and rough RB51 are indicated. Notice that the proteomes of bacteria derived from 2308 background are more closely related in relation to the proteome of unrelated vaccine strain S19.

By this quantitative LC-MS/MS proteomic approach, 167 differentially expressed proteins were identified, of which 25 were Omps. Approximately half of the Omps decreased in abundance, whereas half increased. Notably, expression of five group III Omps decreased whereas five lipoproteins increased in the *bvrS* and *bvrR* mutant strains. By contrast, in the periplasmic space approximately 80% of the 60 differentially expressed proteins augmented in at least one of the avirulent mutants. Many of the periplasmic proteins identified were primarily involved in substrate uptake and transport. Consequently, the uniform increase of this protein class in the *bvrR/bvrS* mutants suggested nutritional stress response, possibly owing to defective outer membrane function as well as carbon metabolism dysfunction (Lamontagne *et al.*, 2007; Dozot *et al.*, 2010; Viadas *et al.*, 2010). This phenomenon is reflected by the lower virulence and higher sensitivity of these mutants to bactericidal substances that primarily attack the outer membrane as well as the slower growth rate in relation to the virulent *B. abortus* counterpart (Martínez-Núñez *et al.*, 2010). As expected, virtually all proteins reverted to wild-type levels in the reconstituted virulent BvrRp strain (Lamontagne *et al.*, 2007). However, some proteins (e.g. VirB and BvrR) have been observed to be augmented in the BvrRp complemented strain (Martínez-Núñez *et al.*, 2010).

As expected, the wide range changes in cell envelope protein expression markedly correlated to the avirulent phenotype of *bvrS* and *bvrR* mutants, suggesting that, at least in this case (but very probably in other cases as well), *Brucella* virulence depends on regulatory networks controlling the cell envelope and metabolism rather than on discrete factors. Based on this, it was proposed that the BvrR/BvrS system could work as a master regulatory system to tune the cell envelope structure and physiology of the bacterium in order to adapt to the environmental changes that occur during the shift from the free-living to the intracellular environment (Martínez-Núñez *et al.*, 2010; see also Chapter 10). This proposal has been further supported by extensive transcriptomic and proteomic analysis (Dozot *et al.*, 2010; Viadas *et al.*, 2010).

The type IV VirB secretion apparatus of *Brucella* is relevant for controlling the intracellular trafficking of the bacterium, and mutants in this system are highly attenuated (Boschiroli *et al.*, 2002; Celli *et al.*, 2003). Therefore, it is not surprising that many efforts to discover proteins transported through the type IV VirB secretion apparatus have been carried out by different laboratories. Presently, only two proteins named VceA and VceC of unknown function have been unambiguously found to be translocated in a VirB-dependent manner (De Jong *et al.*, 2008). In a different proteomic approach, Delpino *et al.* (2009) analysed by 2DGE and MALDI-MS, liquid supernatants (free of bacteria and OMF) from early stationary phase cultures of *B. abortus* VirB10 mutant strain and the wild-type counterpart. In this study, a total of 17 bacterial proteins were identified in supernatant cultures: phosphoserine aminotransferase, aspartate aminotransferase, alkyl hydroperoxide reductase, 6-phosphogluconolactonase, SurA, LIVTA-binding protein, DnaK, choloylglycine hydrolase and cytosol aminopeptidase PNPase were present in supernatant cultures of the virulent *B. abortus* 2308 but absent in those of the *virB* mutants. In addition, five substrate periplasmic transporters and one TRAP Omp were observed in the culture supernatants of both strains. It was proposed that the *virB* genes could be involved, not only in the secretion, but also in the regulation of the expression of some of these proteins. It was also suggested that some of the proteins found in the culture supernatants may not be secretion substrates of the VirB apparatus, but released through other mechanisms. Regarding this, it is relevant that *virB* mutants display cell envelope alterations demonstrated by their higher susceptibility to cationic peptides (Wang *et al.*, 2009; Chaves-Olarte *et al.*, unpublished results), which may also influence the protein secretion pattern. Nevertheless, the proteins identified by this procedure constitute a good approximation and a valuable approach to understand the *Brucella* secretome.

A comparison between the proteomes of extracts obtained by sonication of virulent *B. melitensis* 16 M and the *virB* operon promoter deleted mutant counterpart, grown at middle logarithmic phase under limited stress nutrition and acid shock conditions, were analysed by 2DGE and MALDI-MS (Wang *et al.*, 2009). It was proposed that under this growth conditions the transcription of the *virB* operon was enhanced, incrementing the expression or VirB proteins. When the strains were compared, 59 protein spots were down-regulated and 36 up-regulated in the *virB* mutant, in comparison to the virulent *B. melitensis* 16M. In addition, 19 spots were uniquely expressed in the 16M strain, while 27 were found in the *virB* mutant. From the total number of identified proteins, three were only present in 16M, while three were exclusively observed in the *virB* mutant. Thirty-four were cytoplasmic, nine periplasmic, four Omps, one cell membrane and the rest corresponded to unknown proteins. Several proteins previously involved in attenuation were down-regulated in the *virB* mutant. By RT-PCR analysis, low level of transcription was shown for *vjbR, dnaK, htrA, omp25* and *gntR*. It is difficult to understand, however, how the absence of secretion machinery with a very narrow window of expression *in vitro* would affect the transcription/translation of these proteins. Transcription analysis of *virB* and *vjbR* genes at different growth phases showed that *vjbR* was positively influenced by the VirB system and that in a *vjbR* mutant the expression of *virB* was down-regulated. Based on this, it has been suggested that the *virB* operon is also positively influenced by VjbR (Wang *et al.*, 2009). Therefore, it seems that the VirB and VjbR systems positively influence each other may be forming a positive regulation circuit.

In a complementary study, the proteomics of sonic extracts pelleted by ultracentrifugation from *B. melitensis virB* mutants were compared with those of the 16M virulent strain (Wang *et al.*, 2010). In this analysis a total of 190 and 202 protein spots were detected by 2DGE in 16M and *virB* mutant, respectively. Seventy protein spots were down-regulated and 36 were up-regulated in the *virB* mutant. A total of 45 proteins were identified by MALDI-MS, indicating that many proteins were represented in more than one spot, mainly Omps. From these, 26 proteins were cytoplasmic (three up-regulated and 23 down-regulated), ten Omps (three up-regulated and ten down-regulated), three periplasmic (down-regulated), two inner membrane (down-regulated) and three unknown proteins (down-regulated). The various group III Omps (Omp25, Omp25c, Omp25b and Omp31) were not resolved in single spots but each Omp distributed in several spots in the gel. In general, these Omps were detected in lower amounts in the *virB* mutant. This fact was commensurate with the lower transcription of these genes in the *virB* mutant in relation to the virulent counterpart.

It was determined that the distinct proteomic phenotype of the *virB* mutant was also related to its higher sensitivity to the cationic polymixin B and decreased survival under oxidative, high-salt and high-osmolarity stresses (Wang *et al.*, 2010). These phenotypic characteristics reveal that the *virB* mutants, in addition of lacking functional T4SS, display outer membrane defects that may reflect the avirulent phenotype of this bacterium. These are interesting events, since mutants such as the BvrR/BvrS which also exhibit susceptibility to oxidation and cationic bactericidal molecules linked to outer membrane defects, display some proteomic similarities with the *virB* mutant. Indeed, from 29 common shared proteins, 16 were down-regulated and seven up-regulated in both mutant strains. Furthermore, the profiles of group III Omps and periplasmic proteins between the *virB* mutant and the *bvrR/bvrS* mutants were also similar (Lamontagne *et al.*, 2007; Wang *et al.*, 2009, 2010). In addition, it was observed that the *virB* mutant generated bacterial aggregates at a high cell density and produced exopolysaccharide (Wang *et al.*, 2010), a phenotype that is reminiscent to the *B. melitensis vjbR* mutant (Uzureau *et al.*, 2007) and the *B. melitensis aiiD*-overexpressing strain (Godefroid *et al.*, 2010). Linked to this is the fact that the two-component regulatory system BvrR/BvrS controls the expression of VirB through direct and indirect mechanisms, as well as the expression of the transcriptional activator VjbR (Martínez-Núñez *et al.*, 2010). It is clear, however, that BvrR/BvrS is necessary but not sufficient for VirB expression: even if the BvrR is overexpressed throughout the growth curve, VirB did not constitutively increase at early times. At this phase, probably, the negative regulators acting on the VirB promoter dominate over the positive input added by the BvrR (Sieira *et al.*, 2004; Martínez-Nuñez *et al*, 2010).

Brucella releases quorum sensing *N*-acyl-homoserine lactone molecules to synchronize gene expression within the bacterial population, a phenomenon that depends upon the two LuxR type regulator proteins named VjbR and BabR (also known as BlxR) (Rambow-Larsen *et al.*, 2008; Uzureau *et al.*, 2010; see also Chapter 8). The expression of *vjbR* ensures adequate intracellular parasitism, and mutants in this gene are severely attenuated, demonstrating the critical role that quorum sensing has in *Brucella* virulence. Indeed, the *B. melitensis* autoinducer *N*-acyl-homoserine lactone influences the expression of cell envelope components, such as flagellar molecules, VirB proteins, several Omps and the synthesis and export of exopolysaccharide (Delrue *et al.*, 2005; Uzureau *et al.*, 2010; Wang *et al.*, 2010). Although no specific function has been ascribed to the BabR and its deletion does not generate attenuated brucellae, there seems to be some overlapping between the targets of BabR and VjbR, suggesting a cross-talk between these two regulators (Uzureau *et al.*, 2010). By proteomic analysis of the cytoplasmic fractions (101 proteins) and complementary exploration of the transcripts (338 transcripts) of *B. melitensis* 16M and the corresponding BabR and VjbR mutants, it was demonstrated that close to 5% of the genes (common targets identified by both proteomic and microarray analysis) were influenced by VjbR and/or BabR (Uzureau, *et al.*, 2010). These regulatory proteins directly or indirectly influenced genes coding for products involved in stress response, metabolic adaptation and virulence, indicating that quorum sensing corresponds to a global regulatory phenomenon in *Brucella*. Strikingly, parallels between the targets identified as part of the quorum-sensing regulon and the proteins influenced by the two component regulatory system BvrR/BvrS were found by proteomic analysis: in both, *vjbR* and *bvrR/bvrS Brucella* mutants, the protein profiles seemed to mimic nutrient starvation (Lamontagne *et al.*, 2007; Uzureau *et al.*, 2010). Although significant, this equivalent proteomic profile in both set of mutants was not unanticipated, since it has been demonstrated that the VjbR is also under the influence of the BvrR/BvrS system (Martinez-Nuñez *et al.*, 2010). Taking into account that there is a link between these two systems with the VirB operon (Martinez-Nuñez *et al.*, 2010; Wang at al., 2010), it is concluded that VirB and VjbR influence each other, and that both systems may be under the direct and indirect control of BvrR/BvrS in a regulatory loop that involves the three systems. In addition it seems plausible that the VjbR and BvrR/BvrS regulatory pathways could be also connected through other global starvation sensing mechanisms like the stringent response and/or the phosphotransferase system (PTS), that may be relevant for *Brucella* adaptation during its intracellular cycle (Dozot *et al.*, 2010). As stated before, it may be that the BvrR/BvrS system corresponds to a master regulator of virulence genes that senses when *Brucella* enters the host cell and thus switches on several pathways important to this part of its life cycle.

The comparative and quantitative proteomic analysis of *Brucella* mutants in different systems devoted to virulence, together with the biological characterization of these strains has allowed the definition of common patterns and interactions crucial for the intracellular lifestyle of this bacterium. For instance, it is clear that the control of the cell envelope homeostasis through signalling systems such as BvrR/BvrS and quorum sensing is essential for the bacterial invasion and trafficking in animal host cells. Moreover, a complex interplay between the BvrR/BvrS, VjbR and VirB *Brucella* virulence systems has been partially unravelled (see below).

The *Brucella* proteome under different growth conditions

The first 2DGE comparison of protein patterns of virulent *B. melitensis* 16M subjected to different stress conditions (heat, oxidative and acid stress) was performed by combination of metabolically radioactive label [^{35}S] methionine and [^{35}S] cysteine proteins and N-terminal microsequencing of some of the differentially expressed spots (Texeira-Gomes *et al.*, 2000). By this procedure, close to 680 distinct spots were detected. From this 72 were influenced by heat, 59 by oxidative conditions and 34 by acid stress, sharing between four and eight spots among the different treatments. Nineteen differentially expressed proteins were identified; close to half of them representing proteins that are known to be affected by stress conditions such as DnaK, GroEL, Cu-Zn SOD, Fe/Mn SOD, ClpP, bacterioferritin and malate dehydrogenase. As stated before, these proteins are anticipated to be synthesized in higher amounts according to the expression levels from codon biases predictions in α-Proteobacteria (Karlin *et al.*, 2003). The differences observed in the protein patterns during the various stress conditions suggested that specific regulatory mechanisms were activated to overcome the particular stress response. For instance from the 19 proteins identified, only malate dehydrogenase was expressed at similar level in bacteria either subjected to heat shock or oxidative stresses. Similarly, only the stress protein DnaK displayed the same level of expression in bacteria treated with heat or with acid. One interesting finding was the decreased level of the regulatory protein BvrR after heat shock treatment but not after oxidative or acid stresses. During the early phases of intracellular trafficking it seems that the internalized *Brucella* has to confront acidic conditions within phagosomes (Porte *et al.*, 1999; Jubier-Maurin *et al.*, 2004); nevertheless, at this early times of intracellular life the bacterial expression of BvrR is low, levelling up just after the replication period has been achieved (Lamontagne *et al.*, 2009). This is an indication that the regulatory actions within the intracellular milieu are quite different to those confronted by the bacterium during stress conditions outside the host cell.

The 2DGE analysis has also been successfully used to dissect the putative function of individual *Brucella* proteins and regulatory events. For instance it was demonstrated that the low pH dependent chaperone HdeA produced mainly during the stationary phase growth of *B. abortus* 2308 required the presence of the RNA binding protein Hfq, and that this last protein was also needed for the optimal stationary phase production of the periplasmic Cu-Zn SodC (Valderas *et al.*, 2005). This chain of events may explain why reduced SodC levels could contribute to the attenuation displayed by the *B. abortus* Hfq mutant taking into account that SodC is required for *Brucella* virulence in macrophages (Robertson and Roop, 1999; Gee *et al.*, 2005).

It has been determined that *Brucella* is able to achieve limited growth under microaerobiosis and to survive under strongly oxygen-limited conditions (Loisel-Meyer *et al.*, 2006; Jiménez de Bagüés *et al.*, 2007), a phenomenon that may be reminiscent for an adaptation to the intracellular environment (Lamontagne *et al.*, 2009). In order to investigate the proteomic changes of *B. suis* under low oxygen tension, the bacterium was cultivated under microaerobic and anaerobic denitrifying environments and compared with regular aerobic growth parameters (Al Dahouket al., 2009). Under these two conditions, DNA and protein metabolisms slowed-down, correlating with reduced or minimal bacterial growth. From a total of 1042 spots detected in 2DGE, 37 proteins were differentially expressed between microareobic and aerobic respiratory conditions. From these, 19 proteins were found to be up-regulated while 18 were down-regulated by low oxygen tension. Similarly, when *B. suis* was maintained for two weeks under anaerobic denitrifying environment, 24 proteins were differentially expressed. From these 11 were up-regulated while 13 were down-regulated. Taking into account the number of spots detected, these changes represent only a few variations in the *Brucella* proteome. When compared with aerobic growth, the total number of proteins showing changes of concentration under anaerobic denitrifying conditions (2.3%) was lower than under microaerobiosis (3.5%). However, none of the proteins differentially expressed under microareobic conditions was observed under anaerobic denitrifying cultures and vice versa. This discrepancy suggests that although both conditions involved low oxygen tension, the respective adaptation to each situation seemed to engage a different subset of proteins. However, it is also possible that the experimental system did not have the resolution necessary to comprehensively profile all the proteins involved in both processes. Nevertheless, it seems that *B. suis* adapts to low oxygen conditions by modulating oxidative and non-oxidative energy metabolism, such as up-regulation of glycolysis and denitrification; the latter active under microaerobiosis and anaerobiosis. The reduction of basal metabolism was in line with low or absence of growth of *B. suis*. It seems therefore, that basic metabolic processes may be maintained even when oxygen becomes scarce. Under both microaerobiosis and anaerobiosis, glycolysis and denitrification were favoured. In addition, fatty acid oxidation and possibly citrate fermentation could also contribute to energy production sufficient for survival under anaerobiosis.

Proteomics of intracellular *Brucella*

The first attempt to study the proteome of intracellular *Brucella* was performed after a short period of infection in J774A.1-like macrophages treated with cycloheximide by Lin and Ficht (1995). These investigators used 2DGE to separate proteins from intracellular *B. abortus* that had been metabolically labelled with [^{35}S]methionine. These authors observed relevant changes in the expression of several proteins during stress conditions and intracellular infection by the same *B. abortus* strain. However, the bacterial protein profiles detected during stress were significantly different from those observed during intracellular infections. In addition, some differences in the expression of proteins between the virulent 2308 and the attenuated vaccine S19 strains were observed, a phenomenon that has been reinforced by high-throughput proteomics (Lamontagne *et al.*, 2007, 2009). The highly abundant chaperone GroEL stress response protein was positively identified as one of the main proteins augmented under stress and intracellular infection.

Rafie-Kolpin *et al.* (1996) pursued a very similar approach with radioactively labelled *Brucella* proteins. These authors tested five different stress conditions (H_2O_2, menadione, starvation, heat and low pH) and compared the protein 2DGE profiles with those of *B. abortus* 2308 obtained from bovine macrophages infected for a short period of time and treated with cycloheximide. The identification and further estimation of stress response proteins (DnaK, GroEL, SodC and HTrA) was performed by immunodetection of the individual proteins. By this procedure, close to 147 spots were differentially expressed between bacteria grown in cultures and in cells. Twenty-four spots not detected in cultures were recorded in the *B. abortus* infecting the bovine macrophages. In contrast, approximately 50 protein spots that were expressed in bacteriological culture conditions were repressed by intracellular *Brucella*. The level of expression of 19 protein spots increased while the expression of 54 spots decreases in infected macrophages. Several 'new' protein spots present under the various stress conditions, were absent during regular growth. The overall proteome demonstrated during stress responses was significantly different from the proteome showed by intracellular *Brucella*. Moreover, comparisons between the four stress proteins tested, also revealed that the proteome profile of *B. abortus* inside macrophages after a short period of infection, was different from that obtained under various stringent conditions. In spite of displaying some overlapping, the general conclusion was that the *Brucella* protein response to the intracellular milieu was not a simple summation of the actions of various extracellular environmental stresses.

Cycloheximide does not affect the multiplication of *Brucella* late in its infection cycle, once the bacterium has reached its replicative niche (Detilleux *et al.* 1991). However, this inhibitor of protein synthesis of eukaryotic cells used for studying the protein profile of intracellular bacteria (Buchmeier and Heffron, 1990) may well affect the trafficking of *Brucella* at early times of infection. Indeed, the above protocols that have utilized cycloheximide for studying the proteome of intracellular *Brucella* (Lin *et al.*, 1995; Rafie-Kolpin *et al.*, 1996) have been applied at early time points (2–3 h) in the macrophage infection cycle. At early times some bacteria may display higher stress responses than others, creating heterogeneous protein responses, which eventually may be reproduced in the proteomic (or transcriptomic) analyses of intracellular bacteria (see below). Considering this, protocols for isolating *Brucella* compartments containing live bacteria are probably more informative than protocols that use of protein synthesis inhibitors in infected host cells.

By proteomic analysis in 2DGE, it was determined that VjbR, DnaK, HtrA, Omp25 and GntR in *B. melitensis virB* mutants were down-regulated in relation to 16M (Wang *et al.*, 2009). Therefore, the transcripts of the same proteins were studied by quantitative RT-PCR in these two *B. melitensis* strains after intracellular growth in J774A.1 macrophage like cells (Wang *et al.*, 2009). It was found that the level of transcripts coding for these proteins in the intracellular *virB* mutant were significantly lower as compared with those of the virulent 16M counterpart, mainly after 12 h of infection. However, there are intrinsic difficulties in interpreting these results since it is known that vacuoles containing *virB* mutants readily fuse with lysosomes and then bacteria are digested within phagolysosomes during the first hours of infection. Moreover, the amounts and combination of ampicilin and kanamycin used in the cell cultures to eliminate extracellular *Brucella* have been described to promote the killing of intracellular bacteria (Veale *et al.*, 1975).

The 2DGE proteomic analysis of *B. suis* recovered after 48 h culture from J774 macrophages lysates by centrifugation through 15% sucrose cushion, identified a total of 168 differently expressed proteins (Al Dahouk *et al.*, 2008). The *Brucella* proteins were distinguished and resolved from the macrophage proteins by differential staining with two distinct fluorescent dyes. By this approach, a total of 44 different *Brucella* proteins corresponding to 49 spots were differentially expressed within J774 like macrophages at later times of infection. From these, 20 protein spots were at lower concentrations, indicating that this intracellular adaptation did not correspond to a net increase in bacterial proteins when compared with brucellae cultured in stationary phase. Most of these bacterial proteins were involved in primary metabolism supporting the hypothesis of an adaptive response by quantitative reduction of processes participating in energy, protein, and nucleic acid metabolism. In accordance with other proteomic analysis (Lamontagne *et al.*, 2009), the metabolic adaptation of *Brucella* at this later stage of infection seems to be a key strategy for its survival and replication within host cells.

An alternative comprehensive strategy to study the time course proteome of intracellular *B. abortus* was developed by Lamontagne *et al.*, (2009). However, before reviewing the results of this work, an important event regarding the life and death of intracellular *Brucella* should be clarified. It is well known that at early times of infection the fate of intracellular *Brucella* is marked by one of two different outcomes. In one case, vacuoles containing bacteria that have avoided lysosomal fusion begin to associate with compartments of the endoplasmic reticulum by means of LPS, β-cyclic glucans and hypothetically by effectors released by the type IV VirB secretion system (Detilleux *et al.*, 1990; Celli *et al.*, 2003; Arellano-Reynoso *et al.*, 2005; Den Hartigh *et al.*, 2008; De Jong *et al.*, 2008). Bacteria within these early vacuoles that successfully fuse with endoplasmic reticulum-like compartment, then, proceed to replicate within this multimembranous organelle (Celli *et al.*, 2003). However, vacuoles containing bacteria unable to avoid lysosomal fusion will have to confront harsh conditions within phagolysosomes, which eventually will destroy the ingested bacterium (Celli *et al.*, 2003). It is thus expected that the declining bacteria within lysosomal compartments would display different proteomes than the protein profiles demonstrated by the surviving bacteria that successfully achieved fusions with the endoplasmic reticulum derived compartments (Moreno and Gorvel, 2004). This was the main reason why a gradient centrifugation protocol for separating alive from death bacterial debris was developed (Lamontagne *et al.*, 2009): to avoid the ambiguity of dealing with the mixed *Brucella* populations that would hamper the proteomic analysis of live intracellular bacteria. Considering this, the proteome of intracellular *Brucella* was studied by infecting murine macrophages with virulent *B. abortus* 2308 or vaccine *B. abortus* S19 strains. Then, at different time points (3 h, 20 h and 44 h) after infection, the live intracellular bacteria were retrieved and their proteomics studied and compared using a quantitative LC-based sample fractionation method of complete and unseparated protein bacterial enzymatic digests in combination with extensive quantitative MS/MS analysis (Lamontagne *et al.*, 2009).

By this novel approach, a total of 192 *Brucella* proteins were unambiguously identified to be differentially expressed during the infection time course (Fig. 6.2). Ninety proteins were uniquely differentially expressed by the 2308 strain, 31 proteins by the S19 and 71 proteins by both strains. It was demonstrated that after 3 h of infection, between 91–95% of the intracellular *Brucella* organisms recovered from transitional compartments that did not fuse with lysosomes, were

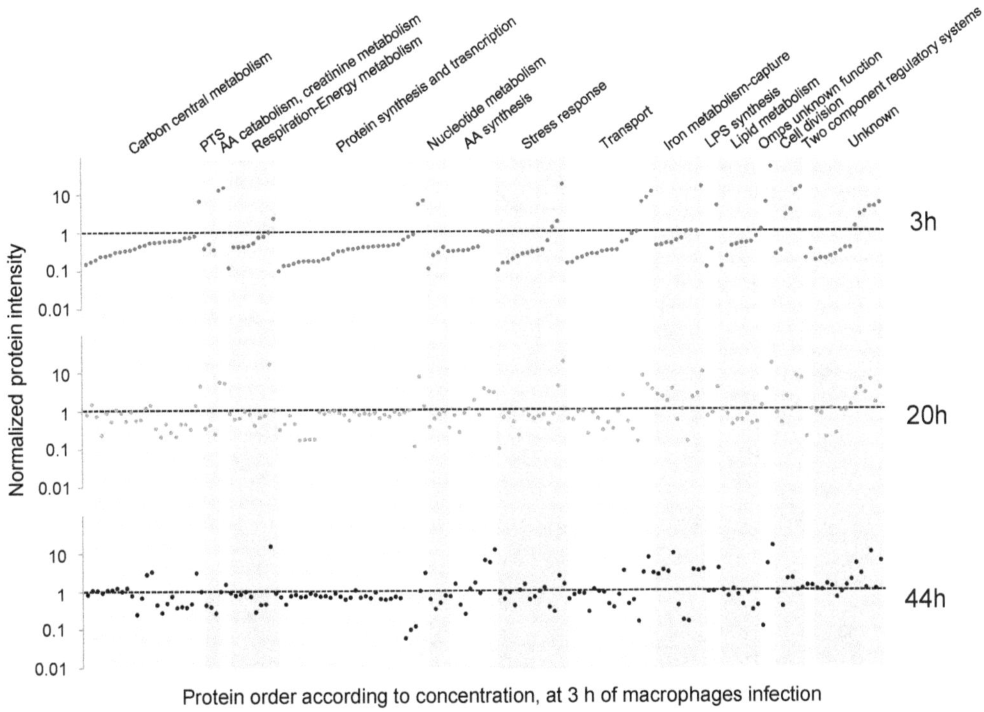

Figure 6.2 Proteins differentially expressed by *B. abortus* 2308 during an intracellular infection. Intensity values per protein were normalized to time zero (corresponding to bacteria grown in liquid culture) and displayed in logarithmic scale per time point (Y-axis). Protein order in the x-axis (represented as a dot) follows the relative concentration values of the *Brucella* proteins at 3h of infection in each functional cluster (alienated and indicated in the upper part of the figure), obtained from the supplementary table in (Lamontagne *et al.*, 2009). For instance, the first protein (represented by the first dot) from left to right in the 'carbon central metabolism' at 3h corresponds to the same protein (first dots) in the subsequent 20h and 44h, and so for. The standard deviation in each point was seldom higher than 10% of each value, and was not included for clarity. Notice that at 44h of infection a large collection of the *Brucella* proteins have approached values close to 0h basal levels.

indeed alive (Lamontagne *et al.*, 2009). Since the proteome of intracellular virulent *Brucella* at 3h of infection demonstrated low levels of bacterial oxidative response proteins, chaperons and other important stress response molecules, it was concluded that this transitional compartment (e.g. early phagosome) did not seem to confront *Brucella* with particularly stressful conditions, such as oxygen radicals or other microbicidals (Fig. 6.3). It has been suggested, however, that within this early compartments *Brucella* has to confront moderate acidic conditions (Jubier-Maurin *et al.*, 2004), a fact that was supported by the low levels of some proteins, such as malate dehydrogenase, which decreases during low pH stress (Teixeira-Gomes *et al.*, 2000).

According to the proteomic analysis, it was proposed that at this early time of infection (3h), the intracellular *Brucella* initiated its adaptation to low oxygen tension type of respiration, responding to the conditions of intracellular compartments. The decrease of periplasmic transporters and the reduction of central carbon metabolic pathways, such as the tricarboxylic acid cycle, also indicated that these systems operated at a restricted level, probably owing to the reduced availability of extracellular substrates within the phagosome. As suggested by the intracellular *Brucella* proteome, the intracellular bacterium switched to anaplerotic routes and to the generation of glutamate by enzymatic conversion of amino acids, for making the necessary intermediates for the reduced

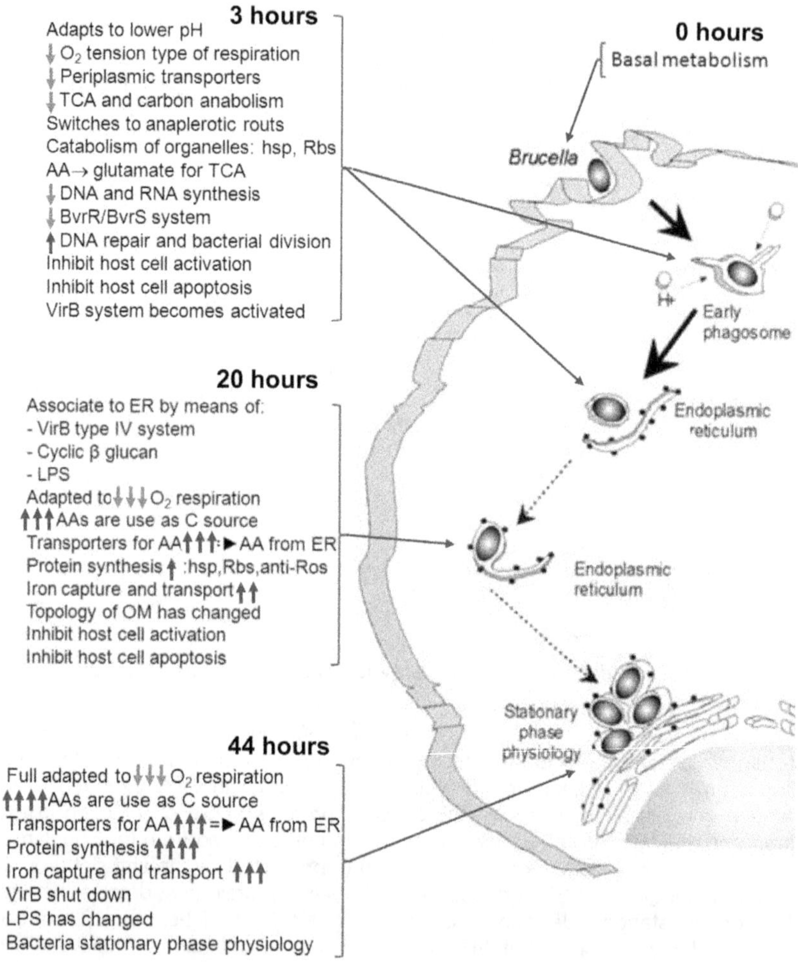

Figure 6.3 Proposal of a model that describes the mayor changes occurring during intracellular trafficking of virulent *Brucella* in macrophages. The 0 h points correspond to the basal level (extracellular *Brucella*) from which all the other changes were recorded at different periods of infection (3, 20 and 44 h). The darker arrows pointing upwards and the lighter arrows pointing downwards, indicate increase or decrease, respectively, in relation to the previous time period. AA, amino acids; TCA, tricarboxylic acid cycle; hsp, stress response proteins; Rbs, ribosomal proteins; C, carbon; ER, endoplasmic reticulum; Ros, reactive oxygen species; OM, outer membrane.

output tricarboxylic acid cycle. The low levels of some intracellular bacterial proteins at this critical time, suggested that several catabolic pathways generating essential substrates by using its own macromolecules were activated, mainly the digestion of ribosomes or abundant stress response proteins (Fig. 6.2). Interestingly, at these early periods of infection, intracellular *Brucella* seemed to restrict the synthesis of proteins and nucleic acid, and to down-regulate the BvrR/BvrS system essential for carbon metabolism and the biogenesis of the cell envelope (Lamontagne *et al.*, 2007; Dozot *et al.*, 2010). In contrast, proteins from other regulatory systems devoted to the control of cell division and DNA repair mechanisms were detected at higher concentrations, and therefore suggested to be activated. Thanks to its stealth strategy, *Brucella* is capable to withstand this early compartment without significant activation of its host cell (Barquero-Calvo *et al.*, 2007, 2009), an event that probably gives the opportunity for the bacterium to activate its VirB type IV secretion system (Boschiroli *et al.*, 2002) and the apoptosis inhibitory mechanisms for maintaining the host cell alive (Gross *et al.*, 2000), and to move to the replication phase.

The proteome of intracellular *B. abortus* after 20 h of infection (Fig. 6.3), was consistent with the proposal that the *Brucella* metabolism has fully adapted to a low oxygen tension type of respiration, by means of the PrrA/PrrB two component regulatory system, while continuing to use amino acids as a carbon source (Lamontagne *et al.*, 2009). Transporters involved in the capture of amino acids and peptides increased, reflecting the availability of these substrates within the endoplasmic reticulum. At this time the proteome suggested that protein and nucleic acid synthesis have partially carried on. In addition, the relative amounts of intracellular *Brucella* chaperone and antioxidant proteins necessary for the adequate folding and damage protection also increased, suggesting that bacterial protein synthesis resumed. Concomitantly, several of the bacterial proteins use to capture and transport iron, essential in many metabolic processes, also increased over time (Fig. 6.4). The amounts of Omps varied in relation with the earlier time points, suggesting adaptation to this new intracellular environment. It is not trivial that at this point of infection, macrophages do not display signs of activation despite the fact that are not refractory to be activated by external influences (Barquero-Calvo *et al.*, 2007).

After 44 h of intracellular life (Fig. 6.3), once *Brucella* has achieved extensive replication within the infected cells (Celli *et al.*, 2003), many of the

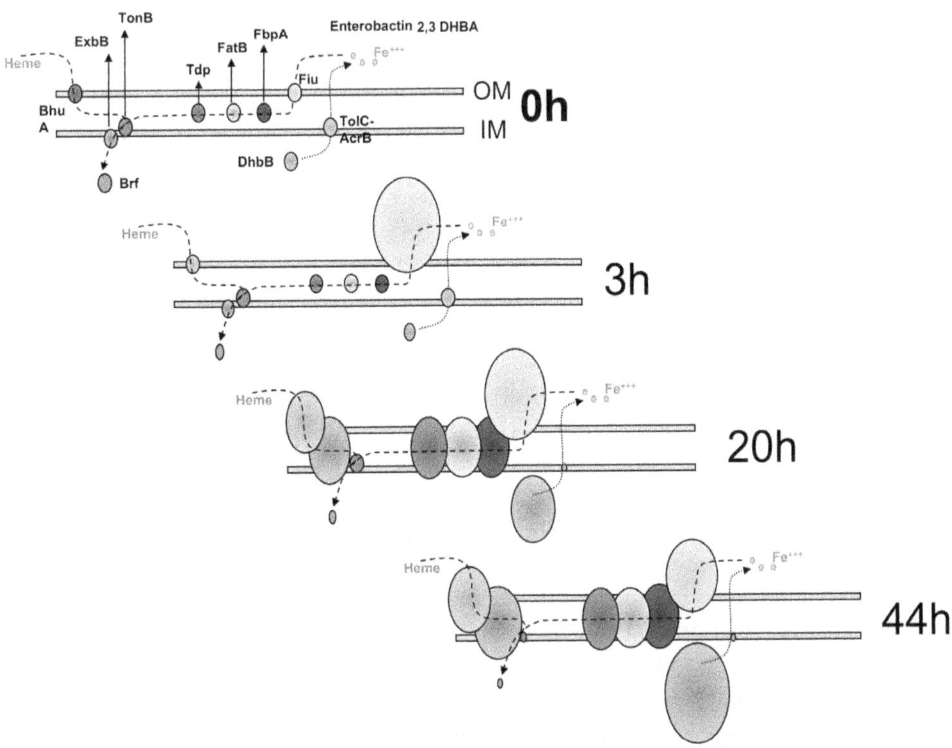

Figure 6.4 A model indicating the variation in concentration of the various proteins involved in iron uptake and transport during different times of macrophage infection. The variation in concentration in each protein depicted as an oval are indicated by the relative size of the ovals in relation to 0 time (normalized values). OM, outer membrane, IM, inner membrane. The names of the proteins correspond to the open reading frames described in Lamontagne *et al.* (2009). Based on the model described by Roop *et al.* (2004).

differentially expressed proteins have reverted to pre-infection expression levels. For instance, all the regulatory systems that increased or decreased earlier in infection have approached pre-infection levels, while others like those devoted to iron uptake remain high (Fig. 6.4). This suggested that the bacterial physiology may be entering a stationary phase type of homeostasis, as proposed previously (Roop *et al.*, 2003; Anderson *et al.*, 2008).

Although the few attenuated *B. abortus* S19 capable to avoid lysosomal fusion seemed to replicate within the endoplasmic reticulum derived BCV, the proteome analysis of this intracellular bacterium revealed significant differences (Lamontagne *et al.*, 2009). For instance the reduction of proteins involved in metabolic and biosynthetic alternatives also occurred in the attenuated strain, but to a far lesser extent. Similarly, changes in proteins involved in the shift to low oxygen tension type of respiration early in infection were also observed in the S19 strain but to lesser extent than the virulent 2308. Finally, later in infection, when the virulent strain re-establishes the proteins involved in protein and nucleic acid synthesis, the attenuated strain was unable to match that level of adjustment. Moreover, S19 does not seem to up-regulate those proteins involved in iron metabolism or those of the cell envelope composition to the same extent as the virulent strain. It seemed that the attenuated strain's inability to sufficiently adjust its metabolism early in infection compromised its capacity to fully exploit the endoplasmic reticulum microenvironment in mid and late infection. It may be, therefore, that effective antimicrobial strategies for combating virulent *B. abortus* infections could be developed by limiting the bacterium's ability to catabolize amino acids or switch to low oxygen tension type of respiration early in infection.

In summary, it is clear that *Brucella* undergoes dramatic changes during its intracellular journey to the final niche, the endoplasmic reticulum derived BCV. These changes reflect the metabolic struggle confronted to avoid the constitutive degradative machinery of the host cell as well as the biochemical adaptations required to adapt and exploit different nutritional conditions and the idiosyncratic intracellular environment. However, it is noticeable that the overall proteomic profile of extracellular logarithmic growing brucellae roughly resembles the corresponding profile of intracellular growing bacterium at later times of infection. This observation indicates that the key for the intracellular lifestyle of *Brucella* lies mainly on its ability to reach the endoplasmic reticulum rather than in adapting to a particular intracellular niche. Indeed, *Brucella* is capable to achieve limited replication in vacuoles of epithelial cells that are different from the endoplasmic reticulum (Chaves-Olarte *et al.*, 2002; J. P. Gorvel, personal communication).

Discussion and conclusions

As in any other emerging field in science, the proteomic studies of *Brucella* have ended up with more questions than answers. Therefore, the straightforward response to the question used as heading of this review 'What have we learned from *Brucella* proteomics?' is simple: very little, in relation to our original expectations! However, our ignorance has to be placed in its right context, since we are dealing with a group of very complex pathogens that are capable of adapting to and living in at least two very different environments. Indeed, we know that *Brucella* seldom leaves its niche within intracellular vesicles, and that the early phagosome containing *Brucella* does not initiate bacterial multiplication with a plethora of nutrients. Thus, the obvious conclusion is that the bacterium must metabolically adapt to the intracellular environment and then manipulate the host cell biology to create conditions for its survival and multiplication. However, we have not been able to identify the *Brucella* components that interact with the host machinery to control these events and the regulatory mechanisms taken place for adapting to the intracellular niche. Therefore, in addition to the identification of the protein changes that occur during the *Brucella* life cycle, it is necessary to understand how proteins interact with each other and with other molecules within and outside the bacterium. We also ignore what are the predominant conditions and topology of vacuolar proteins within the *Brucella* containing compartment during the first hours of infection and the concomitant topological changes occurring on the bacterial surface during trafficking. One fundamental question that rises from proteomic studies corresponds to the nutrients extracted by *Brucella* from host cells and how the intracellular bacterium obtains them for achieving an exponential growth in the second half of the infection cycle. In spite of our efforts, we barely understand the metabolic adaptations carried out by both, the cell host and the bacterium. Last but not least, we do not know how *Brucella* manipulates epithelial cells and macrophages to prolong their life and how the bacteria are released from these cells. All these and other questions related to the *Brucella* lifestyle may be resolved by the use of proteomics, taking into consideration that adequate protocols of organelle and bacterial purification are implemented and the sensitivity of quantitative methods improve.

Proteomics has contributed to identify a number of proteins in the various *Brucella* extracts obtained under different growth conditions. Still, the amount of proteins identified remains comparatively low (5–20%) in relation to the total number of open reading frames (~3200) coded in the genome of this bacterium. A larger proportion of the *Brucella* proteins have been identified by gel-free methods (Lamontagne *et al.*, 2010; Uzureau *et al.*, 2010), a fact that concerns the resolution of the various approaches used. Moreover, a significant proportion of the proteins identified by 2DGE/MS correspond to the same bulk of *Brucella* proteins, most of which have been predicted to have high expression levels (Karlin *et al.*, 2003).

In addition to the limitations already mentioned, there are other factors that might hamper the sensitivity of protein detection in 2DGE. Among the most relevant are: the narrow range of separation of proteins with very low (<10 kDa) or with very high (>150 kDa) molecular weights, proteins with extreme isoelectric points (<4 and >10), strong hydrophobic proteins, minimum level of detection of protein stains and limitation in the retrieval of proteins from the gel. It was not unexpected that most of the *Brucella* proteins identified in the early proteomic works corresponded to relatively abundant soluble molecules with no extreme isoelectric points, such as periplasmic and cytoplasmic proteins, while very few membrane proteins were recognized. Indeed, cytoplasmic and periplasmic soluble proteins are more prone to be extracted and run in the first isoelectric focusing dimension than the abundant hydrophobic membrane proteins. In addition of displaying solubility problems, hydrophobic proteins also show cathodic drift, resulting in an inefficient and non-reproducible separation, commonly generating several protein spots in the gel (Freiburghaus, 1994). Considering this, it has been practical to perform analysis of membrane proteins with purified cell envelope fractions or OMF (Guzmán-Verri *et al.*, 2002; Boigegrain *et al.*, 2004; Lamontagne *et al.*, 2007), using defined electrophoretic or chromatographic protocols. In order to overcome some of the above described drawbacks for the detection of proteins separated in 2DGE, several modification protocols have also been implemented, such as the inclusion of ultrasensitive luminescent ruthenium organic complex dyes and fluorescent-labelling methods (Al Dahouk *et al.*, 2008). In addition, the use of narrower linear pH gradients incorporated in some protocols have allowed better protein separation in 2DGE (Mujer *et al.*, 2002; Wagner *et al.*, 2002; Al Dahouk *et al.*, 2009). Although radioactive label proteins increase the sensitivity in 2DGE, it seems to be unpractical, especially at the time of recovering the proteins from the gel.

Some of the inherent factors that influence the composition of the *Brucella* proteomes set limits on the phenotypic interpretations are the different growth conditions and the bacterium cell cycle. In a comparative proteomic study performed with *O. anthropi* (the closest *Brucella* relative) grown at different times, it was demonstrated that from a total of 131 identified proteins in the soluble fraction (89 at early phase and 95 at late growth phase), 34 were uniquely produced during the early phase of growth, 40 were found to be unique to the late phase, while 55 proteins were detected under both growth conditions (Graham *et al.*, 2007). Close to 97% of the *O. anthropi* proteins identified in this study, have orthologues in the *Brucella* proteome, suggesting that similar features may be observed in the later bacterium. This has already been demonstrated for some *Brucella* proteins such as the HdeA, group III Omps, BvrR, VjbR and the VirB proteins, whose expression varies through the different growth phases (Valderas *et al.*, 2005; Martinez-Nuñez *et al.*, 2010; Wang *et al.*, 2010). Other features that particularly may influence the number of *Brucella* protein spots correspond to the extraction methods and the bacterial strains used. Indeed, it is known that *Brucella* cells are quite resistance to be disrupted and that the membrane components of the cell envelope are not readily extractable by conventional procedures (Moriyón *et al.*, 1983, 1998). Regarding the bacterial strain, separation of protein extracts obtained from smooth bacteria may be more problematic than with extracts obtained from rough strains. As indicated before, the use of rough strains for protein analysis has the advantage over smooth strains that the rough LPS migrates to the bottom of the gel, generating less aggregation and distortion of the protein patterns (Gamazo *et al.*, 1989).

The integration of the data generated through proteomic analysis of mutants in the BvrR/BvrS, VjbR and VirB systems, together with the description of the proteomic changes that *Brucella* undergoes during its intracellular lifestyle, have helped us to propose a unified model in which these three systems interact to allow the successful infection of host cells (Fig. 6.5). This model begins with *Brucella* entering animal cells by means of activating the host Rho GTPases, with the involvement of the factors influenced by the BvrR/BvrS system (Guzmán-Verri *et al.*, 2002; Chaves-Olarte *et al.*, 2003). Once inside the host cell, the bacterium senses its new environment through particular intracellular cues (e.g. acidification of the early phagosome) by means of the sensor BvrS protein, which in turn phosphorylates the regulator BvrR. The activated BvrR, then binds the promoter region of the transcriptional regulator VjbR protein, activating its transcription (Viadas *et al.*, 2010; Martinez Nuñez *et al.*, 2010), as well as its own promoter and promoters of genes involved in the synthesis of outer membrane and periplasmic proteins, *virB*, and LPS biosynthesis genes (Manterola *et al.*, 2005; Lamontagne *et al.*, 2007). At this point, the concentration of homoserine lactone is low, allowing VjbR to bind its regulated promoters. The *virB* operon and the flagellar genes are then activated allowing the secretion of factors (e.g. VceA and VceC) needed for the evasion of the phagolysosomal route and the recruitment of endoplasmic reticulum derived vesicles (Rambow-Larsen *et al.*, 2008; De Jong, 2009). At the same time, the VjbR protein binds to the promoters of the genes that are negatively regulated by this transcriptional regulator. That is, VjbR shuts down protein synthesis, cell division, decreases the transporters at the cell membrane and slows down the metabolism during the intracellular journey to its ER derived replication niche (Lamontagne *et al.*, 2009; Uzureau *et al.*, 2010). A negative feedback is implemented in order to avoid hyper activation of transit responses and the level of BvrR is dramatically diminished (Lamontagne *et al.*, 2009). This down-regulation could be achieved by inhibition of BvrR expression, selective proteolysis or even VirB-mediated secretion of this response regulator (Marcheasini *et al.*, 2004). As *Brucella* creates its endoplasmic reticulum-derived compartment and replicates, the concentration of homoserine lactone starts to rise reaching a threshold concentration able to inactivate the DNA binding activity of VjbR. With BvrR absent and VjbR inactivated, the transcription of VirB ceases being its mission accomplished with the *Brucella* now safe at its replicating niche, the endoplasmic reticulum. All the genes repressed before by VjbR are now reassuming a transcriptional state. Protein synthesis begins again, transporters are launched to the membrane and the division machinery starts operation to allow *Brucella* to replicate exponentially in order to reach hundreds of bacteria filling the endoplasmic reticulum-derived compartment of infected host cells. The continuous increase in intracellular *Brucella* population raises even higher the homoserine lactone concentrations and the bacteria reach a state similar to the phenotype displayed by mutants in VjbR in stationary phase with a dense polysaccharide being secreted and covering the bacterial population. This polysaccharide is either intended to hide the soon-to-be released bacteria from the action of the soluble components of the immune system (antibodies, complement) or aids to the now extracellular bacteria to attach non-infected neighbouring cells to start a new replications cycle, or both (Godefroid *et al.*, 2010).

Although the proteomic analyses have contributed to put forward models on the dynamics that may occur during *Brucella* intracellular life,

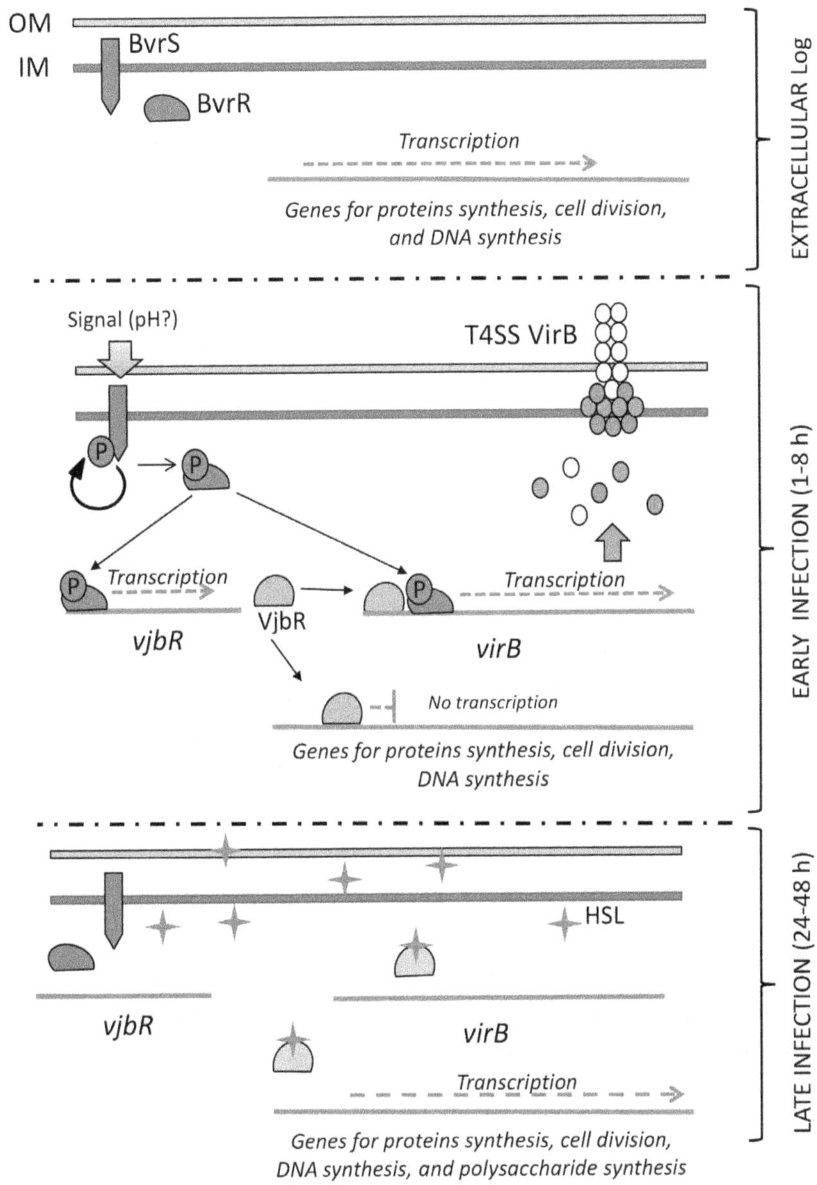

Figure 6.5 Model of regulatory networks among different systems during the adaptation of *Brucella* to the intracellular niche. Extracellular log phase bacteria have the virulence systems in an 'off' state. As soon as the bacterium invades the host cells, the sensor protein (BvrS) of the two component regulatory BvrR/BvrS system detects an intracellular signal (e.g. low pH) and its kinase activity becomes manifest by phosphorylating the regulatory BvrR. Once phosphorylated, BvrR binds the *vjbR* promoter inducing its transcription. Then, both VjbR and BvrR bind the *virB* promoter allowing the assembly of the secretion machinery in the outer membrane. Simultaneously, VjbR stops transcription of genes relevant for bacterial replication. The effectors translocated by the VirB type IV secretion system mediate the generation of the endoplasmic reticulum derived replication niche. Once the bacterium is within this replicative compartment, the concentration of the quorum sensing molecule homoserine lactone (HSL) increases, inactivating the DNA-binding activity of the VjbR. As consequence, *virB* transcription ceases. Concomitantly, DNA and protein synthesis and multiplication restarts and the bacteria multiply.

there are still many pieces in this complex puzzle that are lacking. The fact that the knock out of a single gene influences the overall proteome of the *Brucella* mutants in relation to the parental strains reveals the complexity of the interacting components and the intricacy of this life system. In this regard, it is necessary to complement proteomics with reliable isolation methods as well other high-throughput analytical techniques such as screening systems for the isolation of *Brucella* genes encoding surface exposed and secreted proteins (Marchesini *et al.*, 2004), a high efficiency cloning and expression systems (Ding *et al.*, 2006), yeast two-hybrid screening techniques (Dozot *et al.*, 2010), co-immunoprecipitation methods (Martinez-Nuñez *et al.*, 2010), *in vitro* functionality of protein pairs (Goley *et al.*, 2010) and protein–DNA interactions (De Jong *et al.*, 2008; Martinez-Nuñez *et al.*, 2010). Moreover, more extensive proteome analysis of the *Brucella* containing compartments is necessary, to understand the host cells components that are recruited and interact with the bacterium (Fugier *et al.*, 2009). Concomitantly, all findings have to be confronted with *ex vivo* and *in vivo* models, in order to correlate the magnitude of the molecular changes in relation with the phenotype and the biology of the bacteria.

Acknowledgements

This work was supported by grants from NeTropica, Fondos FS-CONARE, FS-CONARE-equipo, Vicerrectoria de Investigación de la Universidad de Costa Rica, and FI-240-2010 from MICIT/CONICIT of Costa Rica, and by the NIAID/NIH contract HHSN266200400056C.

References

Al Dahouk, S., Nöckler, K., Scholz, H.C., Tomaso, H., Bogumil, R., and Neubauer, H. (2006). Immunoproteomic characterization of *Brucella abortus* 1119-3 preparations used for the serodiagnosis of *Brucella* infections. J. Immunol. Methods *309,* 34–47.

Al Dahouk, S., Jubier-Maurin, V., Scholz, H.C., Tomaso, H., Karges, W., Neubauer, H., and Köhler, S. (2008). Quantitative analysis of the intramacrophagic *Brucella suis* proteome reveals metabolic adaptation to late stage of cellular infection. Proteomics *8,* 3862–3870.

Al Dahouk, S., Loisel-Meyer, S., Scholz, H.C., Tomaso, H., Kersten, M., Harder, A., Neubauer, H., Köhler, S., and Jubier-Maurin, V. (2009). Proteomic analysis of *Brucella suis* under oxygen deficiency reveals flexibility in adaptive expression of various pathways. Proteomics *9,* 3011–3021.

Alvarado, R., Tran, D., Ching, B., and Phinney, B.S. (2010). A comparative study of in-gel digestions using microwave and pressure-accelerated technologies. J. Biomol. Tech. *21,* 148–155.

Anderson, E.S., Paulley, J.T., and Roop, R.M., 2nd. (2008). The AraC-like transcriptional regulator DhbR is required for maximum expression of the 2,3-dihydroxybenzoic acid biosynthesis genes in *Brucella abortus* 2308 in response to iron deprivation. J. Bacteriol. *190,* 1838–1842.

Arellano-Reynoso, B., Lapaque, N., Salcedo, S., Briones, G., Ciocchini, A.E., Ugalde, R., Moreno, E., Moriyón, I. and Gorvel, J.P. (2005). The cyclic β–1,2-glucan is a *Brucella* virulence factor required for intracellular survival. Nat. Immunol. *6,* 618–625.

Barquero-Calvo, E., Chaves-Olarte, E., Weiss, D. S., Guzmán-Verri, C., Chacón-Díaz, C., Rucavado, A., Moriyón, I., and Moreno, E. (2007). *Brucella abortus* uses a stealthy strategy to avoid activation of the innate immune system during the onset of infection. PLoS One *2,* e631.

Barquero-Calvo, E., Conde-Alvarez, R., Chacón-Díaz, C., Quesada-Lobo, L., Martirosyan, A., Guzmán-Verri, C., Iriarte, M., Mancek-Keber, M., Jerala, R., Gorvel, J.P., *et al.* (2009). The differential interaction of *Brucella* and *Ochrobactrum* with innate immunity reveals traits related to the evolution of stealthy pathogens. PLoS One *4,* e5893.

Bohlin, J., Snipen, L., Cloeckaert, A., Lagesen, K., Ussery, D., Kristoffersen, A.B., and Godfroid, J. (2010). Genomic comparisons of *Brucella* spp. and closely related bacteria using base compositional and proteome based methods. BMC Evol. Biol. *10,* 249.

Boigegrain, R.A., Salhi, I., Alvarez-Martinez, M.T., Machold, J., Fedon, Y., Arpagaus, M., Weise, C., Rittig, M., and Rouot, B. (2004). Release of periplasmic proteins of *Brucella suis* upon acidic shock involves the outer membrane protein Omp25. Infect. Immun. *72,* 5693–5703.

Boschiroli, M.L., Ouahrani-Bettache, S., Foulongne, V., Michaux-Charachon, S., Bourg, G., Allardet-Servent, A., Cazevieille, C., Liautard, J. P., Ramuz, M., and O'Callaghan, D. (2002). The *Brucella suis* virB operon is induced intracellularly in macrophages. Proc. Natl. Acad. Sci. U.S.A. *99,* 1544–1549.

Bricker, B.J., Tabatabai, L.B., Judge, B.A., Deyoe, B.L., and Mayfield, J.E. (1990). Cloning, expression, and occurrence of the *Brucella* Cu-Zn superoxide dismutase. Infect. Immun. *58,* 2935–2939.

Brooks-Worrell, B.M., and Splitter, G.A. (1992a). Sodium dodecyl sulfate- and salt-extracted antigens from various *Brucella* species induce proliferation of bovine lymphocytes. Infect. Immun. *60,* 2136–2138.

Brooks-Worrell, B.M., and Splitter, G.A. (1992b). Antigens of *Brucella abortus* S19 immunodominant for bovine lymphocytes as identified by one- and two-dimensional cellular immunoblotting. Infect. Immun. *60,* 2459–2464.

Buchmeier, N.A., and Heffron F. (1990). Induction of *Salmonella* stress proteins upon infection of macrophages. Science *248,* 730–732.

Cañas, B., Pineiro, C., Calvo, E. López-Ferrer, D., and Gallardo, J.M. (2007). Trends in sample preparation for classical and second generation proteomics. J. Chromatogr. A. *1153,* 235–258.

Celli, J., de Chastellier, C., Pizarro-Cerdá, J., Moreno, E., and Gorvel, J.P. (2003). *Brucella* evasion of macrophage killing through VirB-dependent sustained interactions with the endoplasmic reticulum. J. Exp. Med. *198,* 545–556.

Chaves-Olarte, E., Guzmán-Verri, C., Méresse, S., Desjardins, M., Pizarro-Cerdá, J., Badilla, J., Gorvel, J.P., and Moreno, E. (2002). Activation of Rho and Rab GTPases dissociates *Brucella abortus* internalization from intracellular trafficking. Cell. Microbiol. *4,* 663–676.

Chaves-Olarte, E., Guzmán-Verri, C., and Moreno, E. (2003). Expression of the ExoS protein from *Sinorhizobium meliloti* complements biological properties of a *Brucella abortus* mutant lacking the BvrS protein. In: Brucellosis 2003 International Research Conference, 15–17 September 2003. Pamplona: Spain, p. 157.

Cherwonogrodzky, J. W., Dubray, G., Moreno, E., and Mayer, H. (1990). Antigens of *Brucella.* In: Animal Brucellosis, Nielsen, K., and Duncan, B., eds. (CRC Press, Inc: Boca Raton, Florida), pp. 19–64.

Cloeckaert, A., de Wergifosse, P., Dubray, G., and Limet, J.N. (1990). Identification of seven surface-exposed *Brucella* outer membrane proteins by use of monoclonal antibodies: immunogold labeling for electron microscopy and enzyme-linked immunosorbent assay. Infect. Immun. *58,* 3980–3987.

Cloeckaert, A., Verger, J.M., Grayon, M., and Vizcaíno, N. (1996). Molecular and immunological characterization of the major outer membrane proteins of *Brucella.* FEMS Microbiol. Lett. *145,* 1–8.

Cloeckaert, A., Tibor, A., and Zygmunt, M.S. (1999). *Brucella* outer membrane lipoproteins share antigenic determinants with bacteria of the family Rhizobiaceae. Clin. Diagn. Lab. Immunol. *6,* 627–629.

Comerci, D.J., Martínez-Lorenzo, M.J., Sieira, R., Gorvel, J.P., and Ugalde, R.A. (2001). Essential role of the VirB machinery in the maturation of the *Brucella abortus*-containing vacuole. Cell. Microbiol. *3,* 159–168.

Connolly, J.P., Comerci, D., Alefantis, T.G., Walz, A., Quan, M., Chafin, R., Grewal, P., Mujer, C.V., Ugalde, R.A., and DelVecchio, V.G. (2006). Proteomic analysis of *Brucella abortus* cell envelope and identification of immunogenic candidate proteins for vaccine development. Proteomics *6,* 3767–3780.

Cutillas, P.R., and Vanhaesebroeck, B. (2007). Quantitative profile of five murine core proteomes using label-free functional proteomics. Mol. Cell. Proteomics *6,* 1560–1573.

De Groot, M.J., Daran-Lapujade, P., van Breukelen, B., Knijnenburg, T.A., de Hulster, E.A., Reinders, M.J., Pronk, J.T., Heck, A.J., and Slijper, M. (2007). Quantitative proteomics and transcriptomics of anaerobic and aerobic yeast cultures reveals post-transcriptional regulation of key cellular processes. Microbiology *153,* 3864–3878.

De Jong, M.F. Sun, Y.H., den Hartigh, A.B., van Dijl, J.M., and Tsolis, R.M. (2008). Identification of VceA and VceC, two members of the VjbR regulon that are translocated into macrophages by the *Brucella* type IV secretion system. Mol. Microbiol. *70,* 1378–1396.

Delpino, M.V., Comerci, D.J., Wagner, M.A., Eschenbrenner, M., Mujer, C.V., Ugalde, R.A., Fossati, C.A., Baldi, P.C., and Delvecchio, V.G. (2009). Differential composition of culture supernatants from wild-type *Brucella abortus* and its isogenic *virB* mutants. Arch. Microbiol. *191,* 571–581.

Delrue, R.M., Martinez-Lorenzo, M., Lestrate, P., Danese, I., Bielarz, V., Mertens, P., De Bolle, X., Tibor, A., Gorvel, J.P., and Letesson, J.J. (2001). Identification of *Brucella* spp. genes involved in intracellular trafficking. Cell. Microbiol. 3, 487–497.

Delrue, R.M., Deschamps, C., Léonard, S., Nijskens, C., Danese, I., Schaus, J.M., Bonnot, S., Ferooz, J., Tibor, A., De Bolle, X., *et al.* (2005). A quorum-sensing regulator controls expression of both the type IV secretion system and the flagellar apparatus of *Brucella melitensis.* Cell. Microbiol. *7,* 1151–1161.

DelVecchio, V.G., and Mujer, C.V. (2004). Comparative proteomics of *Brucella* species. In: *Brucella:* Molecular and Cellular Biology. López-Goñi, I., and Moriyón, I., eds. (Horizon Scientific Press: United Kingdom), pp. 103–115.

DelVecchio, V.G., Kapatral, V., Redkar, R.J., Patra, G., Mujer, C., Los, T., Ivanova, N., Anderson, I., Bhattacharyya, A., Lykidis, A., *et al.* (2002a). The genome sequence of the facultative intracellular pathogen *Brucella melitensis.* Proc. Natl. Acad. Sci. U.S.A. *99,* 443–448.

DelVecchio, V.G., Wagner, M.A., Eschenbrenner, M., Horn, T.A., Kraycer, J.A., Estock, F., Elzer, and P., Mujer, C.V. (2002b). *Brucella* proteomes: a review. Vet. Microbiol. *90,* 593–603.

Den Hartigh, A.B., Rolán, H.G., de Jong, M.F., and Tsolis, R.M. (2008). VirB3 to VirB6 and VirB8 to VirB11, but not VirB7, are essential for mediating persistence of *Brucella* in the reticuloendothelial system. J. Bacteriol. *190,* 4427–4436.

Detilleux, P.G., Deyoe, B.L., and Cheville, N.F. (1990). Penetration and intracellular growth of *Brucella abortus* in nonphagocytic vells *in vitro.* Infect. Immun. *158,* 2320–2328.

Detilleux, P.G., Deyoe, B.L., and Cheville, N.F. (1991). Effect of endocytic and metabolic inhibitors on the internalization and intracellular growth of *Brucella abortus* in Vero cells. Am. J. Vet. Res. *52,* 1658–1664.

Díaz, R., and Bosseray N. (1974). Estudio de las relaciones antigenicas entre *Yersinia enterocolitica* serotipo IX y otras especies bacterianas Gram-negativas. Microbiol. Esp. (Spain) *27,* 1–14.

Díaz, R., and Chordi, A. (1966). Immunoelectrophoretic analysis of *Brucella suis.* Experientia *22,* 825–826.

Díaz, R., and Dorronsoro, I. (1971). Contribución al diagnóstico serológico de brucelosis y yersiniosis. I.

Utilidad de la reacción de precipitación en gel. Rev. Clin. Esp. *121,* 367–372.

Díaz, R., Jones, L.M., and Wilson, J.B. (1967). Antigenic relationship of *Brucella ovis* and *Brucella melitensis*. J. Bacteriol. *93,* 1262–1268.

Díaz, R., Jones, L.M., and Wilson, J.B. (1968). Antigenic relationship of the gram-negative organism causing canine abortion to smooth and rough brucellae. J. Bacteriol. *95,* 618–624.

Díaz, R., Maraví-Poma, E., and Rivero, A. (1976). Comparison of Counter-Inmunoelectrophoresis with other serological tests in the diagnosis of human brucellosis. Bull. World Health Organ. *64,* 417–424.

Díaz, R., Maraví-Poma, E., Fernández, J.L., García-Merlo, S., and Rivero-Puente, A. (1982). Brucelosis: estudio de 222 casos. Parte IV: diagnóstico de la brucelosis humana. Rev. Clin. Esp. (Spain) *166,* 107–110.

Ding, X.Z., Paulsen, I.T., Bhattacharjee, A.K., Nikolich, M.P., Myers, G., and Hoover, D.L. (2006). A high efficiency cloning and expression system for proteomic analysis. Proteomics *6,* 4038–4046.

Douglas, J.T., Rosenberg, E.Y., Nikaido, H., Verstreate, D.R., and Winter, A.J. (1984). Porins of *Brucella* species. Infect. Immun. *44,* 16–21.

Dozot, M., Poncet, S., Nicolas, C., Copin, R., Bouraoui, H., Mazé, A., Deutscher, J., De Bolle, X., and Letesson, J.J. (2010). Functional characterization of the incomplete phosphotransferase system (PTS) of the intracellular pathogen *Brucella melitensis*. PLoS One 5, e12679.

Duan, X., Young, R., Straubinge,r R.M., Page, B., Cao, J., Wang, H., Yu, H., Canty, J.M., and Qu, J. (2009). A straightforward and highly efficient precipitation/on-pellet digestion procedure coupled to a long gradient nano-LC separation and orbitrap mass spectrometry for label-free expression profiling of the swine heart mitochondrial proteome. J. Proteome Res. *8,* 2838–2850.

Eschenbrenner, M., Horn, T.A., Wagner, M.A., Mujer, C.V., Miller-Scandle, T.L., and DelVecchio, V.G. (2006). Comparative proteome analysis of laboratory grown *Brucella abortus* 2308 and *Brucella melitensis* 16M. J. Proteome Res. *5,* 1731–1740.

Ewalt, D.R., Payeur, J.B., Martin, B.M., Cummins, D.R., and Miller, W.G. (1994). Characteristics of a *Brucella* species from a bottlenose dolphin (*Tursiops truncatus*). J. Vet. Diagn. Invest. *6,* 448–452.

Ficht, T.A., Pei, J., and Kahl-McDonagh, M. (2010). *In vitro* mutagenesis of *Brucella* species. Methods Mol. Biol. *634,* 15–35.

Foulongne, V., Bourg, G., Cazevieille, C., Michaux-Charachon, S., and O'Callaghan, D. (2000). Identification of *Brucella suis* genes affecting intracellular survival in an *in vitro* human macrophage infection model by signature-tagged transposon mutagenesis. Infect. Immun. *68,* 1297–303.

Freer, E., Moreno, E., Moriyón, I., Pizarro-Cerdá, J., Weintraub, A., and Gorvel, J.P (1996). *Brucella*-Salmonella-lipopolysaccharide chimeras are less permeable to hydrophobic probes and more sensitive to cationic peptides and EDTA than their native *Brucella* sp. counterparts. J. Bacteriol. *178,* 5867–5876.

Freiburghaus, A.U. (1994). Two-dimensional polyacrylamide gel electrophoresis of membrane proteins. Mol. Biotechnol. *2,* 281–293.

Fugier, E., Salcedo, S.P., de Chastellier, C., Pophillat, M., Muller, A., Arce-Gorvel, V., Fourquet, P., and Gorvel, J.P. (2009). The glyceraldehyde-3-phosphate dehydrogenase and the small GTPase Rab 2 are crucial for *Brucella* replication. PLoS Pathog. *5,* e1000487.

Gamazo, C., Winter, A.J., Moriyón, I., Riezu-Boj, J.I., Blasco, J.M., and Díaz, R. (1989). Comparative analyses of proteins extracted by hot saline or released spontaneously into outer membrane blebs from field strains of *Brucella ovis* and *Brucella melitensis*. Infect. Immun. *57,* 1419–26.

Gauci, S., Veenhoff, L.M., Heck A.J.R., and Krijgsveld, J. (2009). Orthogonal separation techniques for the characterization of the yeast nuclear proteome. J. Proteome Res. *8,* 3451–3463.

Gee, J.M., Valderas, M.W., Kovach, M.E., Grippe, V.K., Robertson, G.T., Ng, W.L., Richardson, J.M., Winkler, M.E., and Roop, R.M., 2nd. (2005). The *Brucella abortus* Cu,Zn superoxide dismutase is required for optimal resistance to oxidative killing by murine macrophages and wild-type virulence in experimentally infected mice. Infect. Immun. *73,* 2873–2880.

Godefroid, M., Svensson, M.V., Cambier, P., Uzureau, S., Mirabella, A., De Bolle, X., Van Cutsem, P., Widmalm, G., and Letesson, J.J. (2010). *Brucella melitensis* 16M produces a mannan and other extracellular matrix components typical of a biofilm. FEMS Immunol. Med. Microbiol. *59,* 364–377.

Goley, E.D., Dye, N.A., Werner, J.N., Gitai, Z., and Shapiro, L. (2010). Imaging-based identification of a critical regulator of FtsZ protofilament curvature in *Caulobacter.* Mol. Cell. *39,* 975–987.

Graham, R.L., Sharma, M.K., Ternan, N.G., Weatherly, D.B., Tarleton, R.L, and McMullan, G. (2007). A semi-quantitative GeLC-MS analysis of temporal proteome expression in the emerging nosocomial pathogen *Ochrobactrum anthropi*. Genome Biol. *8,* R110

Granvogl, B., Gruber, P., and Eichacker, L.A. (2007). Standardisation of rapid in-gel digestion by mass spectrometry. Proteomics 7, 642–654.

Greiner, M., Verloo, D., and de Massis, F. (2009). Meta-analytical equivalence studies on diagnostic tests for bovine brucellosis allowing assessment of a test against a group of comparative tests. Prev. Vet. Med. *92,* 373–831.

Gross, A., Terraza, A., Ouahrani-Bettache, S., Liautard, J.P., and Dornand, J. (2000). *In vitro Brucella suis* infection prevents the programmed cell death of human monocytic cells. Infect. Immun. *68,* 342–51.

Gupta, R.S., and Mok, A. (2007). Phylogenomics and signature proteins for the alpha proteobacteria and its main groups. BMC Microbiol. *7,* 106.

Guzman-Verri, C., Manterola, L., Sola-Landa, A., Parra, A., Cloeckaert, A., Garin, J., Gorvel, J.P., Moriyón, I., Moreno, E., and López-Goñi, I. (2002). The two-component system BvrR/BvrS essential for *Brucella abortus* virulence regulates the expression of outer membrane proteins with counterparts in members of the Rhizobiaceae. Proc. Natl. Acad. Sci. U. S A. *99,* 12375–12380.

Halling, S.M., Detilleux, P.G., Tatum, F.M., Judge, B.A., and Mayfield, J.E. (1991). Deletion of the BCSP31 gene of *Brucella abortus* by replacement. Infect. Immun. *59,* 3863–3868.

Halling, S.M., Peterson-Burch, B.D., Bricker, B.J., Zuerner, R.L., Qing, Z., Li, L.L., Kapur, V., Alt, D.P., and Olsen, S.C. (2005). Completion of the genome sequence of *Brucella abortus* and comparison to the highly similar genomes of *Brucella melitensis* and *Brucella suis*. J. Bacteriol. *187,* 2715–26.

Hill, R.A., and Cook, D.R. (1994). Protein profiles of *Brucella suis* and *Brucella abortus* in isoelectric focusing and sodium dodecyl sulphate-polyacrylamide gel electrophoresis. Vet. Microbiol. *39,* 25–32.

Jiménez-de-Bagüés, M.P, Loisel-Meyer, S., Liautard, J.P., and Jubier-Maurin, V. (2007). Different roles of the two high-oxygen-affinity terminal oxidases of *Brucella suis*: cytochrome c oxidase, but not ubiquinol oxidase, is required for persistence in mice. Infect. Immun. *75,* 531–535.

Jones, L.M., and Berman, D.T. (1975). Antibody-mediated and delayed-type hypersensitivity reactions to *Brucella* skin test antigens in guinea pigs. Infect. Immun. *11,* 360–364.

Jones, L.M., Díaz, R., and Taylor, A.G. (1973). Characterization of allergens prepared from smooth and rough strains of *Brucella melitensis*. Br. J. Exp. Pathol. *54,* 492–508.

Jubier-Maurin, V,. Loisel, S., Liautard, J. P., and Köhler, S. (2004). The intramacrophagic environment of *Brucella* spp. and their replicative niche In: *Brucella* Molecular and Cellular Biology, López-Goñi, I., and Moriyón, I., eds. (Horizon Scientific Press: United Kingdom), pp. 313–340.

Karlin, S., Barnett, M.J., Campbell, A.M., Fisher, R.F., and Mrazek, J. (2003). Predicting gene expression levels from codon biases in alpha-proteobacterial genomes. Proc. Natl. Acad. Sci. U.S.A. *100,* 7313–7318.

Kim, S., Watarai, M., Kondo, Y., Erdenebaatar, J., Makino, S., and Shirahata, T. (2003). Isolation and characterization of mini-Tn5Km2 insertion mutants of *Brucella abortus* deficient in internalization and intracellular growth in HeLa cells. Infect. Immun. *71,* 3020–3027.

Kittelberger, R., Hilbink, F., Hansen, M.F., Ross, G.P., de Lisle, G.W., Cloeckaert, A., and de Bruyn. J. (1995). Identification and characterization of immunodominant antigens during the course of infection with *Brucella ovis*. J. Vet. Diagn. Invest. *7,* 210–218.

Lai, T.C., Chou, H.C., Chen, Y.W., Lee, T.R., Chan, H.T., Shen, H.H., Lee, W.T., Lin, S.T., Lu, Y.C., Wu, C.L., and Chan, H.L. (2010). Secretomic and proteomic analysis of potential breast cancer markers by two-dimensional differential gel electrophoresis. J. Proteome Res. *9,* 1302–1322.

Lamontagne, J., Butler, H., Chaves-Olarte, E., Hunter, J., Schirm, M., Paquet, C., Tian, M., Kearney, P., Hamaidi, L., Chelsky, D., *et al.* (2007). Extensive cell envelope modulation is associated with virulence in *Brucella abortus*. J. Proteome Res. *6,* 1519–1529.

Lamontagne, J., Forest, A., Marazzo, E., Denis, F., Butler, H., Michaud, J.F., Boucher, L., Pedro, I., Villeneuve, A., Sitnikov, D., *et al.* (2009). Intracellular adaptation of *Brucella abortus*. J. Proteome Res. *8,* 1594–1609.

Lamontagne, J., Béland, M., Forest, A., Côté-Martin, A., Nassif, N., Tomaki, F., Moriyón, I., Moreno, E., and Paramithiotis, E. (2010). Proteomics-based confirmation of protein expression and correction of annotation errors in the *Brucella abortus* genome. BMC Genomics *11,* 300–310.

Latimer, E., Simmers, J., Sriranganathan, N., Roop, R.M., 2nd., Schurig, G.G., and Boyle, S.M. (1992). *Brucella abortus* deficient in copper/zinc superoxide dismutase is virulent in BALB/c mice. Microb. Pathog. *12,* 105–113

Lavín, J.L., Binnewies. T.T., Pisabarro, A.G., Ussery, D.W., García-Lobo, J.M., and Oguiza, J.A. (2010). Differences in two-component signal transduction proteins among the genus *Brucella*: implications for host preference and pathogenesis. Vet. Microbiol. *144,* 478–483.

LeVier, K., Phillips, R.W., Grippe, V.K., Roop, R.M., 2nd., and Walker, G.C. (2000). Similar requirements of a plant symbiont and a mammalian pathogen for prolonged intracellular survival. Science *287,* 2492–2493.

Lin, J., and Ficht, T.A. (1995). Protein synthesis in *Brucella abortus* induced during macrophage infection. Infect. Immun. *63,* 1409–1414.

Lin, J.L., Adams, G., and Ficht, T.A. (1992). Characterization of the heat shock response in *Brucella abortus* and isolation of the genes encoding the GroE heat shock proteins. Infect. Immun. *60,* 2425–2431.

Loisel-Meyer, S., Jiménez-de-Bagüés, M.P., Bassères, E., Dornand, J., Köhler, S., Liautard, J.P., and Jubier-Maurin, V. (2006). Requirement of *norD* for *Brucella suis* virulence in a murine model of *in vitro* and *in vivo* infection. Infect. Immun. *74,* 1973–1976.

López-Goñi, I., Manterola, L., and Pan S.Q. (2004). The *Brucella* bvrS/bvrR and related two-component regulatory systems of the α-2 Proteobacteria: common regulatory strategies of animal and plant pathogens and endosymbionts. In: *Brucella* Molecular and Cellular Biology, López-Goñi, I., and Moriyón, I., eds. (Horizon Scientific Press: United Kingdom), pp. 211–230.

Manterola, L., Moriyón, I., Moreno, E., Sola-Landa, A., Weiss, D.S., Koch, M.H., Howe, J., Brandenburg, K., and López-Goñi, I. (2005). The lipopolysaccharide of *Brucella abortus* BvrS/BvrR mutants contains lipid A modifications and has higher affinity for bactericidal cationic peptides. J. Bacteriol. *187,* 5631–5639.

Manterola, L., Guzman-Verri, C., Chaves-Olarte, E., Barquero-Calvo, E., de Miguel, M. J., Moriyón, I., Grilló, M. J., López-Goñi, I., and Moreno, E. (2007). The BvrR/BvrS-controlled outer membrane proteins Omp3a and Omp3b are not essential for *Brucella abortus* virulence. Infect. Immun. *75,* 4867–4874.

Marchesini, M.I., Ugalde, J.E., Czibener, C., Comerci, D.J., and Ugalde, R.A. (2004). N-terminal-capturing screening system for the isolation of *Brucella abortus* genes encoding surface exposed and secreted proteins. Microb. Pathog. *37*, 95–105.

Martínez-Núñez, C., Altamirano-Silva, P., Alvarado-Guillén, F., Moreno, E., Guzmán-Verri, C., and Chaves-Olarte, E. (2010). The two component system BvrR/BvrS regulates the expression of the Type IV secretion system VirB in *Brucella abortus*. J. Bacteriol. *192*, 5603–5608.

Martín-Martín, A.I., Caro-Hernández, P., Sancho, P., Tejedor, C., Cloeckaert, A., Fernández-Lago, L., and Vizcaíno, N. (2009). Analysis of the occurrence and distribution of the Omp25/Omp31 family of surface proteins in the six classical *Brucella* species. Vet. Microbiol. *137*, 74–82.

Mastroleo, F., Leroy, B., Van, Houdt, R., s' Heeren, C., Mergeay, M., Hendrickx, L., and Wattiez, R. (2009). Shotgun proteome analysis of *Rhodospirillum rubrum* S1H: integrating data from gel-free and gel-based peptides fractionation methods. J. Proteome Res. *8*, 2530–2541.

Meyer, M.E., and Cameron, H.S. (1961). Metabolic characterization of the genus *Brucella*. I. Statistical evaluation of the oxidative rates by which type I of each species can be identified. J. Bacteriol. *82*, 387–395.

Moreno, E. (1998). Genome evolution within the alpha Proteobacteria: why do some bacteria not possess plasmids and others exhibit more than one different chromosome? FEMS Microbiol. Rev. *22*, 255–275.

Moreno, E., and Gorvel, J.P. (2004). Invasion, intracellular trafficking and teplication of *Brucella* organisms in professional and non-professional phagocytes. In: *Brucella*, Molecular and Cellular Biology, López-Goñi, I., and Moriyón, I., eds. (Horizon Scientific Press: United Kingdom), pp. 287–312.

Moreno, E., and Moriyón, I. (2002). *Brucella melitensis*: a nasty bug with hidden credentials for virulence. Proc. Natl. Acad. Sci. U.S.A. *99*, 1–3

Moreno, E., and Moriyón, I. (2006). The genus *Brucella*. In: The Prokaryotes Vol. 5, Part 1, Section 3.1, Dworkin, M., Falkow, S., Rosenberg, E., Schleifer, K.H., and Stackebrant, E., eds. (Springer-Verlag, New York), pp. 315–456.

Moriyón, I., and Berman, D.T. (1983). Isolation, purification, and partial characterization of *Brucella abortus* matrix protein. Infect. Immun. *39*, 394–402.

Moriyón, I., and López-Goñi, I. (1998). Structure and properties of the outer membranes of *Brucella abortus* and *Brucella melitensis*. Int. Microbiol. *1*, 19–26.

Morris, J.A. (1973). The use of polyacrylamide gel electrophoresis in taxonomy of *Brucella*. J. Gen. Microbiol. *76*, 231–237.

Mujer, C.V., Wagner, M.A., Eschenbrenner, M., Horn, T., Kraycer, J.A., Redkar, R., Hagius, S., Elzer, P., and Delvecchio, V.G. (2002). Global analysis of *Brucella melitensis* proteomes. Ann. N. Y. Acad. Sci. *969*, 97–101.

Myers, D.M., Jones, L.M., and Varela-Díaz, V.M. (1972). Studies of antigens for complement fixation and gel diffusion tests in the diagnosis of infections caused by *Brucella ovis* and other *Brucella*. Appl. Microbiol. *23*, 894–902.

Myers, D.M., Varela-Díaz, V.M., and Coltorti, E. (1974). A Comparative sensitivity of gel-diffusion and tube agglutination tests for the detection of *Brucella* canis antibodies in experimentally infected dogs. Appl. Microbiol. *28*, 1–4

O'Callaghan, D., Cazevieille, C., Allardet-Servent, A., Boschiroli, M.L., Bourg, G., Foulongne, V., Frutos, P., Kulakov, Y., and Ramuz, M. (1999). A homologue of the *Agrobacterium tumefaciens* VirB and *Bordetella pertussis* Ptl type IV secretion systems is essential for intracellular survival of *Brucella suis*. Mol. Microbiol. *33*, 1210–1220.

Olitzki, A.L. (1960). The antigenic structure of *Brucellae*. Boll. Inst. Slerotev. Milan. *39*, 97–104.

Patel, V.J., Thalassinos, K., Slade, S.E, Connolly, J.B., Crombie, A., Murrell, J.C., and Scrivens, J.H. (2009). A comparison of labeling and label-free mass spectrometry-based proteomics approaches. J. Proteome Res. *8*, 3752–3759.

Paulsen, I.T., Seshadri, R., Nelson, K.E., Eisen, J.A., Heidelberg, J.F., Read, T.D., Dodson, R.J., Umayam, L., Brinkac, L.M., Beanan, M.J., *et al.* (2002). The *Brucella suis* genome reveals fundamental similarities between animal and plant pathogens and symbionts. Proc. Natl. Acad. Sci. U.S.A. *99*, 13148–13153.

Porte, F, Liautard, J.P., and Köhler, S. (1999). Early acidification of phagosomes containing *Brucella suis* is essential for intracellular survival in murine macrophages. Infect. Immun. 67, 4041–4047.

Rafie-Kolpin, M., Essenberg, R.C., and Wyckoff, J.H. 3rd. (1996). Identification and comparison of macrophage-induced proteins and proteins induced under various stress conditions in *Brucella abortus*. Infect. Immun. *64*, 5274–5283.

Rambow-Larsen, A.A., Rajashekara, G., Petersen, E., and Splitter, G. (1982). Putative quorum-sensing regulator BlxR of *Brucella melitensis* regulates virulence factors including the type IV secretion system and flagella. J. Bacteriol. *190*, 3274–3282.

Raybould, T.J. (1982). Antigens of diagnostic significance in *Brucella abortus*. Can. J. Microbiol. *28*, 557–566.

Robertson, G.T., and Roop, R.M., Jr. (1999). The *Brucella abortus* host factor I (HF-I) protein contributes to stress resistance during stationary phase and is a major determinant of virulence in mice. Mol. Microbiol. 34, 690–700.

Roop, R.M., II, Gee, J.M., Robertson, G.T., Richardson, J.M., Ng, W.L., and Winkler, M.E. (2003). *Brucella* stationary-phase gene expression and virulence. Annu. Rev. Microbiol. *57*, 57–76.

Roop, R.M. II, Bellaire, B.H., Anderson, E., and Paulley, J.T. (2004). Iron metabolism in *Brucella*. In *Brucella:* Molecular and Cellular Biology, López-Goñi, I., and Moriyón, I., eds. (Horizon Scientific Press: United Kingdom), pp. 243–262.

Sánchez, D.O., Zandomeni, R.O., Cravero, S., Verdún, R.E., Pierrou, E., Faccio, P., Díaz, G., Lanzavecchia, S., Agüero, F., Frasch, A.C., *et al.* (2001). Gene discovery through genomic sequencing of *Brucella abortus*. Infect. Immun. *69,* 865–868.

Schurig, G.G., Jones, L.M., Speth, S.L., and Berman, D.T. (1978). Antibody response to antigens distinct from smooth lipopolysaccharide complex in *Brucella* infection. Infect. Immun. *21,* 994–1002.

Schurig, G.G., Pringle, A.T., and Breese, S.S., Jr. (1981). Localization of *Brucella* antigens that elicit a humoral immune response in *Brucella abortus*-infected cells. Infect. Immun. *34,* 1000–1007.

Sieira, R., Comerci, D.J., Sánchez, D.O., and Ugalde, R.A. (2000). A homologue of an operon required for DNA transfer in *Agrobacterium* is required in *Brucella abortus* for virulence and intracellular multiplication. J. Bacteriol. *182,* 4849–4855.

Sieira, R., Comerci, D.J., Pietrasanta, L.I., and Ugalde, R.A. (2004). Integration host factor is involved in transcriptional regulation of the *Brucella abortus virB* operon. Mol. Microbiol. *54,* 808–822.

Sieira, R., Arocena, G.M., Bukata, L., Comerci, D.J., and Ugalde, R.A. (2010). Metabolic control of virulence genes in *Brucella abortus*: HutC coordinates virB expression and the histidine utilization pathway by direct binding to both promoters. J. Bacteriol. *192,* 217–224.

Sola-Landa, A., Pizarro-Cerdá, J., Grilló, M.J., Moreno, E., Moriyón, I., Blasco, J.M., Gorvel, J.P., and López-Goñi, I. (1998). A two-component regulatory system playing a critical role in plant pathogens and endosymbionts is present in *Brucella abortus* and controls cell invasion and virulence. Mol. Microbiol. *29,* 125–138.

Sowa, B.A., Crawford, R.P., Heck, F.C., Williams, J.D., Wu, A.M., Kelly, K.A., and Adams, L.G. (1986). Size, charge and structural heterogeneity of *Brucella abortus* lipopolysaccharides demonstrated by two-dimensional gel electrophoresis. Electrophoresis 7, 283–288.

Sowa, B.A., Kelly, K.A., Ficht, T.A., and Adams, L.G. (1992). Virulence associated proteins of *Brucella abortus* identified by paired two-dimensional gel electrophoretic comparisons of virulent, vaccine and LPS deficient strains. Appl. Theor. Electrophor. *3,* 33–40.

Stemshorn, B.W. (1984). Recent progress in the diagnosis of brucellosis. Dev. Biol. Stand. *56,* 325–340.

Stewart, II., Thomson, T., and Figeys D. (2001). 18O labeling: a tool for proteomics. Rapid Commun. Mass Spectrom. *15,* 2456–2465.

Tabatabai, L.B., and Deyoe, B.L. (1984). Biochemical and biological properties of soluble protein preparations from *Brucella abortus*. Dev. Biol. Stand. *56,* 199–211.

Tatum, F.M., Morfitt, D.C., and Halling, S.M. (1993). Construction of a *Brucella abortus* RecA mutant and its survival in mice. Microb. Pathog. *14,* 177–185.

Teixeira-Gomes, A.P., Cloeckaert, A., Bézard, G., Bowden, R.A., Dubray, G., and Zygmunt, M.S. (1979a). Identification and characterization of *Brucella ovis* immunogenic proteins using two-dimensional electrophoresis and immunoblotting. Electrophoresis *18,* 1491–1497.

Teixeira-Gomes, A.P., Cloeckaert, A., Bézard, G., Dubray, G., and Zygmunt, M.S. (1997b). Mapping and identification of *Brucella melitensis* proteins by two-dimensional electrophoresis and microsequencing. Electrophoresis *18,* 156–62.

Teixeira-Gomes, A.P., Cloeckaert, A., and Zygmunt, M.S. (2000). Characterization of heat, oxidative, and acid stress responses in *Brucella melitensis*. Infect. Immun. *68,* 2954–2961.

Tsolis, R.M. (2002). Comparative genome analysis of the alpha-proteobacteria: relationships between plant and animal pathogens and host specificity. Proc. Natl. Acad. Sci. U.S.A. *99,* 12503–12505.

Tsolis, R.M., Seshadri, R., Santos, R.L., Sangari, F.J., Lobo, J.M., de Jong, M.F., Ren, Q., Myers, G., Brinkac, L.M., Nelson, W.C., *et al.* (2009). Genome degradation in *Brucella ovis* corresponds with narrowing of its host range and tissue tropism. PLoS One *4,* e5519.

Uzureau, S., Godefroid M., Deschamps, C., Lemaire, J., De Bolle, X., and Letesson, J.J. (2007). Mutations of the quorum sensing-dependent regulator VjbR lead to drastic surface modifications in *Brucella melitensis*. J. Bacteriol. *189,* 6035–6047.

Uzureau, S., Lemaire, J., Delaive, E., Dieu, M., Gaigneaux, A., Raes, M., De Bolle, X., and Letesson, J.J. (2010). Global analysis of quorum sensing targets in the intracellular pathogen *Brucella melitensis* 16 M. J. Proteome Res. *9,* 3200–3217.

Valderas, M.W., Alcantara, R.B., Baumgartner, J.E., Bellaire, B.H., Robertson, G.T., Ng, W.L., Richardson, J.M., Winkler, M.E., and Roop, R.M, 2nd. (2005). Role of HdeA in acid resistance and virulence in *Brucella abortus* 2308. Vet. Microbiol. *107,* 307–312.

Veale, D.R., Smith, H., and Witt, K. (1975). Penetration of penicillin into human phagocytes containing gonococci. Lancet. *1,* 306–308.

Velasco, J., Díaz, R., Grilló, M.J., Barberán, M., Marín, C., Blasco, J.M., and Moriyón, I. (1997). Antibody and delayed-type hypersensitivity responses to Ochrobactrum anthropi cytosolic and outer membrane antigens in infections by smooth and rough *Brucella* spp. Clin. Diagn. Lab. Immunol. *4,* 279–84.

Velasco, J., Romero, C., López-Goñi, I., Leiva, J., Díaz, R., and Moriyón, I. (1998). Evaluation of the relatedness of *Brucella* spp. and *Ochrobactrum anthropi* and description of *Ochrobactrum intermedium* sp. nov., a new species with a closer relationship to *Brucella* spp. Int. J. Syst. Bacteriol. *48* Pt 3, 759–768.

Verstreate, D.R., and Winter, A.J. (1984). Comparison of sodium dodecyl sulfate-polyacrylamide gel electrophoresis profiles and antigenic relatedness among outer membrane proteins of 49 *Brucella abortus* strains. Infect. Immun. *46,* 182–187.

Viadas, C., Rodríguez, M.C., Sangari, F.J., Gorvel, J.P., García-Lobo, J.M., and López-Goñi, I. (2010). Transcriptome analysis of the *Brucella abortus* BvrR/BvrS two-component regulatory system. PLoS One *5,* e10216

Vizcaíno, N., Zygmunt, M.S., Verger, J.M., Grayon, M., and Cloeckaert, A. (1997). Localization and characterization of a specific linear epitope of the *Brucella* DnaK protein. FEMS Microbiol. Lett. *154,* 117–122.

Wagner, M,A., Eschenbrenner, M., Horn, T.A., Kraycer, J.A., Mujer, C.V., Hagius, S., Elzer, P., and DelVecchio, V.G. (2002). Global analysis of the *Brucella melitensis* proteome: identification of proteins expressed in laboratory-grown culture. Proteomics *2,* 1047–1060.

Wang, Y., Chen, Z., Qiao, F., Ying, T., Yuan, J., Zhong, Z., Zhou, L., Du, X., Wang, Z., Zhao, J., *et al.* (2009). Comparative proteomics analyses reveal the virB of *B. melitensis* affects expression of intracellular survival related proteins. PLoS One *4,* e5368.

Wang, Y., Chen, Z., Qiao, F., Zhong, Z., Xu, J., Wang, Z., Du, X., Qu, Q., Yuan, J., Jia, L., *et al.* (2010). The type IV secretion system affects the expression of Omp25/Omp31 and the outer membrane properties of *Brucella melitensis*. FEMS Microbiol. Lett. *303,* 92–100.

Wattam, A.R., Williams, K.P., Snyder, E.E., Almeida, N.F., Jr., Shukla, M., Dickerman, A.W., Crasta, O.R., Kenyon, R., Lu, J., Shallom, J.M., *et al.* (2009). Analysis of ten *Brucella* genomes reveals evidence for horizontal gene transfer despite a preferred intracellular lifestyle. J. Bacteriol. *191,* 3569–3579.

Biology and Genetics of the *Brucella* Outer Membrane

7

Nieves Vizcaíno and Axel Cloeckaert

Abstract

The particular characteristics of the *Brucella* outer membrane (OM) are considered to importantly contribute to the biological properties of these bacteria that are able to survive in the hostile environment of phagocytes and cause persistent infections in diverse mammals. This contrasts with the *Brucella*'s closest relatives in the genus *Ochrobactrum* which are only occasionally associated to opportunistic infections. High hydrophobicity, resistance to complement, bactericidal cationic peptides, detergents or EDTA, resistance to macrophage degradation or the protection against the host immune response are characteristics related to the *Brucella* OM. The O-polysaccharide chains of the smooth lipopolysaccharide (S-LPS) are a key component for the virulence of smooth strains but, since naturally rough *B. canis* and *B. ovis* that lack O-chains in the LPS are virulent in their respective natural hosts, other traits must have a significant role in virulence. Heterogeneity in LPS, OM proteins, and phospholipids draw a particular OM for each *Brucella* species (in some instances at the biovar or strain level) with a definite composition and topology. This variability leads to modifications in the antigenic and functional properties that could affect the interaction with the host and be related to the differences in pathogenicity and host tropism among the brucellae.

Introduction

The outer membrane (OM) of many Gram-negative bacteria is the most external layer of the cell constituting an important barrier for survival in hostile environments and an accessible target for the interaction of bacterial pathogens with the host and the defence mechanisms of the immune system. Therefore, the components of the OM can contribute by several mechanisms to the establishment of infection, but can equally be targets for the neutralization of the pathogen by the host.

Brucella evades innate immune mechanisms and is able to establish an intracellular niche inside phagocytes where it can not only survive but also replicate. Many *Brucella* factors, both known and as yet to be identified, contribute to these particular host–pathogen interactions but the specific properties conferred to the bacterium by the atypical composition of its OM, when compared with that of other Gram-negative bacteria, are considered crucial for this behaviour. Moreover, a relative heterogeneity of the OM within the genus *Brucella* probably contributes to the differences of pathogenicity and host preference known to exist among the species of this highly homogeneous genus at the DNA level (Verger *et al.*, 1985).

This chapter reviews the data about the main components of the *Brucella* OM. We highlight the molecular differences that have been detected among species and discuss their relevance for the interaction of the bacterium with the host. They also may have clinical significance since the *Brucella* OM components induce a strong specific immune response in the animal host that constitutes the basis of the diagnostic tools and vaccines commonly used for brucellosis at present and in development.

General structure of the *Brucella* outer membrane

The *Brucella* OM, as that of typical Gram-negative bacteria (Kamio and Nikaido, 1976), is an asymmetrical lipid bilayer composed of lipopolysaccharide (LPS), proteins and phospholipids (PLs) with the LPS molecules and PLs located in the outer and inner leaflet, respectively (Fig. 7.1). Depending on its structure, LPS can be smooth (S-LPS) or rough (R-LPS). S-LPS has a lipid A -of hydrophobic nature that anchors the hydrophilic moiety of LPS into the OM-, the core oligosaccharide and the O-polysaccharide (O-PS) chains that are displayed towards the surrounding environment outside the bacteria. The absence of O-chains is the differential characteristic of R-LPS. Several outer membrane proteins (OMPs) have been detected in the genus *Brucella*, the most extensively studied being several minor lipoproteins and the major OMPs (see below) that are strongly associated with the peptidoglycan (PG) of the cell wall (Fig. 7.1) (Sowa *et al.*, 1991; Cloeckaert *et al.*, 1992b; Vizcaíno *et al.*, 1996). The PLs composition of the *Brucella* OM has not been specifically determined, but the analysis of the whole cell revealed that the main PLs of *B. melitensis* and *B. abortus* are phosphatidylcholine (lecithin), phosphatidylethanolamine (cephalin), diphosphatidylglycerol (cardiolipin) and phosphatidylglycerol (Thiele and Schwinn, 1973).

Although most Gram-negative bacteria share this basic structure, the *Brucella* OM has peculiar physicochemical characteristics giving distinctive traits to the bacterium when compared with other *Proteobacteria*. Some of these traits are considered ancestral and are shared by members of the *Rhizobiaceae* that group with *Brucella* in the α–2 subdivision of the *Proteobacteria* and that also establish parasitic or symbiotic interactions with higher eukaryotic cells (Moreno *et al.*, 1990, 2002). However, other OM traits are particular for *Brucella* and indeed differ from those of its phylogenetically closest relatives of the genus *Ochrobactrum* (Velasco *et al.*, 2000; Barquero-Calvo *et al.*, 2009).

The integrity of many Gram-negative bacteria is dependent on the strong negative charge of their surface that is stabilized by divalent cations and polyamines converting the OM in a selective permeability barrier excluding hydrophobic compounds (Hancock, 1997; Nikaido, 2003). However, a reduced negative charge has been

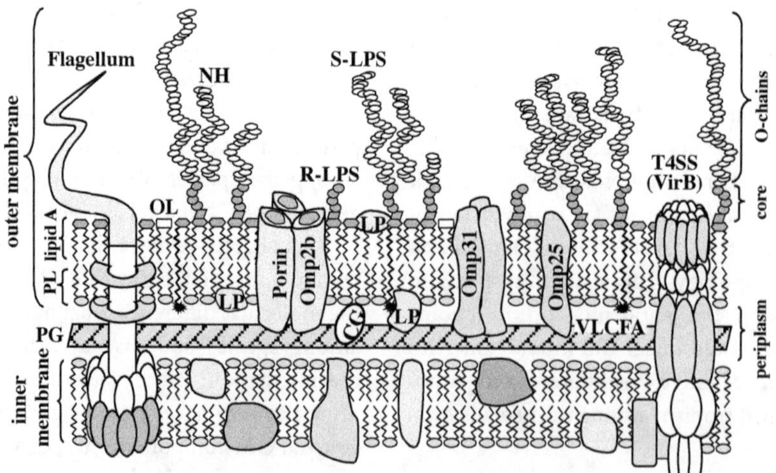

Figure 7.1 Schematic representation of the OM of a smooth *Brucella* strain. CG, cyclic glucan; LP, lipoprotein; NH, native hapten; OL, ornithine lipid; PL, phospholipid; PG, peptidoglycan; R-LPS, rough lipopolysaccharide; S-LPS, smooth lipopolysaccharide; T4SS, type-IV secretion system; VLCFA, very long-chain fatty acids.

reported for the *Brucella* surface (Moriyón and López-Goñi, 1998), which is explained in part by particular LPS characteristics -such as the lower amounts of the acidic sugar 3-deoxy-d-manno-2-octulosonic acid (KDO), the absence of other acidic core sugars or the low proportion of phosphate in the lipid A- and by the presence of positively charged free ornithine lipids that would neutralize the low amount of anionic groups in lipid A (Moriyón and López-Goñi, 1998). Therefore, it is believed that the stability of the *Brucella* OM is related to several particular characteristics: (i) in contrast to many Gram-negative bacteria where phosphatidylethanolamine is the major PL (Goldfine, 1984), the most abundant PL in both *B. melitensis* and *B. abortus* species is phosphatidylcholine (about 37%) (Thiele and Schwinn, 1973), a common PL in eukaryotic cells but rare in prokaryotes and that provides more stability to the lipid bilayer of bacterial membranes (Goldfine, 1984); (ii) the presence of positively charged ornithine lipids (Fig. 7.1), usually coextracted with LPS and thought to locate in the outer leaflet of the OM (Moriyón and López-Goñi, 1998); (iii) the strong association of some OMPs to LPS and PG (Dubray and Bézard, 1980; Winter, 1987; Cloeckaert *et al.*, 1992b; Vizcaíno *et al.*, 1996); and (iv) the presence of very long chain fatty acids (VLCFAs) in the lipid A of LPS (Moreno *et al.*, 1990; Bhat *et al.*, 1991; Velasco *et al.*, 2000) (Fig. 7.1). These VLCFAs – which have also been detected in other members of the α-2 subgroup of *Proteobacteria* (Bhat *et al.*, 1991) – have the potential to span the OM displaying the terminal hydroxyl group in the periplasmic space (Fig. 7.1), what probably favours a strong anchorage of the LPS and the integrity of the OM (Velasco *et al.*, 2000; Iriarte *et al.*, 2004; Barquero-Calvo *et al.*, 2009).

In addition to LPS, another polysaccharide seems to be associated with the OM of smooth *Brucella* strains. The native hapten (NH) is a polysaccharide with an identical composition to that determined for the O-PS of the S-LPS (Aragón *et al.*, 1996) and whose existence as an independent molecule has been questioned for a while. Some works reported indeed that it was an artefact of the extraction protocols that released O-PS chains by hydrolysis of the S-LPS (Zygmunt *et al.*, 1988). However, other works have shown that the acid hydrolysis of S-LPS releases O-PS containing KDO while purified NH lacks this core sugar, which led to the proposal that the NH is a precursor of the S-LPS constituted by the O-chains not linked to the core–lipid A complex (Aragón *et al.*, 1996). Additionally, several evidences support its localization in the OM where it would be anchored through interactions with the S-LPS (Aragón *et al.*, 1996) (Fig. 7.1).

The *virB* operon of *Brucella* spp. (O'Callaghan *et al.*, 1999; Sieira *et al.*, 2000) encodes 12 proteins showing homology with proteins of several type IV secretion systems -including the widely studied *virB* operon of *Agrobacterium tumefaciens*- that presumably form a channel spanning both bacterial membranes and serve to secrete macromolecules across the cell envelope (Álvarez-Martínez and Christie, 2009). The multimolecular cell envelope VirB complex, identified as an important virulence factor in *Brucella* spp., contains some OMPs but, since it will be exhaustively analysed in another chapter of this book, it will not be discussed here. Also other OM related structures, the flagellar apparatus and a matrix-forming exopolysaccharide that have been detected in *Brucella melitensis* under some experimental conditions (Fretin *et al.*, 2005; Uzureau *et al.*, 2007), will be analysed in other chapters owing to their connexion with *virB* and quorum sensing regulation.

Little is known until present about the regulatory mechanisms intervening on the expression of genes involved in the biogenesis and homeostasis of the *Brucella* OM although mutations in two important regulatory systems, the two-component system BvrR/BvrS and the quorum sensing regulator VjbR, have been shown to induce surface modifications altering not only the OMPs pattern (Guzmán-Verri *et al.*, 2002; Lamontagne *et al.*, 2007; Uzureau *et al.*, 2007), the expression of genes encoding OM-related components – such as OMPs, VirB type IV secretion system or the flagellar apparatus- (Delrue *et al.*, 2005; Viadas *et al.*, 2010; Weeks *et al.*, 2010) or the acylation degree of lipid A and LPS acyl-chain fluidity (Manterola *et al.*, 2005), but also the cell properties associated with the OM (Sola-Landa *et al.*, 1998; Manterola *et al.*, 2005; Lamontagne *et al.*, 2007; Uzureau *et al.*, 2007). Both regulatory systems are recognized virulence factors in *Brucella* spp. (Sola-Landa *et al.*, 1998; Delrue *et al.*, 2005) and the attenuation in virulence of their corresponding mutant strains has been related in part with these surface alterations (Manterola *et al.*, 2005; Lamontagne *et al.*, 2007; Uzureau *et al.*, 2007). The heat shock sigma factor RpoH2 (Delory *et al.*, 2006) and the transcriptional regulator BlxR (Rambow-Larsen *et al.*, 2008) were also reported to be involved in the expression of the *virB* operon and the flagellar genes. Additionally, a differential expression profile of several OMP-encoding genes and genes involved in LPS biosynthesis has been found in *B. melitensis* 16M between late logarithmic and stationary growth phases, the most and least invasive growth phase in HeLa cells in culture, respectively (Rossetti *et al.*, 2009). The intramacrophagic proteome of *B. suis* also showed the increased production of several OMPs (Al Dahouk *et al.*, 2008)

OM-related properties of the *Brucella* species

The particular properties of the *Brucella* OM, when compared with related bacteria, have been shown by the behaviour of several strains in the presence of cationic peptides, detergents, EDTA, non-immune serum and hydrophobic compounds (Corbeil *et al.*, 1988; Martínez de Tejada and Moriyón, 1993; Martínez de Tejada *et al.*, 1995; Freer *et al.*, 1996, 1999; Eisenschenk *et al.*, 1999; Velasco *et al.*, 2000; Fernández-Prada *et al.*, 2001; Estein *et al.*, 2004; Barquero-Calvo *et al.*, 2009; Martín-Martín *et al.*, 2010a) and it has been stressed that they account for the distinctive interaction of *Brucella* spp. with the host (Moriyón and López-Goñi, 1998; Velasco *et al.*, 2000; Barquero-Calvo *et al.*, 2009).

Since cationic peptides, such as defensins and cathelicidins, are part of the oxygen-independent bactericidal mechanisms of phagocytes and are also found in mammal tissues and body fluids, it is believed that they perform diverse biological roles in host defence (Jenssen *et al.*, 2006). Accordingly, the unusual resistance of smooth and rough *Brucella* strains to the cationic peptide polymyxin B and other peptides of animal origin (Martínez de Tejada and Moriyón, 1993; Martínez de Tejada *et al.*, 1995; Freer *et al.*, 1996, 1999; Caro-Hernández *et al.*, 2007; Jensen and Halling, 2010; Martín-Martín *et al.*, 2010a) has been proposed as explanation for the inefficiency of the oxygen-independent systems of phagocytes to kill intracellular brucellae (Moriyón and López-Goñi, 1998). Polymyxin B, the most widely studied bactericidal cationic peptide, exerts its antimicrobial activity by an initial permeabilization of the OM – partially through its detergent-like property mediated by the fatty acid tail (Nikaido, 2003) – that allow it to reach the cytoplasmic membrane where it causes leakage of the cytoplasmic components (Vaara, 1992). Because the interaction between cationic peptides and Gram-negative bacteria is thought to occur through electrostatic bonding with negatively charged components of the OM, such as phosphate groups of the LPS lipid A or KDO of the inner core (Vaara, 1992; Nikaido, 2003), the resistance of *Brucella* spp. to cationic peptides has been attributed to the reduced negative charge of its OM (see above) and to a steric hindrance caused by the LPS O-PS chains and core that prevent the access of the cationic peptide to the deeper negatively charged groups of the OM (Moriyón and López-Goñi, 1998; Velasco *et al.*, 2000; Iriarte *et al.*, 2004). The reduced negative charge also accounts for the resistance of *Brucella* spp. to exposure to the chelator EDTA, which contrasts with the high sensitivity of other Gram-negative bacteria that require divalent cations to stabilize the negatively charged OM (Moriyón and Berman, 1982; Freer *et al.*, 1996).

The bactericidal components of non-immune serum – complement, cationic peptides or others – that are part of the innate immunity, constitute another defence mechanism of the host against pathogens when located in an extracellular environment. The activation of the complement cascade by *B. melitensis* and *B. abortus* has been shown to occur through the classic and lectin-mediated pathways without the participation of the alternative pathway (Eisenschenk *et al.*, 1999; Fernández-Prada *et al.*, 2001). Despite the heterogeneity of results, probably derived from the diversity of methodologies used, at least a partial resistance to the bactericidal effect of human, bovine or ovine non-immune serum has been reported for the genus *Brucella* (Corbeil *et al.*, 1988; Allen *et al.*, 1998; Eisenschenk *et al.*, 1999;

Fernández-Prada *et al.*, 2001; Estein *et al.*, 2004; Barquero-Calvo *et al.*, 2007; Caro-Hernández *et al.*, 2007; Martín-Martín *et al.*, 2010a). Several studies have shown that rough strains of *B. abortus* and *B. melitensis* exhibit an increased sensitivity to normal serum when compared with that observed with smooth strains (Corbeil *et al.*, 1988; Allen *et al.*, 1998; Eisenschenk *et al.*, 1999; Fernández-Prada *et al.*, 2001; González *et al.*, 2008). Additionally, the OMPs extracted from *B. abortus* smooth and rough strains were able to bind to the C1q component of bovine complement, while smooth strains did not activate the complement cascade (Eisenschenk *et al.*, 1999). Accordingly, it was proposed that the O-PS chains shield outer membrane components that are targets for complement (Eisenschenk *et al.*, 1999). However, naturally rough *B. ovis* and *B. canis* strains are even more resistant to the bactericidal effect of human serum than some smooth *Brucella* strains (Table 7.1) (Martín-Martín *et al.*, 2010a) and, therefore, other bacterial characteristics, probably related with the OM, must account for this resistance.

Another peculiarity of the *Brucella* OM, consistent with the traits described above regarding the stability of the OM, is its unusual high hydrophobicity that is in agreement with its high permeability to hydrophobic compounds (Martínez de Tejada and Moriyón, 1993; Freer *et al.*, 1999). This contrasts with the OM of other Gram-negative bacteria acting as a selective permeability barrier excluding hydrophobic compounds (Hancock, 1997). This characteristic could be harmful to *Brucella* since certain toxic compounds produced by the host, such as bile salts or other steroid-like molecules, would easily penetrate the cell. However, other mechanisms, such as efflux systems able to extrude the potentially noxious products, could compensate for the increased OM permeability. In this regard, a protein of the multidrug and toxic compound extrusion (MATE) family and a tripartite RND-type efflux pump have been

Table 7.1 Autoagglutination properties and resistance to exposure to 1 mg/ml polymyxin B (Poly B), 1 mg/ml sodium deoxycholate (DOC), 0.1% Triton X-100 (Triton), 10 mg/ml CHAPS, non-immune human serum (Serum), and pH4 of several strains of the six classical *Brucella* species (modified from Martín-Martín *et al.*, 2010a)

	Resistance[2] to exposure to						
Strain (relevant characteristics)[1]	Poly B (80 min)	DOC (80 min)	Triton (80 min)	CHAPS (80 min)	Serum (4 h)	pH 4 (5 h)	Agglutination[3] (48 h)
Bme Rev1 (smooth vaccine)	+++	–	++	+++	+++	+++	–
Bme B115 (rough strain)	+	±	+	+	+++	++	–
Bab 544 (smooth strain)	+	+++	+++	+++	±	+++	–
Bab 2308 (smooth strain)	+	+++	+++	+++	±	++	–
Bab RB51 (rough vaccine)	–	+++	+++	+++	±	+++	+++
Bab 45/20 (rough strain)	–	++	++	±	+	+++	+++
Bab B19 (smooth vaccine)	+	+	+++	–	±	++	–
Bsu 1330 (smooth strain)	+++	±	+++	+++	++	++	–
Bov PA (rough strain)	+++	–	–	+	+++	++	–
Bov Reo198 (rough strain)	+++	–	–	±	++	±	–
Bca RM6/66 (rough strain)	++	++	+++	+++	+++	++	++
Bne 5K33 (smooth strain)	+	±	+	++	±	++	–

[1]Bme, *B. melitensis*; Bab, *B. abortus*; Bsu, *B. suis*; Bov, *B. ovis*; Bca, *B. canis*; Bne, *B. neotomae*. [2]Percentage of survival after exposure to different agents – colony counts in solid medium after exposure (time in parenthesis) versus colony counts after incubation in PBS or heated serum (100% survival); –, 0–1%; ±, 1–20%; +, 20–50%; ++, 50–75%; +++, 75–100% survival. [3]Percentage of optical density at 600 nm after static incubation in culture medium for 48 h when compared with the values at time 0 (100% $OD_{600\,nm}$); –, 95–120% (no auto-agglutination); ++, 25–50%; +++, 0–5% $OD_{600\,nm}$.

found to be functional in *Brucella* spp. (Braibant *et al.*, 2002; Posadas *et al.*, 2007; Martín *et al.*, 2009). The OM component of the *B. suis* 1330 tripartite efflux pump, named BepC and belonging to the TolC family, was found to participate in the efflux of small and relatively hydrophobic compounds (Posadas *et al.*, 2007) and it has been suggested that this efflux pump may contribute to the survival of *Brucella* spp. in the initial stages of orally acquired infections by increasing the resistance to bile components during the intestinal transit of the bacterium (Posadas *et al.*, 2007; Martín *et al.*, 2009).

Additionally, the OM – and particularly the LPS – has been related to the poor phagocytosis of *Brucella*, the weak induction of the proinflammatory response in the host, the prevention of apoptosis in infected macrophages or the interference with the antigen presenting activity of macrophages (Lapaque *et al.*, 2005; Roop *et al.*, 2009), aspects that are developed in another chapter of this book.

Differences in OM-related properties among the *Brucella* species

Members of the genus *Ochrobactrum* are the closest phylogenetic relatives of the genus *Brucella* (Moreno *et al.*, 2002) sharing atypical surface characteristics such as phosphatidylcholine as a major PL, and the presence of VLCFAs and ornithine lipids in the OM (Velasco *et al.*, 2000). Moreover, gene orthologues of *Brucella* virulence factors associated to the cell envelope – *bvrR/bvrS*, *virB*, flagellin and cyclic β1–2 glucan genes, or several OM lipoproteins genes such as *omp10*, *omp16* or *omp19* – have been detected in the genome of *Ochrobactrum anthropi* (Barquero-Calvo *et al.*, 2009), and antigenic similarities have been reported between *Brucella* and *Ochrobactrum* OM components by reactivity with antibodies (Cloeckaert *et al.*, 1999; Barquero-Calvo *et al.*, 2009). However, the OM of *Ochrobactrum* spp. is impermeable to hydrophobic probes and more sensitive to polymyxin B (Velasco *et al.*, 2000), which contrasts with the OM properties defined for the genus *Brucella*. Additionally, *O. anthropi* fixes complement through molecules other than LPS and is more sensitive to the bactericidal effect of bovine non-immune serum than *B. abortus* 2308 (Barquero-Calvo *et al.*, 2009). *Brucella* spp. are animal pathogens inducing chronic infections and *Ochrobactrum* strains are free living bacteria only occasionally associated to opportunistic infections. This different pathogenic behaviour has been associated in part to their OM properties that would reflect an adaptive evolution of the genus *Brucella* towards pathogenicity from a common ancestor (Velasco *et al.*, 2000; Barquero-Calvo *et al.*, 2009).

Also remarkable are the differences found among the *Brucella* species regarding host preference and organ tropism, which contrast with the high homogeneity of this genus at the DNA level (Verger *et al.*, 1985; Foster *et al.*, 2009; Wattam *et al.*, 2009). Several factors probably contribute to these marked differences but the variability in the OM properties evidenced among species and even among strains of the same species is one most probable candidate (Jensen and Halling, 2010; Martín-Martín *et al.*, 2010a). The smooth phenotype is recognized as necessary for virulence of smooth *Brucella* strains (Godfroid *et al.*, 1998; McQuiston *et al.*, 1999; Monreal *et al.*, 2003; González *et al.*, 2008). However, two naturally rough *Brucella* species – *B. canis* and *B. ovis* – are virulent in their respective preferred natural hosts and also in experimental animal models (Caro-Hernández *et al.*, 2007). Additionally, on the contrary to rough mutants derived from smooth virulent strains, *B. canis* and *B. ovis* survive and replicate inside murine macrophages (Martín-Martín *et al.*, 2008, 2010b) and enter the host cell by using cholesterol-rich lipid rafts, thus limiting macrophage activation and favouring intracellular survival and multiplication (Martín-Martín *et al.*, 2010b).

Several properties related to the OM, such as auto-agglutination and resistance to several bactericidal agents mimicking conditions encountered in the host – cationic peptides, detergents, non-immune serum, acid and oxidative stress – were analysed in 12 *Brucella* strains representative of the six classical *Brucella* species and of the smooth and rough phenotype (Martín-Martín *et al.*, 2010a). Important differences among the *Brucella* strains but that could not be associated to the rough or smooth phenotype were revealed

(Martín-Martín *et al.*, 2010a) (Table 7.1). Auto-agglutination in culture medium was only observed in rough strains but some rough strains – *B. melitensis* B115 and *B. ovis* PA and Reo198 – behaved as all *Brucella* smooth strains and did not agglutinate (Table 7.1) (Martín-Martín *et al.*, 2010a). Resistance to the bactericidal effect of polymyxin B was found to be dependent on the presence of O-chains in the LPS but only when comparing strains of the same species. Thus, smooth *B. melitensis* Rev1 was more resistant than rough *B. melitesnsis* B115 and smooth *B. abortus* 2308, 544 and B19 were more resistant than rough *B. abortus* RB51 and 45/20. However, rough *B. ovis* and *B. canis* strains were more resistant to polymyxin B than smooth *B. abortus* and *B. neotomae* strains (Table 7.1) (Martín-Martín *et al.*, 2010a). Assays of susceptibility to several detergents, acid pH and human serum also evidenced intra and interspecies differences not justified by the smooth or rough phenotype and revealed rough *B. canis* RM6/66 as the most resistant strain overall (Table 7.1) (Martín-Martín *et al.*, 2010a).

The next sections review the knowledge about the major components of the *Brucella* surface and their encoding genes, emphasizing the differences among species that could contribute to explain their particular characteristics regarding OM properties, pathogenicity and host tropism. Identification of the most relevant proteins mentioned in the text is given in Table 7.2 for the genome of *B. suis* 1330 (accession numbers NC_004310 and NC_004311).

General characteristics of the *Brucella* LPSs

As previously mentioned, LPS is a major component of the *Brucella* OM and in smooth species occurs as S-LPS and in rough species, such as *B. canis* and *B. ovis*, as R-LPS devoid of the O-PS which is the most exposed part of S-LPS in the OM of smooth organisms. The physicochemical characteristics of the LPS are important contributors to many of the particular properties described for the *Brucella* OM in the first part of this chapter. The LPSs of *Brucella* present biological properties distinct from enterobacterial LPSs (Lapaque *et al.*, 2005; Cardoso *et al.*, 2006), such as low endotoxicity, high resistance to macrophage degradation and protection against immune responses. Taken together, these properties may constitute key virulence mechanisms of *Brucella* spp. for intracellular survival and replication and for persistence inside the animal host. The O-PS is critical in this regard for smooth strains, as rough mutants that lack O-PS do not survive as well as their smooth counterparts in macrophages or in animal models of infection.

This part of the chapter will give an overview on structure, antigenic characteristics and genetics of *Brucella* LPS. Special attention will be paid to the well-known O-PS, although the importance of R-LPS and other OM components must not be neglected considering that naturally rough *B. ovis* and *B. canis* lack O-PS but survive inside phagocytic cells and are virulent in their respective natural hosts and in animal models (Caro-Hernández *et al.*, 2007; Martín-Martín *et al.*, 2008, 2010b).

Structural and antigenic heterogeneity of *Brucella* LPSs

S-LPS, or O-PS, carries the major antigenic determinants involved in the humoral immune responses in animals and humans infected by smooth strains (Fernández-Lago and Díaz, 1986; Díaz-Aparicio *et al.*, 1993; Alonso-Urmeneta *et al.*, 1998), while immunodominant antigens in rough *Brucella* strains are R-LPS and OMPs (Riezu-Boj *et al.*, 1986). Consequently, the O-PS moiety of S-LPS has been shown to be an important protective antigen against smooth *Brucella* infection in the mouse model, both in active immunization experiments (Winter *et al.*, 1988) and in passive immunization with monoclonal antibodies (MAbs) (Montaraz *et al.*, 1986; Limet *et al.*, 1989; Cloeckaert *et al.*, 1992a). MAbs specific for R-LPS epitopes were found to give little or no protection against smooth *B. abortus* infection in mice (Cloeckaert *et al.*, 1993a), whereas one of these MAbs was found highly protective against the naturally rough pathogen *B. ovis* (Bowden *et al.*, 1995b). These differences were in agreement with surface accessibility of R-LPS epitopes in the corresponding smooth or rough pathogens (Bowden *et al.*, 1995a).

Table 7.2 Main genes mentioned in the text, predicted function of the encoded proteins and identification of the ORFs according to the annotation in the genome of *B. suis* 1330 (accession numbers NC_004310 and NC_004311)

Gene	Locus	ORF in the genome of *B. suis* 1330	Role in biosynthesis of	Predicted function or characteristic
LPS genes				
wbkE	*wbk*	BR0540	O-PS	Mannosyltransferase
manA$_{O\text{-}Ag}$	*wbk*	BR0539	Uncertain	Mannose furnishing
manC$_{O\text{-}Ag}$	*wbk*	BR0538	Uncertain	Mannose furnishing
manB$_{O\text{-}Ag}$	*wbk*	BR0537	Uncertain	Phosphomannomutase
wbkA	*wbk*	BR0529	O-PS	Mannosyltransferase
gmd	*wbk*	BR0522	O-PS	GDP-mannose 4,6-dehydratase
per	*wbk*	BR0521	O-PS	Perosamine synthetase
wzm	*wbk*	BR0520	O-PS	ABC transporter
wzt	*wbk*	BR0519	O-PS	ABC transporter
wbkB	*wbk*	BR0518	Uncertain	Unknown
wbkC	*wbk*	BR0517	O-PS	Formylation
wbkF	*wbk*	BR0511	O-PS	Undecaprenyl-glycosyl transferase
wbkD	*wbk*	BR0510	O-PS	Epimerase/dehydratase
wboA	*wbo* (GI-2)	BR0981	O-PS	Manosyltransferase
wboB	*wbo* (GI-2)	BR0982	O-PS	Manosyltransferase
*wa***		BR0615	Core	Glycosyltransferase
pgm		BR0058	Core	Phosphoglucomutase
*manB*core		BRA0348	Core	Phosphomannomutase
OMP genes				
omp2a		BR0637		Porin (experimental evidence); not synthesized in *Brucella* spp.
omp2b		BR0639		Porin (experimental evidence)
omp31		BRA0423		Omp25/Omp31 family. Porin
omp25		BR0701		Omp25/Omp31 family
omp31b		BR1622		Omp25/Omp31 family
omp25b		BR0971		Omp25/Omp31 family
cmp25c		BR0119		Omp25/Omp31 family
omp25d		BR0118		Omp25/Omp31 family
omp22		BR1284		Omp25/Omp31 family
omp10		BRA0077		OM lipoprotein
omp16		BR1695		OM lipoprotein
omp19		BR1930		OM lipoprotein
bepC		BR0945		Efflux pump OMP component
ugpB		BRA0655		Adhesin/invasin SP41

The antigenic determinants involved in serotyping with polyclonal sera are borne by the O-PS. At present, smooth *Brucella* strains are classified into three serotypes, i.e. A^+M^-, A^-M^+, and A^+M^+, according to slide agglutination with A and M monospecific polyclonal sera (Díaz *et al.*, 1968; Alton, 1988). These agglutination results correspond to strains expressing mainly the A (A dominant) or M (M dominant) antigen or both antigens in nearly equivalent amounts, respectively. Additionally, smooth *Brucella* strains share common epitopes on the O-PS with cross-reacting bacteria, of which the most important is *Yersinia enterocolitica* O:9 (Bundle *et al.*, 1984; Caroff *et al.*, 1984; Díaz-Aparicio *et al.*, 1993). By using MAbs a number of epitope specificities on the O-PS have been reported: A, M, and epitopes shared by both A- and M-dominant strains, which have been named common (C) epitopes or C/Y epitopes if also common to *Y. enterocolitica* O:9 (Douglas and Palmer, 1988; Bundle *et al.*, 1989; Meikle *et al.*, 1989; Palmer and Douglas, 1989; Vizcaíno *et al.*, 1991; Cloeckaert *et al.*, 1993b, 1998; Weynants *et al.*, 1997).

The *Brucella* O-PS structure has been described as being constituted by homopolymers of 4,6-dideoxy-4-formamido-α-D-mannopyranose residues. O-PS from A-dominant strains is a linear α-1,2-linked polymer with about 2% α-1,3 linkages, while O-PS from M-dominant strains is a linear polymer of pentasaccharide repeating units containing one α-1,3-linked and four α-1,2-linked monosaccharide residues (Meikle *et al.*, 1989). O-PS from A^+M^+ strains, i.e. *B. melitensis* biovar 3 and *B. suis* biovar 4, contain about 8% and 13% α-1,3 linkages, respectively (Meikle *et al.*, 1989). MAbs specific for the C/Y epitopes probably recognize α-1,2-linked tri- or tetrasaccharides of the O-PS (Bundle *et al.*, 1989). It was also established that antibodies specific for the A epitope require five contiguous α-1,2-linked monosaccharide residues for binding. The α-1,3 linkage should be mainly involved in the structure recognized by MAbs specific for the M epitope since such MAbs fail to react with *Y. enterocolitica* O:9 and their preferential binding to M-dominant O-PS correlates with an increased number of α-1,3-linked monosaccharide residues. Depending on the technique used binding of MAbs specific for the M epitope to some A-dominant strains is consistent with the presence of some α-1,3-linked monosaccharide residues in A-dominant O-PS. The percentage of α-1,3-linked monosaccharide residues and their accessibilities on A-dominant O-PS are probably critical to allow binding of anti-M epitope MAbs. The occurrence and distribution of α-1,3 linkages on the O-PS may contribute largely to epitopic heterogeneity observed among *Brucella* strains. O-PS from *B. suis* biovar 2 strains, which is apparently devoid of C and M epitopes, may lack α-1,3-linked monosaccharide residues in its O chain. Thus, determining the O-PS structure of *B. suis* biovar 2 strains could possibly give interesting information regarding epitope recognition of these MAbs in relation to the presence or absence of α-1,3-linked monosaccharide residues. The European hare (*Lepus capensis*) is the principal host of *B. suis* biovar 2 and can form a reservoir of infection from which pigs, both domestic and wild, may become infected (Alton, 1990). Interestingly, *B. suis* biovar 2 is rarely associated to human infection.

As mentioned above, the OM of smooth *Brucella* strains contains free O-PS, termed NH, at the surface of the OM where it would be anchored through interactions with the LPS O-chain (Moreno *et al.*, 1987; Zygmunt *et al.*, 1988; Aragón *et al.*, 1996). From a practical point of view, NH is also of interest for the serological diagnosis of animal and human brucellosis (Fernández-Lago and Díaz, 1986; Díaz-Aparicio *et al.*, 1993; Alonso-Urmeneta *et al.*, 1998; Marín *et al.*, 1999).

Besides heterogeneity in O-PS structures and antigenic determinants, S-LPSs have also been shown to be heterogeneous in size and banding pattern in sodium dodecyl sulphate/polyacrylamide gel electrophoresis and Western blotting among *Brucella* species, biovars and strains (Dubray and Limet, 1987; Garin-Bastuji *et al.*, 1990; Freer *et al.*, 1995; Baucheron *et al.*, 2002). In addition, a study of Freer *et al.* has shown that *Brucella* LPS preparations are a family of related molecules which display heterogeneity not only at the level of the O polysaccharide, but also at the core oligosaccharide and the lipid A (Freer *et al.*, 1995). In this study reactivity with MAbs suggested that regardless of the lipid A heterogeneity the basic structure of lipid A backbone was maintained. Interestingly, the two-component BvrR/BvrS regulation system involved in regulation of OMPs expression and shown to be essential in *Brucella* virulence has been shown to be also involved in modification of lipid A through acylation (Manterola *et al.*, 2005). Decrease in lipid A acylation in *B. abortus*

BvrR/BvrS mutants would result in higher affinity for bactericidal cationic peptides.

Genetics of *Brucella* O-PS biosynthesis

Several studies aiming to identify *Brucella* virulence factors involved in intracellular survival through transposon mutagenesis have resulted in the identification of several genes involved in *Brucella* O-PS biosynthesis (Allen *et al.*, 1998; Godfroid *et al.*, 1998; McQuiston *et al.*, 1999; Foulongne *et al.*, 2000; Hong *et al.*, 2000; Köhler *et al.*, 2002; González *et al.*, 2008; Paixao *et al.*, 2009; Lacerda *et al.*, 2010). The genes identified were located in two major separate chromosomal loci termed the *wbk* locus and the *wbo* locus (Table 7.2).

Organization and polymorphism at the O-PS *wbk* locus

The *wbk* locus consist of 13 O-PS biosynthetic genes and in terms of gene organization and gene similarity part of this locus is closely related to that found in *Y. enterocolitica* O :9 (Godfroid *et al.*, 2000; Skurnik and Bengoechea, 2003; Skurnik *et al.*, 2007; González *et al.*, 2008), which at least partly explains the common O-PS structure between *Brucella* spp. and *Y. enterocolitica* O :9 and consequent O-PS antigenic relatedness and serological cross-reactions. The genes have a G+C content lower (around 48%) than that typical of *Brucella* spp. (58%) and thus may have been acquired from a species other than *Brucella*. Region *wbk* contains the genes putatively involved in perosamine synthesis, i.e. *gmd* (encoding GDP-mannose 4,6-dehydratase) and *per* (encoding perosamine synthetase)- its formylation (*wbkC*) and polymerization (glycosyltransferases) (*wbkA* and *wbkE*), and genes involved in bactoprenol priming (*wbkD* and *wbkF*) and O-PS translocation (*wzm* and *wzt*). Inactivation of the latter genes results in intracellular accumulation of O-antigen without apparent affecting the integrity of the LPS core part (Godfroid *et al.*, 2000). In addition, *wbk* contains genes, $manA_{O\text{-}Ag}$, $manB_{O\text{-}Ag}$, and $manC_{O\text{-}Ag}$, which may code for the enzymes that furnish mannose, the perosamine precursor. The genes *wbkB*, coding for a putative membrane protein of unknown function (Godfroid *et al.*, 2000), and $manB_{O\text{-}Ag}$ seem not essential for O-PS biosynthesis since their disruption does not generate rough mutants (Godfroid *et al.*, 2000; González *et al.*, 2008).

PCR-restriction fragment length polymorphism (PCR-RFLP) and DNA sequencing have been used to study the genetic diversity of these genes in *Brucella* species and biovars (Cloeckaert *et al.*, 2000; Zygmunt *et al.*, 2009). These studies showed that the *wbk* genes are highly conserved including in the rough *B. ovis* species, although species-specific markers have been detected for several genes and are detailed next.

Polymorphism in $manA_{O\text{-}Ag}$

B. neotomae had a distinct $manA_{O\text{-}Ag}$ restriction pattern consisting of an additional *Ava*II site. *In silico* analysis showed a specific profile for *B. ovis* consisting of a nucleotide substitution (GAA to GGA) at position 497 which modified the ManA C-terminal sequence at amino acid 165 (not shown). Also, a single nucleotide deletion (CAAT to CA-T) was detected at position 738; this frame shift leads to a change in amino acid sequence after position 246. Nucleotide sequence of PCR products from several strains confirmed the deletion in $manA_{O\text{-}Ag}$ as characteristic of *B. ovis*. Since *wbk man* genes are dispensable, the nucleotide substitution and frame shift detected in *B. ovis* $manA_{O\text{-}Ag}$ should not contribute to its R phenotype.

Polymorphism in *wbkD* and *wbkF*

Interestingly, a deletion of about 400 bp, encompassing the *wbkD* and *wbkF* genes, was detected by PCR-RFLP in *B. canis*. Nucleotide sequencing revealed a 351 bp deletion extending from *wbkD* nucleotide 1594 (in BMEI 1426) to *wbkF* nucleotide 918 (in BMEI 1427) that was confirmed in the genome sequence of *B. canis* RM6/66 (ATCC 23365) (Genbank accession numbers CP000872 and CP000873). This deletion might have occurred by a slipped mispairing mechanism, since a direct repeat sequence of 7 bp 'GGATCAT' is present at both sides of the deleted sequence in the other *Brucella* species. It is clear that this deletion has profound consequences in the synthesis of LPS since *wbkF* is essential in O-PS biosynthesis, and *wbkD* seems to be involved in the synthesis of quinovosamine, a sugar that is possibly linking the *Brucella* O-PS to the R-LPS (González *et al.*, 2008). Thus, this double mutation clearly explains the rough phenotype of *B. canis* and is consistent with the absence of quinovosamine in this species (Bowser *et al.*, 1974).

Compared to its homologues in *B. melitensis*, *B. abortus* and *B. suis*, the *B. ovis wbkF* showed a single nucleotide deletion at position 35. Presumably, WbkF is involved in catalysing the transfer of an acetylated aminosugar to undecaprenylphosphate, thus priming this carrier for O-chain polymerization (González *et al.*, 2008). The N-terminal region of the *E. coli* WbkF homologue was found to be necessary for this function and, therefore, it seems probable that the frame-shift in *B. ovis wbkF* produces a non-functional protein, thus explaining in part the R phenotype of this species (Zygmunt *et al.*, 2009).

The other *wbk* genes show no or limited DNA polymorphism which is not species-specific (Cloeckaert *et al.*, 2000; Zygmunt *et al.*, 2009). Interestingly, in strains isolated from marine mammals, i.e. *B. ceti* or *B. pinnipedialis* strains, an IS*711* element was detected disrupting $manB_{O\text{-}Ag}$ (Zygmunt *et al.*, 2009). This observation supports also the dispensability of $manB_{O\text{-}Ag}$ in O-PS biosynthesis since both *B. pinnipedialis* and *B. ceti* species are smooth. As indicated above the *wbk* region has features suggestive of horizontal acquisition whereas other mannose synthesis genes located outside *wbk* such as $manB_{core}$ (and $manC_{core}$) would be older *Brucella* genes necessary for the synthesis of the LPS core oligosaccharide (Monreal *et al.*, 2003; Zygmunt *et al.*, 2009). Accordingly, a drift to dysfunction of the *wbk man* genes may have been made possible by the redundancy created after horizontal acquisition of *wbk*, and the similarity in this regard between *B. ceti* and *B. pinnipedialis* suggests a common ancestor (Zygmunt *et al.*, 2009).

Besides O-PS biosynthetic genes, within the *wbk* locus of *B. melitensis* four entire and one partial insertion sequences (IS) have been identified (Godfroid *et al.*, 2000). IS*Bm1* showed 66% identity with IS*869* of *Agrobacterium tumefaciens* and is a member of the IS5 family. IS*Bm1* is disrupted by an IS*711* element. IS*Bm2* was 60% identical to IS*511* of *Caulobacter crescentus* and is a member of the IS3 family. IS*Bm3* showed 60% identity with IS*481* of *Bordetella pertussis* and is also a member of the IS3 family. Finally, the partial IS*Bm4* is homologous to IS*1248a* of *Paracoccus denitrificans*.

Organization and polymorphism at the O-PS *wbo* locus

Region *wbo* encodes two putative glycosyltransferases (*wboA* and *wboB*) and corresponds to a genomic island of 15.1 kb termed genomic island 2 (GI-2) located near a tRNA for glycine (Rajashekara *et al.*, 2004; Vizcaíno *et al.*, 2004). Genomic islands are often inserted at tRNA loci that presumably act as targets for the integration of foreign DNA. GI-2 has been shown to be absent in *B. ovis* explaining partly its rough phenotype (Rajashekara *et al.*, 2004; Vizcaíno *et al.*, 2004; Tsolis *et al.*, 2009). Among other DNA polymorphisms, a 1.7 kb DNA inversion within GI-2 has been identified in some *B. ceti* strains (Vizcaíno *et al.*, 2004). Two perfect inverted repeats of 5 bp were detected in *B. ceti* flanking the DNA inversion and were suggested to be responsible for the DNA inversion that occurred in *B. ceti*. The left inverted repeat is also conserved in *B. melitensis* 16M and *B. suis* 1330, but it is not the case for the 5-nucleotide region corresponding to the right inverted repeat that differs in both *B. melitensis* 16M and *B. suis* 1330 (Vizcaíno *et al.*, 2004). The most feasible hypothesis to explain the DNA inversion is that the DNA flanking the DNA inversion was initially as that of *B. melitensis* 16M (no perfect inverted repeats). A common ancestor to *B. ceti* strains might have suffered a mutation on the DNA flanking the right side of the DNA inversion that would have originated a 5 bp inverted repeat of the 5 bp flanking the left side of the deletion, thus favouring an inverted repeat-mediated DNA inversion. Another important DNA polymorphism has been identified in rough *B. abortus* vaccine strain RB51 consisting of an IS*711* element disrupting *wboA* (Vemulapalli *et al.*, 1999). However, this vaccine strain probably has other mutations responsible for its rough phenotype since complementation with a functional *wboA* gene resulted in O-antigen synthesis but no change in rough phenotype (Vemulapalli *et al.*, 2000b). Nevertheless, this complementation resulted in enhanced vaccine efficacy of strain RB51 (Vemulapalli *et al.*, 2000b), showing again the importance of O-PS in protection against *Brucella* infection. GI-2 has also been demonstrated to be a major virulence determinant of *B. melitensis,* since deletion of this island resulted in the attenuation of bacterial growth in macrophages and virulence in interferon regulatory factor 1-deficient [IRF-1(–/–)] mice (Rajashekara *et al.*, 2008).

Most recently, GI-2 has been demonstrated to excise spontaneously resulting in rough mutants of *B. abortus, B. melitensis* and *B. suis* (Mancilla *et al.*, 2010; see also Chapter 3). A predicted excised circular intermediate of GI-2 was identified in cultures of these *Brucella* species. Moreover, disruption of a putative phage integrase gene in *B. abortus* GI-2 caused a reduction in O-PS loss rates under conditions promoting smooth-rough dissociation. Thus, this island appears to be an integrative and unstable genetic element contributing to *Brucella* LPS spontaneous smooth to rough dissociation.

The limited genetic diversity observed in O-PS biosynthetic genes could not explain the A or M antigenic profiles explained above since no clear correlation was detected between DNA polymorphism of the genes and occurrence of the different O-PS epitopes in S-LPS of *Brucella* species and biovars. All this evidence suggests that, rather than the presence of α-1,3-specific transferases in the M serotype, there may be subtle variations in the expression of *wboB, wbkA* or *wbkE,* or in the activity of the corresponding glycosyltransferases that lead to the increase in α-1,3 linkages typical of the M and A=M serotypes (Zygmunt *et al.*, 2009).

Genetics of *Brucella* core oligosaccharide and lipid A biosynthesis

Much less is known on the *Brucella* core oligosaccharide part of LPS. It contains 2-keto, 3-deoxyoctulosonic acid, mannose, glucose, glucosamine and quinovosamine but the structure is unknown (Cardoso *et al.*, 2006; González *et al.*, 2008). Thus far, only three genes have been proved to be involved in core synthesis: *pgm* (phosphoglucomutase, a general biosynthetic function), *manB*$_{core}$ (phosphomannomutase) and *wa*** (putative glycosyltransferase) (Monreal *et al.*, 2003; González *et al.*, 2008). Disruption of *manB*$_{core}$ has been shown to generate a deep R-LPS in *B. abortus* and interestingly it has been demonstrated that a complete core is required for rough vaccines to be efficient against *B. abortus* and *B. ovis* infection in the mouse model (Monreal *et al.*, 2003). Genetic diversity of these genes has been recently studied by PCR-RFLP (Zygmunt *et al.*, 2009). Despite using six restriction enzymes, all brucellae displayed the same RFLP pattern for the *manB*$_{core}$ amplicon. *In silico*, the four genomes available showed low polymorphism. A single nucleotide deletion at position 812 was detected in *B. ovis*, which should modify the C-terminal sequence of the protein making unlikely a change severe enough to contribute to the R phenotype. This is also supported by the fact that *B. ovis* R-LPS is not deeply truncated like that of *manB*$_{core}$ mutants (Monreal *et al.*, 2003). Similarly, a low degree of polymorphism was observed in *wa***. With the exception of *B. suis* biovar 2, one *Pst*I pattern was specific of *B. suis*. Biovar 2 also lacked an *Ava*II site, which could be considered as a biovar marker. With *Hinf*1, a pattern was found specific of *B. ovis*.

Little is known about lipid A biosynthesis. The *bacA* mutant of *B. abortus* 2308 showed an important reduction in the amount of lipid A modified with VLCFA, and BacA, an integral inner membrane protein, was suggested to be involved in transport of VLCFA out of the cytoplasm (Ferguson *et al.*, 2004). Additionally, the *B. abortus* 2308 mutant in the BvrS/BvrR two-component regulatory system showed increased amounts of underacylated lipid A (Manterola *et al.*, 2005). Although some other genes putatively involved in the biosynthesis of lipid A have been identified in the *B. suis* 1330 and *B. melitensis* 16 M genome sequences, mutants for these genes have not been described yet, what makes harder to define their role in LPS synthesis (Cardoso *et al.*, 2006).

Major OMPs and related proteins: general characteristics

The major (most abundant) OMPs of *Brucella* spp. were initially identified, in the early 1980s, in the sodium dodecyl sulphate insoluble (SDS-I) cell wall fraction shown to be protective against a *Brucella* infection in the mouse model (Dubray and Bézard, 1980; Dubray, 1981; Dubray and Charriaut, 1983). Two major OMPs were identified in the SDS-I fraction of *B. abortus* and designated as the 36–38 kDa and 25–27 kDa proteins according to their apparent molecular mass in SDS-polyacrylamide gel electrophoresis (PAGE), (Dubray and Bézard, 1980; Dubray and Charriaut, 1983). They were shown to be strongly associated with PG as evidenced by the fact that their visualization by SDS-PAGE was only achieved after releasing by lysozyme digestion (Dubray and Bézard, 1980; Dubray and Charriaut, 1983; Douglas *et al.*, 1984; Sowa *et al.*, 1991). Almost simultaneously, the 36–38 kDa and 25–27 kDa OMPs of *B. abortus* were also detected after extraction with other detergents and named group 2 and group 3 proteins, respectively (Verstreate *et al.*, 1982). Group 1 OMPs, known as Omp89 or Omp1, were determined to be minor proteins of 88–94 kDa (Verstreate *et al.*, 1982). Group 2 OMPs were characterized as bacterial porins by analysing the permeability of proteoliposomes reconstituted from purified proteins and egg phosphatidylcholine (Douglas *et al.*, 1984). An additional major OMP of 31–34 kDa, included in group 3 OMPs, was detected in the SDS-I fraction of *B. melitensis* (Dubray, 1981), while the 36–38 kDa OMP was less represented in this *Brucella* species (Dubray, 1981; Santos *et al.*, 1984).

Subsequently, the gene *omp2b* of *B. abortus* was determined to be responsible for the synthesis of the 36–38 kDa OMP that was then designated Omp2b (Ficht *et al.*, 1988, 1989). The homologous gene *omp2a* was also identified close to *omp2b* in the genome (Ficht *et al.*, 1989). Similarly, the 25–27 kDa and 31–34 kDa OMPs were determined to be encoded by *omp25* and *omp31*, respectively, and were designated Omp25 and Omp31 (de Wergifosse *et al.*, 1995; Vizcaíno *et al.*, 1996), which share 34% amino acid identity (Vizcaíno *et al.*, 1996). Major *Brucella* OMPs, even when synthesized in recombinant *E. coli*, display a characteristic multiple band pattern in SDS-PAGE (Douglas *et al.*, 1984; Cloeckaert *et al.*, 1990, 1996b; Vizcaíno *et al.*, 1996; Paquet *et al.*, 2001; Martín-Martín *et al.*, 2009) and multiple spots differing in molecular mass and isoelectric point have been detected in two-dimensional electrophoresis for each OMP (Eschenbrenner *et al.*, 2002; Guzmán-Verri *et al.*, 2002; Wagner *et al.*, 2002; Connolly *et al.*, 2006). Although the reason for this behaviour remains unknown it has been related to oligomer formation and mainly to the association to PG subunits of several sizes or to LPS or both (Douglas *et al.*, 1984; Sowa *et al.*, 1991; Cloeckaert *et al.*, 1992b; Connolly *et al.*, 2006).

As shown by reactivity with polyclonal or MAbs in immunological techniques such as Western blot, enzyme-linked immunosorbent assay (ELISA), immunoelectron microscopy and flow cytometry, the three major OMPs are exposed on the surface of *Brucella* spp. and are also exported to the OM in recombinant *E. coli* (Cloeckaert *et al.*, 1990, 1991; Marquis and Ficht, 1993; Bowden *et al.*, 1995a,b; Vizcaíno *et al.*, 1996, 2001b). However, as shown by the lower accessibility of OMPs to antibodies in smooth strains when compared with rough *Brucella* strains, the O-polysaccharide chains of LPS cause a steric hindrance reducing the surface exposure of OMPs in smooth strains (Cloeckaert *et al.*, 1990; Bowden *et al.*, 1995a). Accordingly, and although some protective activity against smooth strains has been described for OMP-based DNA vaccines (Cassataro *et al.*, 2005; Commander *et al.*, 2010), the poor interest reported for the *Brucella* major OMPs and their specific MAbs in protecting against infections by smooth strains (Montaraz *et al.*, 1986; Winter *et al.*, 1988; Cloeckaert *et al.*, 1991; Jacques *et al.*, 1992; Bowden *et al.*, 1998; Guilloteau *et al.*, 1999) has been attributed to the low accessibility of OMPs on the bacterial surface (Cloeckaert *et al.*, 2002). Additionally, major OMPs induce moderate antibody responses in *B. abortus* infected cattle and *B. melitensis* infected sheep and, therefore, they are not considered suitable antigens for the serological diagnosis of brucellosis caused by smooth *Brucella* strains (Cloeckaert *et al.*, 2002). This unsuitability was confirmed when recombinant Omp31 from *B. melitensis* was tested in ELISA for the diagnosis of human and animal brucellosis caused by smooth strains (Cassataro *et al.*, 2004). On the contrary, several reports describe the interest of OMPs for the prophylaxis and diagnosis of infections caused by rough *Brucella* strains, such as those of the species *B. canis* and *B. ovis*, where OMPs are highly exposed on the surface (Kittelberger *et al.*, 1998; Estein *et al.*, 2003, 2009; Cassataro *et al.*, 2007).

Whole genome sequencing of *B. melitensis* 16M and *B. suis* 1330 strains in 2002 enabled the identification of five other genes encoding putative OMPs sharing a high level of amino acid identity with Omp25 and/or Omp31 (Salhi *et al.*, 2003; Vizcaíno *et al.*, 2004). As Omp2b, Omp25 and Omp31, these five proteins have a clear signal peptide sequence and a phenylalanine terminal residue that are typical OMP characteristics involved in export to the OM (Ficht *et al.*, 1989; de Wergifosse *et al.*, 1995; Vizcaíno *et al.*, 1996; Salhi *et al.*, 2003). Since then, the seven homologous proteins are known as the Omp25/Omp31 family (Salhi *et al.*, 2003) that has orthologues in several members of the order Rhizobiales which suggests that this OMP family was present before the evolutional divergence from the common ancestor. Orthologues are found in Rhizobiales living in close association with eukaryotic cells either as symbionts or as pathogens, i.e. the four-member RopB family of *Rhizobium leguminosarum* bv. *viciae* (Foreman *et al.*, 2010), the five haemin binding proteins of the Hbp family of *Bartonella quintana* (Minnick *et al.*, 2003) or the three-member RopB homologous family of *A. tumefaciens* (http://depts.washington.edu/agro/genomes/c58/c58homeF.htm). Some of these orthologues have been related with the interaction of bacteria with the eukaryotic host (Jia *et al.*, 2002; Minnick *et al.*, 2003), but they are also found in other Rhizobiales that are mostly free-living bacteria – such as *O. anthropi* (accession numbers NC_009667 and NC_009668).

Important levels of DNA polymorphism have been found at the *omp2* locus among the *Brucella* species and biovars (Ficht *et al.*, 1990, 1996; Cloeckaert *et al.*, 1995, 2001, 2002), a characteristic that has been largely used for the molecular typing of the *Brucella* strains (Ficht *et al.*, 1990, 1996; Cloeckaert *et al.*, 1995, 2001, 2002; Ohishi *et al.*, 2004; Yu and Nielsen, 2010). Although in a lesser extent, DNA polymorphism has also been detected in *omp25* (Cloeckaert *et al.*, 1995), *omp31* (Vizcaíno *et al.*, 1997) and the other members of the *omp25/omp31* family (Vizcaíno *et al.*, 2004). Accordingly, DNA characterization of *omp* genes has become a common test for the description of new *Brucella* species (Foster *et al.*, 2007; Scholz *et al.*, 2008, 2010) and for the molecular typing of *Brucella* strains (Whatmore, 2009).

Major Omp2b protein and homologous Omp2a

Cloning and sequencing of *B. abortus* bv.1 *omp2a* and *omp2b* genes revealed that they share more than 85% sequence identity and that they are oriented in opposite directions separated by 900 bp in the genome (Ficht *et al.*, 1988, 1989). The main difference found between both genes was a deletion of 138 bp in *omp2a* that shortens the encoded protein. Thus, *omp2b* from *B. abortus* bv.1 encodes a protein of apparent molecular mass of 36 kDa while *omp2a* codes for a 33 kDa protein (Ficht *et al.*, 1989). The deletion in *omp2a* was only detected in *B. abortus* biovars 1, 2 and 4 and it was proposed to occur through imperfect direct repeats of 26 bp that were found flanking both sides of the deleted sequence in the other *Brucella* species and biovars (Ficht *et al.*, 1996). Only Omp2b (Omp36 in some works) can be detected in *B. abortus* under laboratory conditions, which is explained by the fact than only one functional promoter, which is oriented to transcribe *omp2b*, is detected in the intergenic *omp2a*–*omp2b* region (Ficht *et al.*, 1989). Synthesis of Omp2a was only achieved in recombinant *E. coli* bearing *omp2a* under the control of the *lacZ* promoter or the *omp2b* promoter, although expression in *B. abortus in vivo* could not be discounted (Ficht *et al.*, 1989), and both recombinant proteins showed characteristics of bacterial porins (Marquis and Ficht, 1993). Omp2b was then identified as the previously described group 2 porin (Douglas *et al.*, 1984; Marquis and Ficht, 1993) and it was suggested that Omp2a forms a larger pore than Omp2b, which could be advantageous for intracellular growth of *Brucella* spp. since the concentration of nutrients may be particularly low within the phagosome (Marquis and Ficht, 1993). However, synthesis of Omp2a in *Brucella* spp. or transcription from *omp2a* has not been reported until present – even using whole proteomic or transcriptomic approaches – in any *Brucella* strain neither in culture medium (Ficht *et al.*,

1989; Wagner *et al.*, 2002; Connolly *et al.*, 2006; Lamontagne *et al.*, 2007; Rossetti *et al.*, 2009) nor in the intramacrophagic environment (Al Dahouk *et al.*, 2008; Lamontagne *et al.*, 2009). Although a recent work describes the detection of Omp2a in *B. abortus* 2308 by liquid chromatography–mass spectrometry of the whole proteome (Lamontagne *et al.*, 2010), the peptide sequences detected are found in both Omp2a and Omp2b of this *Brucella* strain (accession number NC_007618), thus precluding the accurate identification of the corresponding protein that would most probably be Omp2b.

Polymorphism at the *omp2* locus is mostly due to gene conversion phenomena that modify parts of *omp2a* taking *omp2b* as the template and inversely, leading to 'mosaic' structures and a tendency to the homogenization of the two gene sequences (Ficht *et al.*, 1996; Vizcaíno *et al.*, 2000; Cloeckaert *et al.*, 2001, 2002; Paquet *et al.*, 2001). The genetic conversion phenomenon between the two *omp2* genes probably preclude the reconstitution of the phylogenetic tree for *omp2*-based evolutionary analysis but polymorphism at the *omp2* locus has been shown to be useful for the classification of the *Brucella* species depending on their ecological niche and host preference. Thus, an increasing number of *omp2a* and *omp2b* variants has been identified with the description of new *Brucella* species, variants that were in accordance with the differentiation of the genus in species and biovars (Cloeckaert *et al.*, 1995, 2001; Ohishi *et al.*, 2004; Scholz *et al.*, 2008, 2010). PCR-RFLP and DNA sequencing of the *omp2* locus in *Brucella* spp. confirmed the close genetic relatedness between *B. melitensis* and *B. abortus* and between *B. canis* and *B. suis* (Ficht *et al.*, 1996; Vizcaíno *et al.*, 2000; Cloeckaert *et al.*, 2001; Ohishi *et al.*, 2004) and the higher divergence of *B. ovis*, *B. neotomae* and the *Brucella* strains isolated from marine mammals (Ficht *et al.*, 1996; Vizcaíno *et al.*, 2000; Cloeckaert *et al.*, 2001; Ohishi *et al.*, 2004). Marine mammal *Brucella* strains contain specific motifs not found in terrestrial isolates (Foster *et al.*, 2007) and the most recently described species, *B. microti* and *B. inopinata*, also presented distinctive traits at the *omp2* locus (Ohishi *et al.*, 2004; Scholz *et al.*, 2008, 2010).

This high variability in *omp2a* and *omp2b* sequences among the *Brucella* species leads to Omp2b porin natural chimeric variants – differing not only in the amino acid sequence but also in size (Paquet *et al.*, 2001) – consistent with the different protein migration patterns found in Western blot for the *Brucella* species (Ficht *et al.*, 1990; Paquet *et al.*, 2001). Omp2a and Omp2b are predicted to have 16 transmembrane β strands spanning the outer membrane and eight large surface-exposed loops (L1 to L8) (Mobasheri *et al.*, 1997; Paquet *et al.*, 2000, 2001; Cloeckaert *et al.*, 2002). The major sequence differences (insertion and/or deletion) among Omp2b variants were found in the external loops L3 and L5 but minor amino acid substitutions were distributed along the protein from residue 35 to the C-terminus (Paquet *et al.*, 2001). In contrast to most known porins, where loop L3 is involved in constriction of the pore, loop L5 appears to be critical in Omp2b and Omp2a sugar permeability and is thought to participate to the formation of the 'external mouth' of the pore, which serves to pre-screen the solute in many known porins (Paquet *et al.*, 2001; Cloeckaert *et al.*, 2002). The influence of charged residues of loops L3 and L5 in ion conductance and ion selectivity remains to be elucidated (Cloeckaert *et al.*, 2002).

The panel of Omp2b size variants purified from recombinant *E. coli* displayed different permeability properties towards small sugars and the respective *Brucella* strains showed slight antigenic differences in Omp2b (Omp2a is not produced in *Brucella*) as detected by reactivity with specific MAbs in Western blot and ELISA (Paquet *et al.*, 2001). The antigen–antibody reactivity revealed that loops L6, L7 and L8 are part of conformational epitopes highly exposed on the bacterial surface and that show some degree of antigenic variability among the *Brucella* strains analysed (Paquet *et al.*, 2001). This variability in the permeability and antigenic properties of Omp2b in the genus *Brucella* could also be relevant for the interaction with the animal host and have a role in host tropism (Paquet *et al.*, 2001). Additionally, Omp2b has been recently identified as a possible inhibitor of cell apoptosis since it inhibited Bax-induced cell dead in yeast in a model currently used for the study of human anti-apoptotic proteins (Laloux *et al.*, 2010).

Major Omp25 protein

With the exception of *B. ovis* where *omp25* (in some works named *omp3a*) has suffered a deletion of 36 bp near the 3′-end, *omp25* is highly conserved in the genus *Brucella* differing among species in only few single nucleotide substitutions (Cloeckaert *et al.*, 1996a,b, 2002; Vizcaíno *et al.*, 2004). Variations were observed at only 12 positions of the 639-bp long *omp25* when analysing the nucleotide sequence of the reference strains of the six classical *Brucella* species (Cloeckaert *et al.*, 1996b). Also 12 nt positions of a 490 bp *omp25* region were determined to differ among 160 isolates of *Brucella* corresponding to the six classical species infecting terrestrial mammals and *Brucella* strains isolated from marine mammals (Whatmore *et al.*, 2007). Because of the 36-bp DNA deletion, Omp25 was 12 amino acids shorter in *B. ovis* and displayed antigenic differences when compared with the *B. melitensis* Omp25 as determined by the reactivity profile with a panel of anti-Omp25 MAbs (Cloeckaert *et al.*, 1996b). The mature Omp25 protein is predicted to contain eight transmembrane β-strands, connected alternatively by short turns (supposed to be oriented towards the periplasm) or long external loops (Cloeckaert *et al.*, 2002). The deletion in the *B. ovis* Omp25 protein is predicted to be located in the last C-terminal surface-exposed loop (Cloeckaert *et al.*, 2002) and probably accounts for the different antigenic behaviour observed for the *B. ovis* Omp25 protein (Cloeckaert *et al.*, 1996b).

Acidification of the phagosome has been reported to trigger the survival and intracellular replication of *B. suis* inside macrophages (Porte *et al.*, 1999). Omp25 – but not Omp31 – was found to be necessary in *B. suis* for the release of periplasmic proteins at acid pH suggesting a possible contribution of the protein to *B. suis* virulence (Boigegrain *et al.*, 2004). Additionally, a Δ*omp25 B. suis* 1330 mutant, on the contrary to the parental strain, induced the production of tumour necrosis factor alpha (TNF-α) by human macrophages after phagocytosis (Jubier-Maurin *et al.*, 2001). A similar observation was made when *B. suis* 1330 was treated with a MAb specific for an Omp25 surface-exposed epitope. Complementation of the Δ*omp25* mutant with wild type gene reverted this effect since macrophages infected with the complemented strain secreted much less TNF-α than macrophages infected with the parental strain. Accordingly, although the Δ*omp25* mutant of *B. suis* behaved as the parental strain regarding intracellular replication, it was stressed that Omp25 is involved in the negative regulation of TNF-α production upon infection of human macrophages by an undetermined mechanism and could therefore contribute to the virulence of the bacterium (Jubier-Maurin *et al.*, 2001; Dornand *et al.*, 2002). An Omp25-dependent mechanism was also determined to be responsible for the impairment of human dendritic cells infected with *B. suis* in their capacity to secrete TNF-α, which lead to the avoidance of both dendritic cell maturation and interleukin-12 secretion and resulted in the alteration of antigen presentation to naive T cells (Billard *et al.*, 2007). The inhibitory effect of Omp25 in the production of TNF-α is dependent on the phagocyte since it is not observed in murine macrophages, where a rapid and significant production of TNF-α is detected after phagocytosis of *Brucella* (Caron *et al.*, 1996; Dornand *et al.*, 2002; Manterola *et al.*, 2007). It must be noted that, unlike humans, mice are not naturally infected with *Brucella* and in laboratory-induced infection the bacterium is eliminated after several weeks, being TNF-α a key cytokine for the control of infection (Dornand *et al.*, 2002).

When compared with the parental strain, Omp25 is barely represented in the proteome of *B. abortus* 2308 mutants for the two-component regulatory system BvrR/BvrS (Guzmán-Verri *et al.*, 2002; Lamontagne *et al.*, 2007) and its encoding gene is strongly down-regulated in this mutant strain (Viadas *et al.*, 2010). Since these mutants are impaired in intracellular survival in phagocytes and strongly attenuated in virulence in the mouse model (Sola-Landa *et al.*, 1998) it was thought that absence of Omp25 in the bacterial surface could contribute to this behaviour. However, a *bvrR/bvrS* mutant of *B. suis* 1330 showed reduced survival in human macrophages but Omp25 levels similar to those of the parental strain (Salhi *et al.*, 2003). Additionally, studies analysing the role of

Omp25 in intracellular replication and virulence of *Brucella* spp. have provided discrepant results. A role for Omp25 in intracellular replication was not demonstrated in murine or human phagocytes for *B. suis* 1330 (Jubier-Maurin *et al.*, 2001; Dornand *et al.*, 2002; Billard *et al.*, 2007), *B. abortus* 2308 (Manterola *et al.*, 2007) or *B. ovis* PA (Martín-Martín *et al.*, 2008). However, an impaired replication inside late-gestational bovine chorionic trophoblasts and macrophage has been reported for a *B. abortus* 2308 Δ*omp25* mutant (Edmonds *et al.*, 2001). Deletion mutants of *B. abortus* 2308, *B. melitensis* 16M and *B.* ovis LSU99 were shown to be attenuated in the mouse model as determined by the splenic colonization along a 20-week period (Edmonds *et al.*, 2002a). On the contrary, the splenic infection level reached in mice by another Δ*omp25* mutant of *B. abortus* 2308 was equivalent to that of the parental strain (Manterola *et al.*, 2007), and the Δ*omp25* mutant of virulent *B. ovis* PA was found to be also fully virulent in the mouse model (Caro-Hernández *et al.*, 2007). The reason for these discrepancies is not known although it could be related to the experimental models using different inoculation doses and administration routes and, in the case of *B. ovis*, to the use of two strains (LSU99 and PA) that could have a different genetic background. Nevertheless, a role for Omp25 in the virulence of *Brucella* spp. in the natural host cannot be discarded. In this respect, *B. abortus* 2308 and *B. melitensis* 16M Δ*omp25* mutants were attenuated in cattle and goats, respectively, although it must be noted that these mutants were those previously found attenuated in the mouse model (Edmonds *et al.*, 2001, 2002a,b).

Major Omp31 protein

Two main interspecies differences are found in the genus *Brucella* concerning *omp31*: the remarkable absence of the gene in *B. abortus* strains (Vizcaíno *et al.*, 1997, 1999) and the nine nucleotide differences found in *B. ovis* when compared with *omp31* from *B. melitensis* (Vizcaíno *et al.*, 2001b). Absence of *omp31* in *B. abortus* is explained by the deletion of a 25-kb DNA fragment (Vizcaíno *et al.*, 2001a) that, in addition to *omp31*, includes another 21 genes most of which encode hypothetical proteins with significant homology to proteins related to the synthesis of polysaccharides in other bacteria. It has been suggested that these proteins could be involved in the synthesis of an unknown polysaccharide in *Brucella* spp. (Vizcaíno *et al.*, 2001a). The nine nucleotide substitutions of *omp31* in *B. ovis* lead to seven amino acid differences between the *B. ovis* and *B. melitensis* Omp31 proteins, resulting in marked antigenic differences as revealed by reactivity of recombinant *E. coli* synthesizing the respective recombinant proteins with a panel of Omp31-specific MAbs and sera from infected sheep (Vizcaíno *et al.*, 2001b). Omp31 is predicted to be an eight-stranded β-barrel with four surface-exposed loops larger than those in Omp25 (Cloeckaert *et al.*, 2002). The N-terminal surface-exposed loop (L1) corresponds to an immunodominant region in both *B. melitensis* and *B. ovis* (Vizcaíno *et al.*, 1996, 2001b; Cloeckaert *et al.*, 2002), contains three of the seven amino acid differences detected between Omp31 of *B. melitensis* and *B. ovis*, and was proposed to be responsible for the important antigenic differences observed between both proteins (Cloeckaert *et al.*, 2002). The remaining four amino acid substitutions were also located in external loops of Omp31 (Cloeckaert *et al.*, 2002) and another immunodominant region, covering part of loop L4, was detected near the C-terminal region in *B. ovis* Omp31 (Vizcaíno *et al.*, 2001b; Cloeckaert *et al.*, 2002).

Omp31 from *B. melitensis* was shown to form oligomers resistant to SDS denaturation at low temperatures suggesting that it could have porin activity (Vizcaíno *et al.*, 1996). Although this aspect has not been further studied, the possible function of Omp31 as a porin is also supported by the fact that Omp31 is a major OMP in *B. melitensis* while Omp2b porin, as also happens in *B. canis* and *B. ovis*, is a minor OMP (Dubray, 1981; Santos *et al.*, 1984; Gallot-Lavallée *et al.*, 1995). On the contrary, Omp2b porin is a major OMP in *B. abortus* strains (Dubray and Bézard, 1980; Dubray and Charriaut, 1983; Santos *et al.*, 1984), known to lack Omp31 as a consequence of the 25-kb DNA deletion (Vizcaíno *et al.*, 1997, 1999). If a porin activity is supposed for Omp31, its absence from the *B. abortus* OM could have dramatic effects for cell viability unless it was compensated by another protein with similar activity. Thus, the high levels of Omp2b porin detected in *B. abortus* might compensate for the absence of Omp31. Although Omp2b was not included in the analysis, evidence for a compensatory mechanism between the members of the Omp25/Omp31 family in *B.* suis 1330 and *B. ovis* PA mutants has been reported (Salhi *et al.*, 2003; Caro-Hernández *et al.*, 2007; Martín-Martín *et al.*, 2009).

Omp31 shares 31.7% of amino acid identity with the hemin-binding protein HbpA from *Bartonella quintana,* a bacterial erythrocyte parasite with high hemin requirements in order to grow *in vitro* (Carroll *et al.*, 2000). The hemin-binding activity has also been detected for *B. melitensis, B. abortus* and *B. suis* Omp31 which synthesis seems to be increased under iron-limiting conditions in the culture medium (Delpino *et al.*, 2006). Despite the relevance that the ability of Omp31 to bind hemin or its hypothetical porin activity could have for the virulence of *Brucella* spp., absence of Omp31 in *B. ovis* PA does not have a dramatic effect on intracellular survival in murine macrophages (Martín-Martín *et al.*, 2008) or on splenic multiplication in mice (Caro-Hernández *et al.*, 2007). Additionally, *B. abortus* naturally lacks Omp31 but it is virulent in mice and in its normal host, and inactivation of *omp31* in attenuated smooth *B. melitensis* Rev1 does not reduce its virulence in the mouse or sheep models (Cloeckaert *et al.*, 2004; Guilloteau *et al.*, 2006). Interestingly, HbpA is a member of a family of five homologous proteins in *B. quintana,* as it is also the case for the seven members of the *Brucella* spp. Omp25/Omp31 family (Salhi *et al.*, 2003; Vizcaíno *et al.*, 2004). Inactivation of *hbpA* enhances the hemin-binding phenotype of *B. quitana* and increases the expression of other genes of the family, suggesting a compensatory mechanism to counterbalance the absence of HbpA (Minnick *et al.*, 2003) and supporting the similar observations reported for the *Brucella* Omp25/Omp31 family (Salhi *et al.*, 2003; Caro-Hernández *et al.*, 2007; Martín-Martín *et al.*, 2009). A compensatory mechanism between the members of the Omp25/Omp31 family could explain why absence of Omp31, which is a major OMP, does not have drastic effects on the virulence of *Brucella* spp.

Other members of the Omp25/Omp31 family

Omp31b, Omp25b, Omp25c, Omp25d and Omp22, together with Omp25 and Omp31, constitute the Omp25/Omp31 family of *Brucella* spp. (Salhi *et al.*, 2003; Vizcaíno *et al.*, 2004). Mature Omp25b, Omp25c and Omp25d share about 60% identity between them and about 40% identity with Omp25 (Salhi *et al.*, 2003). Omp31b share 67% identity with Omp31 while Omp22 is the less homologous protein (about 25% with the six other members of the family) (Salhi *et al.*, 2003).

The corresponding genes were firstly identified analysing the whole genome and each encoded protein was initially detected in *B. melitensis, B. abortus* or *B. suis* (Guzmán-Verri *et al.*, 2002; Wagner *et al.*, 2002; Salhi *et al.*, 2003). According to DNA data, only *B. suis* and *B. neotomae* would be able to synthesize the seven proteins of the Omp25/Omp31 family (Vizcaíno *et al.*, 2004). However, the simultaneous detection of the seven proteins in a particular *Brucella* strain has never been reported. Nucleotide sequence of *omp25c, omp25d* and *omp22* (*omp3b* in some works) is highly conserved in the genus *Brucella* with a maximum of three nucleotide differences, depending on the strain, when compared with the consensus sequence (Vizcaíno *et al.*, 2004). The most remarkable difference is the substitution of the ATG start codon for an ACG codon in *B. ceti* B14/94 *omp22,* which might prevent the synthesis of the encoded protein (Vizcaíno *et al.*, 2004). Omp25c has been detected in the six classical species of the genus *Brucella,* either by direct identification of the protein (Wagner *et al.*, 2002; Connolly *et al.*, 2006; Lamontagne *et al.*, 2007; Al Dahouk *et al.*, 2008) or by reactivity with antibodies (Martín-Martín *et al.*, 2009), and in *B. ovis* it seems to be as abundant as Omp31 and might also be considered a major OMP (Martín-Martín *et al.*, 2009). Omp22 has been detected in *B. abortus* 2308 (Guzmán-Verri *et al.*, 2002; Connolly *et al.*, 2006), *B. ovis* PA and *B. neotomae* 5K33 (Martín-Martín *et al.*, 2009) while Omp25d has only been detected in *B. abortus* 2308 (Connolly *et al.*, 2006) and in a *B. ovis* PA Δ*omp25d* complemented mutant overexpressing the wild-type gene (Martín-Martín *et al.*, 2009).

Omp31b is probably not synthesized in *B. melitensis* strains where a deletion of 232 bp, including the ribosome binding site and the 5′ end of *omp31b*, was detected in strains belonging to all three biovars (Salhi *et al.*, 2003; Vizcaíno *et al.*, 2004; Lamontagne *et al.*, 2007). Although *omp31b* transcripts have been detected in *B. melitensis* strains (Rossetti *et al.*, 2009; Weeks *et al.*, 2010), they have smaller size than those of the other *Brucella* species (Martín-Martín *et al.*, 2009) and the Omp31b protein they would encode has never been observed in *B. melitensis* while it has been identified in *B. suis*, *B. abortus* and *B. neotomae* by direct detection of the protein or/and reactivity with antibodies (Salhi *et al.*, 2003; Connolly *et al.*, 2006; Lamontagne *et al.*, 2007; Al Dahouk *et al.*, 2008; Martín-Martín *et al.*, 2009). Since one nucleotide substitution creates a premature stop codon in the encoding gene, Omp31b would be truncated in *B. ovis* and *B. canis* losing the C-terminal phenylalanine necessary for efficient translocation of proteins across the OM (Vizcaíno *et al.*, 2004). Detection of Omp31b in these two species has not been reported.

DNA polymorphism studies revealed *omp25b* as the most heterogeneous gene of the *omp25/omp31* family (Vizcaíno *et al.*, 2004). According to the nucleotide sequence, only *B. melitensis*, *B. suis* and *B. melitensis* 16M should be able to synthesize a full-length Omp25b (Vizcaíno *et al.*, 2004). *B. ovis* lacks *omp25b* since it is found in a 15-kb genomic island deleted in this species (Vizcaíno *et al.*, 2004). Most strains of *B. ceti* have an inverted 1.7-kb DNA fragment including the 3′ end of *omp25b*, and *B. pinnipedialis* strains, some *B. ceti* strains, *B. abortus* and *B. canis* have one nucleotide insertion or deletion truncating the encoded protein. Omp25b has been experimentally detected in *B. melitensis* and *B. suis* (Al Dahouk *et al.*, 2008). Increased levels of Omp31, Omp31b, Omp25b and Omp25c has been detected in *B. suis* recovered from infected murine macrophages (Al Dahouk *et al.*, 2008). Additionally, a *B. abortus* 2308 mutant defective in the two-component regulatory system *bvrR/bvrS* is impaired in intracellular survival, is severely attenuated in animal models and exhibits altered OM properties related to modifications of the OM components, among them lower levels of Omp2b porin, Omp31b, Omp25, Omp25c and Omp22 (Sola-Landa *et al.*, 1998; Guzmán-Verri *et al.*, 2002; Lamontagne *et al.*, 2007; Viadas *et al.*, 2010). The *B. melitensis* 16M mutant in the quorum-sensing regulator VjbR is also strongly attenuated in cellular and animal models and exhibits an altered pattern of the members of the Omp25/Omp31 family when incubated in the presence of N-dodecanoyl homoserine lactone signalling molecule (Delrue *et al.*, 2005; Uzureau *et al.*, 2007). All these observations suggest a probable implication of the Omp25/Omp31 family in the virulence of *Brucella* spp. but, although this aspect has not been exhaustively evaluated, the studies performed until present show variable results depending on the *Brucella* strain analysed. Thus, in addition to the discrepancies discussed above regarding Omp25, a *B. abortus* 2308 mutant unable to synthesize Omp22 – which is an under-represented OMP in the attenuated *B. abortus* 2308 *bvrR/bvrS* mutant (Guzmán-Verri *et al.*, 2002) – was not impaired in virulence in cellular or animal models (Manterola *et al.*, 2007), while deletion of *omp22* strongly reduced the intracellular survival in murine macrophages and the splenic colonization in mice of *B. ovis* PA (Caro-Hernández *et al.*, 2007; Martín-Martín *et al.*, 2008). The *B. ovis* PA Δ*omp25d* mutant was also strongly attenuated, although to a lesser extent than the Δ*omp22* mutant. On the contrary, absence of Omp25, Omp31 or Omp25c, the other Omp25/Omp31 proteins synthesized in *B. ovis* since *omp31b* and *omp25b* are defective genes in this species (Vizcaíno *et al.*, 2004; Martín-Martín *et al.*, 2009) did not lead to important effects on virulence in spite of alterations in the OM properties (Caro-Hernández *et al.*, 2007; Martín-Martín *et al.*, 2008). The role in virulence of Omp25/Omp31 proteins, other than Omp25 and Omp22, has not been evaluated yet in other *Brucella* species than *B. ovis*.

The seven Omp25/Omp31 homologous OMPs in *Brucella* spp. might be structural and functionally interchangeable proteins that could have been maintained along evolution as an advantageous characteristic assuring the integrity and properties of the OM and the virulence of the bacterium. This redundancy of OMPs makes difficult to evaluate the role of each individual protein in the biology of *Brucella* since a compensatory mechanism increasing the synthesis of one or several members of the family could compensate for the absence of another. As previously mentioned, evidences in this respect have already been reported in *Brucella* spp. (Salhi *et al.*, 2003; Caro-Hernández *et al.*, 2007; Martín-Martín *et al.*, 2009) similarly to other homologous proteins of related bacteria (Minnick *et al.*, 2003).

Other components of the *Brucella* OM

Three minor surface-exposed OMPs – Omp10, Omp16 and Omp19 – were identified by reactivity with MAbs (Cloeckaert *et al.*, 1990), and determined to be lipoproteins (Tibor *et al.*, 1999) after cloning and sequencing of the respective genes (Tibor *et al.*, 1994, 1996). Omp16 showed an important degree of similarity with three bacterial peptidoglycan-associated lipoproteins (PALs) (Tibor *et al.*, 1994). Poor polymorphism was found in *omp10, omp16* and *omp19* among the *Brucella* species and biovars (Vizcaíno *et al.*, 2000) and orthologues are found in the *O. anthropi* genome (Barquero-Calvo *et al.*, 2009). Additionally, the three lipoproteins share common epitopes with other Rhizobiales, *O. anthropi* being the most antigenically related bacteria (Cloeckaert *et al.*, 1999), and were detected in the 34 *Brucella* strains analysed that were representative of the six classical species of the genus and their biovars (Tibor *et al.*, 1994, 1996). Mutant Δ*omp10* and Δ*omp19* strains of *B. abortus* 544 are attenuated in the mouse model with a lower splenic colonization than the parental strain, although Omp10 is not required for intracellular survival (Tibor *et al.*, 2002) and inactivation of *omp19* in attenuated *B. abortus* RB51 did not increase its attenuation (Vemulapalli *et al.*, 2000a). Lipidated Omp16 and Omp19, but not unlipidated forms, have revealed immunomodulatory properties that could have a role in the progress of infection (Giambartolomei *et al.*, 2004; Zwerdling *et al.*, 2008). Additionally, the protein moiety of Omp16 has recently been identified as a bacterial pathogen-associated molecular pattern (PAMP) able to activate dendritic cells *in vivo* and conferring some degree of protection against a *B. abortus* 544 challenge in mice (Pasquevich *et al.*, 2010). Interestingly, higher levels of Omp19 were detected in the attenuated *bvrR/bvrS* mutant of *B. abortus* 2308 (Lamontagne *et al.*, 2007) as it was also the case for Omp10 and Omp19 in the *B. melitensis* 16M attenuated mutant on the quorum sensing regulator VjbR (Uzureau *et al.*, 2007).

Other minor OMPs have been identified in the genus *Brucella*. They have been less studied regarding their occurrence in the genus *Brucella* although some of them have also been related to virulence. Among these minor OMPs – that also include OMPs related to the flagellar and VirB structures (Fig. 7.1) analysed in other chapters- the following are mentioned as example: (i) several OMPs detected in cell envelope proteomic analysis (Lamontagne *et al.*, 2007); (ii) the TolC-homologous BepC protein suggested to increase the resistance of the bacterium to bile components during intestinal transit and which *B. suis* mutant is impaired for survival in mice (Posadas *et al.*, 2007; Martín *et al.*, 2009); and (iii) the SP41 surface protein, encoded by *ugpB*, that was determined to be important for the internalization of *B. suis* in non-phagocytic epithelial HeLa cells (Castañeda-Roldán *et al.*, 2006).

Phosphatidylcholine and phosphatidylethanolamine, two major PLs in *Brucella* (Thiele and Schwinn, 1973), have been determined to be important for the virulence of *B. abortus* 2308 (Comerci *et al.*, 2006; Conde-Álvarez *et al.*, 2006; Bukata *et al.*, 2008). Percentages of both PLs are similar in *B. melitensis* and *B. abortus* but these two species differ in the abundance of other PLs and in the fatty acid pattern of the OM ornithine lipids (Thiele and Schwinn, 1973), characteristics that could influence the biological properties of the bacteria.

Concluding remarks

Members of the genus *Ochrobactrum* are the closest relatives to the genus *Brucella*, sharing genomic, antigenic and structural similarities (Moreno *et al.*, 2002). Despite these analogies, they strongly differ in the interaction with higher eukaryotic organisms, *Brucella* spp. being pathogenic bacteria able to infect a wide variety of mammals and *Ochrobactrum* organisms being free-living bacteria. Even within the genus *Brucella*, marked differences are found among species regarding pathogenicity and host preference, which contrasts with the homogeneity of the genus at the DNA level (Verger *et al.*, 1985).

The compact structure of the *Brucella* OM and its particular properties have been described as important factors determining the pathogen–host interactions (Velasco *et al.*, 2000; Barquero-Calvo *et al.*, 2009). Beside differential characteristics not necessarily related to the OM – such as species-specific genomic islands, gene inactivations, deletions, or particular gene expression patterns – the *Brucella* species differ in OM-related properties independently of the smooth or rough phenotype (Martín-Martín *et al.*, 2010a), which is suggestive of qualitative and/or quantitative variability of the OM components, other than O-PS, between species. These differences might explain, at least in part, why naturally rough and virulent *B. canis* and *B. ovis* strains are the exceptions to the accepted rule claiming the relevance of O-PS for the virulence of *Brucella* spp. (Allen *et al.*, 1998; Fernández-Prada *et al.*, 2001; Barquero-Calvo *et al.*, 2007). Contrary to rough mutants derived from smooth strains, *B. canis* and *B. ovis*, enter the phagocyte by a lipid-raft-mediated mechanism, survive and replicate inside macrophages and establish persistent infections in their natural hosts and in animal models (Martín-Martín *et al.*, 2008, 2010b). The heterogeneity described in this chapter regarding the main OM components reveals that each *Brucella* species exhibit a particular OM composition and topology leading to different antigenic and functional properties that could influence the interaction with the host. Whether this heterogeneity is involved in the pathogenicity and host tropism characteristics of each *Brucella* species remains to be elucidated.

Acknowledgements

Research at the Departamento de Microbiología y Genética, Universidad de Salamanca, Spain is supported by the Ministerio de Ciencia e Innovación of Spain (Proyect AGL2008-04514-C03-02/GAN).

References

Al Dahouk, S., Jubier-Maurin, V., Scholz, H.C., Tomaso, H., Karges, W., Neubauer, H., and Köhler, S. (2008). Quantitative analysis of the intramacrophagic *Brucella suis* proteome reveals metabolic adaptation to late stage of cellular infection. Proteomics *8*, 3862–3870.

Allen, C.A., Adams, L.G., and Ficht, T.A. (1998). Transposon-derived *Brucella abortus* rough mutants are attenuated and exhibit reduced intracellular survival. Infect. Immun. *66*, 1008–1016.

Alonso-Urmeneta, B., Marín, C., Aragón, V., Blasco, J.M., Díaz, R., and Moriyón, I. (1998). Evaluation of lipopolysaccharides and polysaccharides of different epitopic structures in the indirect enzyme-linked immunosorbent assay for diagnosis of brucellosis in small ruminants and cattle. Clin. Diagn. Lab. Immunol. *5*, 749–754.

Alton, G.G. (1990). *Brucella suis*. In Animal Brucellosis, Nielsen, K., and Duncan, J.R., eds. (CRC Press, Inc.: Boca Raton, FL), pp. 411–422.

Alton, G.G., Jones, L.M., Angus, R.D., and Verger, J.M. (1988). Techniques for the Brucellosis Laboratory (Institut National de la Recherche Agronomique: Paris, France).

Álvarez-Martínez, C.E., and Christie, P.J. (2009). Biological diversity of prokaryotic type IV secretion systems. Microbiol. Mol. Biol. Rev. *73*, 775–808.

Aragón, V., Díaz, R., Moreno, E., and Moriyón, I. (1996). Characterization of *Brucella abortus* and *Brucella melitensis* native haptens as outer membrane O-type polysaccharides independent from the smooth lipopolysaccharide. J. Bacteriol. *178*, 1070–1079.

Barquero-Calvo, E., Chaves-Olarte, E., Weiss, D.S., Guzmán-Verri, C., Chacón-Díaz, C., Rucavado, A., Moriyón, I., and Moreno, E. (2007). *Brucella abortus* uses a stealthy strategy to avoid activation of the innate immune system during the onset of infection. PLoS One 2, e631.

Barquero-Calvo, E., Conde-Álvarez, R., Chacón-Díaz, C., Quesada-Lobo, L., Martirosyan, A., Guzmán-Verri, C., Iriarte, M., Mancek-Keber, M., Jerala, R., Gorvel, J.P., *et al.* (2009). The differential interaction of *Brucella* and *Ochrobactrum* with innate immunity reveals traits related to the evolution of stealthy pathogens. PLoS One 4, e5893.

Baucheron, S., Grayon, M., Zygmunt, M.S., and Cloeckaert, A. (2002). Lipopolysaccharide heterogeneity in *Brucella* strains isolated from marine mammals. Res. Microbiol. *153*, 277–280.

Bhat, U.R., Carlson, R.W., Busch, M., and Mayer, H. (1991). Distribution and phylogenetic significance of 27-hydroxy-octacosanoic acid in lipopolysaccharides from bacteria belonging to the alpha-2 subgroup of *Proteobacteria*. Int. J. Syst. Bacteriol. *41*, 213–217.

Billard, E., Dornand, J., and Gross, A. (2007). *Brucella suis* prevents human dendritic cell maturation and antigen presentation through regulation of tumor necrosis factor alpha secretion. Infect. Immun. *75*, 4980–4989.

Boigegrain, R.A., Salhi, I., Alvarez-Martinez, M.T., Machold, J., Fedon, Y., Arpagaus, M., Weise, C., Rittig, M., and Rouot, B. (2004). Release of periplasmic proteins of *Brucella suis* upon acidic shock involves the outer membrane protein Omp25. Infect. Immun. 72, 5693–5703.

Bowden, R.A., Cloeckaert, A., Zygmunt, M.S., Bernard, S., and Dubray, G. (1995a). Surface exposure of outer membrane protein and lipopolysaccharide epitopes in *Brucella* species studied by enzyme-linked immunosorbent assay and flow cytometry. Infect. Immun. *63*, 3945–3952.

Bowden, R.A., Cloeckaert, A., Zygmunt, M.S., and Dubray, G. (1995b). Outer-membrane protein- and rough lipopolysaccharide-specific monoclonal antibodies protect mice against *Brucella ovis*. J. Med. Microbiol. *43*, 344–347.

Bowden, R.A., Cloeckaert, A., Zygmunt, M.S., and Dubray, G. (1998). Evaluation of immunogenicity and protective activity in BALB/c mice of the 25-kDa major outer-membrane protein of *Brucella melitensis* (Omp25) expressed in *Escherichia coli*. J. Med. Microbiol. *47*, 39–48.

Bowser, D.V., Wheat, R.W., Foster, J.W., and Leong, D. (1974). Occurrence of quinovosamine in lipopolysaccharides of *Brucella* species. Infect. Immun. *9*, 772–774.

Braibant, M., Guilloteau, L., and Zygmunt, M.S. (2002). Functional characterization of *Brucella melitensis* NorMI, an efflux pump belonging to the multidrug and toxic compound extrusion family. Antimicrob. Agents Chemother. *46*, 3050–3053.

Bukata, L., Altabe, S., de Mendoza, D., Ugalde, R.A., and Comerci, D.J. (2008). Phosphatidylethanolamine synthesis is required for optimal virulence of *Brucella abortus*. J. Bacteriol. *190*, 8197–8203.

Bundle, D.R., Gidney, M.A., Perry, M.B., Duncan, J.R., and Cherwonogrodzky, J.W. (1984). Serological confirmation of *Brucella abortus* and *Yersinia enterocolitica* O:9 O-antigens by monoclonal antibodies. Infect. Immun. *46*, 389–393.

Bundle, D.R., Cherwonogrodzky, J.W., Gidney, M.A., Meikle, P.J., Perry, M.B., and Peters, T. (1989). Definition of *Brucella* A and M epitopes by monoclonal typing reagents and synthetic oligosaccharides. Infect. Immun. *57*, 2829–2836.

Cardoso, P.G., Macedo, G.C., Azevedo, V., and Oliveira, S.C. (2006). *Brucella* spp noncanonical LPS: structure, biosynthesis, and interaction with host immune system. Microb Cell Fact *5*, 13.

Caro-Hernández, P., Fernández-Lago, L., de Miguel, M.J., Martín-Martín, A.I., Cloeckaert, A., Grilló, M.J., and Vizcaíno, N. (2007). Role of the Omp25/Omp31 family in outer membrane properties and virulence of *Brucella ovis*. Infect. Immun. *75*, 4050–4061.

Caroff, M., Bundle, D.R., and Perry, M.B. (1984). Structure of the O-chain of the phenol-phase soluble cellular lipopolysaccharide of *Yersinia enterocolitica* serotype O:9. Eur. J. Biochem. *139*, 195–200.

Caron, E., Gross, A., Liautard, J.P., and Dornand, J. (1996). *Brucella* species release a specific, protease-sensitive, inhibitor of TNF-alpha expression, active on human macrophage-like cells. J. Immunol. *156*, 2885–2893.

Carroll, J.A., Coleman, S.A., Smitherman, L.S., and Minnick, M.F. (2000). Hemin-binding surface protein from *Bartonella quintana*. Infect. Immun. *68*, 6750–6757.

Cassataro, J., Pasquevich, K., Bruno, L., Wallach, J.C., Fossati, C.A., and Baldi, P.C. (2004). Antibody reactivity to Omp31 from *Brucella melitensis* in human and animal infections by smooth and rough Brucellae. Clin. Diagn. Lab. Immunol. *11*, 111–114.

Cassataro, J., Pasquevich, K.A., Estein, S.M., Laplagne, D.A., Zwerdling, A., de la Barrera, S., Bowden, R., Fossati, C.A., Giambartolomei, G.H., and Goldbaum, F.A. (2007). A DNA vaccine coding for the chimera BLSOmp31 induced a better degree of protection against *B. ovis* and a similar degree of protection against *B. melitensis* than Rev.1 vaccination. Vaccine *25*, 5958–5967.

Cassataro, J., Velikovsky, C.A., de la Barrera, S., Estein, S.M., Bruno, L., Bowden, R., Pasquevich, K.A., Fossati, C.A., and Giambartolomei, G.H. (2005). A DNA vaccine coding for the *Brucella* outer membrane protein 31 confers protection against *B. melitensis* and *B. ovis* infection by eliciting a specific cytotoxic response. Infect. Immun. *73*, 6537–6546.

Castañeda-Roldán, E.I., Ouahrani-Bettache, S., Saldaña, Z., Avelino, F., Rendón, M.A., Dornand, J., and Girón, J.A. (2006). Characterization of SP41, a surface protein of *Brucella* associated with adherence and invasion of host epithelial cells. Cell. Microbiol. *8*, 1877–1887.

Cloeckaert, A., de Wergifosse, P., Dubray, G., and Limet, J.N. (1990). Identification of seven surface-exposed *Brucella* outer membrane proteins by use of monoclonal antibodies: immunogold labeling for electron microscopy and enzyme-linked immunosorbent assay. Infect. Immun. *58*, 3980–3987.

Cloeckaert, A., Jacques, I., Bosseray, N., Limet, J.N., Bowden, R., Dubray, G., and Plommet, M. (1991). Protection conferred on mice by monoclonal antibodies directed against outer-membrane-protein antigens of *Brucella*. J. Med. Microbiol. *34*, 175–180.

Cloeckaert, A., Jacques, I., de Wergifosse, P., Dubray, G., and Limet, J.N. (1992a). Protection against *Brucella melitensis* or *Brucella abortus* in mice with immunoglobulin G (IgG), IgA, and IgM monoclonal antibodies specific for a common epitope shared by the *Brucella* A and M smooth lipopolysaccharides. Infect. Immun. *60*, 312–315.

Cloeckaert, A., Zygmunt, M.S., de Wergifosse, P., Dubray, G., and Limet, J.N. (1992b). Demonstration of peptidoglycan-associated *Brucella* outer-membrane proteins by use of monoclonal antibodies. J. Gen. Microbiol. *138*, 1543–1550.

Cloeckaert, A., Jacques, I., Bowden, R.A., Dubray, G., and Limet, J.N. (1993a). Monoclonal antibodies to *Brucella* rough lipopolysaccharide: characterization and evaluation of their protective effect against *B. abortus*. Res. Microbiol. *144*, 475–484.

Cloeckaert, A., Zygmunt, M.S., Dubray, G., and Limet, J.N. (1993b). Characterization of O-polysaccharide specific monoclonal antibodies derived from mice infected with the rough *Brucella melitensis* strain B115. J. Gen. Microbiol. *139*, 1551–1556.

Cloeckaert, A., Verger, J.M., Grayon, M., and Grépinet, O. (1995). Restriction site polymorphism of the genes encoding the major 25 kDa and 36 kDa outer-membrane proteins of *Brucella*. Microbiology *141*, 2111–2121.

Cloeckaert, A., Verger, J.M., Grayon, M., and Vizcaíno, N. (1996a). Molecular and immunological characterization of the major outer membrane proteins of *Brucella*. FEMS Microbiol. Lett. *145*, 1–8.

Cloeckaert, A., Verger, J.M., Grayon, M., Zygmunt, M.S., and Grépinet, O. (1996b). Nucleotide sequence and expression of the gene encoding the major 25-kilodalton outer membrane protein of *Brucella ovis*: Evidence for antigenic shift, compared with other *Brucella* species, due to a deletion in the gene. Infect. Immun. *64*, 2047–2055.

Cloeckaert, A., Weynants, V., Godfroid, J., Verger, J.M., Grayon, M., and Zygmunt, M.S. (1998). O-Polysaccharide epitopic heterogeneity at the surface of *Brucella* spp. studied by enzyme-linked immunosorbent assay and flow cytometry. Clin. Diagn. Lab. Immunol. *5*, 862–870.

Cloeckaert, A., Tibor, A., and Zygmunt, M.S. (1999). *Brucella* outer membrane lipoproteins share antigenic determinants with bacteria of the family *Rhizobiaceae*. Clin. Diagn. Lab. Immunol. *6*, 627–629.

Cloeckaert, A., Grayon, M., Verger, J.M., Letesson, J.J., and Godfroid, F. (2000). Conservation of seven genes involved in the biosynthesis of the lipopolysaccharide O-side chain in *Brucella* spp. Res. Microbiol. *151*, 209–216.

Cloeckaert, A., Verger, J.M., Grayon, M., Paquet, J.Y., Garin-Bastuji, B., Foster, G., and Godfroid, J. (2001). Classification of *Brucella* spp. isolated from marine mammals by DNA polymorphism at the *omp2* locus. Microbes Infect. 3, 729–738.

Cloeckaert, A., Vizcaíno, N., Paquet, J.Y., Bowden, R.A., and Elzer, P.H. (2002). Major outer membrane proteins of *Brucella* spp.: past, present and future. Vet. Microbiol. *90*, 229–247.

Cloeckaert, A., Jacques, I., Grilló, M.J., Marín, C.M., Grayon, M., Blasco, J.M., and Verger, J.M. (2004). Development and evaluation as vaccines in mice of *Brucella melitensis* Rev.1 single and double deletion mutants of the *bp26* and *omp31* genes coding for antigens of diagnostic significance in ovine brucellosis. Vaccine 22, 2827–2835.

Comerci, D.J., Altabe, S., de Mendoza, D., and Ugalde, R.A. (2006). *Brucella abortus* synthesizes phosphatidylcholine from choline provided by the host. J. Bacteriol. *188*, 1929–1934.

Commander, N.J., Brewer, J.M., Wren, B.W., Spencer, S.A., Macmillan, A.P., and Stack, J.A. (2010). Liposomal delivery of *p-ialB* and *p-omp25* DNA vaccines improves immunogenicity but fails to provide full protection against *B. melitensis* challenge. Genet. Vaccines Ther. *8*, 5.

Conde-Álvarez, R., Grilló, M.J., Salcedo, S.P., de Miguel, M.J., Fugier, E., Gorvel, J.P., Moriyón, I., and Iriarte, M. (2006). Synthesis of phosphatidylcholine, a typical eukaryotic phospholipid, is necessary for full virulence of the intracellular bacterial parasite *Brucella abortus*. Cell. Microbiol. *8*, 1322–1335.

Connolly, J.P., Comerci, D., Alefantis, T.G., Walz, A., Quan, M., Chafin, R., Grewal, P., Mujer, C.V., Ugalde, R.A., and DelVecchio, V.G. (2006). Proteomic analysis of *Brucella abortus* cell envelope and identification of immunogenic candidate proteins for vaccine development. Proteomics *6*, 3767–3780.

Corbeil, L.B., Blau, K., Inzana, T.J., Nielsen, K.H., Jacobson, R.H., Corbeil, R.R., and Winter, A.J. (1988). Killing of *Brucella abortus* by bovine serum. Infect. Immun. *56*, 3251–3261.

Delory, M., Hallez, R., Letesson, J.J., and De Bolle, X. (2006). An RpoH-like heat shock sigma factor is involved in stress response and virulence in *Brucella melitensis* 16M. J. Bacteriol. *188*, 7707–7710.

Delpino, M.V., Cassataro, J., Fossati, C.A., Goldbaum, F.A., and Baldi, P.C. (2006). *Brucella* outer membrane protein Omp31 is a haemin-binding protein. Microbes Infect. *8*, 1203–1208.

Delrue, R.M., Deschamps, C., Leonard, S., Nijskens, C., Danese, I., Schaus, J.M., Bonnot, S., Ferooz, J., Tibor, A., De Bolle, X., *et al.* (2005). A quorum-sensing regulator controls expression of both the type IV secretion system and the flagellar apparatus of *Brucella melitensis*. Cell. Microbiol. *7*, 1151–1161.

Díaz, R., Jones, L.M., Leong, D., and Wilson, J.B. (1968). Surface antigens of smooth brucellae. J. Bacteriol. *96*, 893–901.

Díaz-Aparicio, E., Aragón, V., Marín, C., Alonso, B., Font, M., Moreno, E., Pérez-Ortiz, S., Blasco, J.M., Díaz, R., and Moriyón, I. (1993). Comparative analysis of *Brucella* serotype A and M and *Yersinia enterocolitica* O:9 polysaccharides for serological diagnosis of brucellosis in cattle, sheep, and goats. J. Clin. Microbiol. *31*, 3136–3141.

Dornand, J., Gross, A., Lafont, V., Liautard, J., Oliaro, J., and Liautard, J.P. (2002). The innate immune response against *Brucella* in humans. Vet. Microbiol. *90*, 383–394.

Douglas, J.T., and Palmer, D.A. (1988). Use of monoclonal antibodies to identify the distribution of A and M epitopes on smooth *Brucella* species. J. Clin. Microbiol. *26*, 1353–1356.

Douglas, J.T., Rosenberg, E.Y., Nikaido, H., Verstreate, D.R., and Winter, A.J. (1984). Porins of *Brucella* species. Infect. Immun. *44*, 16–21.

Dubray, G. (1981). Etude ultrastructurale et biochimique des enveloppes des bactéries du genre *Brucella*. Ph.D. thesis. Université de Paris-Sud, Paris.

Dubray, G., and Bézard, G. (1980). Isolation of three *Brucella abortus* cell-wall antigens protective in murine experimental brucellosis. Ann. Rech. Vet. *11*, 367–373.

Dubray, G., and Charriaut, C. (1983). Evidence of three major polypeptide species and two major polysaccharide species in the *Brucella* outer membrane. Ann. Rech. Vet. *14*, 311–318.

Dubray, G., and Limet, J. (1987). Evidence of heterogeneity of lipopolysaccharides among *Brucella* biovars in relation to A and M specificities. Ann. Inst. Pasteur Microbiol. *138*, 27–37.

Edmonds, M.D., Cloeckaert, A., Booth, N.J., Fulton, W.T., Hagius, S.D., Walker, J.V., and Elzer, P.H. (2001).

Attenuation of a *Brucella abortus* mutant lacking a major 25 kDa outer membrane protein in cattle. Am. J. Vet. Res. *62*, 1461–1466.

Edmonds, M.D., Cloeckaert, A., and Elzer, P.H. (2002a). *Brucella* species lacking the major outer membrane protein Omp25 are attenuated in mice and protect against *Brucella melitensis* and *Brucella ovis*. Vet. Microbiol. *88*, 205–221.

Edmonds, M.D., Cloeckaert, A., Hagius, S.D., Samartino, L.E., Fulton, W.T., Walker, J.V., Enright, F.M., Booth, N.J., and Elzer, P.H. (2002b). Pathogenicity and protective activity in pregnant goats of a *Brucella melitensis* Delta*omp25* deletion mutant. Res. Vet. Sci. *72*, 235–239.

Eisenschenk, F.C., Houle, J.J., and Hoffmann, E.M. (1999). Mechanism of serum resistance among *Brucella abortus* isolates. Vet. Microbiol. *68*, 235–244.

Eschenbrenner, M., Wagner, M.A., Horn, T.A., Kraycer, J.A., Mujer, C.V., Hagius, S., Elzer, P., and DelVecchio, V.G. (2002). Comparative proteome analysis of *Brucella melitensis* vaccine strain Rev 1 and a virulent strain, 16M. J. Bacteriol. *184*, 4962–4970.

Estein, S.M., Cassataro, J., Vizcaíno, N., Zygmunt, M.S., Cloeckaert, A., and Bowden, R.A. (2003). The recombinant Omp31 from *Brucella melitensis* alone or associated with rough lipopolysaccharide induces protection against *Brucella ovis* infection in BALB/c mice. Microbes Infect. *5*, 85–93.

Estein, S.M., Cheves, P.C., Fiorentino, M.A., Cassataro, J., Paolicchi, F.A., and Bowden, R.A. (2004). Immunogenicity of recombinant Omp31 from *Brucella melitensis* in rams and serum bactericidal activity against *B. ovis*. Vet. Microbiol. *102*, 203–213.

Estein, S.M., Fiorentino, M.A., Paolicchi, F.A., Clausse, M., Manazza, J., Cassataro, J., Giambartolomei, G.H., Coria, L.M., Zylberman, V., Fossati, C.A., *et al.* (2009). The polymeric antigen BLSOmp31 confers protection against *Brucella ovis* infection in rams. Vaccine 27, 6704–6711.

Ferguson, G.P., Datta, A., Baumgartner, J., Roop, R.M., 2nd, Carlson, R.W., and Walker, G.C. (2004). Similarity to peroxisomal-membrane protein family reveals that *Sinorhizobium* and *Brucella* BacA affect lipid-A fatty acids. Proc. Natl. Acad. Sci. U.S.A. *101*, 5012–5017.

Fernández-Lago, L., and Díaz, R. (1986). Demonstration of antibodies against *Brucella melitensis* 16M lipopolysaccharide and native hapten in human sera by enzyme-linked immunosorbent assay. J. Clin. Microbiol. 24, 76–80.

Fernández-Prada, C.M., Nikolich, M., Vemulapalli, R., Sriranganathan, N., Boyle, S.M., Schurig, G.G., Hadfield, T.L., and Hoover, D.L. (2001). Deletion of *wboA* enhances activation of the lectin pathway of complement in *Brucella abortus* and *Brucella melitensis*. Infect. Immun. *69*, 4407–4416.

Ficht, T.A., Bearden, S.W., Sowa, B.A., and Adams, L.G. (1988). A 36-kilodalton *Brucella abortus* cell envelope protein is encoded by repeated sequences closely linked in the genomic DNA. Infect. Immun. *56*, 2036–2046.

Ficht, T.A., Bearden, S.W., Sowa, B.A., and Adams, L.G. (1989). DNA sequence and expression of the 36-kilodalton outer membrane protein gene of *Brucella abortus*. Infect. Immun. *57*, 3281–3291.

Ficht, T.A., Bearden, S.W., Sowa, B.A., and Marquis, H. (1990). Genetic variation at the *omp2* porin locus of the brucellae: species-specific markers. Mol. Microbiol. *4*, 1135–1142.

Ficht, T.A., Husseinen, H.S., Derr, J., and Bearden, S.W. (1996). Species-specific sequences at the *omp2* locus of *Brucella* type strains. Int. J. Syst. Bacteriol. *46*, 329–331.

Foreman, D.L., Vanderlinde, E.M., Bay, D.C., and Yost, C.K. (2010). Characterization of a gene family of outer membrane proteins (*ropB*) in *Rhizobium leguminosarum* bv. *viciae* VF39SM and the role of the sensor kinase ChvG in their regulation. J. Bacteriol. *192*, 975–983.

Foster, G., Osterman, B.S., Godfroid, J., Jacques, I., and Cloeckaert, A. (2007). *Brucella ceti* sp. nov. and *Brucella pinnipedialis* sp. nov. for *Brucella* strains with cetaceans and seals as their preferred hosts. Int. J. Syst. Evol. Microbiol. *57*, 2688–2693.

Foster, J.T., Beckstrom-Sternberg, S.M., Pearson, T., Beckstrom-Sternberg, J.S., Chain, P.S., Roberto, F.F., Hnath, J., Brettin, T., and Keim, P. (2009). Whole-genome-based phylogeny and divergence of the genus *Brucella*. J. Bacteriol. *191*, 2864–2870.

Foulongne, V., Bourg, G., Cazevieille, C., Michaux-Charachon, S., and O'Callaghan, D. (2000). Identification of *Brucella suis* genes affecting intracellular survival in an *in vitro* human macrophage infection model by signature-tagged transposon mutagenesis. Infect. Immun. *68*, 1297–1303.

Freer, E., Rojas, N., Weintraub, A., Lindberg, A.A., and Moreno, E. (1995). Heterogeneity of *Brucella abortus* lipopolysaccharides. Res. Microbiol. *146*, 569–578.

Freer, E., Moreno, E., Moriyón, I., Pizarro-Cerdá, J., Weintraub, A., and Gorvel, J.P. (1996). *Brucella-Salmonella* lipopolysaccharide chimeras are less permeable to hydrophobic probes and more sensitive to cationic peptides and EDTA than are their native *Brucella* sp. counterparts. J. Bacteriol. *178*, 5867–5876.

Freer, E., Pizarro-Cerdá, J., Weintraub, A., Bengoechea, J.A., Moriyón, I., Hultenby, K., Gorvel, J.P., and Moreno, E. (1999). The outer membrane of *Brucella ovis* shows increased permeability to hydrophobic probes and is more susceptible to cationic peptides than are the outer membranes of mutant rough *Brucella abortus* strains. Infect. Immun. *67*, 6181–6186.

Fretin, D., Fauconnier, A., Köhler, S., Halling, S., Léonard, S., Nijskens, C., Ferooz, J., Lestrate, P., Delrue, R.M., Danese, I., *et al.* (2005). The sheathed flagellum of *Brucella melitensis* is involved in persistence in a murine model of infection. Cell. Microbiol. 7, 687–698.

Gallot-Lavallée, T., Zygmunt, M.S., Cloeckaert, A., Bézard, G., and Dubray, G. (1995). Growth phase-dependent variations in the outer membrane protein profile of *Brucella melitensis*. Res. Microbiol. *146*, 227–236.

Garin-Bastuji, B., Bowden, R.A., Dubray, G., and Limet, J.N. (1990). Sodium dodecyl sulfate-polyacrylamide gel electrophoresis and immunoblotting analysis of smooth-lipopolysaccharide heterogeneity among *Brucella* biovars related to A and M specificities. J. Clin. Microbiol. *28*, 2169–2174.

Giambartolomei, G.H., Zwerdling, A., Cassataro, J., Bruno, L., Fossati, C.A., and Philipp, M.T. (2004). Lipoproteins, not lipopolysaccharide, are the key mediators of the proinflammatory response elicited by heat-killed *Brucella abortus*. J. Immunol. *173*, 4635–4642.

Godfroid, F., Taminiau, B., Danese, I., Denoel, P., Tibor, A., Weynants, V., Cloeckaert, A., Godfroid, J., and Letesson, J.J. (1998). Identification of the perosamine synthetase gene of *Brucella melitensis* 16M and involvement of lipopolysaccharide O side chain in *Brucella* survival in mice and in macrophages. Infect. Immun. *66*, 5485–5493.

Godfroid, F., Cloeckaert, A., Taminiau, B., Danese, I., Tibor, A., de Bolle, X., Mertens, P., and Letesson, J.J. (2000). Genetic organisation of the lipopolysaccharide O-antigen biosynthesis region of *Brucella melitensis* 16M (wbk). Res. Microbiol. *151*, 655–668.

Goldfine, H. (1984). Bacterial membranes and lipid packing theory. J. Lipid Res. *25*, 1501–1507.

González, D., Grilló, M.J., De Miguel, M.J., Ali, T., Arce-Gorvel, V., Delrue, R.M., Conde-Álvarez, R., Muñoz, P., López-Goñi, I., Iriarte, M., *et al.* (2008). Brucellosis vaccines: assessment of *Brucella melitensis* lipopolysaccharide rough mutants defective in core and O-polysaccharide synthesis and export. PLoS One *3*, e2760.

Guilloteau, L.A., Laroucau, K., Olivier, M., Grilló, M.J., Marín, C.M., Verger, J.M., and Blasco, J.M. (2006). Residual virulence and immunogenicity of CGV26 and CGV2631 *B. melitensis* Rev. 1 deletion mutant strains in sheep after subcutaneous or conjunctival vaccination. Vaccine *24*, 3461–3468.

Guilloteau, L.A., Laroucau, K., Vizcaíno, N., Jacques, I., and Dubray, G. (1999). Immunogenicity of recombinant *Escherichia coli* expressing the *omp31* gene of *Brucella melitensis* in BALB/c mice. Vaccine *17*, 353–361.

Guzmán-Verri, C., Manterola, L., Sola-Landa, A., Parra, A., Cloeckaert, A., Garin, J., Gorvel, J.P., Moriyón, I., Moreno, E., and López-Goñi, I. (2002). The two-component system BvrR/BvrS essential for *Brucella abortus* virulence regulates the expression of outer membrane proteins with counterparts in members of the *Rhizobiaceae*. Proc. Natl. Acad. Sci. U.S.A. *99*, 12375–12380.

Hancock, R.E. (1997). The bacterial outer membrane as a drug barrier. Trends Microbiol. *5*, 37–42.

Hong, P.C., Tsolis, R.M., and Ficht, T.A. (2000). Identification of genes required for chronic persistence of *Brucella abortus* in mice. Infect. Immun. *68*, 4102–4107.

Iriarte, M., González, D., Delrue, R.M., Monreal, D., Conde, R., López-Goñi, I., Letesson, J.J., and Moriyón, I. (2004). *Brucella* lipopolysaccharide: structure, biosynthesis and genetics. In: *Brucella:* Molecular and Cellular Biology, López-Goñi, I., and Moriyón, I., eds. (Horizon Biosciences; Norfolk, UK), pp. 159–191.

Jacques, I., Cloeckaert, A., Limet, J.N., and Dubray, G. (1992). Protection conferred on mice by combinations of monoclonal antibodies directed against outer-membrane proteins or smooth lipopolysaccharide of *Brucella*. J. Med. Microbiol. *37*, 100–103.

Jensen, A.E., and Halling, S.M. (2010). Effect of polymyxin B and environmental conditions on isolation of *Brucella* species and the vaccine strain RB51. Comp. Immunol. Microbiol. Infect. Dis. *33*, 121–131.

Jenssen, H., Hamill, P., and Hancock, R.E. (2006). Peptide antimicrobial agents. Clin. Microbiol. Rev. *19*, 491–511.

Jia, Y.H., Li, L.P., Hou, Q.M., and Pan, S.Q. (2002). An *Agrobacterium* gene involved in tumorigenesis encodes an outer membrane protein exposed on the bacterial cell surface. Gene *284*, 113–124.

Jubier-Maurin, V., Boigegrain, R.A., Cloeckaert, A., Gross, A., Álvarez-Martínez, M.T., Terraza, A., Liautard, J., Köhler, S., Rouot, B., Dornand, J., *et al.* (2001). Major outer membrane protein Omp25 of *Brucella suis* is involved in inhibition of tumor necrosis factor alpha production during infection of human macrophages. Infect. Immun. *69*, 4823–4830.

Kamio, Y., and Nikaido, H. (1976). Outer membrane of *Salmonella typhimurium*: accessibility of phospholipid head groups to phospholipase c and cyanogen bromide activated dextran in the external medium. Biochemistry *15*, 2561–2570.

Kittelberger, R., Diack, D.S., Vizcaíno, N., Zygmunt, M.S., and Cloeckaert, A. (1998). Characterization of an immuno-dominant antigen in *Brucella ovis* and evaluation of its use in an enzyme-linked immunosorbent assay. Vet. Microbiol. *59*, 213–227.

Köhler, S., Foulongne, V., Ouahrani-Bettache, S., Bourg, G., Teyssier, J., Ramuz, M., and Liautard, J.P. (2002). The analysis of the intramacrophagic virulome of *Brucella suis* deciphers the environment encountered by the pathogen inside the macrophage host cell. Proc. Natl. Acad. Sci. U.S.A. *99*, 15711–15716.

Lacerda, T.L., Cardoso, P.G., Augusto de Almeida, L., Camargo, I.L., Afonso, D.A., Trant, C.C., Macedo, G.C., Campos, E., Cravero, S.L., Salcedo, S.P., *et al.* (2010). Inactivation of formyltransferase (wbkC) gene generates a *Brucella abortus* rough strain that is attenuated in macrophages and in mice. Vaccine *28*, 5627–5634.

Laloux, G., Deghelt, M., de Barsy, M., Letesson, J.J., and De Bolle, X. (2010). Identification of the essential *Brucella melitensis* porin Omp2b as a suppressor of bax-induced cell death in yeast in a genome-wide screening. PLoS One *5*, e13274.

Lamontagne, J., Butler, H., Chaves-Olarte, E., Hunter, J., Schirm, M., Paquet, C., Tian, M., Kearney, P., Hamaidi, L., Chelsky, D., *et al.* (2007). Extensive cell envelope modulation is associated with virulence in *Brucella abortus*. J. Proteome Res. *6*, 1519–1529.

Lamontagne, J., Forest, A., Marazzo, E., Denis, F., Butler, H., Michaud, J.F., Boucher, L., Pedro, I., Villeneuve, A., Sitnikov, D., *et al.* (2009). Intracellular adaptation of *Brucella abortus*. J. Proteome Res. *8*, 1594–1609.

Lamontagne, J., Béland, M., Forest, A., Côté-Martin, A., Nassif, N., Tomaki, F., Moriyón, I., Moreno, E., and Paramithiotis, E. (2010). Proteomics-based confirmation of protein expression and correction of annotation errors in the *Brucella abortus* genome. BMC Genomics *11*, 300.

Lapaque, N., Moriyón, I., Moreno, E., and Gorvel, J.P. (2005). *Brucella* lipopolysaccharide acts as a virulence factor. Curr. Opin. Microbiol. *8*, 60–66.

Limet, J.N., Bosseray, N., Garin-Bastuji, B., Dubray, G., and Plommet, M. (1989). Humoral immunity in mice mediated by monoclonal antibodies against the A and M antigens of *Brucella*. J. Med. Microbiol. *30*, 37–43.

McQuiston, J.R., Vemulapalli, R., Inzana, T.J., Schurig, G.G., Sriranganathan, N., Fritzinger, D., Hadfield, T.L., Warren, R.A., Lindler, L.E., Snellings, N., *et al.* (1999). Genetic characterization of a Tn5-disrupted glycosyltransferase gene homolog in *Brucella abortus* and its effect on lipopolysaccharide composition and virulence. Infect. Immun. *67*, 3830–3835.

Mancilla, M., López-Goñi, I., Moriyón, I., and Zárraga, A.M. (2010). Genomic island-2 is an unstable genetic element contributing to *Brucella* lipopolysaccharide spontaneous smooth to rough dissociation. J. Bacteriol. *192*, 6346–6351.

Manterola, L., Moriyón, I., Moreno, E., Sola-Landa, A., Weiss, D.S., Koch, M.H., Howe, J., Brandenburg, K., and López-Goñi, I. (2005). The lipopolysaccharide of *Brucella abortus* BvrS/BvrR mutants contains lipid A modifications and has higher affinity for bactericidal cationic peptides. J. Bacteriol. *187*, 5631–5639.

Manterola, L., Guzmán-Verri, C., Chaves-Olarte, E., Barquero-Calvo, E., de Miguel, M.J., Moriyón, I., Grilló, M.J., López-Goñi, I., and Moreno, E. (2007). BvrR/BvrS-controlled outer membrane proteins Omp3a and Omp3b are not essential for *Brucella abortus* virulence. Infect. Immun. *75*, 4867–4874.

Marín, C.M., Moreno, E., Moriyón, I., Díaz, R., and Blasco, J.M. (1999). Performance of competitive and indirect enzyme-linked immunosorbent assays, gel immunoprecipitation with native hapten polysaccharide, and standard serological tests in diagnosis of sheep brucellosis. Clin. Diagn. Lab. Immunol. *6*, 269–272.

Marquis, H., and Ficht, T.A. (1993). The *omp2* gene locus of *Brucella abortus* encodes two homologous outer membrane proteins with properties characteristic of bacterial porins. Infect. Immun. *61*, 3785–3790.

Martín, F.A., Posadas, D.M., Carrica, M.C., Cravero, S.L., O'Callaghan, D., and Zorreguieta, A. (2009). Interplay between two RND systems mediating antimicrobial resistance in *Brucella suis*. J. Bacteriol. *191*, 2530–2540.

Martín-Martín, A.I., Caro-Hernández, P., Orduña, A., Vizcaíno, N., and Fernández-Lago, L. (2008). Importance of the Omp25/Omp31 family in the internalization and intracellular replication of virulent *B. ovis* in murine macrophages and HeLa cells. Microbes Infect. *10*, 706–710.

Martín-Martín, A.I., Caro-Hernández, P., Sancho, P., Tejedor, C., Cloeckaert, A., Fernández-Lago, L., and Vizcaíno, N. (2009). Analysis of the occurrence and distribution of the Omp25/Omp31 family of surface proteins in the six classical *Brucella* species. Vet. Microbiol. *137*, 74–82.

Martín-Martín, A.I., Sancho, P., Tejedor, C., Fernández-Lago, L., and Vizcaíno, N. (2010a). Differences in the outer membrane-related properties of the six classical *Brucella* species. Vet. J. doi:10.1016/j.tvjl.2010.05.021.

Martín-Martín, A.I., Vizcaíno, N., and Fernández-Lago, L. (2010b). Cholesterol, ganglioside GM_1 and class A scavenger receptor contribute to infection by *Brucella ovis* and *Brucella canis* in murine macrophages. Microbes Infect. *12*, 246–251.

Martínez de Tejada, G., and Moriyón, I. (1993). The outer membranes of *Brucella* spp. are not barriers to hydrophobic permeants. J. Bacteriol. *175*, 5273–5275.

Martínez de Tejada, G., Pizarro-Cerdá, J., Moreno, E., and Moriyón, I. (1995). The outer membranes of *Brucella* spp. are resistant to bactericidal cationic peptides. Infect. Immun. *63*, 3054–3061.

Meikle, P.J., Perry, M.B., Cherwonogrodzky, J.W., and Bundle, D.R. (1989). Fine structure of A and M antigens from *Brucella* biovars. Infect. Immun. *57*, 2820–2828.

Minnick, M.F., Sappington, K.N., Smitherman, L.S., Andersson, S.G., Karlberg, O., and Carroll, J.A. (2003). Five-member gene family of *Bartonella quintana*. Infect. Immun. *71*, 814–821.

Mobasheri, H., Ficht, T.A., Marquis, H., Lea, E.J., and Lakey, J.H. (1997). Brucella Omp2a and Omp2b porins: single channel measurements and topology prediction. FEMS Microbiol. Lett. *155*, 23–30.

Monreal, D., Grilló, M.J., González, D., Marín, C.M., De Miguel, M.J., López-Goñi, I., Blasco, J.M., Cloeckaert, A., and Moriyón, I. (2003). Characterization of *Brucella abortus* O-polysaccharide and core lipopolysaccharide mutants and demonstration that a complete core is required for rough vaccines to be efficient against *Brucella abortus* and *Brucella ovis* in the mouse model. Infect. Immun. *71*, 3261–3271.

Montaraz, J.A., Winter, A.J., Hunter, D.M., Sowa, B.A., Wu, A.M., and Adams, L.G. (1986). Protection against *Brucella abortus* in mice with O-polysaccharide-specific monoclonal antibodies. Infect. Immun. *51*, 961–963.

Moreno, E., Mayer, H., and Moriyón, I. (1987). Characterization of a native polysaccharide hapten from *Brucella melitensis*. Infect. Immun. *55*, 2850–2853.

Moreno, E., Cloeckaert, A., and Moriyón, I. (2002). *Brucella* evolution and taxonomy. Vet. Microbiol. *90*, 209–227.

Moreno, E., Stackebrandt, E., Dorsch, M., Wolters, J., Busch, M., and Mayer, H. (1990). *Brucella abortus* 16S rRNA and lipid A reveal a phylogenetic relationship with members of the alpha-2 subdivision of the class *Proteobacteria*. J. Bacteriol. *172*, 3569–3576.

Moriyón, I., and Berman, D.T. (1982). Effects of nonionic, ionic, and dipolar ionic detergents and EDTA on the *Brucella* cell envelope. J. Bacteriol. *152*, 822–828.

Moriyón, I., and López-Goñi, I. (1998). Structure and properties of the outer membranes of *Brucella abortus* and *Brucella melitensis*. Int. Microbiol. *1*, 19–26.

Nikaido, H. (2003). Molecular basis of bacterial outer membrane permeability revisited. Microbiol. Mol. Biol. Rev. *67*, 593–656.

O'Callaghan, D., Cazevieille, C., Allardet-Servent, A., Boschiroli, M.L., Bourg, G., Foulongne, V., Frutos, P., Kulakov, Y., and Ramuz, M. (1999). A homologue of the *Agrobacterium tumefaciens* VirB and *Bordetella pertussis* Ptl type IV secretion systems is essential for intracellular survival of *Brucella suis*. Mol. Microbiol. *33*, 1210–1220.

Ohishi, K., Takishita, K., Kawato, M., Zenitani, R., Bando, T., Fujise, Y., Goto, Y., Yamamoto, S., and Maruyama, T. (2004). Molecular evidence of new variant *Brucella* in North Pacific common minke whales. Microbes Infect. *6*, 1199–1204.

Paixao, T.A., Roux, C.M., den Hartigh, A.B., Sankaran-Walters, S., Dandekar, S., Santos, R.L., and Tsolis, R.M. (2009). Establishment of systemic *Brucella melitensis* infection through the digestive tract requires urease, the type IV secretion system, and lipopolysaccharide O antigen. Infect. Immun. *77*, 4197–4208.

Palmer, D.A., and Douglas, J.T. (1989). Analysis of *Brucella* lipopolysaccharide with specific and cross-reacting monoclonal antibodies. J. Clin. Microbiol. *27*, 2331–2337.

Paquet, J.Y., Vinals, C., Wouters, J., Letesson, J.J., and Depiereux, E. (2000). Topology prediction of *Brucella abortus* Omp2b and Omp2a porins after critical assessment of transmembrane beta strands prediction by several secondary structure prediction methods. J. Biomol. Struct. Dyn. *17*, 747–757.

Paquet, J.Y., Díaz, M.A., Genevrois, S., Grayon, M., Verger, J.M., de Bolle, X., Lakey, J.H., Letesson, J.J., and Cloeckaert, A. (2001). Molecular, antigenic, and functional analyses of Omp2b porin size variants of *Brucella* spp. J. Bacteriol. *183*, 4839–4847.

Pasquevich, K.A., García Samartino, C., Coria, L.M., Estein, S.M., Zwerdling, A., Ibanez, A.E., Barrionuevo, P., Oliveira, F.S., Carvalho, N.B., Borkowski, J., *et al.* (2010). The protein moiety of *Brucella abortus* outer membrane protein 16 is a new bacterial pathogen-associated molecular pattern that activates dendritic cells *in vivo*, induces a Th1 immune response, and is a promising self-adjuvanting vaccine against systemic and oral acquired brucellosis. J. Immunol. *184*, 5200–5212.

Porte, F., Liautard, J.P., and Köhler, S. (1999). Early acidification of phagosomes containing *Brucella suis* is essential for intracellular survival in murine macrophages. Infect. Immun. *67*, 4041–4047.

Posadas, D.M., Martín, F.A., Sabio y García, J.V., Spera, J.M., Delpino, M.V., Baldi, P., Campos, E., Cravero, S.L., and Zorreguieta, A. (2007). The TolC homologue of *Brucella suis* is involved in resistance to antimicrobial compounds and virulence. Infect. Immun. *75*, 379–389.

Rajashekara, G., Glasner, J.D., Glover, D.A., and Splitter, G.A. (2004). Comparative whole-genome hybridization reveals genomic islands in *Brucella* species. J. Bacteriol. *186*, 5040–5051.

Rajashekara, G., Covert, J., Petersen, E., Eskra, L., and Splitter, G. (2008). Genomic island 2 of *Brucella melitensis* is a major virulence determinant: functional analyses of genomic islands. J. Bacteriol. *190*, 6243–6252.

Rambow-Larsen, A.A., Rajashekara, G., Petersen, E., and Splitter, G. (2008). Putative quorum-sensing regulator BlxR of *Brucella melitensis* regulates virulence factors including the type IV secretion system and flagella. J. Bacteriol. *190*, 3274–3282.

Riezu-Boj, J.I., Moriyón, I., Blasco, J.M., Marín, C.M., and Díaz, R. (1986). Comparison of lipopolysaccharide and outer membrane protein-lipopolysaccharide extracts in an enzyme-linked immunosorbent assay for the diagnosis of *Brucella ovis* infection. J. Clin. Microbiol. *23*, 938–942.

Roop, R.M., 2nd, Gaines, J.M., Anderson, E.S., Caswell, C.C., and Martin, D.W. (2009). Survival of the fittest: how *Brucella* strains adapt to their intracellular niche in the host. Med. Microbiol. Immunol. *198*, 221–238.

Rossetti, C.A., Galindo, C.L., Lawhon, S.D., Garner, H.R., and Adams, L.G. (2009). *Brucella melitensis* global gene expression study provides novel information on growth phase-specific gene regulation with potential insights for understanding *Brucella*:host initial interactions. BMC Microbiol. *9*, 81.

Salhi, I., Boigegrain, R.A., Machold, J., Weise, C., Cloeckaert, A., and Rouot, B. (2003). Characterization of new members of the group 3 outer membrane protein family of *Brucella* spp. Infect. Immun. *71*, 4326–4332.

Santos, J.M., Verstreate, D.R., Perera, V.Y., and Winter, A.J. (1984). Outer membrane proteins from rough strains of four *Brucella* species. Infect. Immun. *46*, 188–194.

Scholz, H.C., Hubalek, Z., Sedlácek, I., Vergnaud, G., Tomaso, H., Al Dahouk, S., Melzer, F., Kämpfer, P., Neubauer, H., Cloeckaert, A., *et al.* (2008). *Brucella microti* sp. nov., isolated from the common vole *Microtus arvalis*. Int. J. Syst. Evol. Microbiol. *58*, 375–382.

Scholz, H.C., Nockler, K., Gollner, C., Bahn, P., Vergnaud, G., Tomaso, H., Al-Dahouk, S., Kampfer, P., Cloeckaert, A., Maquart, M., *et al.* (2010). *Brucella inopinata* sp. nov., isolated from a breast implant infection. Int. J. Syst. Evol. Microbiol. *60*, 801–808.

Sieira, R., Comerci, D.J., Sánchez, D.O., and Ugalde, R.A. (2000). A homologue of an operon required for DNA transfer in *Agrobacterium* is required in *Brucella abortus* for virulence and intracellular multiplication. J. Bacteriol. *182*, 4849–4855.

Skurnik, M., and Bengoechea, J.A. (2003). The biosynthesis and biological role of lipopolysaccharide O-antigens of pathogenic *Yersiniae*. Carbohydr. Res. *338*, 2521–2529.

Skurnik, M., Biedzka-Sarek, M., Lubeck, P.S., Blom, T., Bengoechea, J.A., Pérez-Gutiérrez, C., Ahrens, P., and Hoorfar, J. (2007). Characterization and biological role of the O-polysaccharide gene cluster of *Yersinia enterocolitica* serotype O:9. J. Bacteriol. *189*, 7244–7253.

Sola-Landa, A., Pizarro-Cerdá, J., Grilló, M.J., Moreno, E., Moriyón, I., Blasco, J.M., Gorvel, J.P., and López-Goñi, I. (1998). A two-component regulatory system playing a critical role in plant pathogens and endosymbionts is present in *Brucella abortus* and controls cell invasion and virulence. Mol. Microbiol. *29*, 125–138.

Sowa, B.A., Kelly, K.A., Ficht, T.A., Frey, M., and Adams, L.G. (1991). SDS-soluble and peptidoglycan-bound proteins in the outer membrane–peptidoglycan complex of *Brucella abortus*. Vet. Microbiol. *27*, 351–369.

Thiele, O.W., and Schwinn, G. (1973). The free lipids of *Brucella melitensis* and *Bordetella pertussis*. Eur. J. Biochem. *34*, 333–344.

Tibor, A., Weynants, V., Denoel, P., Lichtfouse, B., De Bolle, X., Saman, E., Limet, J.N., and Letesson, J.J. (1994). Molecular cloning, nucleotide sequence, and occurrence of a 16.5-kilodalton outer membrane protein of *Brucella abortus* with similarity to PAL lipoproteins. Infect. Immun. *62*, 3633–3639.

Tibor, A., Saman, E., de Wergifosse, P., Cloeckaert, A., Limet, J.N., and Letesson, J.J. (1996). Molecular characterization, occurrence, and immunogenicity in infected sheep and cattle of two minor outer membrane proteins of *Brucella abortus*. Infect. Immun. *64*, 100–107.

Tibor, A., Decelle, B., and Letesson, J.J. (1999). Outer membrane proteins Omp10, Omp16, and Omp19 of *Brucella* spp. are lipoproteins. Infect. Immun. *67*, 4960–4962.

Tibor, A., Wansard, V., Bielartz, V., Delrue, R.M., Danese, I., Michel, P., Walravens, K., Godfroid, J., and Letesson, J.J. (2002). Effect of *omp10* or *omp19* deletion on *Brucella abortus* outer membrane properties and virulence in mice. Infect. Immun. *70*, 5540–5546.

Tsolis, R.M., Seshadri, R., Santos, R.L., Sangari, F.J., García Lobo, J.M., de Jong, M.F., Ren, Q., Myers, G., Brinkac, L.M., Nelson, W.C., *et al.* (2009). Genome degradation in *Brucella ovis* corresponds with narrowing of its host range and tissue tropism. PLoS One *4*, e5519.

Uzureau, S., Godefroid, M., Deschamps, C., Lemaire, J., De Bolle, X., and Letesson, J.J. (2007). Mutations of the quorum sensing-dependent regulator VjbR lead to drastic surface modifications in *Brucella melitensis*. J. Bacteriol. *189*, 6035–6047.

Vaara, M. (1992). Agents that increase the permeability of the outer membrane. Microbiol. Rev. *56*, 395–411.

Velasco, J., Bengoechea, J.A., Brandenburg, K., Lindner, B., Seydel, U., González, D., Zahringer, U., Moreno, E., and Moriyón, I. (2000). *Brucella abortus* and its closest phylogenetic relative, *Ochrobactrum* spp., differ in outer membrane permeability and cationic peptide resistance. Infect. Immun. *68*, 3210–3218.

Vemulapalli, R., McQuiston, J.R., Schurig, G.G., Sriranganathan, N., Halling, S.M., and Boyle, S.M. (1999). Identification of an IS711 element interrupting the *wboA* gene of *Brucella abortus* vaccine strain RB51 and a PCR assay to distinguish strain RB51 from other *Brucella* species and strains. Clin. Diagn. Lab. Immunol. *6*, 760–764.

Vemulapalli, R., Cravero, S., Calvert, C.L., Toth, T.E., Sriranganathan, N., Boyle, S.M., Rossetti, O.L., and Schurig, G.G. (2000a). Characterization of specific immune responses of mice inoculated with recombinant vaccinia virus expressing an 18-kilodalton outer membrane protein of *Brucella abortus*. Clin. Diagn. Lab. Immunol. *7*, 114–118.

Vemulapalli, R., He, Y., Buccolo, L.S., Boyle, S.M., Sriranganathan, N., and Schurig, G.G. (2000b). Complementation of *Brucella abortus* RB51 with a functional *wboA* gene results in O-antigen synthesis and enhanced vaccine efficacy but no change in rough phenotype and attenuation. Infect. Immun. *68*, 3927–3932.

Verger, J.M., Grimont, F., Grimont, P.A., and Grayon, M. (1985). *Brucella*, a monospecific genus as shown by deoxyribonucleic-acid hybridization. Int. J. Syst. Bacteriol. *35*, 292–295.

Verstreate, D.R., Creasy, M.T., Caveney, N.T., Baldwin, C.L., Blab, M.W., and Winter, A.J. (1982). Outer membrane proteins of *Brucella abortus*: isolation and characterization. Infect. Immun. *35*, 979–989.

Viadas, C., Rodríguez, M.C., Sangari, F.J., Gorvel, J.P., García-Lobo, J.M., and López-Goñi, I. (2010). Transcriptome analysis of the *Brucella abortus* BvrR/BvrS two-component regulatory system. PLoS One *5*, e10216.

Vizcaíno, N., Chordi, A., and Fernández-Lago, L. (1991). Characterization of smooth *Brucella* lipopolysaccharides and polysaccharides by monoclonal antibodies. Res. Microbiol. *142*, 971–978.

Vizcaíno, N., Cloeckaert, A., Zygmunt, M.S., and Dubray, G. (1996). Cloning, nucleotide sequence, and expression of the *Brucella melitensis omp31* gene coding for an immunogenic major outer membrane protein. Infect. Immun. *64*, 3744–3751.

Vizcaíno, N., Verger, J.M., Grayon, M., Zygmunt, M.S., and Cloeckaert, A. (1997). DNA polymorphism at the *omp-31* locus of *Brucella* spp.: evidence for a large deletion in *Brucella abortus*, and other species-specific markers. Microbiology *143*, 2913–2921.

Vizcaíno, N., Cloeckaert, A., Zygmunt, M.S., and Fernández-Lago, L. (1999). Molecular characterization of a *Brucella* species large DNA fragment deleted in *Brucella abortus* strains: evidence for a locus involved in the synthesis of a polysaccharide. Infect. Immun. *67*, 2700–2712.

Vizcaíno, N., Cloeckaert, A., Verger, J., Grayon, M., and Fernández-Lago, L. (2000). DNA polymorphism in the genus *Brucella*. Microbes Infect. *2*, 1089–1100.

Vizcaíno, N., Cloeckaert, A., Zygmunt, M.S., and Fernández-Lago, L. (2001a). Characterization of a *Brucella* species 25-kilobase DNA fragment deleted from *Brucella abortus* reveals a large gene cluster related to the synthesis of a polysaccharide. Infect. Immun. *69*, 6738–6748.

Vizcaíno, N., Kittelberger, R., Cloeckaert, A., Marín, C.M., and Fernández-Lago, L. (2001b). Minor nucleotide substitutions in the *omp31* gene of *Brucella ovis* result in antigenic differences in the major outer membrane protein that it encodes compared to those of the other *Brucella* species. Infect. Immun. *69*, 7020–7028.

Vizcaíno, N., Caro-Hernández, P., Cloeckaert, A., and Fernández-Lago, L. (2004). DNA polymorphism in the *omp25/omp31* family of *Brucella* spp.: identification of a 1.7-kb inversion in *Brucella cetaceae* and of a 15.1-kb genomic island, absent from *Brucella ovis*, related to the synthesis of smooth lipopolysaccharide. Microbes Infect. *6*, 821–834.

Wagner, M.A., Eschenbrenner, M., Horn, T.A., Kraycer, J.A., Mujer, C.V., Hagius, S., Elzer, P., and DelVecchio, V.G. (2002). Global analysis of the *Brucella melitensis* proteome: Identification of proteins expressed in laboratory-grown culture. Proteomics 2, 1047–1060.

Wattam, A.R., Williams, K.P., Snyder, E.E., Almeida, N.F., Jr., Shukla, M., Dickerman, A.W., Crasta, O.R., Kenyon, R., Lu, J., Shallom, J.M., *et al.* (2009). Analysis of ten *Brucella* genomes reveals evidence for horizontal gene transfer despite a preferred intracellular lifestyle. J. Bacteriol. *191*, 3569–3579.

Weeks, J.N., Galindo, C.L., Drake, K.L., Adams, G.L., Garner, H.R., and Ficht, T.A. (2010). *Brucella melitensis* VjbR and C12-HSL regulons: contributions of the N-dodecanoyl homoserine lactone signaling molecule and LuxR homologue VjbR to gene expression. BMC Microbiol. *10*, 167.

de Wergifosse, P., Lintermans, P., Limet, J.N., and Cloeckaert, A. (1995). Cloning and nucleotide sequence of the gene coding for the major 25-kilodalton outer membrane protein of *Brucella abortus*. J. Bacteriol. *177*, 1911–1914.

Weynants, V., Gilson, D., Cloeckaert, A., Tibor, A., Denoel, P.A., Godfroid, F., Limet, J.N., and Letesson, J.J. (1997). Characterization of smooth lipopolysaccharides and O polysaccharides of *Brucella* species by competition binding assays with monoclonal antibodies. Infect. Immun. *65*, 1939–1943.

Whatmore, A.M. (2009). Current understanding of the genetic diversity of *Brucella*, an expanding genus of zoonotic pathogens. Infect. Genet. Evol. *9*, 1168–1184.

Whatmore, A.M., Perrett, L.L., and MacMillan, A.P. (2007). Characterisation of the genetic diversity of *Brucella* by multilocus sequencing. BMC Microbiol. 7, 34.

Winter, A.J. (1987). Outer membrane proteins of *Brucella*. Ann. Inst. Pasteur Microbiol. *138*, 87–89.

Winter, A.J., Rowe, G.E., Duncan, J.R., Eis, M.J., Widom, J., Ganem, B., and Morein, B. (1988). Effectiveness of natural and synthetic complexes of porin and O polysaccharide as vaccines against *Brucella abortus* in mice. Infect. Immun. *56*, 2808–2817.

Yu, W.L., and Nielsen, K. (2010). Review of detection of *Brucella* spp. by polymerase chain reaction. Croat. Med. J. *51*, 306–313.

Zwerdling, A., Delpino, M.V., Barrionuevo, P., Cassataro, J., Pasquevich, K.A., García Samartino, C., Fossati, C.A., and Giambartolomei, G.H. (2008). *Brucella* lipoproteins mimic dendritic cell maturation induced by *Brucella abortus*. Microbes Infect. *10*, 1346–1354.

Zygmunt, M.S., Dubray, G., Bundle, D.R., and Perry, M.P. (1988). Purified native haptens of *Brucella abortus* B19 and *B. melitensis* 16M reveal the lipopolysaccharide origin of the antigens. Ann. Inst. Pasteur Microbiol. *139*, 421–433.

Zygmunt, M.S., Blasco, J.M., Letesson, J.J., Cloeckaert, A., and Moriyón, I. (2009). DNA polymorphism analysis of *Brucella* lipopolysaccharide genes reveals marked differences in O-polysaccharide biosynthetic genes between smooth and rough *Brucella* species and novel species-specific markers. BMC Microbiol. *9*, 92.

8

Brucella Quorum Sensing: Much More than Sensing Quorum

Matthieu Terwagne, Sophie Bonnot and Jean-Jacques Letesson

Abstract

Quorum sensing is a regulatory system that allows bacteria to coordinate gene expression according to the local population density. Recently, we demonstrated that the virulence of the facultative intracellular bacteria *Brucella* depends on quorum sensing. Similar to other Gram-negative bacteria, *Brucella* quorum sensing utilizes the production and detection of *N*-acyl homoserine lactone as a signal. However, in *Brucella*, *N*-acyl homoserine lactone could serve to monitor the confinement state, a situation in which a single bacterium enclosed in a vacuole can be the quorum. Here, we present a current review covering the intricacies of quorum sensing in *Brucella*, highlighting the abilities of quorum sensing to influence both *Brucella* virulence and metabolism.

Introduction

Quorum sensing (QS) is a regulatory system that allows gene expression to be controlled in response to small diffusible signalling molecules produced and released by bacteria. When a threshold bacterial concentration is reached, signalling molecules modulate the activity of a transcriptional regulator, leading to the induction or repression of target genes. QS derives its name from an original study demonstrating the ability of a bacterial population to synchronize adaptive responses in response to increasing population density (Fuqua *et al.*, 1994). The concentration of a signalling molecule accumulating during population growth directly reflects the number of bacteria, and the threshold concentration is equal to the quorum of bacteria required for the co-ordinated response of the overall population. Indeed, QS regulates functions requiring the action of multiple bacteria, such as swarming motility in several species (for review see Daniels *et al.*, 2004), bioluminescence in *Vibrio fischeri* (Stevens and Greenberg, 1997) and virulence in *Pseudomonas aeruginosa* (Kohler *et al.*, 2009; Rumbaugh *et al.*, 2009).

Nevertheless, as signalling molecule accumulation can be affected by abiotic and biotic parameters, it has been argued that in addition to monitoring population density, QS also monitors bacterial spatial distribution (Alberghini *et al.*, 2009) and environmental factors such as bacterial confinement (Boedicker *et al.*, 2009; Carnes *et al.*, 2010), pH (Decho *et al.*, 2009) and diffusion rate (Redfield, 2002). Consequently, in particular cases, QS has been renamed to emphasize ecological context; further, it has been shown that the principles of QS can also be applied to individual bacteria (for review see Platt and Fuqua, 2010).

Recently, we demonstrated that the virulence of the facultative intracellular bacteria *Brucella* depends on such a regulatory system, based on the production of *N*-acyl homoserine lactone (AHL) as a signal, as in many others Gram-negative bacteria (Delrue *et al.*, 2005). *Brucella* is the aetiological agent of brucellosis, a worldwide zoonosis that remains a public health concern in endemic countries. Although information detailing the infectious cycle within its natural mammalian host is lacking, it is commonly assumed that during an infection *Brucella* spends most of its time inside phagocytic cells (for review see Roop *et al.*, 2009). Its impressive ability to subvert intracellular trafficking pathway to promote its survival and replication in cells has led authors to refer to *Brucella* as a 'facultatively extracellular intracellular parasite' (Moreno and Moriyón, 2002). Throughout its entire intracellular life, *Brucella* is enclosed in a membrane-bound compartment (Celli *et al.*, 2003). Although it remains possible that AHLs cross the membranes of *Brucella*-containing vacuoles and can be sensed by neighbouring bacteria in highly infected cells, in intracellular *Brucella* infection it seems probable that QS regulation is not a matter of social behaviour. Instead, AHLs could monitor the confinement state, a situation in which a single bacterium can be the quorum (Carnes *et al.*, 2010). Here, we review *Brucella* QS, highlighting what is known about each of the actors in this system.

The partners

As a general principle, AHL-based QS depends on three main actors: (i) AHL, the signal molecule or autoinducer; (ii) the enzyme required for the AHL synthesis; and (iii) a transcriptional regulator which recognizes the autoinducer and modulates target gene expression. In some cases, a fourth player that is an AHL degrading enzyme, adds an additional layer of complexity by preventing signal molecule accumulation.

The QS regulator optimally recognizes AHL produced by its cognate AHL synthase, thus guaranteeing specificity. While most cognate pairs of AHL synthase and QS regulator are genetically linked, exceptions exist where the pairs are distantly encoded on the chromosome. Additionally, in Proteobacteria, QS regulators that lack a cognate AHL synthase exist; these QS regulators are termed 'orphan' or 'solo' (Patankar and Gonzalez, 2009; Subramoni and Venturi, 2009). Here, each of these four distinct components will be addressed, starting with a brief introduction of the general features and then focusing on what is known or suspected in *Brucella*.

The signal

N-acyl homoserine lactone (AHL)

N-acyl homoserine lactone molecules are comprised of a homoserine lactone ring (with a L stereochemistry) and an acyl chain of variable length. According to the specific AHL, the acyl chain is between 4 and 18 carbons and may contain C3 substitutions (either a hydroxyl or a carbonyl group). Some AHLs have one or two double bonds in the acyl chain. The specificity of the signal molecule for its cognate receptor (see below) resides in the length of the acyl chain and any substitution(s) within it (Parsek *et al.*, 1999). A bacterium can produce different types of AHL, and multiple types of bacteria can produce the same AHL.

The amphiphilic nature of AHLs allows them to freely diffuse through both bacterial (Kaplan and Greenberg, 1985) and eukaryotic membranes (Williams *et al.*, 2004). Nevertheless, long chain AHLs could necessitate an efflux pump to be secreted (Pearson *et al.*, 1999). To our knowledge, no specific entrance mechanism has ever been described.

Brucella signalling molecules

In the late 1990s, the increasing awareness of the ability of QS to regulate bacterial virulence prompted us to investigate whether *Brucella* produced AHLs. Screening for AHLs was conducted using biosensors that contained both a *luxR* homologue and a target QS-driven promoter fused to a reporter gene such as *luxABCDE*; these were maintained in *Escherichia coli*, which does not produce endogenous AHLs, or in a bacterial strain mutated for the synthase gene that only exhibits a QS-dependent phenotype when exogenous AHLs are detected.

On solid 2YT rich media, sensor strains cross streaked with *B. melitensis* did not detect AHLs, meaning that either they were not produced under the conditions tested or that they were produced at too low of a concentration for detection. Only concentrated extracts prepared from *B. melitensis* harvested at the end of the exponential phase in defined RPMI 1640 medium demonstrated dose-dependent activation of the biosensor (Taminiau *et al.*, 2002). High-performance liquid chromatography (HPLC) fractionation, thin-layer chromatography (TLC) analysis and finally mass spectrometry (MS) identified the major active molecule as an *N*-dodecanoylhomoserine lactone (C_{12}-HSL). Although a second active minor component was also detected by HPLC and TLC analysis and presumptively identified as 3oxo-C_{12}-HSL, it could not be analysed further by MS (Taminiau *et al.*, 2002). This was the first description of an AHL produced by an intracellular pathogen and the first identification of C_{12}-HSL as a natural product. Previously, C_{12}-HSL was only known as a synthetic molecule.

The low concentration of AHLs detected in *Brucella* culture supernatants is of note. Indeed, the final characterization of these signal molecules necessitated 18 litres of culture media. Several hypothesis can be put forward to explain this fact: (i) the hydrophobicity of the long chain AHLs limits their diffusion and increases their interaction with bacterial membranes, which lowers their presence in the culture medium (Pearson *et al.*, 1999; Schaefer *et al.*, 2002); (ii) *in vitro* AHL production does not correlate with *in vivo* production, as seen in *P. aeruginosa* (Erickson *et al.*, 2002) where some signals are absent *in vitro*; and (iii) given the enclosed vacuolar space of the intracellular niche of *Brucella*, the bacteria would not need to produce a large amount of AHL and could control its synthesis and/or stability (see 'the AHL-degrading enzymes', below).

AHL synthase

The LuxI, LuxM/AinS and HdtS families

Three families of AHL synthases have been described to date: LuxI (Parsek *et al.*, 1999), LuxM/AinS (Bassler *et al.*, 1993) and HdtS (Laue *et al.*, 2000). These families differ both at the sequence level and at the level of the substrate used during the synthesis.

The LuxI family is the most common (Case *et al.*, 2008) and has been identified even in non-cultured bacteria (Hao *et al.*, 2010). The heterologous expression of *luxI* homologues is both necessary and sufficient to induce AHL synthesis (Engebrecht and Silverman, 1984). AHL synthesis occurs through a unique reaction (Parsek *et al.*, 1999) that requires S-adenosylmethionine (SAM) and fatty acid biosynthetic precursors provided as acylated-acyl carrier protein (ACP) conjugates (Schaefer *et al.*, 1996). LuxI enzymes use SAM, not as methyl donor, but as provider of the homoserine lactone ring. In brief, binding of SAM by LuxI initiates the reaction. Subsequently, acyl-ACP binds to the enzyme complex, followed by amide bond formation and the release of holo-ACP. Next, lactonization of the homoserine ring occurs, and the product is liberated. In a final step, 5′-methylthioadenosine (5′-MTA) is released (Parsek *et al.*, 1999).

Amino acid alignments of LuxI-type proteins reveal that the amino-terminal portions of LuxI-type proteins are the most conserved. The carboxy terminus is more divergent, suggesting that this part of the molecule may provide for the recognition of different acyl chains on precursor acyl-ACPs (Hanzelka *et al.*, 1997). Crystallographic analysis of AHL-synthases (EsaI from *Erwinia carotovora* and LasI from *P. aeruginosa*) demonstrated that the geometry of a hydrophobic cavity confers the specificity for an acyl chain of a given length (Watson *et al.*, 2002; Gould *et al.*, 2004).

Two other AHL-producing enzymes, called AinS and LuxM, were described in *Vibrio fischeri* and *Vibrio harveyi*, respectively (Bassler *et al.*, 1993; Hanzelka *et al.*, 1999). They form the AinS/LuxM family of AHL synthases. AinS and LuxM do not share any similarities with LuxI family proteins. Heterologous expression at least of *ainS* induces C8-HSL production. Although AinS and LuxM appear to use the same precursors (SAM and acyl-ACP) for AHL synthesis (Hanzelka *et al.*, 1999), in contrast to LuxI type proteins, AinS can use either acyl-ACP or acyl CoA as acyl chain donors.

A third type of AHL-synthase called HdtS lacking homology with either the LuxI or AinS/LuxM families has been described in *Pseudomonas fluorescens* (Laue *et al.*, 2000). However, this finding remains controversial (Cullinane *et al.*, 2005).

Brucella AHL-synthase: the frustrating quest for the Holy Grail

The search for the presumptive AHL-synthase was initiated prior to the publication of *Brucella* genome sequence. Although the synthase genes identified at the time exhibited low sequence similarities, we undertook a homology-based search in *B. melitensis*. However, both PCR using degenerate primers and Southern blot analysis with labelled *traI* and *raiI* probes from either *Agrobacterium tumefaciens* or *Rhizobium etli* were unsuccessful (unpublished data, B. Taminiau).

Next, we utilized a functional screen based on *E. coli* sensor strains transformed with a *B. abortus* genomic library. If a genomic fragment of *B. abortus* contained a synthase gene, its heterologous expression would induce light production by the sensor strain. In total, more than 6000 clones were screened. Only one open reading frame (ORF) was able to restore light production in the sensor strain. This ORF encoded a product that shared 43% identity and 59% similarity with CepR, a response regulator of the LuxR family from *Burkholderia cepacia*. This *B. abortus* gene product was called BabR (*B. abortus* regulator). Either alone or following heterodimerization with LuxR present in the sensor strain, BabR was able to bind to the *luxI* promoter and induce light production. BabR will be described in more detail in the following section.

The availability of the *B. melitensis* genome (DelVecchio *et al.*, 2002) did not allow the identification of the synthase gene because no gene homologous to *luxI*, *luxM* or *luxS* was found. A *Brucella* ORF called *hdtS* encodes a protein sharing 20% identity and 54% similarity with HdtS from *P. fluorescens* (Laue *et al.*, 2000). However, expression of *Brucella hdtS* in an *E. coli* sensor strain did not induce AHL production, and similarly, deletion of *B. melitensis hdtS* did not prevent AHL production (unpublished data, Ch. Dechamps).

Recently, the Ficht laboratory reported a limited functional screen (Weeks *et al.*, 2010) in which they identified fifteen candidate genes whose products are suspected to interact with the AHL metabolic precursors SAM or acyl-ACP. Further, the candidate gene's expression was altered by addition of C_{12}-HSL or by mutation of another QS regulator (VjbR see below). However, expression of these candidates was tested in an *E. coli* sensor strain, and they all failed to induce the sensor.

Despite these investigations, the pathway for autoinducer synthesis in *Brucella* remains elusive and the synthase has yet to be identified. Consequently, as an alternative to an intrinsic production of AHL by *Brucella,* several authors have proposed that the QS regulators (VjbR and BabR, see below) mediate 'inter-kingdom' communication through the detection of signalling molecules from its eukaryotic host (Arocena *et al.*, 2010), similar to the OxyR regulator of *Xanthomonas oryzae* (Ferluga and Venturi, 2009).

Although we cannot completely exclude this hypothesis, we believe that *Brucella* itself produces the autoinducer for the following reasons:

- C_{12}-HSL and 3-oxoC_{12}-HSL have been identified in supernatants from *in vitro Brucella* cultures.
- AHL production (or at least molecules that can activate *P. aeruginosa* regulator LasR and that can be destroyed by AHL acylase) has been demonstrated inside *Brucella* cells whether grown *in vitro* or in a eukaryotic host cell (manuscript in preparation).
- Mutations of critical residues in VjbR known to be involved in AHL binding in homologous QS regulators affect regulator function (Uzureau *et al.*, 2007).

On this basis, our preferred hypothesis is that *Brucella* realizes AHL synthesis through a novel mode of synthesis, which is probably still connected to fatty acid and amino acid metabolism.

The regulator(s)

General features

Transcriptional regulators that are AHL-binding are all homologous to the LuxR regulator, first discovered in *V. fisheri* (Engebrecht and Silverman, 1984). LuxR family QS regulators are approximately 250 amino acids in size and share between 18% and 25% end-to-end sequence identity. LuxR-type regulators comprise two functional modules: a N-terminal AHL-binding domain (regulatory domain) and a C-terminal DNA-binding domain (activator domain) (Choi and Greenberg, 1991; Hanzelka and Greenberg, 1995). Binding of AHL by LuxR-type regulators leads to conformational changes that modulate their transcriptional activity.

Most LuxR-type regulators are transcriptional activators in the presence of AHL and recruit RNA polymerase to the target promoter. For example, LuxR (*V. fischeri*), TraR (*A. tumefaciens*),

LasR and RhlR (*P. aeruginosa*) form homodimers that specifically bind their target sequences when AHL reaches the threshold concentration (Zhu and Winans, 2001; Lamb *et al.*, 2003; Schuster *et al.*, 2004b; Urbanowski *et al.*, 2004). However, some LuxR-type regulators bound to target DNA behave as transcriptional repressors. This repression is relieved in the presence of AHL either by inhibiting DNA-binding activity (e.g. EsaR of *Pantotea stwerwatii*) or by changing the multimeric state of the regulator (e.g. CarR of *E. carotovora*) (Nasser *et al.*, 1998; Horng *et al.*, 2002; Minogue *et al.*, 2002).

Biochemically, LuxR-type regulators exist in three classes. Classes 1 and 2 require AHL for folding while Class 3 does not. Further, while Class 1 regulators exhibit high affinities for their ligands, Class 2 and 3 regulators bind their ligands reversibly. Examples of Class 1, 2 and 3 regulators are LasR, LuxR and EsaR, respectively.

The AHL-binding domain

This domain covers two thirds of the N-terminal and contains both an AHL-binding domain (between aa 79 and 127 as defined in *V. fischeri*) and a multimerization domain (between aa 116 and 161) (Choi and Greenberg, 1992). Genetic and structural analysis of TraR (*A. tumefaciens*) and other LuxR-type regulators (LuxR (Finney *et al.*, 2002), LasR (Kiratisin *et al.*, 2002) and RhlR (Lamb *et al.*, 2003) have identified six residues in the N-terminal domain that are conserved in at least 95% of the regulators and are involved in cognate AHL binding. In reference to TraR, the residues are as follows: W57, Y61, D70, P71, W85 and G113. The remaining portion of the regulator is more variable, probably reflecting the necessity of accommodating AHL diversity (Yao *et al.*, 2006; Bottomley *et al.*, 2007). The stoichiometric ratio between the AHL and the LuxR regulators is 1/1 (Welch *et al.*, 2000; Zhu and Winans, 2001; Schuster *et al.*, 2004b).

The DNA-binding domain

The final C-terminal of LuxR-type regulators contains a DNA-binding motif belonging to the LuxR superfamily. Additionally, the distal portion of the regulator may also contain an activation domain involved in interactions with the initiation complexes. The HTH motif that defines the LuxR type superfamily is called GerE, and it is found not only in the QS regulator associated with the AHL-binding domain but also in two-component regulators associated with the response regulator domain. In general, the DNA-binding domains are much more conserved between different LuxR-type regulators as compared with the AHL-binding domains. Three amino acids (E178, L182 and G188) are found in more than 95% of the regulators.

A palindromic sequence (20-nucleotide inverted repeat) centred 44 nucleotides upstream of the transcription start site of the reference luminescence operon (*V. fischeri*), which is recognized by the LuxR protein, was named the *lux* box (Fuqua and Winans, 1994). Similar palindromic sequences were identified for other LuxR-type regulators such as TraR and LasR (Fuqua *et al.*, 1996; Lee *et al.*, 2006). Despite this, it is possible that not all QS-controlled genes have a recognizable palindromic motif upstream of their promoter (Fuqua *et al.*, 1996).

The QS regulators of *Brucella*

Only two *bona fide* QS regulators containing both the C-terminal HTH domain GerE and a canonical AHL-binding domain in their N-terminus (Zhang *et al.*, 2002) have been identified in all available *Brucella* genomes. Three others ORFs have a C-terminal GerE domain but belong to the two-component regulator family (BMEI1583, BMEI1607 and BMEII0051). While the gene encoding the synthase is often in close proximity to the genes encoding their cognate regulators, this does not appear to be the case for *Brucella* QS regulators. Thus, they have been classified as orphan or solo QS regulators (Patankar and Gonzalez, 2009). These regulators are called VjbR and BabR and are described here.

VjbR (BMEII1116)

This regulator was identified in a systematic screen of a transpositional mutant library searching for *B. melitensis* mutants unable to replicate in HeLa cells (Delrue, 2002). In addition to a panel of type IV secretion system (*virB*) mutants, this screen picked up a QS regulator mutant. This mutant exhibited similar intracellular behaviour to the *virB* mutants, with the bacteria unable to reach its replicative vacuole. For this reason, this regulator was called VjbR for 'Vacuolar Jacking *Brucella* Regulator'. This mutant was also clearly attenuated in other cellular infection models and in a murine infection model (Delrue *et al.*, 2005; Weeks *et al.*, 2010).

The genomic location of *vjbR* is also interesting; it lies downstream of a TetR family gene and is in the immediate vicinity of the *flhB* and *fliG* flagellar genes. This peculiar organization is conserved in both *Sinorhizobium meliloti* and *Mesorhizobium loti*; however, in these species, the TetR regulator is replaced by the VisN QS regulator (Sourjik *et al.*, 2000).

The amino acid sequence of VjbR is highly conserved (> 99% identity) in all available *Brucella* genomes; this also true for the C-terminal DNA-binding domain of the VjbR homologues of *Ochrobactrum* spp. Although the N-terminal domain is more divergent in *Ochrobactrum* spp., they still represent the best non-*Brucella* homologue. The increased divergence probably reflects subtle differences in the nature of their AHL ligands. Curiously, in VjbR, only the first four 'consensus amino acids' (W57, Y61, D70 and P71) described as being involved in AHL binding are present; the last two residues are missing (W85 and G113). The aspartate in position 70 in the TraR regulator (D70) corresponds to the aspartate in position 82 (D82) in VjbR.

VjbR is required for *virB* expression (see below) and intracellular replication (Delrue *et al.*, 2005). C_{12}-HSL decreases VirB8 production by repressing transcription from the *virB* promoter (P*virB*) (Taminiau *et al.*, 2002); further, addition of exogenous C_{12}-HSL impairs *Brucella's* ability to reach its replicative niche. These observations suggest that the repressor effect of C_{12}-HSL on the P*virB* promoter could be linked to its inhibitory effect on the VjbR regulator. This hypothesis is supported by the fact that two *vjbR* alleles, *vjbR*(D82A) and *vjbR*(Δ1–180), which are mutated in AHL-binding domain and are unresponsive to signal molecules constitutively activate P*virB* (Uzureau *et al.*, 2007). Further indirect evidence that VjbR mediates C_{12}-HSL-induced effects comes from observations that several VjbR targets are affected by C_{12}-HSL even in the absence of *babR* (Uzureau *et al.*, 2010). Additionally, supplemental of exogenous C_{12}-HSL in the culture medium mimics *vjbR* deletion. Recently, Ugalde's group (Arocena *et al.*, 2010) obtained direct proof of C_{12}-HSL-VjbR binding by demonstrating that C_{12}-HSL dissociates VjbR from P*virB*, impairing VjbR-induced DNAseI protection and DNAse I hypersensitivity sites. Thus, in contrast to others LuxR-type regulators described until now, VjbR is a transcriptional activator in the absence of AHLs.

With regard the ability of VjbR to bind DNA, conflicting results have been reported. Using EMSA, de Jong identified an 18 bp degenerated palindromic sequence centred at position −37 relative to the transcriptional start site of the *virB* promoter (de Jong *et al.*, 2008). Similar sequences were identified in the intergenic region between *virB1* and *virB2* and upstream of the *tetR* promoter; in total, they identified 144 *Brucella* promoters that contain these putative binding sites. Additionally, 15 of the 24 promoters that were transcriptionally fused to *lacZ* and tested in *E. coli* were induced when *vjbR* was expressed. The predicted consensus of the *vjbR* box is reported as ATCCGCGATATCGCGAT. None of the identified promoters contained the consensus sequence; rather, they contained a degenerate sequence with more than four substitutions. However, as noted by Arocena *et al.* (Arocena *et al.*, 2010), binding of VjbR to this predicted *vjbR box* on the P*virB* would sterically interfere with the binding of RNA polymerase to the −35 position and therefore impair promoter activation.

Using a DNAseI protection assay on P*virB*, Arocena *et al.* (2010) identified a 30 bp protected region centred at position −94. This region contains a 9 bp motif, GCCCCCTCA, which is similar to the half site of the palindromic sequence recognized by a LuxR-type regulator of *Mesorhizobium tianshanense*. However, they were unable to detect any shifted bands after probing with P*virB* by EMSA, reflecting low interaction between VjbR and P*virB*. Further, while the 9 bp sequence was found upstream of many ORFs in the *Brucella* genome, none were demonstrated to bind VjbR or located upstream of loci known to be up-regulated by VjbR. They conclude that *Brucella* lacks 'high' affinity binding sites for VjbR, suggesting a weak binding and a subsequent rapid dissociation when AHLs are present.

Binding of VjbR to P*virB* and the intergenic *virB1-virB2* region was confirmed *in vivo* by chromatin immunoprecipitation (ChIP) (Uzureau *et al.*, 2010). Direct VjbR-DNA binding has also been found in the promoter region of *omp25b*, *omp36*, BMEII0590 and BMEII0734 both encoding for components of ABC transporters (specific for sugars and oligopeptides, respectively), BMEI0668 coding for a putative calcium binding protein and BMEI0030 coding for a hypothetical protein. Analysis of the upstream regions of ORFs directly bound by VjbR failed to identify a motif recognized by this regulator, most probably reflecting poor conservation of VjbR binding sites.

BabR (BMEI1758)

As described above, a second QS regulator of *Brucella*, BabR (for *Brucella abortus* regulator) was discovered by a functional screen designed to identify the AHL synthase (Taminiau *et al.*, 2002; Letesson and De Bolle, 2004). Surprisingly, this regulator was re-discovered and renamed BlxR by Rambow-Larsen *et al.* (2008). In contrast to the *vjbR* mutant, virulence attenuation was not detected for the *babR* mutant either in cellular models or in mice (Letesson and De Bolle, 2004; Weeks *et al.*, 2010).

The *babR* gene is surrounded by two genes whose products are involved in amino acid metabolism, an omega amino acid pyruvate amino transferase and a 5-methyltetrahydrofolate-homocysteine methyltransferase. This genomic organization is found only in *Brucella* spp. and *Ochrobactrum* spp.; it is not present in other alpha-proteobacteria.

The amino acid sequence of BabR is also highly conserved in all available *Brucella* genomes. *Ochrobactrum* is the best non-*Brucella* homologue with 100% identity in its DNA-binding domain; however, it is less conserved in the putative AHL-binding region. BabR possesses the five conserved residues predicted to be involved in AHL binding. As C_{12}-HSL still exerts an effect on several target genes in a Δ*vjbR* background, it has been suggested that BabR can also detect C_{12}-HSL (Uzureau *et al.*, 2010) since it is the only other defined *Brucella* QS regulator with a *bona fide* AHL-binding domain. However, no data are currently available on the DNA-binding ability of the BabR regulator.

Cross-talk between VjbR and BabR

LuxR-type regulators are often described as regulating their own expression, and when multiple regulators are present in the same bacteria they are often involved in hierarchical transcriptional regulation (Venturi, 2006). A combined transcriptional and proteomic analysis of the *B. melitensis* QS regulon also pointed to probable cross-talk between the two *Brucella* QS regulators (Uzureau *et al.*, 2010) because they shared numerous target genes and acted in opposite manners on more than half of these genes.

Two studies have proposed that VjbR activates its own transcription. First, Rambow-Larsen *et al.* used *lux* transcriptional fusion with the *vjbR* promoter to quantify its expression in both *B. melitensis* 16M wild type and in *vjbR* deleted backgrounds (Rambow-Larsen *et al.*, 2008). Second, de Jong *et al.* (2008) used a P*vjbR-lacZ* transcriptional fusion in an heterologous host *E. coli* co-expressing *vjbR*. Despite these indications, the studies suffer from drawbacks including, for de Jong study, possible discrepancies in the transcriptional machinery between alpha-proteobacteria and *E. coli*. Further, as the reporter gene was plasmid-born in both studies, a multicopy effect with the possible titration of the available regulator could have occurred.

More recently, data gathered by Uzureau *et al.* and Weeks *et al.* using transcriptomic approaches and qRT-PCR suggest the following model of cross talk between the regulators (Uzureau *et al.*, 2010; Weeks *et al.*, 2010). We hypothesize that VjbR is an activator of *babR* expression and that BabR is a repressor of *vjbR* expression. In addition, it was shown that C_{12}-HSL repressed VjbR expression in the parental strain and the Δ*babR* background. Conversely, C_{12}-HSL activated *babR* expression in both the wild type and the Δ*vjbR* backgrounds. Whether this hierarchy involves the direct binding of the QS regulators to the promoters of interest remains to be substantiated. However, it is certainly modulated according to environmental conditions; Weeks *et* *al.* reported differences according to the growth phase (Uzureau *et al.*, 2010; Weeks *et al.*, 2010). The model presented in Fig. 8.1 does not take into account the possibility of a BabR–VjbR heterodimer.

The AHL-degrading enzymes

General features

Quorum quenching is a term that encompasses all processes able to interfere with QS (Dong *et al.*, 2001). Among these mechanisms, the modification or hydrolysis of AHLs has been described in numerous organisms (Uroz and Heinonsalo, 2008) and is the most deeply investigated.

The chemical structure of AHLs allows their degradation by four different mechanisms. Two mechanisms, using either a lactonase or a decarboxylase, inactivate the molecule by opening the lactone ring. The remaining mechanisms result in the cleavage of the molecule into a homoserine lactone ring and a free acyl chain by either an acylase or a deaminase (Dong and Zhang, 2005). To date, only acylase and lactonase activity have been demonstrated. In addition to these degradative enzymes, the modifying enzymes AHL-oxidase and AHL-oxidoreductase exist.

AHL lactonases

These enzymes catalyse the hydrolytic opening of the homoserine lactone ring, which results in a homoserine that is unable to act as a signal (Dong *et al.*, 2000). This reaction resembles natural lactonization that opens the homoserine lactone ring in basic pH conditions; similarly, it can also be reversed by acidification.

The first enzyme with this activity was described in *Bacillus* spp. (Dong *et al.*, 2000). Today, there are two different families of lactonases in the prokaryotes:

1. The first, AiiA, is the most studied and is found in numerous Gram-positive and -negative bacteria (Uroz and Heinonsalo, 2008). These enzymes are metalloproteases related to β-lactamases (Dong *et al.*, 2001); they require two Zn^{2+} ions to be functional (Kim *et al.*, 2005). As their activity is independent of the length of or substitutions within the acyl chain, they are able to degrade a large panel of AHLs. Lactonases are not secreted.
2. A second AHL-lactonase family was found in *Rhodococcus erythropolis*; to date, this family has not been found outside of the *Rhodococcus* genus (Uroz *et al.*, 2008). This enzyme, QsdA, while also requiring two Zn^{2+} ions, is not related to the AiiA family but to phosphotriesterase.
3. Recently, a new AHL lactonase, AidH, was described in *Ochrobactrum* spp. This enzyme is unrelated to any other described AHL-degrading enzyme, and its activity requires Mn^{2+} (Mei *et al.* 2010).

Importantly, AHL-lactonase activity is also present in eukaryotic organisms (Chun *et al.*,

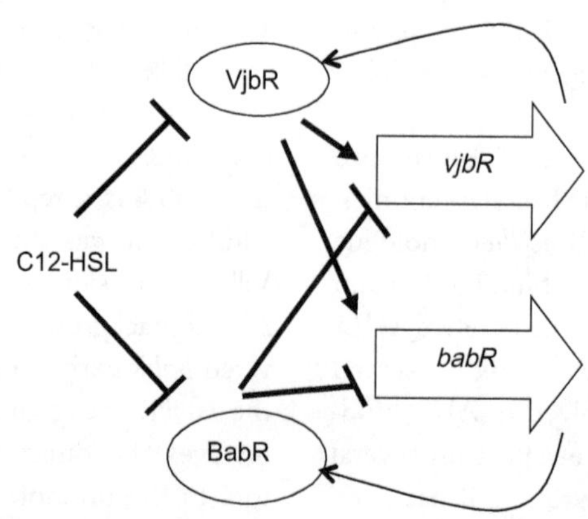

Figure 8.1 Model of the cross-talk between the two QS regulators of *Brucella*.

2004). These enzymes, called paraoxonases (PONs), are present in the serum and at the surface of certain cell types in mammals (Yang *et al.*, 2005); however, their physiological roles are not known. They might play a role in lipid metabolism and protection against the development of atherosclerosis. It has also been suggested that they have anti-oxidative activities. Regardless of their eukaryotic roles, they are able to hydrolyse the lactone rings of AHLs (Draganov *et al.*, 2005) and can therefore interfere with the QS of pathogens such *P. aeruginosa* (Ozer *et al.*, 2005). Consequently, PONs could be important in host defences against pathogenic bacteria (Czajkowski and Jafra, 2009).

AHL -acylases

These enzymes catalyse irreversible AHL degradation by hydrolysing the amide bond, resulting in a separate homoserine lactone ring and acyl chain (Leadbetter and Greenberg, 2000). Around 10 different AHL acylases have been described to date, the most studied being AiiD of *Ralstonia eutropha* (Lin *et al.*, 2003). These enzymes, similar to AiiA, are present in both Gram-positive and negative bacteria.

Usually, four domains can be distinguished in their amino-acid sequences: a signal peptide, an α-subunit, a linker and a β-subunit. The enzyme is a proenzyme that must be cleaved by autoproteolysis to yield the mature form with two or more subunits; generally, active enzymes have a small α-subunit (around 20 kDa) and a large β-subunit (around 60 kDa) (Lin *et al.*, 2003). Currently, the AHL-acylase from *Streptomyces* spp. is the only AHL-acylase known to be secreted, all other AHL-acylases act within the bacterial cell (Park *et al.*, 2005). As opposed to AHL-lactonases that inactivate all types of AHLs, AHL-acylases can exhibit certain specificities. For example, PvdQ and QuiP, two AHL-acylases from *P. aeruginosa*, are able to degrade 3-oxo-C_{12}-HSL but not C_4-HSL (Huang *et al.*, 2003; Sio *et al.*, 2006).

The only reported eukaryotic AHL-acylase activity was described in pig kidney tissue. The porcine kidney acylase I was demonstrated to degrade AHLs with acyl chains with four or eight carbons (Xu *et al.*, 2003).

AHL oxidase and AHL reductase

To date, two enzymes with either AHL oxidase or AHL reductase activity have been described in *Bacillus megaterium* and *Rhodococcus erythropolis*, respectively (Uroz *et al.*, 2005; Chowdhary *et al.*, 2007). Contrary to the previously described quorum quenching enzymes, the AHL molecule is not degraded but is chemically modified in this case so that its recognition by its cognate receptor is impaired to perturb QS (Uroz *et al.*, 2005).

AHL-degrading enzymes in *Brucella*

A bioinformatic screen of *Brucella* genomes was carried out to identify homologues of the two classes of AHL-degrading enzymes. No AHL -lactonase homologues were identified. Using the gene encoding the AHL-acylase AiiD from *R. eutropha* (Lin *et al.*, 2003) as a template, we found a homologue with 20% identity and 44% similarity at the amino acid level. From the 30 genomes tested, all but the five genomes from *B. melitensis* strains contained a gene encoding a 761 amino acid protein (with an identity of more than 98%). However, the corresponding ORF in *B. melitensis* is interrupted by a stop codon. As other well known AHL-acylases, the functional prediction for these enzymes in *Brucella* is penicillin-acylase or amidase (Lin *et al.*, 2003; Park *et al.*, 2005; Sio *et al.*, 2006).

To test the ability of this putative acylase to inactivate AHLs and to determine its potential specificity, a bioassay was performed with various AHLs (C4-HSL, C8-HSL, 3-oxo-C8-HSL and C12-HSL). When expressed in *E. coli* under the control of P*lac*, the putative acylase genes from either *B. melitensis* (BMEII0211 and BMEII0212) or *B. suis* (BRA1089) were able to degrade all AHLs tested regardless of length or the substitution of the C3 carbon of the acyl chain. Additionally, the *B. melitensis* acylase activity was comparable to the *B. suis* acylase activity in spite of the presence of the stop codon. These enzymes can therefore be expected to act as broad-range quorum quenchers. In addition, the acylase gene is QS regulated; its expression is dependent on both VjbR and BabR and is induced specifically by C12-HSL (manuscript in preparation).

Mutants of *B. melitensis* were constructed either by deleting the BMEII0211-BMEII0212 genes or by overexpressing *BRA1089* (Godefroid *et al.*, 2010). Neither the acylase deletion mutant, nor the overexpression mutant (Godefroid *et al.*, 2010) showed a defect in virulence when tested in cellular models (HeLa, bovine macrophages) or in a mouse model (splenic CFU count 1 week after IP infection). Although no virulence attenuation was detected in the *Brucella* strain overexpressing *aiiD*, these bacteria exhibit a clumping phenotype and produced an exopolysaccharide (Godefroid *et al.*, 2010). These phenotypes mimic the phenotype observed by Uzureau *et al.* for QS-deficient mutants (Uzureau *et al.*, 2007).

The targets

QS regulon

Several years ago, QS targets in *Brucella* were confined to the two virulence-related surface structures T4SS and flagellum and to certain outer membrane proteins (Uzureau *et al.*, 2007). Since 2008, four analyses aiming to identify the QS regulon in *Brucella* have been published (de Jong *et al.*, 2008; Rambow-Larsen *et al.*, 2008; Uzureau *et al.*, 2010; Weeks *et al.*, 2010). The consensus drawn from these studies is that QS is a global regulator affecting the expression of a high number of target genes/proteins; a summary is presented in Table 8.1. This is consistent with other QS regulons such as the *P. aeruginosa* QS regulon (Schuster *et al.*, 2003; Wagner *et al.*, 2003, 2004). Despite some discrepancies, *Brucella* QS regulon studies indicate that both BabR and VjbR are able to regulate gene expression depending on C_{12}-HSL (Rambow-Larsen *et al.*, 2008; Uzureau *et al.*, 2010; Weeks *et al.*, 2010). However, to date, only a limited number of genes have been identified as direct VjbR targets (Table 8.2). These data sets constitute the first QS regulon of an intracellular pathogen, and despite being generated from *in vitro* culture experiments, they enable speculation on the role of QS during host cell infection.

The virB operon

The *virB* operon was the first identified target of the *Brucella* QS system. The VirB type IV secretion system is essential for reaching *Brucella*'s replicative niche and is one of *Brucella*'s major virulence factors. Among the numerous signals able to regulate *virB* operon expression, population density appears to play a major role. Indeed, *virB* is highly induced inside the host cell and in the late exponential phase in bacteriological media (Sieira *et al.*, 2000; Boschiroli *et al.*, 2002; Fretin *et al.*, 2005). As *virB* is regulated during the

Table 8.1 Number of QS targets identified

Study	Technique	Comparison	Growth phase	No. of targets
de Jong *et al.* (2008)	Promoter activation by VjbR in heterologous host	–	–	15
Rambow-Larsen *et al.* (2008)	Mini-microarray (289 genes)	ΔbabR/wt	late log phase	36
Uzureau *et al.* (2010)	2D-DIGE combined to microarray	ΔbabR/wt	Mid log phase	80
	2D-DIGE combined to microarray	ΔvjbR/wt	Mid log phase	113
Weeks *et al.* (2010)	Microarray	ΔvjbR/wt	Log phase	202
	Microarray	ΔvjbR+C12-HSL/ΔvjbR	Log phase	34
	Microarray	ΔvjbR/wt	Stationary phase	229
	Microarray	ΔvjbR+C12-HSL/ΔvjbR	Stationary phase	48
	Microarray	wt+C12-HSL/wt	Log phase	322
	Microarray	wt+C12-HSL/wt	Stationary phase	122

Table 8.2 Direct VjbR targets

ORF N° BME	Putative function or name	I	II	III
I0030	Hypothetical protein			
I0076	Inorganic pyrophosphatase			
I0222	Carbonic anhydrase			
I0668	Hypothetical calcium binding protein			
I0948	VceC			
I1020	Molybdopterine-guanine dinucleotide biosynthesis protein A			
I1351	Penicillin binding protein 1A			
I1836	Nicotinate phophorybosyl transferase			
I1007	25kDa Outer membrane immunogenic precursor (Omp25b)			
I1305	Porin (Omp36)			
II0025	Attachment mediating protein VirB1			
II0026	Attachment mediating protein VirB2			
II0151	Flagellar M-ring protein FliF			
II0187	Hypothetical cytosolic protein			
II0390	LysR family transcriptional regulator			
II0590	Sugar-binding periplasmic protein			
II0734	Periplasmic oligopeptide-binding periplasmic protein			
II1116	LuxR family transcriptional regulator (VjbR)			
II1117	TetR family transcriptional regulator			

I, light grey, tested in a heterologous host data from De Jong *et al.* (2008) also validated by Weeks *et al.* (2010).

II, dark grey, tested in Brucella data from ChIP (Uzureau *et al.*, 2010).

III, black, tested in vitro data from Arocena *et al.* (2010).

Horizontal stripes, tested but negative.

growth phase, the effect of C_{12}-HSL on its expression was studied, revealing that C_{12}-HSL exerts an inhibitory effect on *virB* expression (Taminiau *et al.*, 2002). Several years later, it was shown that the deletion of *vjbR* has the same impact on *virB* expression as the addition of C_{12}-HSL into a *Brucella* culture. Thus, it was hypothesized VjbR is responsible for mediating AHL-induced repression of *virB* (Delrue *et al.*, 2005). As described above, this hypothesis was confirmed later both *in vitro* and during cellular infection by the use of a *B. melitensis* strain expressing a mutated allele of *vjbR* unable to bind AHL (Uzureau *et al.*, 2007). Further, studies have demonstrated that the regulation of *virB* by VjbR occurs by direct binding of the regulator to *virB* promoter region both *in vitro* (de Jong *et al.*, 2008; Arocena *et al.*, 2010) and *in vivo* (Uzureau et *al.*, 2010).

Recently, global transcriptome analyses of Δ*babR* and/or Δ*vjbR* strains gave contradictory results regarding the effect of BabR on *virB*. While *virB* genes appeared to be activated by BabR in the study from Rambow-Larsen *et al.* (2008), they appeared to be repressed by BabR in the study from Uzureau *et al.* (2007). This could, in part, be explained by differences in experimental design. However, the recent analysis of QS targets by Weeks *et al.* (2010) and our unpublished results show no attenuation of the *babR* mutant in a cellular model of infection and only slight attenuation in a mouse model. These findings are not compatible with a positive effect of BabR on *virB* gene expression.

The flagellar genes

While described as non-motile, *Brucella* possess 31 genes encoding flagellar proteins (Letesson *et al.*, 2002), and a sheathed flagellum has been observed under the defined conditions of early log phase (Fretin *et al.*, 2005). Although, the flagellar structure is not necessary in cellular infection models, it is important in maintaining infection in mice (Fretin *et al.*, 2005).

Two major observations led to the study of flagellar genes as potential QS targets in *Brucella*. First, the expression of flagellar genes is tightly regulated during the growth phase (Delrue *et al.*, 2005). Second, in *S. meliloti*, the VisN and VisR QS regulators are at the top of the flagellar cascade (Sourjik *et al.*, 2000) and are located in the flagellar locus. Bioinformatic analysis reveals a strong gene synteny between *S. meliloti* and *B. melitensis* flagellar loci, where VjbR and VisR are in the same configuration.

A study by Delrue *et al.* demonstrated that similar to the *virB* operon, expression of the flagellar gene *fliF* and production of the protein FlgE, are dependent on VjbR. Further, addition of C_{12}-HSL to the culture media has the same effect on these targets as the deletion of *vjbR* (Delrue *et al.*, 2005). These results were strengthened by a recent transcriptomic study aiming to identify VjbR and C_{12}-HSL targets in *B. melitensis* (Weeks *et al.*, 2010). Mini-microarray analyses by Rambow-Larson *et al.* (2008) revealed that BabR (BlxR) exerts a positive effect on the expression of the flagellar genes *flhB* and *flhJ* and a negative effect on the expression of chemotaxis genes *motB* and *motD*.

Conclusions

In summary, *Brucella* spp. possess a non-classical QS regulatory system for the following reasons.

1. They lack a classical AHL-synthase but produce low concentration of C_{12}-HSL and probably 3oxo-C_{12}-HSL.
2. C_{12}-HSL exerts an effect both on BabR and VjbR.
3. VjbR is the first QS regulator described as an activator 'silenced' by AHLs (at least for the virB operon and flagellar genes).
4. Through VjbR, QS is involved in the intracellular survival of *B. melitensis* immediately following host cell infection when there are few non-replicating bacteria moving to the replicative vacuole.

In addition to these peculiar features, the QS system of *Brucella* spp. displays properties shared by QS systems in other bacteria.

1. Direct or indirect regulation of a large proportion of the genome.
2. Extreme fine-tuning of the actors, as illustrated by the cross-talk between the two regulators and the existence of a QS-induced AHL-acylase.

In addition to its modulation by its principal components, QS circuitry and its regulation are far more complex than initially believed. An intricate array of transcriptional or post-transcriptional regulators integrates both environmental cues and internal physiological signals to control QS circuitry. This has been predominately deciphered in model organisms such as *P. aeruginosa*, *V. harveyi* and *V. cholerae* (Reimmann *et al.*, 2002; Schuster *et al.*, 2004a; Kay, 2006; Venturi, 2006; Bejerano-Sagie and Xavier, 2007; Hammer and Bassler, 2007). As the study of *Brucella* QS regulation is still in its infancy, we do not deal with these aspects in details. In addition, excellent reviews have been recently published on this complex topic (Boyer and Wisniewski-Dye, 2009; Ng and Bassler, 2009).

However, we highlight several of the links with other global regulatory systems that have emerged recently. For example, recent transcriptomic and proteomic analyses suggest that *vjbR* and *babR* mutants have a strong impact on genes involved in central metabolism (TCA cycle and glycolysis). Interestingly, the same target genes are repressed by VjbR and activated by BabR, suggesting that QS could have a global reorganization effect on central metabolic processes. When in a host cell vacuole, sensing 'quorum' for *Brucella* could mean sensing limited diffusion due to space limitations. This corresponds to 'starvation sensing', which is more significant than sensing quorum. It can be suggested that in addition to its known effects on 'virulence determinants,' QS is directly or indirectly involved in adjusting the metabolism of *Brucella*. Indeed, by slowing down *Brucella*'s basic metabolism, QS (through VjbR) would prevent multiplication until the replicative compartment is reached. Afterwards, the BabR regulator could play a role in reactivating the basal metabolism. A similar proposal was made for BvrR/BvrS two-components system. Both BvrR/BvrS and QS could contribute to the adaptation of the metabolic network during the nutrient shift faced by *Brucella* during its intracellular trafficking. As BvrR has an activating effect on *vjbR* transcription, these two regulatory systems appear to be connected (see also Chapter 10). Whether this link is direct or indirect, or even whether it is through other global starvation sensing mechanisms like the stringent response and/or the phosphotransferase system (Dozot *et al.*, 2010) remains to be determined.

Acknowledgements

Part of this work has been funded by grants from FRFC N° 2452110 and ARC N° Convention N° D8/13-015.

References

Alberghini, S., Polone, E., Corich, V., Carlot, M., Seno, F., Trovato, A., and Squartini, A. (2009). Consequences of relative cellular positioning on quorum sensing and bacterial cell-to-cell communication. FEMS Microbiol. Lett. *292*, 149–161.

Arocena, G.M., Sieira, R., Comerci, D.J., and Ugalde, R.A. (2010). Identification of the quorum-sensing target DNA sequence and N-Acyl homoserine lactone responsiveness of the *Brucella abortus virB* promoter. J. Bacteriol. *192*, 3434–3440.

Bassler, B.L., Wright, M., Showalter, R.E., and Silverman, M R. (1993). Intercellular signalling in *Vibrio harveyi*: sequence and function of genes regulating expression of luminescence. Mol. Microbiol. *9*, 773–786.

Bejerano-Sagie, M., and Xavier, K.B. (2007). The role of small RNAs in quorum sensing. Curr. Opin. Microbiol. *10*, 189–198.

Boedicker, J.Q., Vincent, M.E., and Ismagilov, R.F. (2009). Microfluidic confinement of single cells of bacteria in small volumes initiates high-density behavior of quorum sensing and growth and reveals its variability. Angew. Chem. Int. Ed. Engl. *48*, 5908–5911.

Boschiroli, M.L., Ouahrani-Bettache, S., Foulongne, V., Michaux-Charachon, S., Bourg, G., Allardet-Servent, A., Cazevieille, C., Liautard, J. P., Ramuz, M., and O'Callaghan, D. (2002). The *Brucella suis virB* operon is induced intracellularly in macrophages. Proc. Natl. Acad. Sci. U.S.A. *99*, 1544–1549.

Bottomley, M.J., Muraglia, E., Bazzo, R., and Carfi, A. (2007). Molecular insights into quorum sensing in the human pathogen *Pseudomonas aeruginosa* from the structure of the virulence regulator LasR bound to its autoinducer. J. Biol. Chem. *282*, 13592–13600.

Boyer, M., and Wisniewski-Dye, F. (2009). Cell-cell signalling in bacteria: not simply a matter of quorum. FEMS Microbiol. Ecol. *70*, 1–19.

Carnes, E.C., Lopez, D.M., Donegan, N.P., Cheung, A., Gresham, H., Timmins, G.S., and Brinker, C.J. (2010). Confinement-induced quorum sensing of individual *Staphylococcus aureus* bacteria. Nat. Chem. Biol. *6*, 41–45.

Case, R.J., Labbate, M., and Kjelleberg, S. (2008). AHL-driven quorum-sensing circuits: their frequency and function among the Proteobacteria. Isme. J. *2*, 345–349.

Celli, J., de Chastellier, C., Franchini, D.M., Pizarro-Cerda, J., Moreno, E., and Gorvel, J.P. (2003). *Brucella* evades macrophage killing via VirB-dependent sustained interactions with the endoplasmic reticulum. J. Exp. Med. *198*, 545–556.

Choi, S.H., and Greenberg, E.P. (1991). The C-terminal region of the *Vibrio fischeri* LuxR protein contains an inducer-independent lux gene activating domain. Proc. Natl. Acad. Sci. U.S.A. *88*, 11115–11119.

Choi, S.H., and Greenberg, E.P. (1992). Genetic dissection of DNA binding and luminescence gene activation by the *Vibrio fischeri* LuxR protein. J. Bacteriol. *174*, 4064–4069.

Chowdhary, P.K., Keshavan, N., Nguyen, H.Q., Peterson, J.A., Gonzalez, J.E., and Haines, D.C. (2007). *Bacillus megaterium* CYP102A1 oxidation of acyl homoserine lactones and acyl homoserines. Biochemistry *46*, 14429–14437.

Chun, C.K., Ozer, E.A. Welsh, M.J. Zabner, J., and Greenberg, E.P. (2004). Inactivation of a *Pseudomonas aeruginosa* quorum-sensing signal by human airway epithelia. Proc. Natl. Acad. Sci. U.S.A. *101*, 3587–3590.

Cullinane, M., Baysse, C., Morrissey, J.P., and O'Gara, F. (2005). Identification of two lysophosphatidic acid acyltransferase genes with overlapping function in *Pseudomonas fluorescens*. Microbiology *151*, 3071–3080.

Czajkowski, R., and Jafra, S. (2009). Quenching of acyl-homoserine lactone-dependent quorum sensing by enzymatic disruption of signal molecules. Acta Biochim. Pol. *56*, 1–16.

Daniels, R., Vanderleyden, J., and Michiels, J. (2004). Quorum sensing and swarming migration in bacteria. FEMS Microbiol. Rev. *28*, 261–289.

Decho, A.W., Visscher, P.T., Ferry, J., Kawaguchi, T., He, L., Przekop, K.M., Norman, R.S., and Reid, R P. (2009). Autoinducers extracted from microbial mats reveal a surprising diversity of N-acylhomoserine lactones (AHLs) and abundance changes that may relate to diel pH. Environ. Microbiol. *11*, 409–420.

Delrue, R.-M. (2002). Contribution à l'analyse des mécanismes moléculaires impliqués dans le trafic intracellulaire de *Brucella melitensis* 16M (Namur, Presses Universitaires de Namur).

Delrue, R.M., Deschamps, C., Leonard, S., Nijskens, C., Danese, I., Schaus, J.M., Bonnot, S., Ferooz, J., Tibor, A., De Bolle, X., and Letesson, J.J. (2005). A quorum-sensing regulator controls expression of both the type IV secretion system and the flagellar apparatus of *Brucella melitensis*. Cell. Microbiol. *7*, 1151–1161.

DelVecchio, V.G., Kapatral, V., Redkar, R.J., Patra, G., Mujer, C., Los, T., Ivanova, N., Anderson, I., Bhattacharyya, A., Lykidis, A., *et al.* (2002). The genome sequence of the facultative intracellular pathogen *Brucella melitensis*. Proc. Natl. Acad. Sci. U.S.A. *99*, 443–448.

Dong, Y.H., and Zhang, L.H. (2005). Quorum sensing and quorum-quenching enzymes. J. Microbiol. *43*, 101–109.

Dong, Y.H., Xu, J.L., Li, X.Z., and Zhang, L.H. (2000). AiiA, an enzyme that inactivates the acylhomoserine lactone quorum-sensing signal and attenuates the virulence of *Erwinia carotovora*. Proc. Natl. Acad. Sci. U.S.A. *97*, 3526–3531.

Dong, Y.H., Wang, L.H., Xu, J.L., Zhang, H.B., Zhang, X.F., and Zhang, L.H. (2001). Quenching quorum-sensing-dependent bacterial infection by an N-acyl homoserine lactonase. Nature *411*, 813–817.

Dozot, M., Poncet, S., Nicolas, C., Copin, R., Bouraoui, H., Maze, A., Deutscher, J., De Bolle, X., and Letesson, J.J. (2010) Functional characterization of the incomplete phosphotransferase system (PTS) of the intracellular pathogen *Brucella melitensis*. PLoS One *5*, e12679.

Draganov, D.I., Teiber, J.F., Speelman, A., Osawa, Y., Sunahara, R., and La Du, B.N. (2005). Human paraoxonases (PON1, PON2, and PON3) are lactonases with overlapping and distinct substrate specificities. J. Lipid. Res. *46*, 1239–1247.

Engebrecht, J., and Silverman, M. (1984). Identification of genes and gene products necessary for bacterial bioluminescence. Proc. Natl. Acad. Sci. U.S.A. *81*, 4154–4158.

Erickson, D.L., Nsereko, V.L., Morgavi, D.P., Selinger, L.B., Rode, L.M., and Beauchemin, K.A. (2002). Evidence of quorum sensing in the rumen ecosystem: detection of N-acyl homoserine lactone autoinducers in ruminal contents. Can. J. Microbiol. *48*, 374–378.

Ferluga, S., and Venturi, V. (2009). OryR is a LuxR-family protein involved in interkingdom signaling between pathogenic *Xanthomonas oryzae* pv. *oryzae* and rice. J. Bacteriol. *191*, 890–897.

Finney, A.H., Blick, R.J., Murakami, K., Ishihama, A., and Stevens, A.M. (2002). Role of the C-terminal domain of the alpha subunit of RNA polymerase in LuxR-dependent transcriptional activation of the lux operon during quorum sensing. J. Bacteriol. *184*, 4520–4528.

Fretin, D., Fauconnier, A., Kohler, S., Halling, S., Leonard, S., Nijskens, C., Ferooz, J., Lestrate, P., Delrue, R.M., Danese, I., *et al.* (2005). The sheathed flagellum of *Brucella melitensis* is involved in persistence in a murine model of infection. Cell. Microbiol. *7*, 687–698.

Fuqua, W.C., and Winans, S.C. (1994). A LuxR-LuxI type regulatory system activates *Agrobacterium* Ti plasmid conjugal transfer in the presence of a plant tumor metabolite. J. Bacteriol. *176*, 2796–2806.

Fuqua, W.C., Winans, S.C., and Greenberg, E.P. (1994). Quorum sensing in bacteria: the LuxR-LuxI family of cell density-responsive transcriptional regulators. J. Bacteriol. *176*, 269–275.

Fuqua, C., Winans, S.C., and Greenberg, E.P. (1996). Census and consensus in bacterial ecosystems: the LuxR-LuxI family of quorum-sensing transcriptional regulators. Annu. Rev. Microbiol. *50*, 727–751.

Godefroid, M., Svensson, M.V., Cambier, P., Uzureau, S., Mirabella, A., De Bolle, X., Van Cutsem, P., Widmalm, G., and Letesson, J.J. (2010). *Brucella melitensis* 16M produces a mannan and other extracellular matrix components typical of a biofilm. FEMS Immunol. Med. Microbiol. *59*, 364–377.

Gould, T.A., Schweizer, H.P., and Churchill, M.E. (2004). Structure of the *Pseudomonas aeruginosa* acyl-homoserinelactone synthase LasI. Mol. Microbiol. *53*, 1135–1146.

Hammer, B.K., and Bassler, B.L. (2007). Regulatory small RNAs circumvent the conventional quorum sensing pathway in pandemic *Vibrio cholerae*. Proc. Natl. Acad. Sci. U.S.A. *104*, 11145–11149.

Hanzelka, B.L., and Greenberg, E.P. (1995). Evidence that the N-terminal region of the *Vibrio fischeri* LuxR protein constitutes an autoinducer-binding domain. J. Bacteriol. *177*, 815–817.

Hanzelka, B.L., Stevens, A.M., Parsek, M.R., Crone, T.J., and Greenberg, E.P. (1997). Mutational analysis of the *Vibrio fischeri* LuxI polypeptide: critical regions of an autoinducer synthase. J. Bacteriol. *179*, 4882–4887.

Hanzelka, B.L., Parsek, M.R., Val, D.L., Dunlap, P.V., Cronan, J.E., Jr., and Greenberg, E.P. (1999). Acylhomoserine lactone synthase activity of the *Vibrio fischeri* AinS protein. J. Bacteriol. *181*, 5766–5770.

Hao, Y., Winans, S.C., Glick, B.R., and Charles, T.C. (2010) Identification and characterization of new LuxR/LuxI-type quorum sensing systems from metagenomic libraries. Environ. Microbiol. *12*, 105–117.

Horng, Y.T., Deng, S.C., Daykin, M., Soo, P.C., Wei, J.R., Luh, K.T., Ho, S.W., Swift, S., Lai, H.C., and Williams, P. (2002). The LuxR family protein SpnR functions as a negative regulator of N-acylhomoserine lactone-dependent quorum sensing in *Serratia marcescens*. Mol. Microbiol. *45*, 1655–1671.

Huang, J.J., Han, J.I., Zhang, L.H., and Leadbetter, J.R. (2003). Utilization of acyl-homoserine lactone quorum signals for growth by a soil pseudomonad and *Pseudomonas aeruginosa* PAO1. Appl. Environ. Microbiol. *69*, 5941–5949.

de Jong, M.F., Sun, Y.H., den Hartigh, A.B., van Dijl, J.M., and Tsolis, R.M. (2008). Identification of VceA and VceC, two members of the VjbR regulon that are translocated into macrophages by the *Brucella* type IV secretion system. Mol. Microbiol. *70*, 1378–1396.

Kaplan, H.B., and Greenberg, E.P. (1985). Diffusion of autoinducer is involved in regulation of the *Vibrio fischeri* luminescence system. J. Bacteriol. *163*, 1210–1214.

Kay, E., Reimmann, C. and Haas D. (2006). Small RNAs in bacterial cell-cell communication. Microbes Infect. *1*, 63–69.

Kim, M.H., Choi, W.C., Kang, H.O., Lee, J.S., Kang, B.S., Kim, K.J., Derewenda, Z.S., Oh, T.K., Lee, C.H., and Lee, J.K. (2005). The molecular structure and catalytic mechanism of a quorum-quenching N-acyl-L-homoserine lactone hydrolase. Proc. Natl. Acad. Sci. U.S.A. *102*, 17606–17611.

Kiratisin, P., Tucker, K.D., and Passador, L. (2002). LasR, a transcriptional activator of *Pseudomonas aeruginosa* virulence genes, functions as a multimer. J. Bacteriol. *184*, 4912–4919.

Kohler, T., Buckling, A., and van Delden, C. (2009). Cooperation and virulence of clinical *Pseudomonas aeruginosa* populations. Proc. Natl. Acad. Sci. U.S.A. *106*, 6339–6344.

Lamb, J.R., Patel, H., Montminy, T., Wagner, V.E., and Iglewski, B.H. (2003). Functional domains of the RhlR transcriptional regulator of *Pseudomonas aeruginosa*. J. Bacteriol. *185*, 7129–7139.

Laue, B.E., Jiang, Y., Chhabra, S.R., Jacob, S., Stewart, G.S., Hardman, A., Downie, J.A., O'Gara, F., and Williams, P. (2000). The biocontrol strain *Pseudomonas fluorescens* F113 produces the *Rhizobium* small bacteriocin, N-(3-hydroxy-7-cis-tetradecenoyl)homoserine lactone, via HdtS, a putative novel N-acylhomoserine lactone synthase. Microbiology *146*, 2469–2480.

Leadbetter, J.R., and Greenberg, E.P. (2000). Metabolism of acyl-homoserine lactone quorum-sensing signals by *Variovorax paradoxus*. J. Bacteriol. *182*, 6921–6926.

Lee, J.H., Lequette, Y., and Greenberg, E.P. (2006). Activity of purified QscR, a *Pseudomonas aeruginosa* orphan quorum-sensing transcription factor. Mol. Microbiol. *59*, 602–609.

Letesson, J.J., Lestrate, P., Delrue, R.M., Danese, I., Bellefontaine, F., Fretin, D., Taminiau, B., Tibor, A., Dricot, A., Deschamps, C., *et al.* (2002). Fun stories about *Brucella*: the 'furtive nasty bug'. Vet. Microbiol. *90*, 317–328.

Letesson, J.J. and De Bolle, X. (2004). Brucella virulence: a matter of control. In: Brucella: Molecular and Cellular Biology. Lopez-Goni, I., and Moriyon, I., eds. (Horizon Biosciences: Norfolk England), pp. 117–159.

Lin, Y.H., Xu, J.L., Hu, J., Wang, L.H., Ong, S.L., Leadbetter, J.R., and Zhang, L.H. (2003). Acyl-homoserine lactone acylase from *Ralstonia* strain XJ12B represents a novel and potent class of quorum-quenching enzymes. Mol. Microbiol. 47, 849–860.

Mei, G.Y., Yan, X.X., Turak, A., Luo, Z.Q., and Zhang, L Q. (2010). AidH, an alpha/beta-hydrolase fold family member from an *Ochrobactrum sp.* strain, is a novel N-acylhomoserine lactonase. Appl. Environ. Microbiol. 76, 4933–4942.

Minogue, T.D., Wehland-von Trebra, M., Bernhard, F., and von Bodman, S.B. (2002). The autoregulatory role of EsaR, a quorum-sensing regulator in *Pantoea stewartii ssp. stewartii*: evidence for a repressor function. Mol. Microbiol. *44*, 1625–1635.

Moreno, E., and Moriyón, I. (2002). *Brucella melitensis*: a nasty bug with hidden credentials for virulence. Proc. Natl. Acad. Sci. U.S.A. *99*, 1–3.

Nasser, W., Bouillant, M.L., Salmond, G., and Reverchon, S. (1998). Characterization of the *Erwinia chrysanthemi* expI-expR locus directing the synthesis of two N-acyl-homoserine lactone signal molecules. Mol. Microbiol. *29*, 1391–1405.

Ng, W.L., and Bassler, B.L. (2009). Bacterial quorum-sensing network architectures. Annu. Rev. Genet. *43*, 197–222.

Ozer, E.A., Pezzulo, A., Shih, D.M., Chun, C., Furlong, C., Lusis, A.J., Greenberg, E.P., and Zabner, J. (2005). Human and murine paraoxonase 1 are host modulators of *Pseudomonas aeruginosa* quorum-sensing. FEMS Microbiol. Lett. *253*, 29–37.

Park, S.Y., Kang, H.O., Jang, H.S., Lee, J.K., Koo, B.T., and Yum, D.Y. (2005). Identification of extracellular N-acylhomoserine lactone acylase from a *Streptomyces sp.* and its application to quorum quenching. Appl. Environ. Microbiol. *71*, 2632–2641.

Parsek, M.R., Val, D.L., Hanzelka, B.L., Cronan, J.E., Jr., and Greenberg, E.P. (1999). Acyl homoserine-lactone quorum-sensing signal generation. Proc. Natl. Acad. Sci. U.S.A. *96*, 4360–4365.

Patankar, A.V., and Gonzalez, J.E. (2009). An orphan LuxR homolog of *Sinorhizobium meliloti* affects stress adaptation and competition for nodulation. Appl. Environ. Microbiol. 75, 946–955.

Pearson, J.P., Van Delden, C., and Iglewski, B.H. (1999). Active efflux and diffusion are involved in transport of *Pseudomonas aeruginosa* cell-to-cell signals. J. Bacteriol. *181*, 1203–1210.

Platt, T.G., and Fuqua, C. (2010). What's in a name? The semantics of quorum sensing. Trends Microbiol. *18*, 383–387.

Rambow-Larsen, A.A., Rajashekara, G., Petersen, E., and Splitter, G. (2008). Putative quorum-sensing regulator BlxR of *Brucella melitensis* regulates virulence factors including the type IV secretion system and flagella. J. Bacteriol. *190*, 3274–3282.

Redfield, R.J. (2002). Is quorum sensing a side effect of diffusion sensing? Trends Microbiol. *10*, 365–370.

Reimmann, C., Ginet, N., Michel, L., Keel, C., Michaux, P., Krishnapillai, V., Zala, M., Heurlier, K., Triandafillu, K., Harms, H., *et al.* (2002). Genetically programmed autoinducer destruction reduces virulence gene expression and swarming motility in *Pseudomonas aeruginosa* PAO1. Microbiology *148*, 923–932.

Roop, R.M., 2nd, Gaines, J.M., Anderson, E.S., Caswell, C.C., and Martin, D.W. (2009). Survival of the fittest: how *Brucella* strains adapt to their intracellular niche in the host. Med. Microbiol. Immunol. *198*, 221–238.

Rumbaugh, K.P., Diggle, S.P., Watters, C.M., Ross-Gillespie, A., Griffin, A.S., and West, S.A. (2009). Quorum sensing and the social evolution of bacterial virulence. Curr. Biol. *19*, 341–345.

Schaefer, A.L., Hanzelka, B.L., Eberhard, A., and Greenberg, E.P. (1996). Quorum sensing in *Vibrio fischeri*: probing autoinducer–LuxR interactions with autoinducer analogs. J. Bacteriol. *178*, 2897–2901.

Schaefer, A.L., Taylor, T.A., Beatty, J.T., and Greenberg, E.P. (2002). Long-chain acyl-homoserine lactone quorum-sensing regulation of *Rhodobacter capsulatus* gene transfer agent production. J. Bacteriol. *184*, 6515–6521.

Schuster, M., Lostroh, C. P., Ogi, T., and Greenberg, E.P. (2003). Identification, timing, and signal specificity of *Pseudomonas aeruginosa* quorum-controlled genes: a transcriptome analysis. J. Bacteriol. *185*, 2066–2079.

Schuster, M., Hawkins, A.C., Harwood, C.S., and Greenberg, E.P. (2004a). The *Pseudomonas aeruginosa* RpoS regulon and its relationship to quorum sensing. Mol. Microbiol. *51*, 973–985.

Schuster, M., Urbanowski, M.L., and Greenberg, E.P. (2004b). Promoter specificity in *Pseudomonas aeruginosa* quorum sensing revealed by DNA binding of purified LasR. Proc. Natl. Acad. Sci. U.S.A. *101*, 15833–15839.

Sieira, R., Comerci, D.J., Sanchez, D.O., and Ugalde, R.A. (2000). A homologue of an operon required for DNA transfer in *Agrobacterium* is required in *Brucella abortus* for virulence and intracellular multiplication. J. Bacteriol. *182*, 4849–4855.

Sio, C.F., Otten, L.G., Cool, R.H., Diggle, S.P., Braun, P.G., Bos, R., Daykin, M., Camara, M., Williams, P., and Quax, W.J. (2006). Quorum quenching by an N-acyl-homoserine lactone acylase from *Pseudomonas aeruginosa* PAO1. Infect. Immun. *74*, 1673–1682.

Sourjik, V., Muschler, P., Scharf, B., and Schmitt, R. (2000). VisN and VisR are global regulators of chemotaxis, flagellar, and motility genes in *Sinorhizobium (Rhizobium) meliloti*. J. Bacteriol. *182*, 782–788.

Stevens, A.M., and Greenberg, E.P. (1997). Quorum sensing in *Vibrio fischeri*: essential elements for activation of the luminescence genes. J. Bacteriol. *179*, 557–562.

Subramoni, S., and Venturi, V. (2009). LuxR-family 'solos': bachelor sensors/regulators of signalling molecules. Microbiology *155*, 1377–1385.

Taminiau, B., Daykin, M., Swift, S., Boschiroli, M.L., Tibor, A., Lestrate, P., De Bolle, X., O'Callaghan, D., Williams, P., and Letesson, J.J. (2002). Identification of a quorum-sensing signal molecule in the facultative intracellular pathogen *Brucella melitensis*. Infect. Immun. *70*, 3004–3011.

Urbanowski, M.L., Lostroh, C.P., and Greenberg, E.P. (2004). Reversible acyl-homoserine lactone binding to purified *Vibrio fischeri* LuxR protein. J. Bacteriol. *186*, 631–637.

Uroz, S., Chhabra, S.R., Camara, M., Williams, P., Oger, P., and Dessaux, Y. (2005). N-Acylhomoserine lactone quorum-sensing molecules are modified and degraded by *Rhodococcus erythropolis* W2 by both amidolytic and novel oxidoreductase activities. Microbiology *151*, 3313–3322.

Uroz, S., and Heinonsalo, J. (2008). Degradation of N-acyl homoserine lactone quorum sensing signal molecules by forest root-associated fungi. FEMS Microbiol. Ecol. *65*, 271–278.

Uroz, S., Oger, P.M., Chapelle, E., Adeline, M.T., Faure, D., and Dessaux, Y. (2008). A *Rhodococcus qsdA*-encoded enzyme defines a novel class of large-spectrum quorum-quenching lactonases. Appl. Environ. Microbiol. *74*, 1357–1366.

Uzureau, S., Godefroid, M., Deschamps, C., Lemaire, J., De Bolle, X., and Letesson, J.J. (2007). Mutations of the Quorum Sensing-dependent regulator VjbR lead to drastic surface modifications in *Brucella melitensis*. J. Bacteriol. 189(16):6035–6047.

Uzureau, S., Lemaire, J., Delaive, E., Dieu, M., Gaigneaux, A., Raes, M., De Bolle, X., and Letesson, J.J. (2010). Global analysis of quorum sensing targets in the intracellular pathogen *Brucella melitensis* 16 M. J. Proteome Res. *9*, 3200–3217.

Venturi, V. (2006). Regulation of quorum sensing in Pseudomonas. FEMS Microbiol. Rev. *30*, 274–291.

Wagner, V.E., Bushnell, D., Passador, L., Brooks, A.I., and Iglewski, B.H. (2003). Microarray analysis of *Pseudomonas aeruginosa* quorum-sensing regulons: effects of growth phase and environment. J. Bacteriol. *185*, 2080–2095.

Watson, W.T., Minogue, T.D., Val, D.L., von Bodman, S.B., and Churchill, M.E. (2002). Structural basis and specificity of acyl-homoserine lactone signal production in bacterial quorum sensing. Mol. Cell. *9*, 685–694.

Weeks, J.N., Galindo, C.L., Drake, K.L., Adams, G.L., Garner, H.R., and Ficht, T.A. (2010). *Brucella melitensis* VjbR and C12-HSL regulons: contributions of the N-dodecanoyl homoserine lactone signaling molecule and LuxR homologue VjbR to gene expression. BMC Microbiol. *10*, 167.

Welch, M., Todd, D.E., Whitehead, N.A., McGowan, S.J., Bycroft, B.W., and Salmond, G.P. (2000). N-acyl homoserine lactone binding to the CarR receptor determines quorum-sensing specificity in *Erwinia*. Embo J. *19*, 631–641.

Williams, S.C., Patterson, E.K., Carty, N.L., Griswold, J.A., Hamood, A.N., and Rumbaugh, K.P. (2004). *Pseudomonas aeruginosa* autoinducer enters and functions in mammalian cells. J. Bacteriol. *186*, 2281–2287.

Xu, F., Byun, T., Deussen, H.J., and Duke, K.R. (2003). Degradation of N-acylhomoserine lactones, the bacterial quorum-sensing molecules, by acylase. J. Biotechnol. *101*, 89–96.

Yang, F., Wang, L.H., Wang, J., Dong, Y.H., Hu, J.Y., and Zhang, L.H. (2005). Quorum quenching enzyme activity is widely conserved in the sera of mammalian species. FEBS Lett. *579*, 3713–3717.

Yao, Y., Martinez-Yamout, M.A., Dickerson, T.J., Brogan, A.P., Wright, P.E., and Dyson, H.J. (2006). Structure of the *Escherichia coli* quorum sensing protein SdiA: activation of the folding switch by acyl homoserine lactones. J. Mol. Biol. *355*, 262–273.

Zhang, R.G., Pappas, T., Brace, J.L., Miller, P.C., Oulmassov, T., Molyneaux, J.M., Anderson, J.C., Bashkin, J.K., Winans, S.C., and Joachimiak, A. (2002). Structure of a bacterial quorum-sensing transcription factor complexed with pheromone and DNA. Nature *417*, 971–974.

Zhu, J., and Winans, S.C. (2001). The quorum-sensing transcriptional regulator TraR requires its cognate signaling ligand for protein folding, protease resistance, and dimerization. Proc. Natl. Acad. Sci. U.S.A. *98*, 1507–1512.

Metal Acquisition by *Brucella* Strains

9

R. Martin Roop II, Eric Anderson, Jenifer Ojeda, David Martinson, Evan Menscher and Daniel W. Martin

Abstract

Like most other living organisms, *Brucella* strains require a variety of metals as micronutrients to serve as co-factors and structural components of enzymes and other cellular proteins. Genetic studies have shown that the efficient transport of iron, magnesium, manganese, and zinc are essential for the virulence of *Brucella* strains in experimental and natural hosts. This chapter will review what is presently known about the metal acquisition and homeostasis systems in *Brucella* strains.

Introduction

Metals play critical roles in cellular metabolism (Waldron and Robinson, 2009). Some metals, such as iron and copper, are very redox active and function as co-factors in components of electron-transport chains and other proteins that undergo oxidation–reduction reactions. Iron is also a constituent of haem, which performs a wide variety of important biological functions. Other metals such as manganese, magnesium, zinc, nickel and cobalt play important catalytic and/or structural roles in the biological activity of cellular proteins. Although multiple metals are required by living organisms, many of these metals can be toxic if they accumulate beyond the levels needed to maintain their normal physiologic roles. Consequently, all living organisms that have been studied possess homeostasis systems that ensure they only accumulate the levels of metals they need (Summers, 2009).

Early studies aimed at developing improved growth media for *Brucella* strains determined that iron, manganese and magnesium represent essential micronutrients for these strains when they are cultivated *in vitro* (ZoBell and Meyer, 1932; McCullough *et al.*, 1947; Sanders *et al.*, 1953; Waring *et al.*, 1953; Evenson and Gerhardt, 1955; Gerhardt 1958). Studies performed much later with genetically defined mutants showed that efficient iron (Bellaire *et al.*, 2003a; Paulley *et al.*, 2007), manganese (Anderson *et al.*, 2009a), magnesium (Lestrate *et al.*, 2000; Lavigne *et al.*, 2005) and zinc (Kim *et al.*, 2003, 2004; Yang *et al.*, 2006) transport systems are required for the wild-type virulence of *Brucella* strains in their mammalian hosts.

Iron

Most bacteria require 0.4–1 μM iron for normal growth *in vitro* (Lengeler *et al.*, 1999). Early studies by Waring *et al.* (1953) showed that *B. suis* PS-1 had a similar need for iron, requiring > 0.5 μM iron for growth and displaying optimal growth at iron concentrations > 1.8 μM. Because in nature *Brucella* strains are found almost exclusively associated with a mammalian host (Moreno and Moriyón, 2002; Roop *et al.*, 2009), acquiring sufficient iron to meet their physiologic needs is particularly difficult. Iron that is not incorporated into host tissue components is tightly bound to host iron-binding proteins such as transferrin and lactoferrin in the extracellular spaces, and by ferritin within host cells (Griffiths, 1999). This tight sequestration of iron in mammals serves not only to prevent iron toxicity in the host, but it also limits the availability of this micronutrient to invading microbes. Indeed, iron restriction plays important roles in both the innate and acquired immune responses in the host. The hypoferremic response elicited during the early stages of the host response to infection, which traps iron in the intracellular iron stores, further exacerbates the iron deprivation that pathogens face in the extracellular environment (Weinberg, 1995; Nemeth *et al.*, 2004; Weiss, 2005). Upon the induction of acquired immunity, IFN-γ activation of host macrophages enhances iron restriction in the phagosomal compartment, in part through the reduced production of transferrin receptors on the surface of activated macrophages which reduces the flux of iron through these phagocytes (Byrd and Horwitz, 1989). IFN-activation also stimulates the activity of the natural resistance-associated macrophage protein 1 (Nramp1), which incorporates itself into the phagosomal membrane and pumps divalent cations (including Fe^{2+}) out of this compartment (Cellier *et al.*, 2007).

The capacity of brucellae to gain access to the host's iron stores is critical to their virulence, since these bacteria maintain primarily an intracellular existence in infected mammals. Within their natural hosts, *Brucella* strains predominantly occupy two host cell types, macrophages and placental trophoblasts (Roop *et al.*, 2009). Both of these cell types play central roles in iron metabolism within the host. Macrophages scavenge aged erythrocytes and haptoglobin–haemoglobin complexes from the circulation and in this manner perform a critical role in the recycling of iron (Bratosin *et al.* 1998). Placental trophoblasts serve as a gateway for shuttling iron and other nutrients from the mother to the developing fetus (Schroder, 1995). Experimental evidence suggests that the iron needs of the brucellae differ during their survival and replication in these two host cell types and accordingly, different iron acquisition systems appear to be used by the brucellae to meet their iron requirements within these host cells (Bellaire *et al.*, 1999, 2003a,b; González-Carreró *et al.*, 2002; Parent *et al.*, 2002; Danese *et al.*, 2004; Paulley *et al.*, 2007). Whether these differences are a reflection of the physiologic state and growth kinetics of the brucellae during residence in these host cells, or the types of iron available in these two host cell types, is presently unknown.

Siderophore-mediated Fe^{3+} transport

Many bacteria produce low molecular weight iron chelators known as siderophores and excrete them into the surrounding environment when they are faced with iron deprivation (Chu *et al.*, 2010). Only two siderophores, the catechol 2,3-dihydroxybenzoic acid (2,3-DHBA) (López-Goñi *et al.*, 1992) and a more complex, 2,3-DHBA based siderophore known as brucebactin (González-Carreró *et al.*, 2002) have been described in *Brucella* strains. Brucebactin appears to be very unstable *in vitro*, and consequently, the structure of this siderophore is presently unknown. Analysis of mutants has shown that the enzymes encoded by the *dhbCEBA* operon are required for the production of 2,3-DHBA (Bellaire *et al.*, 1999, 2003a) and the divergently transcribed *vibH* gene is required for the conversion of this monocatechol into brucebactin (González-Carreró *et al.*, 2002; Bellaire *et al.*, 2003a) (Table 9.1). Just downstream of the *dhbCEBA* operon, but transcribed independently, is the *entD* gene whose product has also been proposed to be involved in the conversion of 2,3-DHBA into brucebactin (Bellaire *et al.*, 2003a). As is the case for most bacterial siderophore biosynthesis genes, transcription of the *dhbC* gene is repressed during growth of *B. abortus* 2308 under iron replete conditions (Bellaire *et al.*, 2003a); however, the iron responsiveness of *vibH* or *entD* has not been examined.

Siderophores with bound Fe^{3+} are typically imported into Gram-negative bacteria by specific transport systems consisting of a TonB-dependent outer membrane protein that transports the ferrisiderophore complex across the outer membrane, and a periplasmic binding protein (PBP)-dependent ABC transporter that ferries this complex across the cytoplasmic membrane (Noinaj *et al.*, 2010). *Brucella* strains produce homologues of the TonB-dependent outer membrane proteins Fiu (Hantke *et al.*, 1990) and Cir (Griggs *et al.*, 1987) that transport Fe^{3+} bound to 2,3-DHBA and 2,3-DHBA-serine in *E. coli* (Table 9.1 and Fig. 9.1). Preliminary studies with

Table 9.1 Designations and conservation of *Brucella abortus* 2308 genes predicted to be involved in metal homeostasis

Gene	Designation	Predicted function	Conservation
Iron acquisition and storage			
vibH	BAB2_0016	Siderophore biosynthesis[a]	Conserved[b]
dhbC[c]	BAB2_0015	Siderophore biosynthesis[a]	Pseudogene in *B. canis* ATCC 23365
dhbE	BAB2_0014	Siderophore biosynthesis[a]	Conserved
dhbB	BAB2_0013	Siderophore biosynthesis[a]	Pseudogene in *B. suis* ATCC 23445
dhbA	BAB2_0012	Siderophore biosynthesis	Conserved
entD	BAB2_0011	Siderophore biosynthesis	Pseudogene in *B. abortus* 9-941 and *B. melitensis* ATCC 23457
dhbR	BAB2_1152	Regulation of siderophore; biosynthesis genes[a]	Conserved
fiu	BAB2_0233	Ferrisiderophore uptake[a]	Pseudogene in *B. canis* ATCC 23365
cir	BAB1_1367	Ferrisiderophore uptake[a]	Conserved
fepB	BAB1_1366	Ferrisiderophore uptake	Conserved
fepD	BAB1_1365	Ferrisiderophore uptake	Conserved
fepC	BAB1_1364	Ferrisiderophore uptake	Conserved
fatB	BAB2_0564	Ferrisiderophore uptake[a]	Conserved
fatD	BAB2_0563	Ferrisiderophore uptake	Conserved
fatC	BAB2_0562	Ferrisiderophore uptake[a]	Conserved
fatE	BAB2_0561	Ferrisiderophore uptake	Conserved
bhuA	BAB2_1150	Haem uptake[a]	Conserved
bhuT	BAB2_0483	Haem uptake	Conserved
bhuU	BAB2_0484	Haem uptake	Conserved
bhuV	BAB2_0482	Haem uptake	Pseudogene in *B. suis* ATCC 23445
omp22[d]	BAB1_1302	Haem binding	Conserved
omp25	BAB1_0722	Haem binding	Conserved
omp25[c]	BAB1_0116	Haem binding	Conserved
omp25[d]	BAB1_0115	Haem binding	Conserved
omp31[b]	BAB1_1639	Haem binding	Not present in *B. melitensis*, *canis* and *ovis* strains
exbB	BAB1_1679	Ferrisiderophore and haem uptake[a]	Conserved
exbD	BAB1_1680	Ferrisiderophore and haem uptake	Conserved
tonB	BAB1_1681	Ferrisiderophore and haem uptake	Conserved
sfuB-1	BAB2_0538	Fe^{3+} uptake	Pseudogene in *B. microti* CCM 4915
sfuA-1	BAB2_0539	Fe^{3+} uptake	Conserved
sfuC-1	BAB2_0540	Fe^{3+} uptake	Conserved

Table 9.1 continued

Gene	Designation	Predicted function	Conservation
sfuA-2	BAB2_0519	Fe^{3+} uptake	Conserved
sfuB-2	BAB2_0520	Fe^{3+} uptake	Conserved
sfuC-2	BAB2_0521	Fe^{3+} uptake	Conserved
tpd	BAB2_0840	Fe^{2+} uptakea	Conserved
Hypothetical	BAB2_0839	Fe^{2+} uptake	Conserved
ftr1	BAB2_0838	Fe^{2+} uptake	Conserved
polyf	BAB2_0837	Fe^{2+} uptake	Conserved
bhuO	BAB2_0677	Haem degradation/Fe^{2+} release[a]	Pseudogene in *B. ovis* ATCC 25840
bfr	BAB2_0675	Iron storage/detoxification[a]	Conserved
dps	BAB1_2150	Iron detoxification	Conserved
irr	BAB1_2175	Iron-responsive regulation of iron metabolism genes[a]	Conserved
rirA	BAB2_0678	Iron-responsive regulation of iron metabolism genes[a]	Conserved
Manganese acquisition			
mntH	BAB1_1460	Mn^{2+} uptake[a]	Conserved
mur	BAB1_1668	Mn-responsive regulation of mntH[a]	Conserved
Magnesium acquisition			
mgtB	BAB2_0036	Mg^{2+} uptake[a]	Conserved
mgtC	BAB2_0039	Mg^{2+} uptake[a]	Conserved
mgtE	BAB2_0360	Mg^{2+} uptake	Conserved
corA	BAB1_0583	Mg^{2+} uptake	Conserved
Zinc acquisition			
znuA	BAB2_1079	Zn^{2+} uptake[a]	Conserved
znuC	BAB2_1080	Zn^{2+} uptake	Conserved
znuB	BAB2_1081	Zn^2+ uptake	Conserved
zur	BAB2_1082	Zinc-responsive regulation of *znuA*, *znuB*, and *znuC*	Conserved
Nickel acquisition			
nikA	BAB2_0433	Ni^{2+} uptake[a]	Pseudogenes in *B. abortus* 2308
	BAB2_0434		*B. abortus* 9-941 and *B. abortus* S19
nikB	BAB2_0435	Ni^{2+} uptake	Conserved
nikC	BAB2_0436	Ni^{2+} uptake	Conserved
nikD	BAB2_0437	Ni^{2+} uptake	Pseudogene in *B. ovis* ATCC 25840
nikE	BAB2_0438	Ni^{2+} uptake	Conserved
nikK	BAB1_1384	Ni^{2+} uptake	Conserved

Table 9.1 continued

Gene	Designation	Predicted function	Conservation
nikM	BAB1_1385	Ni^{2+} uptake	Conserved
nikL	BAB1_1386	Ni^{2+} uptake	Conserved
nikQ	BAB1_1387	Ni^{2+} uptake	Conserved
nikO	BAB1_1388	Ni^{2+} uptakea	Conserved
nikR	BAB2_0432	Nickel-responsive regulation of *nikA*, *nikB*, *nikC*, *nikD* and *nikE*	Conserved
Cobalt acquisition			
cbtA	BAB1_1329	Co^{2+} uptake	Pseudogene in *B. abortus* 2308
cbtB	BAB1_1330	Co^{2+} uptake	Conserved
Metal detoxification			
copA	BAB2_0221	Cu^{1+} efflux	Pseudogene in *B. abortus* 2308
zntA	BAB1_2019	Zn^{2+} efflux	Conserved
rcnA	BAB2_0094	Ni^{2+} and Co^{2+} efflux	Conserved
cueO	BAB2_0534	Oxidation of Cu^{1+} to Cu^{2+}	Conserved
copZ	BAB1_0960	Cu^{1+} chaperone	Conserved
ureE	BAB1_0301	Ni^{2+} chaperone	Conserved
cueR	BAB1_0222	Copper-responsive regulation of *copA*	Conserved
zntR	BAB1_2018	Zinc-responsive regulation of *zntA*	Conserved

[a]Experimental data reported supporting predicted gene product function in *Brucella*. See text for details.

[b]Conserved=homologous genes present in the genome sequences of *B. abortus* 9–941, *B. abortus* S19, *B. melitensis* 16M, *B. melitensis* ATCC 23457, *B. melitensis* Rev1, *B. suis* 1330, *B. suis* ATCC 23445, *B. canis* ATCC, 23365, *B. ovis* ATCC 25840, *B. neotomae* 5K33, *B. microti* CCM 4915, *B. ceti* B1/94, *B. pinnipedialis* B2/94 and *B. inopinata* B01.

[c]Genes predicted to be located in operons (e.g. *dhbC*, *E*, *B*, and *A*) are clustered together in this table.

[d]Only the *omp25*/*omp31* genes present in *B. abortus* 2308 are listed in this table. See Vizcaino *et al*. (2004;2009) for comprehensive descriptions of the conservation of these genes among *Brucella* strains.

genetically defined mutants indicate that the *Brucella* Cir and Fiu homologues are required for wild-type iron transport in *B. abortus* 2308 (Anderson and Paulley, unpublished data), but the specificity of these transporters for 2,3-DHBA or brucebactin has not been determined. *Brucella* strains also possess an iron transport system homologous to the FatBCDE complex that transports Fe^{3+}-anguibactin across the cytoplasmic membrane in *Vibrio anguillarum* (Köster *et al.*, 1991). *B. abortus fatB* (González-Carreró *et al.*, 2003) and *B. melitensis fatC* (Danese *et al.*, 2004) mutants cannot use brucebactin and 2,3-DHBA, respectively, as iron sources. In addition, genes predicted to encode a homologue of the FepBCD complex that transports Fe^{3+}-enterochelin across the cytoplasmic membrane in *E. coli* (Shea and McIntosh, 1991; Stephens *et al.*, 1995) are located immediately downstream of the *Brucella* *cir* gene, but whether or not the corresponding gene products play a role in iron transport has not been determined.

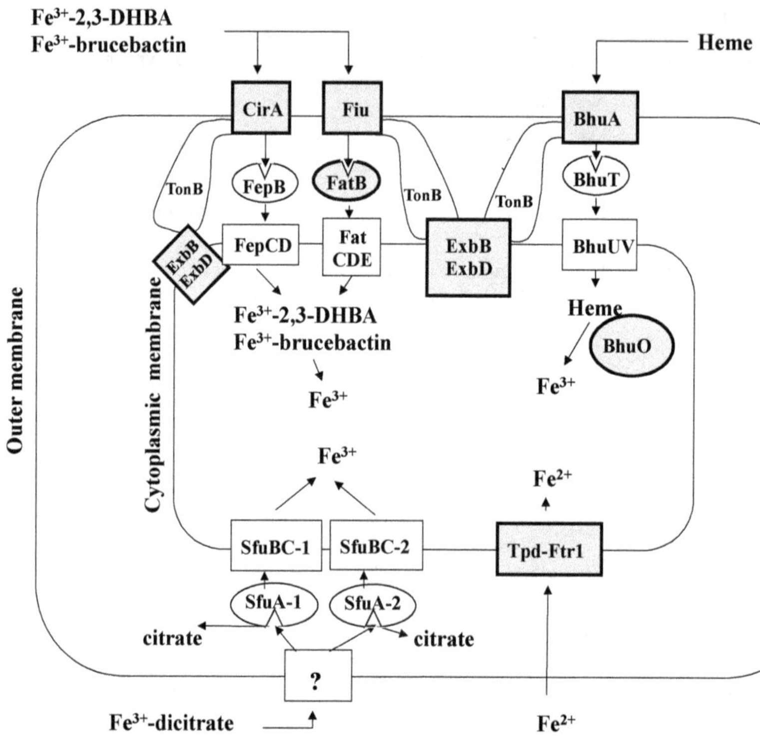

Figure 9.1 Iron acquisition systems in *Brucella*. Proteins or protein complexes for which there are experimental data supporting their proposed function are shown in grey with bold outlines. See text for details.

Independent studies have shown that siderophore-mediated iron transport is not required for the wild-type virulence of *B. abortus* (Bellaire *et al.*, 1999; González-Carreró *et al.*, 2002; Parent *et al.*, 2002) and *B. melitensis* (Danese *et al.*, 2004) strains in experimentally infected mice. Likewise, the *B. abortus dhbC* mutant BHB1 does not display attenuation in the human monocytic cell line THP-1 (B. Bellaire, unpublished data). In contrast, *B. abortus dhbC, cir* and *fiu* mutants exhibit significant attenuation in pregnant cattle (Bellaire *et al.*, 2003a; Anderson, Paulley, Hagius and Elzer, unpublished data) and a *B. abortus dhbC* mutant is also attenuated in pregnant goats (Elzer and Roop, unpublished data). The basis for the divergent virulence phenotypes exhibited by these *Brucella* siderophore biosynthesis and transport mutants in mice and ruminants is unknown, but it has been proposed that this difference may be linked to the inability of these mutants to efficiently use the four-carbon sugar erythritol as a carbon source (Bellaire *et al.*, 2003b).

Erythritol is the preferred carbon source for many *Brucella* strains (McCullough and Beal, 1951; Williams *et al.*, 1964; Anderson and Smith, 1965; Sperry and Robertson, 1975), and the gravid ruminant reproductive tract is rich in erythritol during the latter stages of pregnancy (Smith *et al.*, 1962; Samartino and Enright, 1993). Consequently, it has been postulated that the metabolism of erythritol fuels the rapid proliferation of the brucellae in ruminant placental trophoblasts that leads to abortion and extensive colonization of the reproductive tract (Smith *et al.*, 1962). *In vitro* studies indicate that *B. abortus* 2308 requires more iron when it is metabolizing erythritol than it does when utilizing other carbon and energy sources, and that siderophore production plays an important role in meeting this increased need for iron (Bellaire *et al.*, 2003b).

It is important to note that considerable variation exists in the ability of different *Brucella* strains to produce 2,3-DHBA (and presumably brucebactin as well) in response to iron deprivation (López-Goñi and Moriyón, 1995). As noted in Table 9.1, surveys of available genome sequences indicates that 2,3-DHBA/brucebactin biosynthesis and transport genes are not conserved in all *Brucella* strains, including *B. abortus* 9-941 and *B. melitensis* ATCC 23457. Consequently, it will be important to determine if the link between siderophore production and transport and virulence in pregnant ruminants established for *B. abortus* 2308 extends to *B. melitensis* and other *B. abortus* strains in their natural ruminant hosts. Studies with *Brucella* strains in their non-ruminant natural hosts will also shed more light on the relationship between siderophore production and transport, erythritol metabolism and virulence.

Recent proteomic studies examining the patterns of gene expression by *B. abortus* 2308 during residence in the murine macrophage like cell line RAW264.7 showed that the siderophore biosynthesis gene *dhbB* and transport genes *fatB* and *fiu* exhibit increased expression in the intracellular environment (Lamontagne *et al.*, 2009). As noted previously, multiple studies have shown that siderophore production and transport are not essential for the virulence of this strain in mice (Bellaire *et al.*, 1999; González- Carreró *et al.*, 2002; Parent *et al.*, 2002). The induction of *dhbB*, *fatB* and *fiu* in *B. abortus* 2308 observed in these studies however supports the contention that the brucellae encounter iron deprivation during their intracellular replication in host macrophages.

Haem transport

Brucella strains can transport haem (Almirón *et al.*, 2001) and use it as an iron source in *in vitro* assays (Bellaire *et al.*, 2001; Danese, 2001; Paulley *et al.*, 2007). Macrophages play a critical role in iron recycling in mammals by degrading senescent erythrocytes (Bratosin *et al.*, 1998). The haem released during this process is either incorporated into cellular proteins or transported to the endoplasmic reticulum where it is broken down and the iron released is sequestered in the cellular iron stores or released back into the circulation (Taketani *et al.*, 2005). The so-called 'replicative vacuole' within which the brucellae ultimately reside in macrophages is derived through the sustained interactions of the *Brucella*-containing phagosomes with the host cell endoplasmic reticulum (Celli *et al.*, 2003). Similarly, ruminant placental trophoblasts ingest erythrocytes from the maternal circulation and degrade these cells as a mechanism for providing iron to the developing fetus (Anderson *et al.*, 1986). Thus, haem may be a relevant iron source for the brucellae in the two main intracellular niches they inhabit in their mammalian hosts.

The *Brucella* BhuA protein (Table 9.1 and Fig. 9.1) is a TonB-dependent outer membrane haem transporter that shares extensive amino identity with other bacterial haem transporters including the *Shigella* ShuA (Mills and Payne, 1997) and *Yersinia* HmuR proteins (Thompson *et al.*, 1999). A *B. abortus bhuA* mutant cannot use haem as an iron source in *in vitro* assays, and can establish, but not maintain, spleen infections in experimentally infected mice (Paulley *et al.*, 2007). Such a phenotype is consistent with this mutant being unable to efficiently utilize haem as an iron source once it has reached the replicative vacuole in host macrophages.

Haem transporters in Gram-negative bacteria typically consist of a TonB-dependent transporter such as BhuA that moves haem across the outer membrane, and a periplasmic binding protein (PBP)-dependent ABC type transporter that shuttles the haem across the cytoplasmic membrane (Wandersman, 2010). In many cases, the genes encoding these components reside in the same genetic locus and are often co-regulated. *Brucella* strains appear to be atypical in this regard as the *bhuA* gene appears to be either monocistronic, or transcribed with an upstream gene of unknown function (Anderson and Martinson, unpublished data). But it is not located in close proximity to genes predicted to encode components of a PBP-dependent haem transporter. The genes designated as BAB2_0483-0485 in the *B. abortus* 2308 genome (which are located almost 500 kb away from *bhuA* on chromosome II) are predicted to encode homologues of the ShuT, U and V proteins which form a complex that transports haem across the cytoplasmic membrane of *Shigella dysenteriae* (Wyckoff *et al.*, 1998) and the HmuT, U and V proteins which perform a similar function in *Sinorhizobium meliloti* (Cuív *et al.*, 2008), a close phylogenetic relative of the brucellae. We have given the *Brucella shuT, U,* and *V* homologues the provisional designations *bhuT, U,* and *V* (Table 9.1 and Fig. 9.1). But whether or not these genes play a role in haem transport remains to be experimentally determined.

Although haem is of extreme biological importance to cells, it can also be toxic (Anzaldi and Skaar, 2010). Hence, living cells (including bacteria) are not thought to maintain a pool of haem that is not-incorporated into cellular proteins. Haem oxygenases, which degrade haem into biliverdin and CO and release Fe^{2+}, play an important role in the detoxification of unused haem in cells. These enzymes also allow bacteria to release iron from haem once it has been transported into the cytoplasm (Schmitt, 1997). Biochemical studies have shown that the product of the gene designated BMEII0706 in the *B. melitensis* 16M genome is a haem oxygenase (Puri and O'Brian, 2006). We have given this protein the designation BhuO (Table 9.1 and Fig. 9.1) and it will be important to determine what role BhuO plays in: (a) the capacity of *Brucella* strains to use haem an iron source; and (b) the capacity of these strains to avoid haem toxicity.

The *Brucella* Omp25/Omp31 class of outer membrane proteins are homologues of the haem binding proteins HbpA, HbpB, HbpC, HbpD and HbpE of *Bartonella* (Minnick *et al.*, 2003). The Hbp proteins bind haem to the surface of *Bartonella* strains, which are naturally occurring haem auxotrophs (Sander *et al.*, 2000). Whether or not the Hbps are actively involved in haem acquisition by *Bartonella* strains during residence in the mammalian host, however, is unknown (Minnick and Battisti, 2009). The *B. melitensis* Omp31 protein binds haem (Delpino *et al.*, 2006) and elevated production of this protein is observed in *B. melitensis* 16M in response to iron deprivation. Because some of the *Brucella* Omp25/Omp31 proteins have been linked to virulence (Edmonds *et al.*, 2001, 2002a,b; Caro-Hernández *et al.*, 2007; Martín-Martín *et al.*, 2008), and as noted above, haem appears to be an importance source of iron for the brucellae in the host, it will be imperative to gain a better understanding of what role, if any, the *Brucella* Omp25/Omp31 proteins play in haem acquisition. One possibility is that the Omp25/Omp31 proteins capture and hold haem on the surface of *Brucella* strains for subsequent transport by BhuA. This would be a similar scenario to that proposed for the Hbps and HutA (a BhuA homologue), in *Bartonella* (Battisti and Minnick, 2009; Parrow *et al.*, 2009).

ExbB, ExbD and TonB

The so-called TonB-dependent outer membrane proteins that transport ferrisiderophores and haem across the outer membrane of Gram-negative bacteria (Noinaj *et al.*, 2010) get their name from the fact that their activity requires energy from the proton motive force generated at the cytoplasmic membrane by cellular metabolism, and this energy is transferred to these proteins by the ExbB/ExbD/TonB complex (Postle and Larsen, 2007). *Brucella exbB* mutants cannot use 2,3-DHBA or haem as iron sources in *in vitro* assays (Danese *et al.*, 2004; Paulley and Anderson, unpublished data). These experimental findings suggest that like their counterparts in other bacteria, the *Brucella* Cir, Fiu and BhuA require ExbB/ExbD/TonB (Table 9.1 and Fig. 9.1), and presumably energy from the proton motive force, for their activity.

Transport of Fe^{3+} and Fe^{2+}

At physiologic pH (e.g. 7) under aerobic conditions, iron is present predominately as Fe^{3+}, which readily forms insoluble iron hydroxides under aqueous conditions (Crichton, 2009). Consequently, the level of 'free' Fe^{3+} available in the tissues of mammals has been predicted to be very low (Griffiths, 1999). The bacterial Sfu-, Fbp-, Hit-, Yfu- or Afu-type ABC transporters transport Fe^{3+} from the environment across the cytoplasmic membrane, although it is not clear whether these systems transport 'free' Fe^{3+} or Fe^{3+} complexed with small chelators such as citrate (Braun *et al.*, 1998). *Brucella* strains possess two sets of genes [BAB2_0538–0540 (which we will designate the *sfu-1* locus) and BAB2_0519–521 (which we will designate the *sfu-2* locus) in the *B. abortus* 2308 genome sequence] (Table 9.1 and Fig. 9.1) predicted to encode Fe^{3+} transporters homologous to the *Serratia marcescens* SfuABC complex (Angerer *et al.*, 1992). Elevated expression of the *sfuA-1* gene in *B. abortus* 2308 during intracellular replication in the murine RAW264.7 macrophage-like cell line has been detected in two independent studies (Eskra *et al.*, 2001; Lamontagne *et al.*, 2009). Bison infected with *B. abortus* strains also produce antibodies reactive with SfuA-1 (Lowry *et al.*, 2010). These experimental findings indicate that the Sfu-1 complex is produced by *Brucella* strains during intracellular replication in the host, but a mutational analysis of the contributions of the *Brucella sfu-1* and *sfu-2* loci to iron acquisition and/or virulence has not been reported.

During the early stages of their interactions with host macrophages, the brucellae reside in acidified compartments known as the early and intermediate *Brucella*-containing vacuoles (BCVs) (Celli *et al.*, 2003). The intraluminal pH of these compartments is <5, conditions under which Fe^{2+} would be present and soluble (Crichton, 2009), and could potentially serve as a readily available iron source. The genes designated as BAB2_0837–0840 in the *B. abortus* 2308 genome are predicted to encode a Fe^{2+} transporter (Table 9.1 and Fig. 9.1) homologous to the CupII-type transporters recently described in a variety of other bacteria (Rajasekaran *et al.*, 2010). These transporters consist of a 19 kDa periplasmic Fe^{2+} binding protein (known as Tpd or P19), a periplasmic copper oxidase (cupredoxin) that converts Fe^{2+} to Fe^{3+}, an Ftr1-type high affinity iron permease that transports Fe^{2+} (after it has been oxidized to Fe^{3+}) across the cytoplasmic membrane, and a polyferredoxin that is thought to maintain the redox balance of the transporter. The *tpd-ftr1* locus exhibits maximal expression in response to iron deprivation in *B. abortus* 2308, and an isogenic *tpd* mutant displays increased sensitivity to the iron chelator EDDHA compared with the parental strain (Anderson *et al.*, 2010). Although these experimental findings support a role for the Tpd–Ftr1 complex in iron transport, its specificity for Fe^{2+} has not been verified. A *B. abortus tpd* mutant is extremely attenuated in cultured murine macrophages and experimentally infected mice (Anderson *et al.*, 2010), thus it will

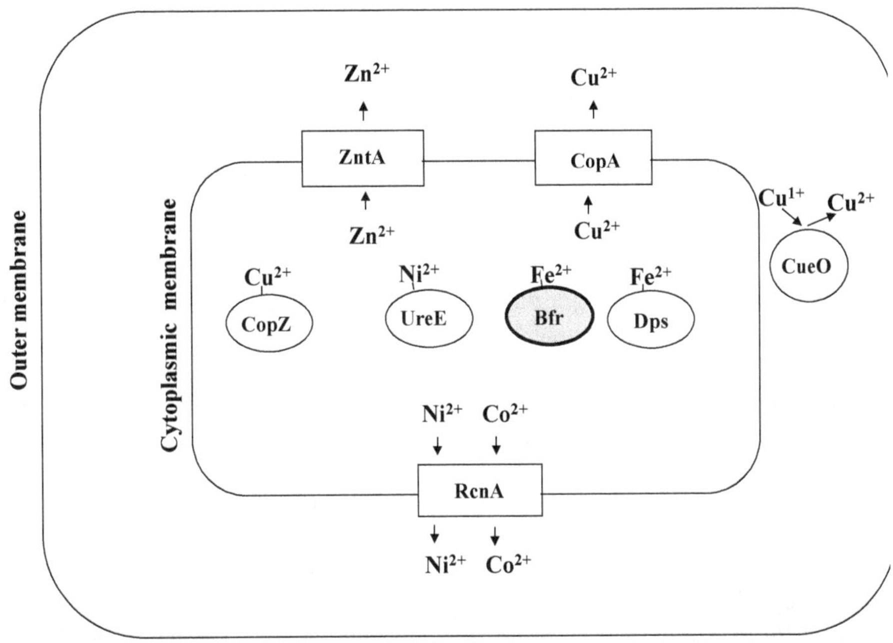

Figure 9.2 Cobalt, iron, copper, magnesium, nickel and zinc detoxification systems in *Brucella*. Proteins or protein complexes for which there is experimental data supporting their proposed function are shown in grey with bold outlines. See text for details.

be important to confirm that Tpd-Ftr1 is an iron transporter; and determine whether or not it is Fe^{2+}-specific.

Iron storage

Iron is an essential micronutrient for most bacteria, but the accumulation of too much intracellular iron can be toxic. One way that bacteria prevent iron toxicity is via the activity of iron storage proteins such as bacterioferritin (Bfr) (Bou-Abdallah, 2010) and Dps (Chiancone and Ceci, 2010). Both of these proteins have been shown to bind excess intracellular iron in a manner that prevents it from participating in the production of reactive oxygen species via Fenton chemistry. *Brucella* strains produce both Bfr and Dps (Table 9.1 and Fig. 9.2). Considering the role that has been proposed for oxidative killing in the brucellacidal activity of host macrophages (Jiang *et al.*, 1993), the iron-storage and detoxification activities of Bfr and Dps would be expected to play important roles in preventing oxidative damage in *Brucella* strains during their residence in the host. However, neither *B. melitensis* nor *B. abortus bfr* mutants exhibit attenuation in mice, human monocyte-derived macrophages or eukaryotic cell lines (Denoel *et al.*, 1997; Almirón and Ugalde, 2010). One possible explanation for these experimental findings is that Bfr and Dps play overlapping and compensatory roles in preventing iron-mediated ROS toxicity in *Brucella* strains. A comparative analysis of the phenotypes of *bfr* and *dps* mutants and a *bfr dps* double mutant would directly address this question.

Regulation of iron acquisition genes

Another way that bacteria protect themselves from iron toxicity is by expressing iron acquisition genes only when they are iron-deprived. Homologues of the *Bradyrhizobium* iron response regulator Irr (Hamza *et al.*, 1998) and the rhizobial iron regulator RirA (Todd *et al.*, 2002) (Table 9.1) appear to controlling the iron-responsive expression of the *Brucella* iron uptake genes that have been studied to date.

Irr has been best studied in *B. japonicum*, where it represses the expression of genes whose products require iron for their activity (e.g. haem biosynthesis genes), and activates the expression of genes whose products participate in iron acquisition (Small *et al.*, 2009). The stability of the *B. japonicum* Irr protein is modulated by cellular iron levels. When sufficient iron is available to allow haem biosynthesis, Irr is degraded. Iron-responsive degradation of Irr requires its direct interaction with ferrochelatase (Qi and O'Brian, 2002), which allows Irr to coordinate the activity of the haem biosynthetic pathways with the cellular availability of iron.

Irr participates in the iron-responsive regulation of the siderophore biosynthesis gene *dhbC* (Martínez *et al.*, 2006) and the haem transport gene *bhuA* in *B. abortus* 2308 (Anderson *et al.*, 2009b). In both cases, a direct interaction between the Irr protein and the promoters of these genes has been demonstrated in electrophoretic mobility shift assays. In addition, a computational analysis of the *B. melitensis* 16M genome sequence searching for predicted Irr binding sites (also known as 'iron control elements' or 'ICE'; Rudolph *et al.*, 2006) has identified several other iron metabolism genes that may be regulated by Irr including *tpd*, *bfr* and *rirA* (see below) (Rodionov *et al.*, 2006a). Although experimental evidence suggests that the same link between cellular iron levels, haem biosynthesis and Irr stability that is observed in *B. japonicum* also exists in *Brucella* (Martínez *et al.*, 2005), the *Brucella* Irr lacks the haem regulatory motif (GCPWHD) that is critical for the iron-responsive degradation of the *B. japonicum* Irr (Qi *et al.*, 1999; Singleton *et al.*, 2010). Thus, it will be important to determine whether the *Brucella* Irr is actively degraded in response to increasing levels of intracellular iron [as is the case with the *B. japonicum* Irr (Qi and O'Brian, 2002)], or whether haem binding to the *Brucella* Irr inhibits its DNA binding activity in the absence of active degradation (as has recently been reported for the *Rhizobium leguminosarum* Irr; Singleton *et al.*, 2010).

The rhizobial iron regulator RirA was identified via a genetic screen seeking mutants that exhibit constitutive expression of iron acquisition genes during growth under iron-replete conditions (Todd *et al.*, 2002). RirA is a member of the Rrf2 family of transcriptional regulators. These transcriptional regulators contain Fe–S clusters and the redox status of these clusters controls their activity (Tucker *et al.*, 2010). Although genetic studies have clearly shown a role for RirA in the repression of iron acquisition genes in *R. leguminosarum* (Todd *et al.*, 2002), *S. meliloti* (Chao *et al.*, 2005) and *Agrobacterium tumefaciens* (Ngok-Ngam *et al.*, 2009), and predicted consensus RirA binding sites have been identified upstream of RirA-regulated genes (Yeoman *et al.*, 2004; Rodionov *et al.*, 2006), direct interaction of the RirA protein with the promoters of RirA-regulated genes has not been demonstrated *in vitro*.

Brucella strains possess a RirA homologue (Table 9.1), and preliminary studies with a *B. abortus rirA* mutant suggest that RirA not only plays an important role in regulating *Brucella* iron metabolism genes, but that this regulator also plays a critical role in virulence (Anderson *et al.*, 2004). The bioinformatics study performed by Rodionov *et al.* (2006a) indicates that putative RirA binding sites (also known as 'iron-responsive operators or IROs'; Yeoman *et al.*, 2004) are present upstream of the *bhuA, dhbC, sfuA, fatB, exbB* and *rirA* genes in *B. melitensis* 16M. Experimental evidence indicates that RirA is required for the iron-responsive expression of the *bhuA* gene in *B. abortus* 2308 (Paulley *et al.*, 2007b), but the precise nature of the regulatory link between RirA and *bhuA* is presently unknown.

In addition to controlling the expression of their iron acquisition genes in response to cellular iron levels, bacteria have also evolved regulatory networks that allow them to produce specific iron transport systems only when the cognate iron source is available in the external environment. The activity of the *Bordetella* AlcR serves as a paradigm for this type of regulation (Brickman and Armstrong, 2009). The genes involved in the biosynthesis and transport of the siderophore alcaligin in *Bordetella* are repressed by the ferric uptake regulator Fur during growth under iron-replete conditions, and 'de-repressed' when these strains are subjected to iron deprivation.

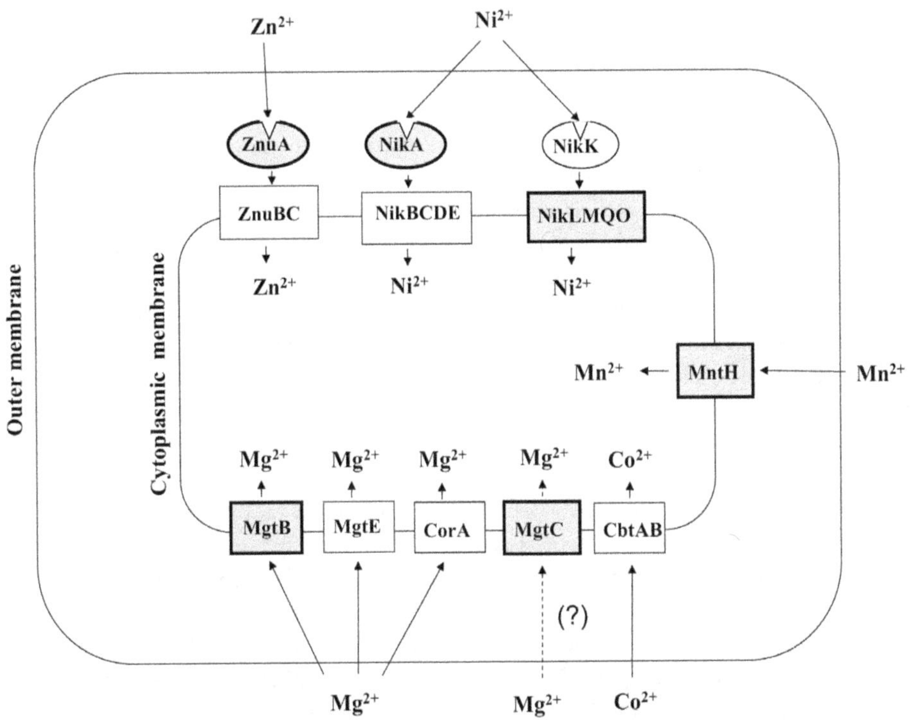

Figure 9.3 Cobalt, magnesium, manganese, nickel and zinc acquisition systems in *Brucella*. Proteins or protein complexes for which there is experimental data supporting their proposed function are shown in grey with bold outlines. See text for details.

Maximum expression of these genes does not occur, however, unless the AraC-type transcriptional activator AlcR binds Fe^{3+}-alcaligin, which serves as a co-activator, allowing AlcR to drive *alcABCDE* and *fauA* transcription (Brickman *et al.*, 2001). Experimental evidence indicates that DhbR (Table 9.1) functions in a similar manner in *B. abortus* 2308. The *dhbCEBA* genes in the 2,3-dihydroxybenzoic acid/brucebactin biosynthesis operon are only expressed when this strain is subjected to iron deprivation (Bellaire *et al.*, 2003a), but maximum expression of these genes does not occur unless DhbR is present and the strain is actively producing and transporting 2,3-DHBA (Anderson *et al.*, 2008). It is presently unknown whether or not the siderophore transport genes (e.g. *fiu*, *cir*, and *fatBCDE*) are subject to DhbR regulation in *B. abortus* 2308.

Manganese

Manganese serves as an important co-factor for a variety of bacterial proteins (Papp-Wallace and Maguire, 2006) and is an essential micronutrient for *Brucella* strains (Sanders *et al.*, 1953; Evenson and Gerhardt, 1955). Mammals also sequester Mn^{2+} as a defence against invading pathogens (Kehl-Fie and Skaar, 2010), and the capacity of the brucellae to successfully compete for this metal in the host plays a role in their virulence (Anderson *et al.*, 2009a). An analysis of the currently available genome sequences indicates that *Brucella* strains produce a single high affinity manganese transporter, MntH (Table 9.1 and Fig. 9.3), and a phenotypic evaluation of a *B. abortus mntH* mutant supports this proposition (Anderson *et al.*, 2009a). Like other described bacterial MntH proteins (Kehres *et al.*, 2000), the *Brucella* MntH appears to be specific for Mn^{2+}. Because they are proton-symporters, bacterial MntH-type Mn^{2+} transporters exhibit their maximum activity at acidic pH, which makes MntH well suited for Mn^{2+} transport by *Brucella* strains in the early and intermediate BCVs in host macrophages.

A *B. abortus mntH* mutant exhibits extreme attenuation in cultured murine macrophages and experimentally infected mice (Anderson *et al*, 2009a). These experimental findings indicate that the acquisition of sufficient levels of cellular Mn^{2+} plays a crucial role in allowing *Brucella* strains to successfully adapt to their intracellular niche in the host. The manganese co-factored superoxide dismutase A (SodA) (Sriranganathan *et al.*, 1991) and the (*p*)ppGpp synthetase/hydrolase Rsh (Dozot *et al.*, 2006) are two cellular proteins whose activities are probably negatively impacted by inefficient Mn^{2+} acquisition in a *Brucella mntH* mutant. Indeed, the *B. abortus mntH* mutant MWV15 produces reduced SodA activity compared with the parental 2308 strain, and exhibits increased sensitivity to both O_2^- and H_2O_2 in *in vitro* assays (Anderson *et al.*, 2009a). However, the attenuation exhibited by the *B. abortus mntH* mutant is much more intense than that displayed by an isogenic *sodA* mutant (Baumgartner *et al.*, 2010). *Brucella* strains produce a single bifunctional (*p*)ppGpp synthetase/hydrolase (designated Rsh) and the enzymatic activity of this family of bacterial proteins is Mn^{2+}-dependent (Papp-Wallace and Maguire, 2006). Rsh activity is required for the induction of the stringent response (Magnusson *et al.*, 2005), and thus inefficient Mn^{2+} acquisition could compromise the ability of the *B. abortus mntH* mutant to successfully adapt to the nutrient deprivation encountered during intracellular residence in host macrophages (Roop *et al.*, 2009). Rsh has also been shown to be important for proper expression of the *virB* operon in *B. melitensis* 16M and *B. suis* 1330 (Dozot *et al.*, 2006). Hence, improper expression of the genes encoding the Type IV secretion machinery could also be contributing to the attenuation exhibited by the *B. abortus mntH* mutant.

Manganese is less redox active than iron, and recent studies performed in *E. coli* have shown that this bacterium elevates expression of *mntH* in response to oxidative stress, and substitutes Mn^{2+} for Fe^{2+} in cellular enzymes as a nonspecific mechanism for protecting these enzymes from oxidative damage as a result of Fenton chemistry (Anjem *et al.*, 2009). Exposure of *B. abortus* 2308 to H_2O_2 *in vitro* also results in elevated *mntH* expression (E. Menscher, unpublished data), suggesting that *Brucella* strains may also be able to employ Mn^{2+} substitution for Fe^{2+} in cellular enzymes as a mechanism to avoid ROS toxicity.

Mur

In many bacteria, the ferric uptake regulator (Fur) controls the expression of iron acquisition genes (Lee and Helmann, 2007). Although *Brucella* strains possess a Fur homologue that can function as an iron-responsive repressor in an *E. coli fur* mutant (Phillips *et al.*, 1996), further studies with this protein have revealed that it does not function in the same manner in *B. abortus* 2308. Rather, this protein appears to be a functional homologue of the manganese-responsive regulator Mur (Table 9.1), first described in *R. leguminosarum* (Díaz-Mireles *et al.*, 2004). Experimental evidence indicates that Mur represses *mntH* expression in *B. abortus* 2308 in response to increasing concentrations of manganese (Anderson *et al.*, 2009a), and the bioinformatics analysis reported by Rodionov *et al.* (2006a) suggests that *mntH* and *irr* may be the only *Brucella* genes subject to regulation by Mur. This latter observation is interesting considering the fact that manganese stabilizes the Irr protein (Puri *et al.*, 2010) and Mur regulates *irr* transcription in *B. japonicum* (Hohle and O'Brian, 2010). These findings suggest that integrated regulation of iron and manganese metabolism genes may be occurring at multiple levels in the α-proteobacteria.

Magnesium

Magnesium is another metal that was found to be required for the growth of *Brucella* strains *in vitro* (Evenson and Gerhardt, 1955), and more recent studies indicate that magnesium transport plays an important role in the physiological adaptation of the brucellae to their intracellular niche in the host (Lestrate *et al.*, 2000; Lavigne *et al.*, 2005). A survey of the available *Brucella* genome sequences indicates that these strains possess three magnesium transporters – MgtB, MgtE and CorA (Table 9.1 and Fig. 9.3). MgtE and CorA are gated channels, and MgtB is a P-type ATPase, that transport Mg^{2+} across the cytoplasmic membrane (Moomaw and Maguire, 2008). A *B. melitensis mgtB* mutant is not reported to exhibit growth restriction in response to magnesium deprivation *in vitro*, but this strain displays significant attenuation in mice (Lestrate *et al.*, 2000).

The precise role that the MgtC protein (Table 9.1 and Fig. 9.3) plays in magnesium transport is unclear (Günzel *et al.*, 2006; Alix and Blanc-Potard, 2007); however, a *B. suis mgtC* mutant grows poorly in a magnesium-restricted growth medium (Lavigne *et al.*, 2005). More importantly, this mutant exhibits significant attenuation in the murine J774 macrophage-like cell line and this attenuation can be partially relieved by the addition of $MgCl_2$ to the tissue culture medium.

Zinc

Although an absolute requirement for zinc was not uncovered in the early studies of the metal requirements of *Brucella* strains during *in vitro* cultivation (McCullough *et al*, 1947), studies by two different groups have clearly shown that zinc transport plays a critical role in the virulence of these strains in HeLa cells (Kim *et al.*, 2003, 2004), murine macrophages (Kim *et al.*, 2004; Yang *et al.*, 2006) and experimentally infected mice (Yang *et al.*, 2006). These experimental findings further support the proposition that the sequestration of zinc is an important host defence against microbial infection (Kehl-Fie and Skaar, 2010). *Brucella* strains produce a single high affinity zinc transporter consisting of a periplasmic zinc binding protein (ZnuA) and an associated ABC-type transporter (ZnuB and C) (Table 9.1 and Fig. 9.3). A gene predicted to encode a homologue of the zinc-responsive regulator Zur (Patzer and Hantke, 1998) resides next to the *znuABC* locus in *Brucella* strains, but studies evaluating the functionality of the *Brucella* Zur homologue, or whether or not the *znuABC* locus is regulated in response to environmental levels of zinc in *Brucella* strains, have not been reported.

Nickel

Nickel has been shown to be an important cofactor for only a limited number of proteins (Li and Zamble, 2009). Notably, one of these is the enzyme urease which is required for the wild-type virulence of *B. abortus* 2308 and *B. suis* 1330 when these strains are administered to mice via the oral route (Bandara *et al.*, 2007; Sangari *et al.*, 2007). It is thought that the acid-neutralizing capacity of urease allows the brucellae to resist the very low pH they encounter during passage through the stomach and gastrointestinal tract after ingestion. Two nickel transporters, an ABC-type transporter designated NikABCDE (Jubier-Maurin *et al.*, 2001) and an energy coupling factor (ECF)-type transporter designated NikKMLQO (Sangari *et al.*, 2010) have been described in *Brucella* strains (Table 9.1 and Fig. 9.3), and *B. suis nikA* and *B. abortus nikO* mutants exhibit reduced urease activity. *nikA* is a pseudogene in *B. abortus* 2308, and it has been proposed that the activity of the NikKMLQO transporter allows this strain to compensate for the absence of NikABCDE and acquire sufficient nickel to synthesize a functional urease (Sangari *et al.*, 2010). As these authors point out, the fact that the *nikK, M, L, Q* and *O* genes are co-transcribed with those in the *ure2* locus in *B. abortus* 2308 could explain why the *ure2* locus is maintained and actively transcribed in this strain even though these genes themselves do not encode an active urease.

Because nickel is an integral component of urease, it seems probable that nickel transport plays an important role in the virulence of *Brucella* strains. This proposition is supported by the fact that *nikA* expression is elevated during the intracellular replication of *B. suis* 1330 in the murine macrophage-like cell line J774, but is not supported by the observation that an isogenic *B. suis nikA* mutant displays wild-type virulence in the human monocytic cell line THP-1 (Jubier-Maurin *et al.*, 2001). Although one possible explanation for these findings is the proposed redundant activity of the two *Brucella* nickel transporters (Sangari *et al.*, 2010), it is important to note that urease activity is not required for the survival and replication of *Brucella* strains in cultured macrophages or their virulence in mice infected via the intraperitoneal route (Bandara *et al.*, 2007; Sangari *et al.*, 2007). Consequently, a definitive evaluation of the relative contributions of the NikABCDE and NikKMLQO transporters to virulence will require the examination of both single and double mutants in an appropriate animal model (i.e. mice infected via the oral route).

A gene encoding a homologue of the nickel-responsive transcriptional regulator NikR (de Pina *et al.*, 1999) resides adjacent to the *nikABCDE* operon (Table 9.1), and *nikA* expression is elevated in *B. suis* 1330 when this strain is cultivated in the presence of the chelator EDTA (Jubier-Maurin *et al.*, 2001). But whether or not *nikA* is being regulated in a nickel-responsive manner, or NikR is responsible for this regulation, has not been determined.

Cobalt

Cobalt is an integral component of cobalamin (also known as vitamin B_{12}), which serves as a co-factor for a variety of cellular enzymes including the class II ribonucleotide reductase (RNR) encoded by the *nrdJ* gene in *S. meliloti* (Taga and Walker, 2010). This enzyme plays an essential role in the establishment of the symbiotic state between *S. meliloti* and its plant host, purportedly because this particular type of RNR is resistant to inactivation by reactive oxygen and nitrogen species. Given the parallels that exist between the intracellular lifestyles of the rhizobia and the brucellae during their interactions with their eukaryotic hosts (Levier *et al.*, 2000; Batut *et al.*, 2004), NrdJ probably also plays an important role in the survival and replication of *Brucella* strains in host macrophages.

Genes encoding several potential cobalt transporters can be found in the *Brucella* genome sequences. Bacterial Nik/CbiMNQO-type transporters can transport either cobalt or nickel (Rodionov *et al.*, 2003). However, they are usually specific for one or the other metal, and their specificity can usually be predicted based upon the location of the corresponding genes in close proximity to genes involved in the biosynthesis of proteins that require nickel or cobalt for their activity. Thus, although the *Brucella nikK, L, M, Q,* and *O* homologues (Table 9.1) are also *cbiK, L, M, Q,* and *O* homologues, these genes have been experimentally linked to nickel transport and urease activity (Sangari *et al.*, 2010). The *Brucella cbtA* and *cbtB* genes (Table 9.1 and Fig. 9.3) (Rodionov, *et al.*, 2003), on the other hand, are located in a cluster of genes (e.g. *cobU* and *cobQ*) predicted be involved in cobalamin biosynthesis, and thus CbtAB is most probably a cobalt-specific transporter. In some cases, the magnesium transporters CorA and MgtE can also transport cobalt (Gibson *et al.*, 1991; Merino *et al.*, 2001; Pfeiffer *et al.* 2002), and the activity of these 'secondary' Co^{2+} transporters may be especially important in *B. abortus* 2308, where *cbtA* is annotated as a pseudogene (Table 9.1).

With regard to cobalamin itself, it is also important to note that *cir* (Table 9.1) has been annotated in some *Brucella* genome sequences as encoding a homologue of the *E. coli* TonB-dependent cobalamin transporter BtuB (Kadner and Liggins, 1973), and a predicted cobalamin riboswitch resides upstream of this gene (Nahvi *et al.*, 2004). Although as noted in a previous section, a *B. abortus cir* mutant exhibits a phenotype suggesting that this genetic locus is involved in iron transport, the capacity of this mutant to transport siderophores has not been directly examined. Consequently, it will be important to experimentally determine whether the *Brucella cir-fepBDC* genes are involved in siderophore or cobalamin transport.

Zinc, copper, nickel and cobalt efflux systems

Copper, zinc, nickel and cobalt perform important structural and catalytic functions in cellular proteins (Summers, 2009). However, owing to their propensity to incorrectly substitute for other metals such as iron and manganese in cellular proteins leading to the inactivation of these proteins (Macomber and Imlay, 2009; Waldron and Robinson, 2009), bacteria produce active efflux systems to maintain the cellular levels of copper, zinc, nickel and cobalt at non-toxic levels (Silver and Phung 2005). They also produce metallochaperones to sequester these cations (Silver and Phung 2005). In addition, some Gram-negative bacteria produce a periplasmic multicopper oxidase CueO that detoxifies copper by converting Cu^{1+} to the less toxic Cu^{2+} (Singh *et al.*, 2004).

Copper, cobalt and nickel toxicity have been reported for *Brucella* strains in the literature (McCullough *et al.*, 1947; Altenbern *et al.*, 1959; Anderson *et al.*, 2009a). Genes predicted to encode homologues of the zinc exporter ZntA (Beard *et al.*, 1997; Rensing *et al.*, 1997), the copper exporter CopA (Rensing *et al.*, 2000), and the nickel and cobalt exporter RcnA (Rodrique *et al.*, 2005) can be found in *Brucella* genome sequences (Table 9.1 and Fig. 9.2), and the *Brucella zntA* and *copA* are flanked by genes predicted to encode homologues of the zinc-responsive regulator ZntR (Brocklehurst *et al.*, 1999) and the copper-responsive regulator CueR (Outten *et al.*, 2000), respectively. The brucellae also possess homologues of the copper chaperone CopZ (Rosenzweig and O'Halloran, 2000) and the nickel chaperone UreE (Colpas *et al.*, 1999), and the gene immediately upstream of *sodC* in *Brucella* strains is predicted to encode a CueO homologue (Table 9.1 and Fig. 9.2). Metal efflux systems (particularly those that export copper) (Francis and Thomas, 1997; Schwan *et al.*, 2005, Stähler *et al.*, 2006; Zhang and Rainey, 2007; Ward *et al.*, 2010) and CueO (Achard *et al.*, 2010) have been linked to bacterial virulence, and it seems probable that these metal 'detoxifiers' make significant contributions to the intracellular lifestyle of *Brucella* strains as well.

Conclusions

Experimental evidence clearly shows that metal homeostasis plays a critical role in the virulence of *Brucella* strains in their mammalian hosts. Table 9.1 provides a list of the genes in *B. abortus* 2308 whose predicted products share significant amino acid identity with bacterial proteins involved in the acquisition, storage, and detoxification of biologically beneficial metals, or regulation of the genes encoding these proteins. Diagrams depicting the proposed activity and cellular localization of the *Brucella* proteins involved in metal acquisition, storage and detoxification can be found in Figs. 9.1, 9.2 and 9.3. As noted in this chapter, the postulated function of some of these *Brucella* gene products has been experimentally verified, but the role that the others play in metal homeostasis remains to be determined. It is hoped that after reading this chapter, the reader will appreciate that gaining a better understanding of the function of the *Brucella* gene products involved in metal homeostasis will provide us with important insights regarding how these bacteria successfully adapt to their intracellular niches and produce disease in their mammalian hosts.

Acknowledgements

Work on iron and manganese acquisition by *Brucella* strains in the laboratory of RMR has been funded by grants from the National Institute of Allergy and Infectious Disease and the United States Department of Agriculture.

References

Achard, M.E.S., Tree, J.J, Holden, J.A., Simpfendorfer, K.R., Wijburg, O.L.C., Strugnell, R.A., Schembri, M.A., Sweet, M.J., Jennings, M.P., and McEwan, A.G. (2010). The multicopper-ion oxidase CueO of *Salmonella enterica* serovar Typhimurium is required for systemic virulence. Infect. Immun. *78*, 2312–2319.

Alix, E., and Blanc-Potard, J.-B. (2007). MgtC: a key player in intramacrophage survival. Trends Microbiol. *15*, 252–256.

Almirón, M., and Ugalde, R.A. (2010). Iron homeostasis in *Brucella abortus*: the role of bacterioferritin. J. Microbiol. *48*, 668–673.

Almirón, M., Martínez, M., Sanjuan, N., and Ugalde, R.A. (2001). Ferrochelatase is present in *Brucella abortus* and is critical for its intracellular survival and virulence. Infect. Immun. *69*, 6225–6230.

Altenbern, R.A., Williams, D.R., and Ginoza, H.S. (1959). Effect of cobalt on population changes in *Brucella abortus*. J. Bacteriol. *77*, 509.

Anderson, E.S., Paulley, J.T., Todd, J.D., Johnston, A.W.B., and Roop R.M. II. (2004). The *Brucella abortus* rhizobial iron regulator (RirA) homolog is necessary for virulence in the murine model of infection. Proceedings of the 47th Annual Brucellosis Research Conference, Abst. 16, p. 23.

Anderson, E.S., Paulley, J.T., and Roop, R.M. II. (2008). The AraC-like transcriptional regulator DhbR is required for maximum expression of the 2,3-dihydroxybenzoic acid biosynthesis genes in *Brucella abortus* 2308 in response to iron deprivation. J. Bacteriol. *190*, 1838–1842.

Anderson, E.S., Paulley, J.T., Gaines, J.M., Valderas, M.W., Martin, D.W., Menscher, E., Brown, T.D., Burns, C.S., and Roop, R.M. II. (2009a). The manganese transporter MntH is a critical virulence determinant for *Brucella abortus* 2308 in experimentally infected mice. Infect. Immun. *77*, 3466–3474.

Anderson, E.S., Paulley, J.T., Gaines, J.M., Martinson, D.A., Steele, K.H., and Roop. R.M. II. (2009b). The *B. abortus* Irr is necessary for iron-responsive regulation of the gene encoding the haem transporter BhuA. Proceedings of the 62nd Annual Brucellosis Research Conference, Abstr. 16, p. 25

Anderson, E.S., Paulley, J.T., Gaines, J.M., Baumgartner, J.E., Martin, D.W., Roop, R. M. II. (2010). Identification of a putative high affinity ferrous iron (Fe^{++}) transporter in *Brucella abortus* 2308 that plays a critical role in virulence. Proc. Biometals Abstr. P-05, p. 52.

Anderson, J.D., and Smith, H. (1965). The metabolism of erythritol by *Brucella abortus*. J. Gen. Microbiol. *38*, 109–124.

Anderson, T.D., Cheville, N.F., and Meador, V.P. (1986). Pathogenesis of placentitis in the goat inoculated with *Brucella abortus*. II. Ultrastructural studies. Vet. Pathol. *23*, 227–239.

Angerer A., Klupp, B., and Braun, V. (1992). Iron transport systems of *Serratia marcescens*. J. Bacteriol. *174*, 1378–1387.

Anjem, A., Varghese, S., and Imlay, J.A. (2009). Manganese import is a key element of the OxyR response to hydrogen peroxide in *Escherichia coli*. Mol. Microbiol. *72*, 844–858.

Anzaldi, L.L., and Skaar, E.P. (2010). Overcoming the heme paradox: heme toxicity and tolerance in bacterial pathogens. Infect. Immun. *78*, 4977–4989.

Bandara, A.B., Contreras, A., Contreras-Rodriguez, A., Martins, A.M., Dobrean, V., Poff-Reichow, S., Rajasekaran, P., Sriranganathan, N., Schurig, G.G., and Boyle, S.M. (2007). *Brucella suis* urease encoded by *ure1* but not *ure2* is necessary for intestinal infection of BALB/c mice. BMC Microbiol. *7*, 57

Batut, J., Andersson, S.G.E., and O'Callaghan, D. (2004). The evolution of chronic infection strategies in the α-proteobacteria. Nat.Rev. Microbiol. *2*, 933–945.

Baumgartner, J.E., Anderson, E.S., Roop, R.M. II, and Martin, D. (2010). The *Brucella abortus* manganese co-factored superoxide dismutase, SodA, is required for the detoxification of endogenous superoxide and virulence in the mouse model of infection. Proceedings of the 54th Annual Wind River Conference. Prok. Biol., p. 52.

Beard, S.J., Hashim, R., Membrillo-Hernández, J., Hughes, M.N., and Poole, R.K. (1997). Zinc (II) tolerance in *Escherichia coli* K-12, evidence that the *zntA* gene (o732) encodes a cation transport ATPase. Mol. Microbiol. *25*, 883–891.

Bellaire, B.H., Elzer, P.H., Baldwin, C.L., and Roop, R.M. II. (1999). The siderophore 2,3-dihydroxybenzoic acid is not required for virulence of *Brucella abortus* in BALB/c mice. Infect. Immun. *67*, 2615–2618.

Bellaire, B. H., Elzer, P.H., Hagius, S., Walker, J., Baldwin, C.L., and Roop, R. M. II. (2003a). Genetic organization and iron-responsive regulation of the *Brucella abortus* 2,3-dihydroxybenzoic acid biosynthesis operon, a cluster of genes required for wild-type virulence in pregnant cattle. Infect. Immun. *71*, 1794–1803.

Bellaire, B. H., Elzer, P.H., Baldwin, C.L., and Roop, R.M. II. (2003b). Production of the siderophore 2,3-dihydroxybenzoic acid is required for wild-type growth of *Brucella abortus* in the presence of erythritol under low-iron conditions *in vitro*. Infect. Immun. *71*, 2927–2932.

Bou-Abdallah, F. (2010). The iron redox and hydrolysis chemistry of the ferritins. Biochim. Biophys. Acta *1800*, 719–731.

Bratosin, D., Mazurier, J., Tissier, J.P., Estaquier, J., Huart, J.J., Amiesen, J.C., Aminoff, D., and Montreuil, J. (1998). Cellular and molecular mechanisms of senescent erythrocyte phagocytosis by macrophages. Biochimie *80*, 173–195.

Braun, V., Hantke, K, and Köster, W. (1998). Bacterial iron transport: mechanisms, genetics and regulation.

In: Metal Ions in Biological Systems, Vol. 35, Sigel, A., and Sigel, H., eds. (Marcel Dekker: New York, NY), pp. 67–145.

Brickman, T.J., and Armstrong, S.K. (2009). Temporal signaling and differential expression of *Bordetella* iron transport systems: the role of ferrimones and positive regulators. Biometals 22, 33–41.

Brickman, T.J., Kang, H.Y., and Armstrong, S.K. (2001). Transcriptional activation of *Bordetella* alcaligin siderophore genes requires the AlcR regulator with alcaligin as inducer. J. Bacteriol. *183*, 483–489.

Brocklehurst, K.R., Hobman, J. L., Lawley, B., Blank, L., Marshall, S.J., Brown, N.L., and Morby. A.P. (1999). ZntR is a Zn(II)-responsive MerR-like transcriptional regulator of *zntA* in *Escherichia coli*. Mol. Microbiol. *31*, 893–902.

Byrd, T.F., and Horwitz, M.A. (1989). Interferon gamma-activated human monocytes down-regulate transferrin receptors and inhibit intracellular multiplication of *Legionella pneumophila* by limiting the availability of iron. J. Clin. Invest. *83*, 1457–1465.

Caro-Hernández, P., Fernández-Lago, L., de Miguel, M., Martín-Martín, A.I., Cloeckaert, A., Grilló, M., and Vizcaíno, N. (2007). Role of the Omp25/Omp31 family in outer membrane properties and virulence of *Brucella ovis*. Infect. Immun. 75, 4050–4061.

Celli, J., de Chastellier, C., Franchini, D.M., Pizarro-Cerdá, J., Moreno, E., and Gorvel, J.-P. (2003). *Brucella* evades macrophage killing via VirB-dependent sustained interactions with the endoplasmic reticulum. J. Exp. Med. *198*, 545–556.

Cellier, M.F., Courville, P., and Campion, C. (2007). Nramp1 phagocyte intracellular metal withdrawal defense. Microbes Infect. *9*, 1662–1670.

Chao, T.-C., Buhrmester, J., Hansmeier, N., Pühler, A., and Weidner, S. (2005). Role of the regulatory gene *rirA* in the transcriptional response of *Sinorhizobium meliloti* to iron limitation. Appl. Environ. Microbiol. *71*, 5969–5982.

Chiancone, E., and Ceci, P. (2010). The multifaceted capacity of Dps proteins to combat bacterial stress conditions: detoxification of iron and hydrogen peroxide and DNA binding. Biochim. Biophys. Acta 1800, 798–805.

Chu, B.C., Garcia-Herrero, A., Johanson, T.H., Krewulak, K.D., Lau, C.K., Peacock, R.S., Slavinskaya, Z., and Vogel, H.J. (2010). Siderophore uptake in bacteria and the battle for iron with the host; a bird's eye view. Biometals 23, 601–611.

Colpas, G.J., Brayman, T.G., Ming, L.J., and Hausinger, R.P. (1999). Identification of metal-binding residues in the *Klebsiella aerogenes* urease nickel metallochaperone, UreE. Biochemistry 38, 4078–4088.

Crichton, R.R. (2009). Iron Metabolism – From Molecular Mechanisms to Clinical Consequences, 3rd ed. (John Wiley and Sons: West Sussex, UK).

Cuív, P., Keogh, D., Clarke, P., and O'Connell, M. (2008). The *hmuUV* genes of *Sinorhizobium meliloti* 2011 encode the permease and ATPase components of an ABC transport system for the utilization of both haem and the hydroxamate siderophores, ferrichrome and ferrioxamine B. Mol. Microbiol. *70*, 1261–1273.

Danese, I. (2001). Contribution à l'étude de l'assimilation du fer chez *Brucella melitensis* 16 M. Doctoral dissertation, Facultes Universitaires Notre-Dame de la Paix, Namur.

Danese, I., Haine, V., Delrue, R.-M., Tibor, A., Lestrate, P., Stevaux, O., Mertens, P., Paquet, J.-Y., Godfroid, J., De Bolle, X., *et al.* (2004). The Ton system, an ABC transporter, and a universally conserved GTPase are involved in iron utilization by *Brucella melitensis* 16M. Infect. Immun. 72, 5783–5790.

Delpino, M. V., Cassataro, J., Fossati, C.A., Goldbaum, F.A., and Baldi, P.C. (2006). *Brucella* outer membrane protein Omp31 is a haemin-binding protein. Microbes Infect. *8*, 1203–1208.

Denoel, P. A., Crawford, R.M., Zygmunt, M.S., Tibor, A., Weynants, V.E., Godfroid, F., Hoover, D. L., and Letesson, J.-J. (1997). Survival of a bacterioferritin mutant of *Brucella melitensis* 16M in human monocyte-derived macrophages. Infect. Immun. *65*, 4337–4340.

De Pina, K., Desjardin, V., Mandrand-Berthelot, M.-A., Giordano, G., and Wu, L.-F. (1999). Isolation and characterization of the *nikR* gene encoding a nickel-responsive regulator in *Escherichia coli*. J. Bacteriol. *181*, 670–674.

Díaz-Mireles, E., Wexler, M., Sawers, G., Bellini, D., Todd, J.D., and Johnston, A.W.B. (2004). The Fur-like protein Mur of *Rhizobium leguminosarum* is a Mn^{2+}-responsive transcriptional regulator. Microbiology *150*, 1447–1456.

Dozot, M., Boigegrain, R.-A., Delrue, R.-M., Hallez, R., Ouahrani-Bettache, S., Danese, I., Letesson, J.-J., De Bolle, X., and Köhler, S. (2006). The stringent response regulator Rsh is required for *Brucella melitensis* and *Brucella suis* virulence, and for expression of the type IV secretion system *virB*. Mol. Microbiol. *8*, 1791–1802.

Edmonds, M.D., Cloeckaert, A., Booth, N.J., Fulton, W.T., Hagius, S.D., Walker, J.V., and Elzer, P.H. (2001). Attenuation of a *Brucella abortus* mutant lacking a major 25 kDa outer membrane protein in cattle. Am. J. Vet. Res. *62*, 1461–1466.

Edmonds, M.D., Cloeckaert, A., and Elzer, P.H. (2002a). *Brucella* species lacking the major outer membrane protein Omp25 are attenuated in mice and protect against *Brucella melitensis* and *Brucella ovis*. Vet. Microbiol. *88*, 205–221.

Edmonds, M.D., Cloeckaert, A., Hagius, S.D., Samartino, L.E., Fulton, W. T., Walker, J.V., Enright, R.M., Booth, N.J., and Elzer, P.H. (2002b). Pathogenicity and protective activity in pregnant goats of a *Brucella melitensis* Δ*omp25* deletion mutant. Res. Vet. Sci. 72, 235–239.

Eskra, L., Canavessi, A., Carey, M., and Splitter, G. (2001). *Brucella* abortus genes identified following constitutive growth and macrophage infection. Infect. Immun. *69*, 7736–7742.

Evenson, M.A., and Gerhardt, P. (1955). Nutrition of brucellae: utilization of iron, magnesium and manganese for growth. Proc. Soc. Exp. Biol. Med. *89*, 678–680.

Francis, M.S., and Thomas, C.J. (1997). Mutants in the CtpA copper transporting P-type ATPase reduce virulence of *Listeria monocytogenes*. Microb. Pathogen. *22*, 67–78.

Gerhardt, P. (1958). The nutrition of brucellae. Microbiol. Rev. *22*, 81–98.

Gibson, M.M., Bagga, D.A., Miller, C. G., and Maguire, M. E. (1991). Magnesium transport in *Salmonella typhimurium*: the influence of new mutations conferring Co^{2+} resistance on the CorA Mg2+ transport system. Mol. Microbiol. *5*, 2753–2762.

González-Carreró, M.I., Sangari, F.J., Agüero, J., and García-Lobo, J.M. (2002). *Brucella abortus* strain 2308 produces brucebactin, a highly efficient catecholic siderophore. Microbiology *148*, 353–360.

González-Carreró, M.I., Sangari, F.J., and García-Lobo, J.M. (2003). Identification and characterization of the brucebactin transport system. Proceedings of the Brucellosis 2003 Intern. Res. Conf., Abstr. no. 112, *p.* 153.

Griffiths, E. (1999). Iron in biological systems. In: Iron and Infection. Molecular, Physiological and Clinical A, Bullen, J.J., and Griffiths, E., eds. (John Wiley and Sons: New York, NY), pp. 1–16.

Griggs, D.W., Tharp, B.A., and Konisky, J. (1987). Cloning and promoter identification of the iron-regulated *cir* gene of *Escherichia coli*. J. Bacteriol. *169*, 5343–5352.

Günzel, D., Kucharski, L.M., Kehres, D.G., Romero, M.F., and Maguire, M.E. (2006). The MgtC virulence factor of *Salmonella enterica* serovar Typhimurium activates Na^+, K^+-ATPase. J. Bacteriol. *188*, 5586–5594.

Hamza, I., Chauhan, S., Hassett, R., and O'Brian, M.R. (1998). The bacterial Irr protein is required for coordination of heme biosynthesis with iron availability. J. Biol. Chem. *273*, 21669–21674.

Hantke, K. (1990). Dihydroxybenzoylserine – a siderophore for *E. coli*. FEMS Microbiol. Lett. *55*, 5–8.

Jiang, X.L., Leonard, B., Benson, R., and Baldwin, C.L. (1993). Macrophage control of *Brucella abortus*: role of reactive oxygen intermediates and nitric oxide. Cell. Immunol. *151*, 309–319.

Jubier-Maurin, V., Rodrigue, A., Ouahrani-Bettache, S., Layssac, M., Mandrand-Berthelot, M.-A., Köhler, S., and Liautard, J.-P. (2001). Identification of the *nik* gene cluster of *Brucella suis*: regulation and contribution to urease activity. J. Bacteriol. *183*, 426–434.

Kadner, R.J., and Liggins, G. L. (1973). Transport of vitamin B_{12} in *Escherichia coli*: genetic studies. J. Bacteriol. *115*, 514–521.

Kehl-Fie, T.E., and Skaar, E.P. (2010). Nutritional immunity beyond iron: a role for manganese and zinc. Curr. Opin. Chem. Biol. *14*, 218–224.

Kehres, D.G., Zaharik, M.L., Finlay, B.B., and Maguire, M.E. (2000). The NRAMP proteins of *Salmonella typhimurium* and *Escherichia coli* are selective manganese transporters involved in the response to reactive oxygen. Mol. Microbiol. *36*, 1085–1100.

Kim, S., Watarai, M., Kondo, Y., Erdenebaatar, J., Makino, S., and Shirahata, T. (2003). Isolation and characterization of mini-Tn5km2 insertion mutants of *Brucella abortus* deficient in internalization and intracellular growth in HeLa cells. Infect. Immun. *71*, 3020–3027.

Kim, S., Watanabe, K., Shirahata, T., and Watarai, M. (2004). Zinc uptake system (*znuA* locus) of *Brucella abortus* is essential for intracellular survival and virulence in mice. J. Vet. Med. Sci. *66*, 1059–1063.

Köster, W.L., Actis, L.A., Waldbeser, L.S., Tomalsky, M.E., and Crosa, J.H. (1991). Molecular characterization of the iron transport system mediated by the pJM1 plasmid in *Vibrio anguillarum*. J. Biol. Chem. *266*, 23829–23833.

Lamontagne, J., Forest, A., Marazzo, E., Denis, F., Butler, H., Michaud, J.-F., Boucher, L., Pedro, I., Villeneuve, A., Sitnikov, D., *et al.* (2009). Intracellular adaptation of *Brucella abortus*. J. Proteome Res. *8*, 1594–1609.

Lavigne, J.-P., O'Callaghan, D., and Blanc-Potard, A.-B. (2005). Requirement of MgtC for *Brucella suis* intramacrophage growth: a potential mechanism shared by *Salmonella enterica* and *Mycobacterium tuberculosis* for adaptation to a low-Mg^{2+} environment. Infect. Immun. *73*, 3160–3163.

Lee, J.W., and Helmann, J.D. (2007). Functional specialization within the Fur family of metalloregulators. Biometals 20, 485–499.

Lengeler, J. W., Drews, G., and Schlegel, H. G. (1999). Biology of the Prokaryotes. (Georg Thieme Verlag: Stuttgart, Germany).

Lestrate, P., Delrue, R.-M., Danese, I., Didembourg, C., Taminau, B., Mertens, P., De Bolle, X., Tibor, A., Tang, C.M., and Letesson, J.-J. (2000). Identification and characterization of *in vivo* attenuated mutants of *Brucella melitensis*. Mol. Microbiol. *38*, 543–551.

LeVier, K., Phillips, R.W., Grippe, V.K., Roop, R.M. II, and Walker, G.W. (2000). Similar requirements of a plant symbiont and a mammalian pathogen for prolonged intracellular survival. Science *287*, 2492–2493.

Li, Y., and Zamble, D.R. (2009). Nickel homeostasis and nickel regulation: an overview. Chem. Rev. *109*, 4617–4643.

López-Goñi, I., and Moriyón, I. (1995). Production of 2,3-dihydroxybenzoic acid by *Brucella* species. Curr. Microbiol. *31*, 291–293.

López-Goñi, I., Moriyón, I., and Neilands, J.B. (1992). Identification of 2,3-dihydroxybenzoic acid as a *Brucella abortus* siderophore. Infect. Immun. *60*, 4496–4503.

Lowry, J.E., Goodridge, L., Vernati, G., Fluegel, A.M., Edwards, W. H., and Andrews, G. P. (2010). Identification of *Brucella abortus* genes in elk (Cervus elaphus) using *in vivo*-induced antigen technology (IVIAT) reveals novel markers of infection. Vet. Microbiol. *142*, 367–372.

McCullough, N.B., and Beal, G.A. (1951). Growth and manometric studies on carbohydrate utilization of *Brucella*. J. Infect. Dis. *89*, 266–271.

McCullough, W.G., Mills, R.C., Herbst, E.J., Roessler, W.G., and Brewer, C.R. (1947). Studies on the nutritional requirements of *Brucella suis*. J. Bacteriol. *53*, 5–15.

Macomber, L., and Imlay, J.A. (2009). The iron–sulfur clusters of dehydratases are primary intracellular targets of copper toxicity. Proc. Natl. Acad. Sci. U.S.A. *106*, 8344–8349.

Magnusson, L. U., Farewell, A., and Nyström, T. (2005). ppGpp: a global regulator in *Escherichia coli*. Trends Microbiol. *13*, 236–242.

Martín-Martín, A.I., Caro-Hernández, P., Orduña, A., Vizcaíno, N., and Fernández-Lago, L. (2008). Importance of the Omp25/Omp31 family in the internalization and intracellular replication of virulent *Brucella ovis* in murine macrophages and HeLa cells. Microbes Infect. *10*, 706–710.

Martín-Martín, A.I., Caro-Hernández, P., Sancho, P., Tejedor, C., Cloeckaert, A., Fernández-Lago, L., and Vizcaíno, N. (2009). Analysis of the occurrence and distribution of the Omp25/Omp31 family of surface proteins in the six classical *Brucella* species. Vet. Microbiol. *137*, 74–82.

Martínez, M., Ugalde, R.A., and Almirón, M. (2005). Dimeric *Brucella abortus* Irr protein controls its own expression and binds haem. Microbiology *151*, 3427–3433.

Martínez, M., Ugalde, R.A., and Almirón, M. (2006). Irr regulates brucebactin and 2,3-dihydroxybenzoic acid biosynthesis, and is implicated in the oxidative stress resistance and intracellular survival of *Brucella abortus*. Microbiology *152*, 2591–2598.

Merino, S., Gavin, R., Altaribba, M., Izquierdo, L., Maguire, M.E., and Tomás, J. M. (2001). The MgtE Mg^{2+} transport protein is involved in *Aeromonas hydrophila* adherence. FEMS Microbiol. Lett. *198*, 189–195.

Mills, M., and Payne, S.M. (1997). Identification of *shuA*, the gene encoding the heme receptor of *Shigella dysenteriae*, and analysis of invasion and intracellular multiplication of a *shuA* mutant. Infect. Immun. *65*, 5358–5363.

Minnick, M.F., and Battisti, J. M. (2009). Pestilence, persistence and pathogenicity: infectious strategies of *Bartonella*. Future Microbiol. 4, 743–758.

Minnick, M.F., Sappington, K.N., Smitherman, L.S., Andersson, S.G.E., Karlberg, O., and Carroll, J. A. (2003). Five-member gene family in *Bartonella quintana*. Infect. Immun. *71*, 814–821.

Moomaw, A.S., and Maguire, M.E. (2008). The unique nature of Mg^{2+} channels. Physiology *23*, 275–285.

Moreno, E., and Moriyón, I. (2002). *Brucella melitensis*: a nasty bug with hidden credentials for virulence. Proc. Natl. Acad. Sci. U.S.A. *99*, 1–3.

Nahvi, A., Barrick, J.E., and Breaker, R.R. (2004). Coenzyme B_{12} riboswitches are widespread genetic control elements in prokaryotes. Nucleic Acids Res. *32*, 143–150.

Nemeth, E., Tuttle, M.S., Powelson, J., Vaughn, M.B., Donovan, A., Ward, D. M., Ganz, T., and Kaplan, J. (2004). Hepcidin regulates cellular iron efflux by binding to ferroportin and inducing its internalization. Science *306*, 2090–2093.

Ngok-Ngam, P., Ruangkiattikul, N., Mahavihakanont, A., Virgem, S.S., Sukchawalit, R., and Mongkolsuk, S. (2009). Roles of *Agrobacterium tumefaciens* RirA in iron regulation, oxidative stress response, and virulence. J. Bacteriol. *191*, 2083–2090.

Noinaj, N., Guillier, M., Barnard, T.J., and Buchanan, S.K. (2010). TonB-dependent transporters: regulation, structure and function. Annu. Rev. Microbiol. *64*, 43–60.

Outten, F.W., Outten, C.E., Hale, J., and O'Halloran, T.V. (2000). Transcriptional activation of an *Escherichia coli* copper efflux regulon by the chromosomal MerR homologue, CueR. J. Biol. Chem. *275*, 31–24–31029.

Ozenberger, B.A., Nahlik, M.S., and McIntosh, M.A. (1987). Genetic organization of multiple *fep* genes encoding ferric enterobactin transport functions in *Escherichia coli*. J. Bacteriol. *169*, 3638–3646.

Papp-Wallace, K.M., and Maguire, M.E. (2006). Manganese transport and the role of manganese in virulence. Annu. Rev. Microbiol. *60*, 187–209.

Parent, M.A., Bellaire, B.H., Murphy, E.A., Roop, R. M. II, Elzer, P.H., and Baldwin, C. L. (2002). *Brucella abortus* siderophore 2,3-dihydroxybenzoic acid (2,3-DHBA) facilitates intracellular survival of the bacteria. Microb. Pathogen. *32*, 239–248.

Parrow, N.L., Abbott, J., Lockwood, A.R., Battisti, J. M., and Minnick, M. F. (2009). Function, regulation, and transcriptional organization of the hemin utilization locus of *Bartonella quintana*. Infect. Immun. *77*, 307–316.

Patzer, S.I., and Hantke, K. (1998). The ZnuABC high-affinity zinc uptake system and its regulator Zur in *Escherichia coli*. Mol. Microbiol. *28*, 1199–1210.

Paulley, J.T. (2007b). Production of BhuA by *Brucella abortus* is required for hemin utilization and virulence and is dependent on the transcriptional regulators *RirA* and *ChrA*. Doctoral dissertation, East Carolina University.

Paulley, J.T., Anderson, E.S., and Roop, R.M. II. (2007a). *Brucella abortus* requires the heme transporter BhuA for maintenance of chronic infection in BALB/c mice. Infect. Immun. *75*, 5248–5254.

Pfeiffer, J., Guhl, J., Waldner, B., Kist, M., and Bereswill, S. (2002). Magnesium uptake by CorA is essential for viability of the gastric pathogen *Helicobacter pylori*. Infect. Immun. *70*, 3930–3934.

Phillips, R.W., Seastone, D.J., Norton, D.D., Farris, M.A., Kovach, M.E., Elzer, P.H., Baldwin, C.L., and Roop, R.M. II. (1996). Molecular cloning of the *Brucella abortus* iron uptake regulation (*fur*) gene. Abstr. 96th Annu. Meet. Amer. Soc. Microbiol., Abstr. D-24, *p*. 245.

Postle, K., and Larsen, R.A. (2007). TonB-dependent energy transduction between outer and cytoplasmic membranes. Biometals *20*, 453–465.

Puri, S., and O'Brian, M.R. (2006). The *hmuQ* and *hmuD* genes from *Bradyrhizobium japonicum* encode heme-degrading enzymes. J. Bacteriol. *188*, 6476–6482.

Puri, S., Hohle, T.H., and O'Brian, M.R. (2010). Control of bacterial iron homeostasis by manganese. Proc. Natl. Acad. Sci. U.S.A. *107*, 10691–10695.

Qi, Z., and O'Brian, M.R. (2002). Interaction between the bacterial iron response regulator and ferrochelatase mediates genetic control of heme biosynthesis. Mol. Cell. *9*, 155–162.

Qi, Z., Hamza, I., and O'Brian, M.R. (1999). Heme is an effector molecule for iron-dependent degradation of the bacterial iron response regulator (Irr) protein. Proc. Natl. Acad. Sci. U.S.A. *96*, 13056–13061.

Rajasekaran, M.B., Nilapwar, S., Andrews, S.C., and Watson, K.A. (2010). EfeO-cupredoxins: major new members of the cupredoxin superfamily with roles in bacterial iron transport. Biometals *23*, 1–17.

Rensing, C., Mitra, B., and Rosen, B.P. (1997). The *zntA* gene of *Escherichia coli* encodes a Zn(II)-translocating P-type ATPase. Proc. Natl. Acad. U.S.A. *94*, 14326–14331.

Rensing, C., Fan, B., Sharma, R., Mitra, B., and Rosen, B.P. (2000). CopA: an *Escherichia coli* Cu(I)-translocating P-type ATPase. Proc. Natl. Acad. Sci. U.S.A. *97*, 652–656.

Rodionov, D.A., Vitreschak, A.G., Mironov, A.A., and Gelfand, M.S. (2003). Comparative genomics of the vitamin B_{12} metabolism and regulation in prokaryotes. J. Biol. Chem. *278*, 41148–41159.

Rodionov, D.A., Gelfand, M.S., Todd, J.D., Curson, A.R.J., and Johnston, A.W.B. (2006a). Computational reconstruction of iron- and manganese-responsive transcriptional networks in α-proteobacteria. PLoS Comput. Biol. *2*, e163.

Rodionov, D.A., Hebbeln, P., Gelfand, M.S., and Eitinger, T. (2006b). Comparative and functional genomic analysis of prokaryotic nickel and cobalt uptake transporters: evidence for a novel group of ATP-binding cassette transporters. J. Bacteriol. *188*, 317–327.

Rodrique, A., Effantin, G., and Mandrand-Berthelot, M.-A. (2005). Identification of *rcnA* (*yohM*), a nickel and cobalt resistance gene in *Escherichia coli*. J. Bacteriol. *187*, 2912–2916.

Roop, R.M. II, Gaines, J.M., Anderson, E.S., Caswell, C.C., and Martin, D.W. (2009). Survival of the fittest: how *Brucella* strains adapt to their intracellular niche. Med. Microbiol. Immunol. *198*, 221–238.

Rosenzweig, A.C., and O'Halloran, T.V. (2000). Structure and chemistry of the copper chaperone proteins. Curr. Opin. Chem. Biol. *4*, 140–147.

Rudolph, G., Semini, G., Hauser, F., Lindenmann, A., Friberg, M., Hennecke, H., and Fischer, H.-M. (2006). The iron control element, acting in positive and negative control of the iron-regulated *Bradyrhizobium japonicum* genes, is a target for the Irr protein. J. Bacteriol. *188*, 733–744.

Samartino, L.E., and Enright, F.M. (1993). Pathogenesis of abortion of bovine brucellosis. Comp. Immunol. Microbiol. Infect. Dis. *16*, 95–101.

Sander, A., Kretzer, S., Bredt, W., Oberle, K., and Bereswill, S. (2000). Hemin-dependent growth and hemin binding of *Bartonella henselae*. FEMS Microbiol. Lett. *189*, 55–59.

Sanders, T.H., Higuchi, K., and Brewer, C.R. (1953). Studies on the nutrition of *Brucella melitensis*. J. Bacteriol. *66*, 294–299.

Sangari, F.J., Seoane, A., Rodríguez, M.C., Agüero, J., and García-Lobo, J. M. (2007). Characterization of the urease operon of *Brucella abortus* and assessment of its role in virulence of the bacterium. Infect. Immun. *75*, 774–780.

Sangari, F.J., Cayón, A.M., Seoane, A., and García-Lobo, J.M. (2010). *Brucella abortus ure2* region contains an acid-activated urea transporter and a nickel transport system. BMC Microbiol. *10*, 107.

Schmitt, M.P. (1997). Utilization of host iron sources by *Corynebacterium diphtheriae*: identification of a gene whose product is homologous to eukaryotic heme oxygenases and is required for acquisition of iron from heme and hemoglobin. J. Bacteriol. *179*, 838–845.

Schroder, H.J. (1995). Comparative aspects of placental exchange functions. Eur. J. Obstet. Gynecol. Reprod. Biol. *63*, 81–90.

Schwan, W.R., Warrener, P., Keunz, E., Stover, C.K., and Folger, K.R. (2005). Mutations in the *cueA* gene encoding the copper homeostasis P-type ATPase reduce the pathogenicity of *Pseudomonas aeruginosa* in mice. Int. J. Med. Microbiol. *295*, 237–242.

Shea, C.M., and McIntosh, M.A. (1991). Nucleotide sequence and genetic organization of the ferric enterobactin transport system: homology to other periplasmic binding protein-dependent systems in *Escherichia coli*. Mol. Microbiol. *5*, 1647–1654.

Silver, S., and Phung, L.T. (2005). A bacterial view of the periodic table: genes and proteins for toxic inorganic ions. J. Ind. Microbiol. Biotechnol. *32*, 587–605.

Singh, S.K., Grass, G., Rensing, C., and Montfort, W.R. (2004). Cuprous oxidase activity of CueO from *Escherichia coli*. J. Bacteriol. *186*, 7815–7817.

Singleton, C., White, G.F., Todd, J.D., Marritt, S.J., Cheesman, M.R., Johnston, A.W.B., and Le Brun, N.E. (2010). Heme-responsive DNA binding by the global iron regulator Irr from *Bradyrhizobium leguminosarum*. J. Biol. Chem. *285*, 16023–16031.

Small, S.K., Puri, S., and O'Brian, M.R. (2009). Heme-dependent metalloregulation by the iron response regulator (Irr) protein in *Rhizobium* and other alpha-proteobacteria. Biometals *22*, 89–97.

Smith, H., Williams, A.E., Pearce, J.H., Keppie, J., Harris-Smith, P.W., Fitz-George, R.B., and Witt, K. (1962). Foetal erythritol: a cause of the localization of *Brucella abortus* in bovine contagious abortion. Nature *193*, 47–49.

Sperry, J.F., and Robertson, D.C. (1975). Erythritol catabolism by *Brucella abortus*. J. Bacteriol. *121*, 619–630.

Sriranganathan, N., Boyle, S.M., Schurig, G., and Misra, H. (1991). Superoxide dismutases of virulent and avirulent strains of *Brucella abortus*. Vet. Microbiol. *26*, 359–366.

Stähler, F.N., Odenbreit, S., Haas, R., Wilrich, J., Van Vliet, A.H.M., Kusters, J.G., Kist, M., and Bereswill, S. (2006). The novel *Helicobacter pylori* CznABC metal efflux pump is required for cadmium, zinc and nickel resistance, modulation and gastric colonization. Infect. Immun. *74*, 3845–3852.

Stephens, D.L., Choe, M.D., and Earhart, C.F. (1995). *Escherichia coli* periplasmic protein FepB binds ferrienterobactin. Microbiology *141*, 1647–1654.

Summers, A.O. (2009). Damage control: defenses against toxic metals and metalloids. Curr. Opin. Microbiol. *12*, 138–144.

Taga, M.E., and Walker, G.C. (2010). *Sinorhizobium meliloti* requires a cobalamin-dependent ribonucleotide reductase for symbiosis with its plant host. Mol. Plant. Microbe Inter. *23*, 1643–1654.

Taketani, S. (2005). Acquisition, mobilization and utilization of cellular iron and heme; endless findings and growing evidence of tight regulation. Tohoku J. Exp. Med. *205*, 297–318.

Thompson, J.M., Jones, H.A., and Perry, R.D. (1999). Molecular characterization of the hemin uptake locus (*hmu*) from *Yersinia pestis* and analysis of *hmu* mutants for hemin and hemoprotein utilization. Infect. Immun. *67*, 3879–3892.

Todd, J.D., Wexler, M., Sawers, G., Yeoman, K.H., Poole, P.S. and Johnston, A.W.B. (2002). RirA, an iron-responsive regulator in the symbiotic bacterium *Rhizobium leguminosarum*. Microbiology *148*, 4059–4071.

Tucker, N. P., Le Brun, N.E., Dixon, R., and Hutchings, M.I. (2009). There's NO stopping NsrR, a global regulator of the bacterial NO stress response. Trends Microbiol. *18*, 149–156.

Vizcaíno, N., Caro-Hernández, P., Cloeckaert, A., and Fernández-Lago, L. (2004). DNA polymorphism in the *omp25*/*omp31* family of *Brucella* spp.: identification of a 1.7-kb inversion in *Brucella cetaceae* and of a 15.1-kb genomic island, absent from *Brucella ovis*, related to the synthesis of smooth lipopolysaccharide. Microbes Infect. *6*, 821–834.

Waldron, K.J., and Robinson, N.J. (2009). How do bacterial cells ensure that metalloproteins get the correct metal? Nat. Rev. Microbiol. *6*, 25–35.

Wandersman, C. (2010). Haem uptake and iron extraction by bacteria. In: Iron Uptake and Homeostasis in Microorganisms, Cornelis, P., and Andrews, S.C., eds. (Caister Academic Press: Norfolk, UK), pp. 17–35

Ward, S.K., Abomoelak, B., Hoye, E.A., Steinberg, H., and Talaat, A.M. (2010). CtpV: a putative copper exporter required for full virulence of *Mycobacterium tuberculosis*. Mol. Microbiol. *77*, 1096–1110.

Waring, W.S., Elberg, S.S., Schneider, P., and Green, W. (1953). The role of iron in the biology of *Brucella suis*. I. Growth and nutrition. J. Bacteriol. *66*, 82–91.

Weiss, G. (2005). Modification of iron regulation by the inflammatory response. Best Prac. Res. Clin. Haematol. *18*, 183–201.

Williams, A.E., Keppie, J., and Smith, H. (1964). The relation of erythritol usage to virulence in the *Brucellas*. J. Gen. Microbiol. *37*, 285–292.

Wyckoff, E.E., Duncan, D., Torres, A.G., Mills, M., and Payne, S.M. (1998). Structure of the *Shigella dysenteriae* haem transport locus and its phylogenetic distribution in enteric bacteria. Mol. Microbiol. *28*, 1139–1152.

Yang, X., Becker, T., Walters, N., and Pascual, D.W. (2006). Deletion of *znuA* virulence factor attenuates *Brucella abortus* and confers protection against wild-type challenge. Infect. Immun. *74*, 3874–3879.

Yeoman, K.H., Curson, A.R.J., Todd, J.D., Sawers, G., and Johnston, A.W.B. (2004). Evidence that the *Rhizobium* regulatory protein RirA binds to *cis*-acting iron-responsive operators (IROs) at promoters of some Fe-regulated genes. Microbiology *150*, 4065–4074.

Zhang, X.X., and Rainey, P.B. (2007). The role of a P1-type ATPase from *Pseudomonas fluorescens* SBW25 in copper homeostasis and plant colonization. Mol. Plant Microbe Interact. *20*, 581–588.

ZoBell, C.E., and Meyer, K.F. (1932). Metabolism studies on the *Brucella* group. VIII. Nutrient requirements in synthetic mediums. J. Infect. Dis. *51*, 344–360.

The Two-component System BvrR/BvrS: a Master Regulator of *Brucella* Virulence

10

Ignacio López-Goñi

Abstract

Recent results from proteomic and transcriptomic analyses show that the two component system BvrR/BvrS of *Brucella* is an global regulator capable of interacting with other regulators, by controlling the synthesis of components of the cell envelope and outer membrane, metabolism of carbon and nitrogen, and of other genes related to the virulence of the bacteria. In this way, the bacteria may sense when they enter a eukaryotic host cell and thus control the expression of a set of genes which allows them to develop during this intracellular phase of their life cycle.

Bacterial two-component regulation systems

Bacteria must modulate the expression of their genes in response to changes in environmental conditions. Signal transduction systems allow bacteria to sense their environment to transform stimuli through their membrane and to achieve the appropriate response. These systems are based on reversible protein phosphorylation, the most common of which are the so-called two-component systems (TCS, Fig. 10.1). One of the components of these systems acts as an environmental sensor. It is frequently located in the cytoplasmic membrane and is capable of transmitting a signal to a second component, a cytoplasmic regulator, which controls changes in gene expression (Hoch *et al.*, 1995). The sensor protein is usually a histidine kinase which reacts to a specific environmental stimulus by autophosphorylation of a conserved histidine residue through an ATP-dependent process. Once the sensor is activated, it transfers its phosphate group to a conserved aspartate residue of the regulator, which generates a cascade of events leading to the activation of factors which interact with specific DNA sequences, thus modulating the expression of certain genes. In this way, bacterial TCS control the response to a great variety of chemical or physical signals that regulate functions such as cell division, communication, development, sporulation, taxis and response to stress or virulence (Hoch *et al.*, 1995). Furthermore, 'cross reactions' between different components of different TCS, between the sensor of one system and the regulator of another, have been observed which allow coordinated responses to be mounted to different stimuli (Hoch *et al.*, 1995).

Two-component systems in *Brucella*

Brucella is an intracellular facultative pathogen which exists in many different environments, both in the exterior and in intracellular medium. The expression of its genes must therefore adapt to the different media in which it is found. The comparative analysis of the complete sequences of the genome of *Brucella* has shown the existence of 22 histidine kinases sensors and 24 regulators which could form part of possible TCS, spread over both chromosomes: 18 histidine kinases and 15 regulators on chromosome I and four histidine kinases and nine regulators on chromosome II (Lavin *et al.*, 2010). However, those histidine kinases and regulators which are grouped on

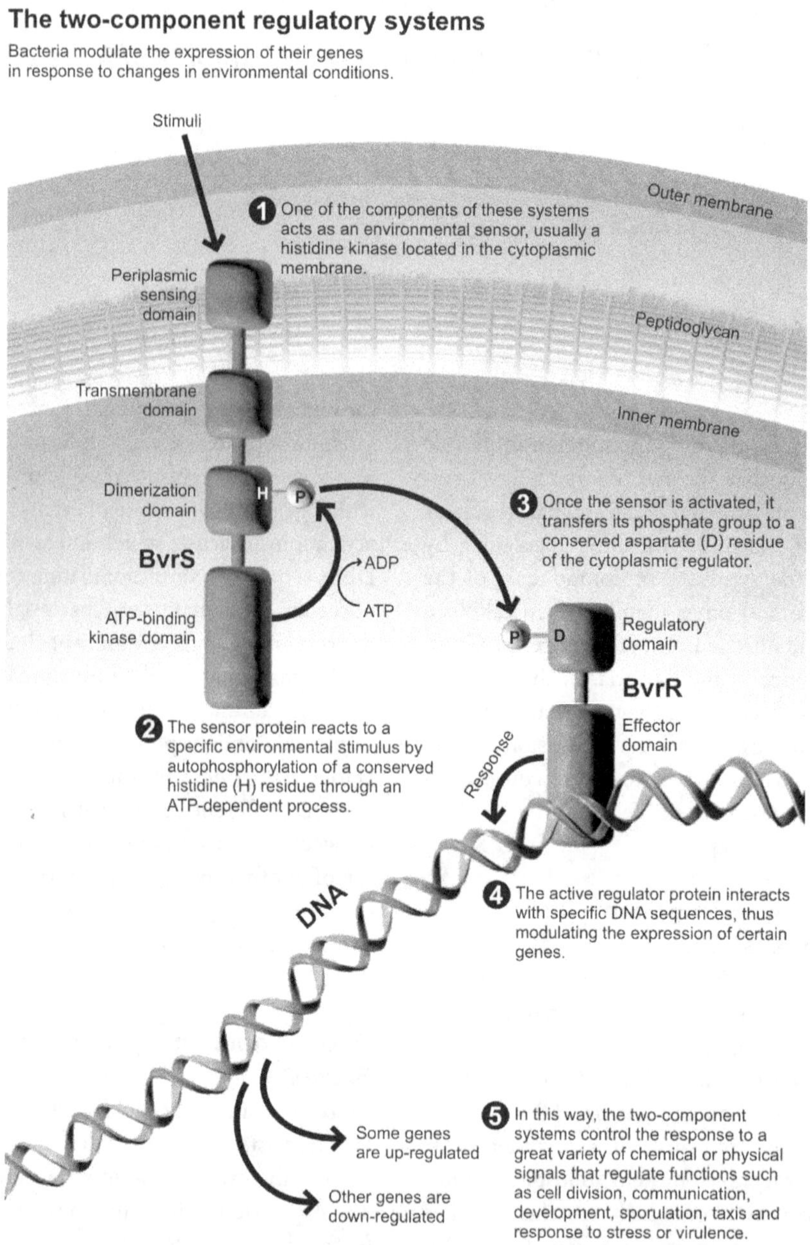

Figure 10.1 The bacterial two-component regulatory systems.

adjacent genes or which are involved in an identified phosphorylation system are considered to be part of a functional TCS. Thus, in *Brucella*, only 15 pairs of histidine kinases and regulators seem to form part of TCS (Lavin *et al.*, 2010). These TCS are very well conserved in the different species of the *Brucella* genus (Lavin *et al.*, 2010). One of the first TCS described in *Brucella* is the FeuP/FeuQ (Dorrell *et al.*, 1998), a homologue to the system involved in the control of iron assimilation in *Rhizobium leguminosarum*. The FeuP mutant showed no abnormalities in growth in iron-free media nor multiplication defects in either macrophages or mice (Dorrell *et al.*, 1998). In contrast, the FeuQ mutant in *B. melitensis* did present attenuation in both mice and cells (Lestrate *et al.*, 2003). Another TCS identified in *B. suis* is the NtrC/NtrB system (Dorrell *et al.*, 1999), which plays a part in nitrogen metabolism. In *Klebsiella pneumoniae* this system regulates the urease gene. The mutant in the *ntrC* gene in *B. suis* grows in the same fashion as the parental strain but presents a reduction in metabolism in certain amino acid sources and no attenuation is apparent in macrophages but they were less virulent in mice (Dorrell *et al.*, 1999). Letesson *et al.* have proposed that this NtrC sensor could respond to intracellular signals (Letesson *et al.*, 2004). Another TCS described in *Brucella* is the NtrY/NtrX system which could be involved in glutamine synthesis, a process which is vital for the intracellular survival of *Brucella*. Thus, the intracellular survival in human macrophages of a mutant in the *ntrY* gene in *B suis* was diminished (Foulongne *et al.*, 2000).

The BvrR/BvrS system in *Brucella*

One of the best characterized TCS is the BvrR/BvrS system, which was identified for the first time in *B. abortus* (Bvr: *Brucella* virulence related). Unlike other Gram-negative bacteria, *Brucella* is exceptionally resistant to bactericidal cationic peptides, whose action is to destabilize the cell envelope (see Chapter 7 for more details). The BvrR/BvrS system was discovered when a group of *B. abortus* mutants sensitive to this type of substances was being analysed (Sola-Landa *et al.*, 1998). The BvrR/BvrS system is well conserved in the different species of the *Brucella* genus (Lavin *et al.*, 2010). With the aid of PCR techniques using different specific primers, it was demonstrated that the *bvrR* and *bvrS* genes were present in all the biovars of all the species of *Brucella*, including *B. ceti* and *B. pinnipedialis* (López-Goñi *et al.*, 2002). This has been also confirmed by genome sequencing projects.

Molecular analysis (Sola-Landa *et al.*, 1998; López-Goñi *et al.*, 2002) has shown that the BvrS sensor is a protein of 601 amino acids associated with the internal membrane, with two transmembrane domains at its N-terminal end between which is found a periplasmic region involved in the detection of environmental signals. Its C-terminal end has a dimerization and self-phosphorylation domain and another domain with ATPase activity, with highly conserved sequences which are typical of this type of sensor proteins. Furthermore, the BvrR protein, with 237 amino acids, is a regulator of the OmpR family which has a regulating domain conserved at its N-terminal end and another effector domain at the C-terminal end. In the regulating domain are found aspartate and lysine residues which are very well conserved in this type of protein which interact with the phosphate groups. The effector domain has a DNA binding motif.

The BvrR/BvrS system is related to the virulence of *Brucella*

The mutants in this system are highly attenuated. These *B. abortus* mutants are probably the most attenuated of those described to date in the literature. These mutants are sensitive to polycationic peptides (such as polymyxin, melittin and polylysine) and to the bactericidal action of non-immune sera from both human and cattle. While the virulent parental strain *B. abortus* 2308 replicates efficiently in mouse models persisting at high levels for over more than 6 weeks, the *bvrR* and *bvrS* mutants are rapidly eliminated from the animal in less than 12 days (Sola-Landa *et al.*, 1998). The mutants are also incapable of replicating in macrophages and epithelial cells (Sola-Landa *et al.*, 1998). Studies using confocal microscopy have shown that these deficiencies are due to defects both in the invasive phase and intracellular replication. Furthermore, unlike the parental strain, the *bvrS* mutant is incapable of recruiting the small GTPases necessary for the polymerization of actin and penetration into cells, which shows that the BvrR/BvrS system is essential for activation of cellular and molecular determinants necessary for invasion (Guzmán-Verri *et al.*, 2001). Furthermore, once inside the cell, these mutants are also incapable of inhibiting fusion with the lysosome and are rapidly destroyed in the phagolysosomes (Sola-Landa *et al.*, 1998). These results suggest that, inside the cell, the BvrR/BvrS system also controls vacuolar maturation and intracellular traffic. In this regard, it has been demonstrated that *bvrR/bvrS* are essential genes for intramacrophagic multiplication of *B. suis* (Köhler *et al.*, 2002)

The severe attenuation of the mutants in the BvrR/BvrS system together with their smooth nature (see below) has led to suggestions for their potential use as live vaccines. Thus, experiments in mice have shown that both *bvrR* and *bvrS* mutants are capable of protecting against infection with the virulent *B. abortus* 2308 strain to levels similar to those obtained with the *B. abortus* S19 strain used as vaccine (Grilló *et al.*, 2006). It has also been suggested that, given the characteristically low endotoxicity of the *Brucella* lipopolysaccharide (LPS), these mutants represent a very attractive option for the development of new vaccines against human brucellosis (López-Goñi *et al.*, 2002).

The BvrR/BvrS system modulates the structure of the cell envelope

The mutants in this system were not only more sensitive to bactericidal polycationic peptides but also to other agents which destabilize the cell envelope such as surfactants and detergents and also exhibited increased levels of hydrophobicity of the bacterial surface (Sola-Landa *et al.*, 1998). As these properties are related to the integrity of the outer membrane, this would suggest that the BrvR/BvrS system must in some way modulate the structure of the cell envelope.

Consequently, a detailed characterization of the cell envelope of the *bvrR* and *bvrS* mutants of *B. abortus* was carried out. Although no differences were detected in the cyclic glucans, the composition of free lipids and the native hapten polysaccharide, the first analyses using two-dimensional electrophoresis clearly revealed differences in outer membrane proteins between the parental strain and the mutants (Guzmán-Verri, *et al.*, 2002). More specifically, it was demonstrated that BvrR/BvrS system controlled the transcription of at least two outer membrane proteins, Omp25 (Omp3a) and Omp22 (Omp3b) (Guzmán-Verri, *et al.*, 2002). Subsequently, a complete proteomic analysis of outer membrane fragments revealed differences in expression for 167 proteins, which included mainly outer membrane proteins (in total, 12 outer membrane assigned proteins), lipoproteins (such as Omp19) and periplasmic proteins related to the transport of substances (Lamontagne *et al.*, 2007). As for the LPS, a principal component of the outer membrane, no abnormalities were found at the level of O chains. As in the parental strain, the *bvrR* and *bvrS* mutants had a smooth LPS. However, the chromatographic pattern of the lipid A and its properties were different, which indicated that the BvrR/BvrS system could in some way control the degree of lipid A acylation of the LPS (Manterola *et al.*, 2005). In addition, the expression of two putative phosphatases involved in lipid A modifications (*lpxE*1and 2) was down-regulated in the *bvrR* and *bvrS* mutants (Conde, 2008).

Full transcriptomic analysis of the BvrR/BvrS system of *B. abortus*

The BvrR/BvrS system is very well conserved within the group of *α-Proteobacteria* both in the gene sequence and the loci in which they are found: homologous systems have been found in *Rhizobium* spp., *Mesorhizobium* spp., *Sinorhizobium* spp., *Bradyrhizobium* spp., *Agrobacterium* spp., *Bartonella* spp. and others (López-Goñi *et al.*, 2004). In the majority of cases, the TCS is flanked by the *pckA* (phosphoenolpyruvate carboxykinase) gene and the *hprK* (HprK/P), *ptsM* ($EIIA^{Man}$) and *ptsO* (NPr) genes of the PTS phosphotransferase system (López-Goñi *et al.*, 2004). Among the systems homologous to BvrR/BvrS which have received most attention are the ExoS/ChvI systems of *Sinorhizobium* (Osteras *et al.*, 1995; Cheng *et al.*, 1998) and ChvG/ChvI de *Agrobacterium* (Charles *et al.*, 1993) involved in the establishment of symbiosis and pathogenicity in plants.

The ChvG/ChvI system of *Agrobacterium* regulates the expression of the *vir* genes, encoding the VirB Type IV secretion system, proteins involved in the processing of the T-DNA and other protein effectors translocated into plant cells to mediate tumorigenesis (Gelvin, 2006). Other virulence genes are also controlled, such as the *aopB* gene which encodes an outer membrane protein involved in virulence (Li *et al.*, 2002).

The ExoS/ChvI system of *Sinorhizobium* controls the expression of the flagellum and the production of succinoglycans, both necessary for the invasion of leguminous plants (Cheng *et al.*, 1998). Recently it has been demonstrated that the BatR/BatS system of the intracellular pathogen *Bartonella henselae*, a homologue of BvrR/BvrS, regulates the expression of several virulence genes, among which is also found the VirB secretion system (Quebatte *et al.*, 2010). Given all of this, it was logical to conclude that the BvrR/BvrS system should control more genes than those related with outer membrane proteins and the structure of the LPS. Furthermore, a similar finding was suggested by the proteomic analyses described above (Lamontagne *et al.*, 2007). With the aim of determining the set of genes regulated by the BvrR/BvrS system, a full transcriptomic analysis was performed using microarray technology comparing the transcription levels of the parental strain *B. abortus* 2308 with the mutant in the *bvrR* regulator (Viadas *et al.*, 2010). A total of 127 were found to be expressed differentially in comparison with the parental strain. Of all of these, 30 genes were annotated as hypothetical proteins with no further information. Of the genes with a known function, 59 were overexpressed and 38 were underexpressed in the *bvrR* mutant. Those genes involved in cellular processes and metabolism were found to be better represented in the list of overexpressed genes. In contrast, those related to membrane transport were found mostly in the list of genes with lower expression in the mutant.

The detailed study of the genes affected revealed that, in general, the BvrR/BvrS system was related to various functions: (i) genes directly involved with the cell envelope and outer membrane, (ii) genes related to the metabolism of carbon and nitrogen, (iii) other regulatory genes, and (iv) genes directly related to bacterial virulence (Fig. 10.2).

Among the genes involved in the cell envelope which were differentially expressed in the *bvrR* mutant were: genes that encoded outer membrane proteins (*omp25a*, *omp25d*), lipoprotein genes, periplasmic chaperone genes (such as *htpX*, *clpA*, and *clpB*), genes for the synthesis of the LPS (glycosyl transferase), genes for the biosynthesis of fatty acids (such as *fabG*, *fabF*, *fadD*, and *cfa*) and genes for the biosynthesis of the flagellum (*fliM*, *flgJ*, *motB*).

Various genes related to metabolism were also expressed differentially. This included twelve genes that encoded ABC-type transport systems (such as *matE*, *dctM*, *fhuD*, *yadH*, *oppA*, *oppC*, *oppD*, *potC*, *ssuB*, *ugpC*), ten genes related to carbohydrate, amino acid and fatty acid metabolism and five related to nitrogen metabolism. More specifically, the first gluconeogenesis enzyme (*pckA*), four genes involved in the Krebs cycle and the metabolism of pyruvate (*fumB*, *lpdA*, pyruvate dehydrogenase, acetyl-CoA acetyltransferase), three genes of amino acid and fatty acid metabolism (aldehyde dehydrogenase, hydroxymethylglutaryl-CoA lyase, among others), and two gene of benzoate degradation were all included. Furthermore, the complete operon of the maltose transport system (*malK*, *malG*, *malF*, *malE*, *iclR*,...) was less expressed in the mutant. Amongst nitrogen metabolism genes, the expression of nitrite, nitric oxide and nitrous oxide reductases (*nirK*, *norC*, *nosZ*) and two deaminases (glutaminase and guanine deaminase) were also found to be differentially expressed in the *bvrR* mutant. These data suggest that the BvrR/BvrS system directly controls elements involved in adjusting the carbohydrate and nitrogen metabolism in *Brucella*, probably by allowing rapid and coordinated nutritional changes necessary for adaptation to intracellular life. Recent

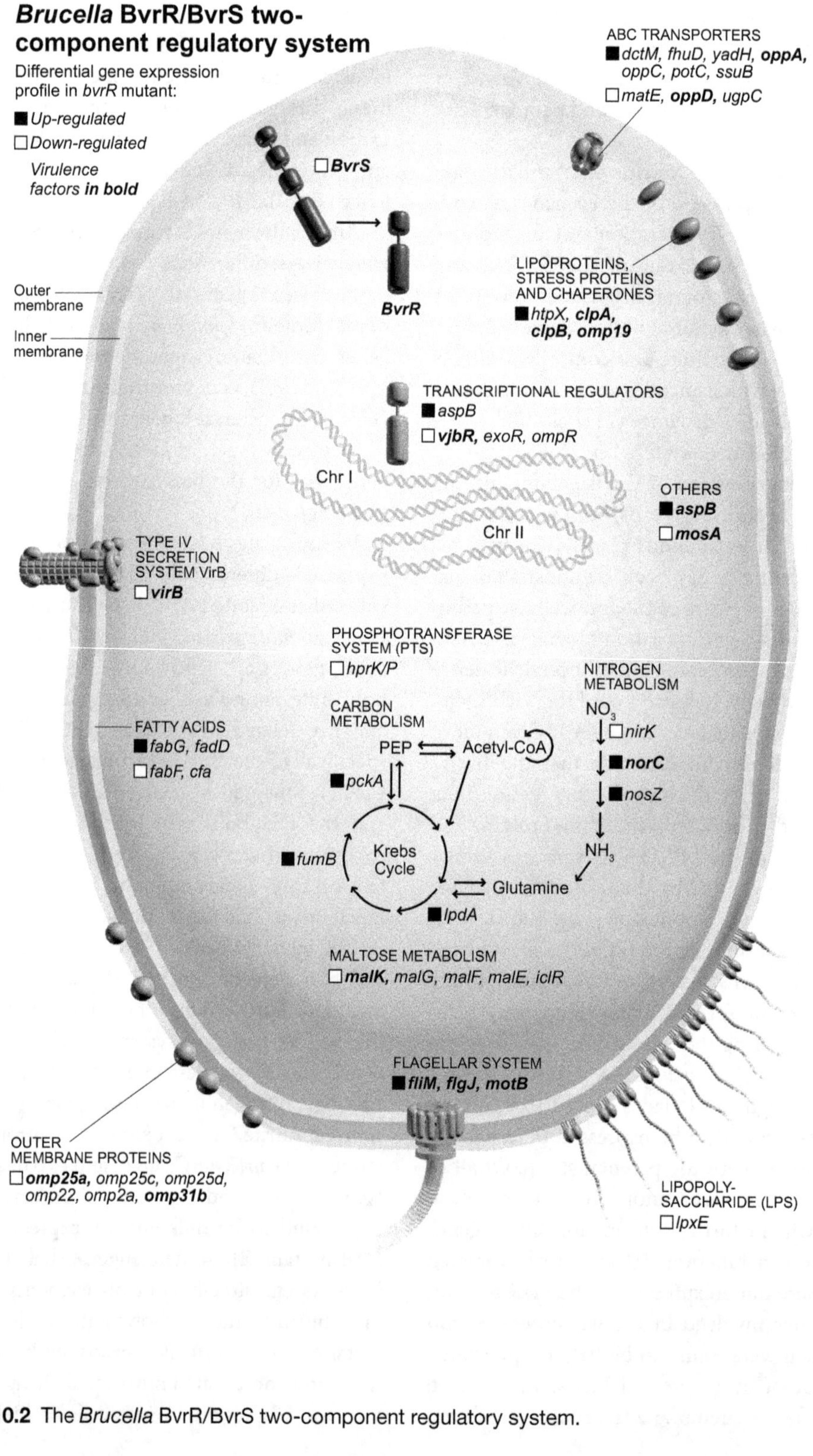

Figure 10.2 The *Brucella* BvrR/BvrS two-component regulatory system.

data published by Letesson's group support this hypothesis. These authors have demonstrated that the phosphotransferase system (PTS) of *B. melitensis* has functions regulating the nitrogen/carbon balance in the bacteria, which it is related to its virulence and is transcribed conjointly with the BvrR/BvrS system (Dozot *et al.*, 2010). In agreement with these results, the data from the transcriptomic analysis using microarrays also showed that the expression of the *hprK* gene of the PTS system was diminished in the *bvrR* mutant (Viadas *et al.*, 2010).

Differential expression was also found for seven transcriptional regulators, among which were found the *vjbR*, *ompR* and *exoR* genes. VjbR is a regulator related to quorum-sensing, a genetic control mechanism dependent on cellular density and mediated by signalling molecules (discussed in Chapter 8). The VjbR regulator has been suggested as a mediator of the effect of the C_{12}-HSL signalling molecule. In *B. melitensis*, the mutant in this regulator presents great attenuation both in cellular infections and mice (Delrue *et al.*, 2005), and directly activates the genes of the VirB secretion and flagellum systems. Furthermore, the VjbR regulator controls the synthesis of exopolysaccharides and the production of certain outer membrane proteins (Uzureau *et al.*, 2007). These data show that the VjbR regulator, as is the BvrR/BvrsR system, is involved in the control of the composition of the cell envelope and virulence. Furthermore, recently it has been shown that the expression of VjbR is directly dependent on the BvrR/BvrS system and that this system also regulates the expression of VirB at the transcriptional level (Martínez-Nuñez *et al.*, 2010). The regulation of VirB expression in *Brucella* is a very complex process both in culture media and inside cells. In culture, the proteins and transcripts of VirB are only detected transitorily at the end of the exponential growth phase (Sieira *et al.*, 2000; Boschiroli *et al.*, 2002; Delrue *et al.*, 2005), whilst inside cells, VirB is only expressed in the first hours of infection (Sieira *et al.*, 2004). Therefore, VirB seems to be somehow regulated by VjbR and the BvrR/BvrS system.

The VjbR regulator also induces the expression of the *ompR* gene, the regulator of the two-component OmpR/EnvZ system (DeJong *et al.*, 2008), which is inhibited in the *bvrR* mutant (Viadas *et al.*, 2010). In *E. coli*, this OmpR/EnvZ system controls the transcription of outer membrane porins, amino acid metabolism, flagellar synthesis and nutrient transport (Oshima, *et al.*, 2002). As noted above, microarray analyses also showed that at least three flagellar genes were over-expressed in the *bvrR* mutant, paralleling the regulation of flagellum biosynthesis and modifications in LPS by *exoR* in *S. meliloti*. On the basis of all these comparisons, and bearing in mind that all these regulators (BvrR/BvrS, VjbR, OmpR and ExoR) are involved in the control of the composition and structure of the cell envelope and are necessary for virulence, it can be concluded that in *Brucella* some type of cross-talk or communication exists between all of them and that this is key for survival of the bacteria.

Finally, although the changes in expression with implications in the cell envelope, metabolism and other regulators could explain by themselves the loss of virulence in mutants of the BvrR/BvrS system, transcriptome analysis detected at least ten specific genes which had previously been associated with the virulence of *Brucella* (Lestrate *et al.*, 2000; Delrue *et al.*, 2005; Haine *et al.*, 2006; Loisel-Meyer *et al.*, 2006). Among these are included the *vjbR, motB, malK, norC, oppA, aspB, mosA* genes and three genes which encode hypothetical proteins. For example, the *malK* mutant of *B. melitensis* and the *aspB* mutant of *B. suis* exhibit attenuation in infection in cellular models and the mutants of *B. melitensis* in *vjbR, motB, oppA, mosA* exhibit attenuation in both models of cellular infection and mice (Lestrate *et al.*, 2000; Delrue *et al.*, 2005).

In conclusion, the recent results from proteome and transcriptome analysis show that the BvrR/BvrS system of *Brucella* is a master regulator capable of interacting with other regulators, by controlling the synthesis of components of the cell envelope, metabolism and virulence.

Acknowledgements

Most of the results present in this review have been obtained by research projects supported by the Spanish Ministry of Science (BIO2005-04985 and AGL2008-04514).

References

Boschiroli, M.L., Ouahrani-Bettache, S., Foulongne, V., Michaux-Charachon, S., Bourg, G., Allardet-Servent, A., Cazevieille, C., Liautard, J.P., Ramuz, M., and O'Callaghan, D. (2002). The *Brucella suis virB* operon is induced intracellularly in macrophages. Proc. Natl. Acad. Sci. U.S.A. *99*, 1544–1549.

Charles, T.C., and Nester, E.W. (1993). A chromosomally encoded two-component sensory transduction system is required for virulence of *Agrobacterium tumefaciens*. J. Bacteriol. *175*, 6614–6625.

Cheng, H.P., and Walker, G.C. (1998). Succinoglycan production by *Rhizobium meliloti* is regulated through the ExoS-ChvI two-component regulatory system. J. Bacteriol. *180*, 20–26.

Conde, R. (2008). A study on the role of phosphatidylcholine and inner lipopolysaccharide sections in *Brucella* virulence. Doctoral dissertation, School of Science, University of Navarra, Pamplona, Spain.

De Jong, M.F., Sun, Y.H., den Hartigh, A.B., van Dijl, J.M., and Tsolis, R.M. (2008). Identification of VceA and VceC, two members of the VjbR regulon that are translocated into macrophages by the *Brucella* type IV secretion system. Mol. Microbiol. *70*, 1378–1396.

Delrue, R.M., Deschamps, C., Leonard, S., Nijskens, C., Danese, I., Schaus, J.M., Bonnat, S., Ferooz, J., Tibor, A., de Bolle, X., *et al.* (2005) A quorum-sensing regulator controls expression of both the type IV secretion system and the flagellar apparatus of *Brucella melitensis*. Cell. Microbiol. *7*, 1151–1161.

Dorrell, N., Spencer, S., Foulongne, V., Guigue-Talet, P., O'Callaghan, D., and Wren, B.W. (1998). Identification, cloning and initial characterisation of FeuPQ in *Brucella suis*: a new sub-family of two-component regulatory systems. FEMS Microbiol. Lett. *162*, 143–150.

Dorrell, N., Guigue, T.P., Spencer, S., Foulongne, V., O'Callaghan, D., and Wren, B.W. (1999). Investigation into the role of the response regulator NtrC in the metabolism and virulence of *Brucella suis*. Microb. Pathog. *27*, 1–11.

Dozot, M., Poncet, S., Nicolas, C., Copin, R., Bouraoui, H., Mazé, A., Deutscher, J., De Bolle, X., and Letesson, J.J. (2010). Functional characterization of the incomplete phosphotransferase system (PTS) of the intracellular pathogen *Brucella melitensis*. PLoS One *5*, e12679.

Foulongne, V., Bourg, G., Cazevieille, C., Michaux-Charachon, S., and O'Callaghan, D. (2000). Identification of *Brucella suis* genes affecting intracellular survival in an *in vitro* human macrophage infection model by signature-tagged transposon mutagenesis. Infect. Immun. *68*, 1297–1303.

Gelvin, S.B. (2006). *Agrobacterium* virulence gene induction. Methods Mol. Biol. *343*, 77–84.

Grilló, M.J., Manterola, L., de Miguel, M.J., Muñoz, P.M., Blasco, J.M., Moriyón, I., and López-Goñi, I. (2006). Increases of efficacy as vaccine against *Brucella abortus* infection in mice by simultaneous inoculation with avirulent smooth *bvrS/bvrR* and rough *wbkA* mutants. Vaccine *24*, 2910–2916.

Guzmán-Verri, C., Chaves-Olarte, E., von Eichel-Streiber, C., López-Goñi, I., Thelestam, M., Arvidson, S., Gorvel, J.P., and Moreno, E. (2001). GTPases of the Rho subfamily are required for *Brucella abortus* internalization in nonprofessional phagocytes: direct activation of Cdc42. J. Biol. Chem. *276*, 44435–44443.

Guzmán-Verri, C., Manterola, L., Sola-Landa, A., Parra, A., Cloeckaert, A., Garin, J., Gorvel, J.P., Moriyón, I., Moreno, E., and López-Goñi, I. (2002). The two-component system BvrR/BvrS essential for *Brucella abortus* virulence regulates the expression of outer membrane proteins with counterparts in members of the *Rhizobiaceae*. Proc. Natl. Acad. Sci. U.S.A. *99*, 12375–12380.

Haine, V., Dozot, M., Dornand, J., Letesson, J.J., and de Bolle, X. (2006). *NnrA* is required for full virulence and regulates several *Brucella melitensis* denitrification genes. J. Bacteriol. *188*, 1615–1619.

Hoch, J.A., and Silhavy, T.J., eds. (1995). Two-component Signal Transduction. (ASM Press: Washington, D.C.).

Kohler, S., Foulongne, V., Ouahrani-Bettache, S., Bourg, G., Teyssier, J., Ramuz, M., and Liautard, J.P. (2002). The analysis of the intramacrophagic virulome of *Brucella suis* deciphers the environment encountered by the pathogen inside the macrophage host cell. Proc. Natl. Acad. Sci. U.S.A. *99*, 15711–15716.

Lamontagne, J., Butler, H., Chaves-Olarte, E., Hunter, J., Schirm, M., Paquet, C., Tian, M., Kearney, P., Hamaidi, L., Chelsky, D., *et al.* (2007). Extensive cell envelope modulation is associated with virulence in *Brucella abortus*. J. Proteome Res. *6*, 1519–1529.

Lavín, J.L., Binnewies, T.T., Pisabarro, A.G., Ussery, D.W., García-Lobo, J.M., and Oguiza, J.A. (2010). Differences in two-component signal transduction proteins among the genus *Brucella*: implications for host preference and pathogenesis. Vet. Microbiol. *144*, 478–483.

Lestrate, P., Delrue, R.M., Danese, I., Didembourg, C., Taminiau, B., Mertens, P., de Bolle, X., Tibor, A., Tang, C.M., and Letesson, J.J. (2000). Identification and characterization of *in vivo* attenuated mutants of *Brucella melitensis*. Mol. Microbiol. *38*, 543–551.

Lestrate, P., Dricot, A., Delrue, R.M., Lambert, C., Martinelli, V., De Bolle, X., Letesson, J.J., and Tibor, A. (2003). Attenuated signature-tagged mutagenesis mutants of *Brucella melitensis* identified during the acute phase of infection in mice. Infect. Immun. *71*, 7053–7060.

Letesson, J.J., and De Bolle, X. (2004). *Brucella* virulence: a matter of control. In: *Brucella*: Molecular and Cellular Biology, López-Goñi, I., and Moriyón, I., eds. (Horizon Bioscience: Norfolk, England), pp. 117–158.

Li, L., Jia, Y., Hou, Q., Charles, T.C., Nester, E.W., and Pan, S.Q. (2002). A global pH sensor: *Agrobacterium* sensor protein ChvG regulates acid-inducible genes on its two chromosomes and Ti plasmid. Proc. Natl. Acad. Sci. U.S.A. *99*, 12369–12374.

Loisel-Meyer, S., Jiménez de Bagues, M.P., Basseres, E., Dornand, J., Kohler, S., Liautard, J.P., and Jubier-Maurin, V. (2006). Requirement of *norD* for *Brucella suis* virulence in a murine model of *in vitro* and *in vivo* infection. Infect. Immun. *74*, 1973–1976.

López-Goñi, I., Guzmán-Verri, C., Manterola, L., Sola-Landa, A., Moriyón, I., and Moreno, E. (2002). Regulation of *Brucella* virulence by the two-component system BvrR/BvrS. Vet. Microbiol. *90*, 329–339.

López-Goñi, I., Manterola, L., and Pan, S.Q. (2004). The *Brucella* BvrS/BvrR and related two-component regulatory systems of the α–2 Proteobacteria: common regulatory strategies of animal and plant pathogens and endosymbionts. In: *Brucella*: Molecular and Cellular Biology, López-Goñi, I., and Moriyón, I., eds. (Horizon Bioscience: Norfolk, England), pp. 213–230.

Manterola, L., Moriyón, I., Moreno, E., Sola-Landa, A., Weiss, D.S., Koch, M.H., Howe, J., Brandenburg, K., and López-Goñi, I. (2005). The lipopolysaccharide of *Brucella abortus* BvrS/BvrR mutants contains lipid A modifications and has higher affinity for bactericidal cationic peptides. J. Bacteriol. *187*, 5631–5639.

Martínez-Núñez, C., Altamirano-Silva, P., Alvarado-Guillén, F., Moreno, E., Guzmán-Verri, C., and Chaves-Olarte, E. (2010). The two component system BvrR/BvrS regulates the expression of the Type IV secretion system VirB in *Brucella abortus*. J. Bacteriol. *192*, 5603–5608.

Oshima, T. (2002). Transcriptome analysis of all two-component regulatory system mutants of *Escherichia coli* K-12. Mol. Microbiol. *46*, 281–291.

Osteras, M., Stanley, J., and Finan, T.M. (1995). Identification of *Rhizobium*-specific intergenic mosaic elements within an essential two-component regulatory system of *Rhizobium* species. J. Bacteriol. *177*, 5485–5494.

Quebatte, M., Dehio, M., Tropel, D., Basler, A., Toller, I., Raddatz, G., Engel, P., Huser, S., Schein, H., Lindroos, H.L., Andersson, S.G., and Dehio, C. (2010). The BatR/BatS two-component regulatory system controls the adaptive response of *Bartonella henselae* during human endothelial cell infection. J. Bacteriol. *192*, 3352–3367.

Sieira, R., Comerci, D.J., Sánchez, D.O., and Ugalde, R.A. (2000). A homologue of an operon required for DNA transfer in *Agrobacterium* is required in *Brucella abortus* for virulence and intracellular multiplication. J. Bacteriol. *182*, 4849–4855.

Sieira, R., Comerci, D.J., Pietrasanta, L.I., and Ugalde, R.A. (2004). Integration host factor is involved in transcriptional regulation of the *Brucella abortus* virB operon. Mol. Microbiol. *54*, 808–822.

Sola-Landa, A., Pizarro-Cerdá, J., Grilló, M.J., Moreno, E., Moriyón, I., Blasco, J.M., Gorvel, J.P., and López-Goñi, I. (1998). A two-component regulatory system playing a critical role in plant pathogens and endosymbionts is present in *Brucella abortus* and controls cell invasion and virulence. Mol. Microbiol. *29*, 125–138.

Uzureau, S., Godefroid, M., Deschamps, C., Lemaire, J., de Bolle, X., and Letesson, J.J. (2007). Mutations of the quorum sensing-dependent regulator VjbR lead to drastic surface modifications in *Brucella melitensis*. J. Bacteriol. *189*, 6035–6047.

Viadas, C., Rodríguez, M.C., Sangari, F.J., Gorvel, J.P., García-Lobo, J.M-, and López-Goñi, I. (2010). Transcriptome analysis of the Brucella abortus BvrR/BvrS two-component regulatory system. PLoS One 5, e10216.

The *Brucella* VirB Type IV Secretion System

11

Renee M. Tsolis and David O'Callaghan

Abstract

The type IV secretion system (T4SS) encoded by the *virB* operon is a key pathogenicity determinant that is highly conserved among *Brucella* species, and is one of the few 'classical' virulence factors identified to date. Since its first description in 1999, work on the VirB T4SS has focused on its role in intracellular survival and persistence, its architecture and assembly in the cell envelope, on the signals that induce its expression and function within the infected host cell, on identification of T4SS substrates, and on its effect on the host response to infection. In this chapter, we review the advances in our understanding of structure/function relationships of the T4SS, and of its function in infection by *Brucella* species.

Introduction

The *Brucella suis virB* operon was first identified as a Tn*blaM* mutant during an early screen for virulence factors (O'Callaghan *et al.*, 1999). Sequencing of a cosmid clone carrying the region showed the presence of an 11 kb region encoding homologues of the *Agrobacterium* VirB proteins and a twelfth protein homologous to the OprF lipoprotein from *Pseudomonas syringae*. Insertion mutants in the *virB* operon were found to be highly attenuated in both macrophages and HeLa cells. Southern blotting showed that the region was present in the genomes of all the known *Brucella* species and the role of the *B. abortus* VirB system in virulence was reported soon after (Sieira *et al.*, 2000). The VirB system has since been shown to be a key virulence factor for *Brucella*. Mutants have been shown to be attenuated in cell culture models, mice and in goats (O'Callaghan *et al.*, 1999; Hong *et al.*, 2000; Zygmunt *et al.*, 2006; den Hartigh *et al.*, 2008).

Type IV secretion and bacterial virulence

Bacterial secretion systems are prokaryotic nanomachines that selectively transport macromolecules across their envelope. T4SS are multiprotein complexes composed of up to 12 proteins named VirB1 to VirB11 plus VirD4 in *A. tumefaciens,* harbouring the prototype of the family of type IV transporters. The complex forms a channel across the bacterial envelope for macromolecule secretion, and in many cases, an extended appendage, or pilus, is also observed. T4SS are very versatile; they translocate proteins and DNA–protein complexes between cells as well as serving to import DNA–protein complexes. First described as being required for conjugation of plasmids in Gram-negative bacteria and translocation of T-DNA complexes from *Agrobacterium* to plant cells (Lessl and Lanka, 1994), they have more recently been shown to be used by many bacterial pathogens and symbionts to translocate protein virulence factors into the cells of infected hosts (Alvarez-Martínez and Christie, 2009). These virulence proteins, or effectors, modulate the host cell biology to allow the bacteria to avoid host defences and create an intracellular niche in which they can survive and replicate.

Structure function analysis of the *Brucella* T4SS

Most of our ideas concerning the structure and function of T4SS are based on the *A. tumefaciens* VirB paradigm. The *Brucella* VirB T4SS is composed of 11 different proteins VirB1–VirB11 which show sequence similarity to the *Agrobacterium* VirB proteins (O'Callaghan *et al.*, 1999). The VirB proteins can be functionally subdivided in three different groups. VirB4, VirB11 and VirD4 are inner membrane ATPases with a large cytoplasmic domain and they are believed to provide the energy required for T4SS assembly and for the translocation of effectors. VirB2 and VirB5 form an extracellular bacterial appendage believed to interact with the host. Finally, VirB3 and VirB6-VirB10 are believed to form a channel-like structure spanning both the inner and outer membranes of the bacteria, allowing the effectors to be translocated from the bacterial cytosol directly to the extracellular space or into the host cytoplasm. The two major differences between the *Agrobacterium* and *Brucella* T4SS are the presence of an extra gene in the *Brucella virB* operon, *virB12*, encoding an outer membrane lipoprotein and the absence of a gene encoding a homologue of the VirD4 coupling protein. Homologues of VirB12 are encoded on conjugative plasmids suggesting that the *Brucella virB* operon was acquired by horizontal gene transfer. There are conflicting results concerning its role in virulence (O'Callaghan *et al.*, 1999; Sun *et al.*, 2005); however, it has been shown to be a highly immunogenic protein in human and animal infections (Rolan *et al.*, 2008; Liang *et al.*, 2010). The VirD4 coupling protein plays an essential role in T4SS function by recruiting the protein substrates to the transporter. For example, in *Agrobacterium* bimolecular fluorescence complementation and immunofluorescence microscopy have been used to show polar localization and interaction between VirD4 and the VirE2 effector protein (Atmakuri *et al.*, 2003). The lack of a coupling protein in *Brucella* opens the question of how this T4SS actually functions: Does another protein act as a coupling protein, or does *Brucella* use an alternative, coupling protein-independent mechanism for secretion as seen with *Bordetella pertussis* (Covacci and Rappuoli, 1993)?

The ATPases form hexameric rings

The crystal structures of the TrwB coupling protein from the IncW plasmid R388 and of *B. suis* VirB11 show that both proteins form hexameric rings (Gomis-Ruth *et al.*, 2001; Hare *et al.*, 2006). A double ring structure is formed by the N- and C-terminal domains of each monomer. Analysis of the *Helicobacter pylori* VirB11 homologue (HP0525) showed that it is a dynamic protein and that nucleotide binding locks it into a compact, symmetrical structure (Yeo *et al.*, 2000; Savvides *et al.*, 2003). It was also suggested that nucleotide binding mediates structural changes that could drive either substrate translocation or T4SS assembly. Although the first to be studied, HP0525 is in fact an atypical VirB11 protein since it has a large domain swap compared with most others, including that of *B. suis* (Hare *et al.*, 2006). This domain swap is caused by differences in the length of the sequence between the N- and C-terminal domains and leads to dramatic changes in the subunit–subunit interface and in the nucleotide-binding sites. Despite this, hexamer formation and function is not affected. The ATPase activity of HP0525 was used as a screening target for the identification of novel antimicrobial molecules that were found to inhibit *H. pylori* T4SS function in infection models (Hilleringmann *et al.*, 2006).

The T4SS pilus

The pili of most T4SS are thought to be multimers of the small VirB2 protein, which has been shown to be processed and circularize (Eisenbrandt *et al.*, 1999; Lai *et al.*, 2002). Although the *Helicobacter* Cag pilus is reported to be sheathed with the VirB10 homologue (Rohde *et al.*, 2003), in most T4SS the only decoration is the VirB5 protein. The crystal structure of TraC, the VirB5 homologue from the conjugative plasmid pKM101 was determined and residues important for both recognition by specific phages and conjugation of pKM101 were identified (Yeo *et al.*, 2003). Aly *et al* (Aly and Baron, 2007) found that the *Agrobacterium* VirB5 protein decorates the tip of the pilus and that the short, non-structured C-terminal tail of the protein was essential for incorporation of VirB5 into the pilus. The presence of VirB5 at the bacterial surface suggests that it may be involved in interaction with host proteins as shown for the *Agrobacterium* VirB2 protein (Hwang and Gelvin, 2004). Work with CagL, the *H. pylori* VirB5 homologue, has shown that it interacts with host cell integrins and that this interaction is important for both translocation of the CagA effector protein and the initiation of signalling cascades in the infected cell (Kwok *et al.*, 2007; Backert *et al.*, 2008; Jiménez-Soto *et al.*, 2009). While a pilus structure has not yet been visualized on the surface of *Brucella* spp., expression of VirB5 and VirB8 has been detected by immunofluorescence in a rough (*manB*) mutant over expressing a subset of the *Brucella* VirB proteins (Carle *et al.*, 2006). Recent, unpublished data from the O'Callaghan laboratory suggest that components predicted to form the *Brucella* VirB pilus can interact with host proteins and that this is important for the infection process.

Understanding the role of the VirB8 protein and the translocation channel

Among the proteins thought to form the membrane-spanning channel, VirB8 has been shown to play a key role in the assembly of the T4SS. In *Agrobacterium* it was found that VirB8 acts as a nucleation centre, recruiting VirB9 and VirB10 into clusters in the outer membrane and localizing VirB proteins at the cell pole (Kumar and Das, 2001). More recently, using high-resolution fluorescence microscopy, it was shown to form helical arrays around the cells (Aguilar *et al.*, 2010). VirB8 is a protein spanning the bacterial inner membrane with a short cytoplasmic N-terminus and a single hydrophobic transmembrane domain. The carboxy-terminal moiety of the protein, consisting of 172 amino acids, is believed to be entirely periplasmic. Recently, the three-dimensional structures of the periplasmic domains of VirB8 from *B. suis* and *A. tumefaciens* have been determined (Terradot *et al.*, 2005; Bailey *et al.*, 2006). Both are predicted to form homodimers. Based on these structural data, in collaboration with the Baron laboratory we used site-directed mutagenesis to create changes in residues of the periplasmic part of *B. suis* VirB8 that were predicted to play a role in its dimerization or interactions with VirB4 or VirB10. When used to complement a *B. suis* strain with a non-polar deletion of *virB8*, we found that variants that were affected in dimerization or in interactions with VirB4 or VirB10 were unable to restore virulence (Paschos *et al.*, 2006).

When crystal and cryo-electron microscopic structures of the membrane-spanning channel were reported by the Waksman laboratory (Bayliss *et al.*, 2007; Chandran *et al.*, 2009; Fronzes *et al.*, 2009; Waksman and Fronzes, 2010), it became apparent that the channel was composed of VirB7, VirB9 and VirB10. Surprisingly, VirB8 was not present. This membrane-spanning core consists of 14 copies of each of the proteins forming two layers that insert in the inner and outer membranes, respectively. The structure was double-walled with VirB10 spanning both the inner and the outer membranes. The Baron laboratory has further studied the role of VirB8 in the assembly of the *Brucella* T4SS (Sivanesan *et al.*, 2010). Data from ELISA, circular dichroism, and surface plasmon resonance assays with soluble periplasmic domains showed that VirB8 forms a stable complex with VirB9 and VirB10, which, together with the finding that VirB8 is absent from the core, suggests that VirB8 recruits these proteins to the T4SS assembly site. Using the bacterial two hybrid system (BTH) (Karimova *et al.*, 2005) with a bicistronic expression construct that allows co-expression of other proteins, this study also showed that VirB9 stimulated VirB8 dimerization. Further, VirB9 partially inhibited the VirB10-VirB10 and VirB9–VirB10 interactions and, taken together, these data suggested a dynamic model for the assembly of the core T4SS.

The finding that dimerization of VirB8 is essential for T4SS assembly allowed the design of a BTH-system based high-throughput screen for potential antimicrobials targeting VirB8 dimerization (Paschos *et al.*, 2011) (discussed in detail in Chapter 12). To further investigate the role of VirB8 dimerization, we constructed chimeric VirB8 proteins, swapping the periplasmic and cytoplasmic/transmembrane domains of *B. suis* VirB8 with those of TraJ, the VirB8 homologue encoded on the broad host range plasmid pSB102 (Patey *et al.*, 2006). The proteins that contained the TraJ cytoplasmic/transmembrane domains were unable to restore virulence to a *virB8* mutant and had a dominant negative effect on the virulence of wild type *B. suis* when expressed *in trans*. Bacterial two hybrid assays showed that the cytoplasmic/transmembrane domain of TraJ forms much stronger interactions with both VirB8 and TraJ than that of VirB8 (Bourg *et al.*, 2009). This indicates that the dimers and heterodimers containing the TraJ cytoplasmic/transmembrane domains are much more tightly associated than the VirB8 cytoplasmic/transmembrane domains and this probably explains the dominant negative effect. It is now clear that T4SS are dynamic structures and that VirB8 dimers must be flexible and recent data from a structure–function analysis of the dimer interface of *Agrobacterium* VirB8 support this notion (Sivanesan and Baron, 2010). Whether this flexibility is required during T4SS assembly or represents a gate keeping function, keeping the pore 'closed' to stop leakage and then 'open' transiently to allow and regulate substrate passage remains to be investigated.

Signals and regulators that modulate expression of the T4SS

Multiple regulatory proteins integrate environmental signals to coordinate expression of the *virB* genes in the infected host. Some of these proteins have been shown to act directly on the *virB* promoter, whereas the role of others is based on genetic evidence (Table 11.1). While the signals in the host that induce expression, assembly and activity of the T4SS are incompletely understood, *in vitro* studies have identified nutrient restriction and acidic pH as being important inducers of *virB* gene expression. In rich media [tryptic soy broth (TSB), pH 7.0] *in vitro*, the *Brucella* species differ in expression of the VirB proteins: while in *B. suis*, the VirB proteins are undetectable, variable low levels of expression were observed in *B. abortus* and *B. melitensis* grown in TSB (Rouot *et al.*, 2003). However, shifting cultures to acidic minimal medium induces expression of the *virB* genes in *B. abortus, B. melitensis, B. canis* and *B.*

Table 11.1 *Brucella* regulators affecting expression of the *virB* genes

Regulator	Signal	Promoter binding demonstrated	Activity: activation/ Repression	Reference
IHF	Nutrient limitation?	+	A	Sieira *et al.* (2004)
VjbR	QS	+	A	Jong *et al.* (2008);, Arocena *et al.* (2010), de Uzureau *et al.* (2010), Weeks *et al.* (2010)
HutC	Metabolic?	+	A	Sieira *et al.* (2010)
BvrR/S	?	+	A	Martínez-Nuñez *et al.* (2010)
BabR	QS	–	A/R	Rambow-Larsen *et al.* (2008), Uzureau *et al.* (2010)
Rsh	Nutrient limitation	–	A	Dozot *et al.* (2006)
PTS	C and N metabolism	–	A	Dozot *et al.* (2010)
ArsR6	?	–	A	Haine *et al.* (2005)
AraC2	?	–	R	Haine *et al.* (2005)
AraC8	?	–	A	Haine *et al.* (2005)
DeoR1	?	–	A	Haine *et al.* (2005)
GntR4	?	–	A	Haine *et al.*, (2005)

suis (Rouot *et al.*, 2003). Acidic minimal medium appears to provide some signals present within the *Brucella*-containing phagosome, which acidifies to a pH of 4.0–4.5 after bacterial uptake (Porte *et al.*, 1999). Phagosomal acidification is absolutely required for induction of *virB* gene expression, since neutralization of phagosomal pH prevents both expression of the *virB* genes and intracellular multiplication of *Brucella* (Boschiroli *et al.*, 2002). The regulator that induces or derepresses *virB* expression in response to acidic conditions is not yet known.

Nutrient limitation is a second signal that induces expression of the *virB* genes (Boschiroli *et al.*, 2002), and three regulators have been implicated in linking nutrient status to expression of the T4SS, the stringent response regulator Rsh, integration host factor (IHF), and the histidine utilization regulator HutC (Sieira *et al.*, 2010). Rsh is RelA/SpoT homologue that synthesizes ppGpp in response to nutrient deprivation (Dozot *et al.*, 2006). The requirement of Rsh for induction of *virB* gene expression suggests that the acidified vacuole containing *Brucella* is a nutrient-poor environment, which is in agreement with previous reports (Kohler *et al.*, 2002).

A second global regulator found to modulate *virB* gene expression is integration host factor (IHF). The α-subunit of IHF was the first regulatory protein found to bind directly to the *virB* promoter (Sieira *et al.*, 2004). DNAse protection experiments showed that IHF binds a 51-bp region that extends from 188 to 137 bp upstream of *virB1*. IHF binding is important for activation of *virB* transcription, since a reporter strain lacking the IHF binding site exhibited reduced expression of the *virB* genes in neutral minimal medium, but not in acidic minimal medium (Sieira *et al.*, 2004). These results suggest that IHF may be a second sensor of nutrient deprivation inside the host cell. Thus, IHF may play a similar role in *Brucella* as in *Salmonella enterica* serotype Typhimurium where it co-ordinately regulates expression of stationary phase/nutrient deprivation-responsive genes and virulence genes (Mangan *et al.*, 2006). Since IHF binding induces bending of the *virB* upstream region, it is possible that this conformational change facilitates binding of additional regulators to the *virB* promoter.

In addition to nutrient deprivation and acidic pH, quorum sensing provides another important signal that modulates expression of the *virB* genes. Two LuxR family regulators have been shown to modulate expression of the *virB* genes, VjbR and BabR (also known as BlxR) (Delrue *et al.*, 2005; Rambow-Larsen *et al.*, 2008; Uzureau *et al.*, 2010). For a more detailed discussion of quorum-sensing regulators in *Brucella* see Chapter 8. VjbR is an activator of *virB* transcription, an activity that is counteracted by the small molecule C_{12}-homoserine lactone (C_{12}-HSL) (Taminiau *et al.*, 2002). In the case of *Brucella* this down-regulation by C_{12}-HSL may occur in the replicative niche, in which bacteria are found in a tightly apposed vacuole. This spatial restriction of this intracellular compartment could allow C_{12}-HSL to accumulate to a level that inhibits binding of VjbR to the *virB* promoter.

VjbR is a direct activator of the *virB* genes and binds the region upstream of *virB1* (de Jong *et al.*, 2008; Arocena *et al.*, 2010; Uzureau *et al.*, 2010). However, different studies have reported different binding sites for VjbR in the *virB* operon (Fig. 11.1). While the study by Arocena *et al* (2010) used DNAse I protection to demonstrate a unique binding site for VjbR in the *virB1* upstream region, the work of Uzureau *et al* (Uzureau *et al.*, 2010) using chromatin immunoprecipitation (ChIP), showed an additional binding for VjbR in the *virB1-virB2* intergenic region. A study by de Jong *et al* (2008) using both electrophoretic mobility shift assays (EMSA), found VjbR binding both in the promoter region and the *virB1-virB2* intergenic region. The different binding sites identified in these studies might reflect the participation in DNA binding of *Brucella* cofactors that are absent from the *in vitro* binding assays but would be present in the ChIP studies or, alternatively, it could result from the different VjbR protein preparations used in the different laboratories. Regulation of *virB* by VjbR is critical for infection, since a *vjbR* mutant is highly attenuated in both cellular and animal infection models (Delrue *et al.*, 2005).

A second quorum sensing regulator, BabR (BlxR), has been shown to regulate the *virB* genes, but while one study showed that BabR activates *virB* gene expression, possibly via an effect on expression of *vjbR* (Rambow-Larsen *et al.*, 2008),

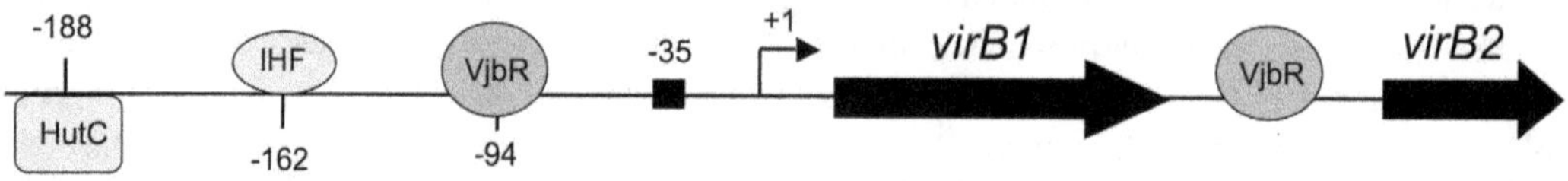

Figure 11.1 Map of the 5′ end of the *virB* operon indicating binding sites for regulators shown to bind directly in the operon. For references, see Table 11.1 and the text.

two studies showed that BabR antagonizes VjbR in regulating expression of the *virB* locus (Uzureau *et al.*, 2010; Weeks *et al.*, 2010).

The two-component regulatory system BvrR/BvrS, which is a global regulator of cell envelope-associate functions in *Brucella* (Guzmán-Verri *et al.*, 2002; Viadas *et al.*, 2010), was recently implicated as a direct regulator of the *virB* genes (Martínez-Nuñez *et al.*, 2010). BvrR/BvrS regulates the *virB* genes both directly via binding the *virB* promoter, and indirectly via an effect on expression of *vjbR*. The host-associated signal that coordinates *virB* expression with cell envelope-associated functions such as LPS biosynthesis and outer membrane protein expression is not yet known. However, in addition to its role in regulating the composition of the *Brucella* cell envelope, BvrR and BvrS were recently shown to be transcriptionally linked to the phosphotransferase (PTS) system, which together with quorum sensing, has a major impact on central metabolism (Dozot *et al.*, 2010; Uzureau *et al.*, 2010). Since PTS mutants have impaired expression of the *virB* genes, it is probable that uptake of *Brucella* spp. by the host cell leads to a change in nutrient availability, and that activation of *virB* expression in response to the host environment is coordinated by multiple systems, including BvrR/BvrS, quorum sensing and the PTS (Dozot *et al.*, 2010; Martínez-Nuñez *et al.*, 2010; Uzureau *et al.*, 2010).

While we are beginning to understand the signals and regulators involved in transcriptional regulation of the *virB* genes, there are some major gaps in our knowledge of post-transcriptional regulation, assembly of the T4SS, and secretion of substrates. Since it has been possible to induce expression of *virB* genes, but not secretion of substrates, in laboratory medium there are probably additional signals specific to the intracellular niche that are essential for assembly of the T4SS complex and translocation of the T4SS substrates into the host cell.

The VirB T4SS controls maturation of the *Brucella*-containing vacuole

As with most intracellular pathogens, *Brucella* modulates the biology of the cell in which it resides to block cell defences and create a novel intracellular niche in which it replicates. After entry into host cells, the first step in *Brucella* intracellular survival is to avoid fusion of its membrane-bound compartment, the *Brucella*-containing vacuole (BCV), with lysosomes thus blocking the first line of cellular defence. This is partly due to the entry into cells via lipid rafts. The next step is to build its replication niche, which is achieved by capturing endoplasmic reticulum (ER) derived vesicles in a similar way to that described for *Legionella* (Hubber and Roy, 2010). These steps are described in detail in recent reviews (Celli, 2006).

The VirB system has been found to be essential for much of this process. The first indications came from a study by Comerci *et al.* (2001), who found that a polar *virB10* mutant was unable to avoid fusion with lysosomes and was rapidly killed. Surprisingly, the non-polar *virB10* mutant was not killed but recycled to cell surface; the reasons for this observation are still not clear, especially considering the key role that VirB10 has in the T4SS structure (Chandran *et al.*, 2009; Fronzes *et al.*, 2009). Delrue *et al* (2001) conducted a transposon mutagenesis screen to identify mutants with altered trafficking, and all but one of the mutants that they identified affected the *virB* locus, and one the *vjbR* gene discussed above, which underlines the important function of VirB.

Watarai and colleagues (Kim *et al.*, 2002; Watarai *et al.*, 2002a,b) suggested that the *Brucella* VirB system is required for entry through lipid rafts. However, these studies were performed with *B. abortus,* which expresses the VirB system constitutively at low levels in rich medium such as TSB. They did not address how and whether the VirB system of other *Brucella* spp., such as *B. suis,* in which expression is tightly regulated (Rouot *et al*, 2003), might play a role in cell entry. Boschiroli *et al* (2002) found that the *B. suis virB* operon is induced intracellularly in response to phagosome acidification. The Watarai lab also reported that the cellular Prion protein was involved in *Brucella* entry (Watarai *et al.*, 2003, 2004), however, other laboratories have not been able to repeat these results (Fontes *et al.*, 2005).

A functional VirB system is essential for the establishment of the replication niche. The replication niche is characterized by the presence of ER markers such as Calnexin on the vacuole membrane and the absence of the lysosomal-associated membrane glycoprotein 1 (LAMP-1), although this is transiently acquired in the early stages after uptake and then excluded during maturation (Pizarro-Cerdá *et al.*, 1998a,b; Starr *et al.*, 2008). Celli (Celli, 2006; Celli *et al.*, 2003, 2005) showed that *Brucella* creates its replication niche by sustained interactions with ER-derived vesicles and that this process is VirB dependent. This is reminiscent to the Dot/Icm T4SS-dependent maturation of the intracellular replication niche of *Legionella pneumophila* (Isberg *et al.*, 2009). Here, effectors translocated by the T4SS allow *Legionella* to create its replication niche by capturing vesicles exiting the ER on their way to the Golgi by recruiting the ADP-ribosylation factor 1 (Arf1), which controls the assembly of the COPI coat complex that mediates the retrograde transport of retrograde transport of ER-resident proteins (Nagai *et al.*, 2002). ARF1/COPI-dependent vesicular transport is sensitive to brefeldin A. Brefeldin A has no effect on *Brucella* infection suggesting that this process is not targeted in the maturation of the BCV. Celli *et al.* (2005) showed that *Brucella* in fact intercept the vesicles at an earlier stage in their route from the ER in a process that is dependent on the small GTPase Sar1, which controls the assembly of COPII complexes on ER membranes. The bacterial factors responsible for this reaction are still not known.

Substrates of the T4SS

To date, only two substrates of the T4SS have been reported, VceA and VceC (de Jong *et al.*, 2008). These proteins were identified in a screen for genes co-regulated with the *virB* genes. A conserved motif overlapping the −35 region of the *virB* promoter and the *virB1-virB2* intergenic region was used to search the *B. suis* 1330 and *B. abortus* 2308 genomes for possible co-regulated genes. Expression of a subset of these genes, including *vceA* and *vceC,* could be activated in a heterologous *E. coli* system by inducing expression of *vjbR.* VirB-dependent translocation of VceA and VceC was demonstrated by generation of translational fusions to TEM-1 beta-lactamase. Translocation of TEM-1::VceA and TEM-1::VceC fusion proteins could be detected starting at 7–9 h after infection, which was consistent with a requirement for intracellular induction of the *virB* genes by phagosomal acidification (Boschiroli *et al.*, 2002). VceC contains positively charged amino acids in its C-terminus that are required for translocation into the host cell (de Jong *et al.*, 2008) suggesting the conservation of C-terminal secretion signals similar to other T4SS (Vergunst *et al.*, 2005). *B. suis* VceC could be translocated into host cells after heterologous expression in *L. pneumophila,* suggesting some conservation of the secretion mechanism between *Brucella* and *Legionella.* However, neither VceC nor VceA share any similarity to any of the described *L. pneumophila* T4SS effectors, or to any protein of known function in bacteria, suggesting that their function within the infected cell may be novel. Recently, four Brucella putative effectors (BPE), BPE123, BPE275, BPE005 and BPE043, encoded by *BAB1_0123, BAB1_1275, BAB1_2005,* and *BAB1_1043,* respectively, were identified in *B. abortus* (Marchesini *et al.*, in press). Most of these T4SS substrates contain eukaryotic-like domains, suggesting an interaction with proteins in the infected host cell. Like VceA and VceC, their intracellular targets and roles in evasion of lysosomal degradation remain to be discovered.

Proinflammatory responses associated with the VirB T4SS

While cellular models have allowed the dissection of how the T4SS mediates interaction with subcellular compartments, less is known about the *in vivo* consequences of these interactions. Animal models, primarily the mouse, have shed some light on how the host orchestrates responses of different cell types to *Brucella* infection and the role of the T4SS in eliciting these responses. While host responses to Gram-negative bacteria are driven primarily by pathogen-associated molecular patterns (PAMPs) such as LPS (Rosenberger *et al.*, 2000), in the case of *Brucella*, the low agonist activity of its PAMPs results in only a mild inflammatory response to infection. A study of transcriptional responses to *B. abortus* and *B. melitensis* infection of the spleen showed that the T4SS is the major signal sensed by the host immune system (Roux *et al.*, 2007). Within 2–3 days after intraperitoneal infection of mice, *B. abortus* and *B. melitensis* elicit production of interferon gamma (IFN-γ) by T cells and NK cells (Copin *et al.*, 2007; Rolan and Tsolis, 2008). Interestingly, this response is dependent on the T4SS, since mice infected with *virB* mutants did not respond to infection with production of IFN-γ (Rolan and Tsolis, 2008). This reduced response was not the result of lower bacterial numbers in this case, since early after intraperitoneal infection, *virB* mutants were present in similar numbers to their wild type controls.

The T4SS affects the host response to *Brucella* infection

As discussed above, *B. abortus*, *B. suis* and *B. melitensis virB* mutants are defective in intracellular survival *in vitro* and are more rapidly cleared from liver and spleen in the mouse infection model. After inoculation via the intraperitoneal route, *virB* mutants can initially disseminate from the inoculation site and colonize, and are able to persist briefly (3–5 days) at wild type levels before their numbers start to decrease (Rolan and Tsolis, 2007). It should be pointed out in this context that this initial ability of *virB* mutants to colonize and persist is at odds with lack of ability to survive intracellularly in cellular infection models (O'Callaghan *et al.*, 1999; Sieira *et al.*, 2000; Delrue *et al.*, 2001; Celli *et al.*, 2003), if one assumes that immediately after infection, bacteria enter an intracellular niche. Interestingly, at 1 day after intragastric inoculation, a non-significant trend was observed for reduced dissemination of a *virB2* mutant to the spleen and mesenteric lymph nodes (Paixao *et al.*, 2009), suggesting that the mechanism for dissemination from the digestive tract and from the peritoneal cavity may have different requirements for the T4SS. Whether this is the result of bacteria being in an intracellular vs. extracellular niche, or whether the inoculated bacteria disseminate from the site of infection in different cell types remains to be determined.

During the early part of the infection, the T4SS is important for induction of inflammatory responses in the spleen and liver (Roux *et al.*, 2007). Transcriptional profiling of infected mouse spleens showed a mild inflammatory response to infection with wild type *B. abortus* or *B. melitensis*, which was characterized by induction of interferon gamma and interferon gamma-inducible genes. However, despite colonizing at the same level early during infection, *virB* mutants fail to elicit any inflammatory response, suggesting that the innate immune response is 'blind' to the presence of a *virB* mutant. On the cellular level, CD4 T cells of mice infected with *virB* mutants exhibit reduced interferon gamma production compared with levels in mice infected with wild type *B. abortus* (Rolan and Tsolis, 2008). The reduced interferon gamma levels, together with reduced persistence of the mutant bacteria, probably also contributes to the shift in IgG isotypes observed in the two groups of mice later in infection: mice inoculated with wild type *B. abortus* produce more IgG2 than mice inoculated with *virB* mutants, while both groups of mice produce similar levels of IgG1. Overall, this suggests that interferon gamma elicited early in the immune response by the T4SS contributes to Th1 polarization of the immune response.

Because of the role of the T4SS in persistence within the mononuclear phagocyte system, it has been difficult to tease apart its role in persistence from effects on the immune response during persistent colonization. However, for reasons that are incompletely understood, mice lacking natural antibody, such as T and B cell-deficient *Rag1*⁻/⁻ mice and B-cell-deficient *Igh6*⁻/⁻ mice, can rescue the persistence defect of *virB* mutants (Rolan and Tsolis, 2007). This permitted an examination of whether the T4SS contributes to granulomatous inflammation in the spleen, a hallmark of persistent *Brucella* infection. Histomorphometric examination of the spleens of the *Igh6*⁻/⁻ mice showed wild type *B. abortus* was able to induce formation of granulomas, while granulomas were absent from spleens of the *virB* mutant-infected mice (Rolan *et al.*, 2009). Thus, the T4SS contributes to the formation of granulomas in mice, probably by initiating a mild inflammatory response in infected cells. Along these lines, Pei and colleagues reported that necrotic cell death, a proinflammatory stimulus induced by rough mutant *B. abortus* in cultured macrophages, was dependent on the T4SS (Pei *et al.*, 2008). Mechanistically, this suggests that an effector or PAMP that is translocated into the cytosol via the T4SS could trigger innate immune pathways leading to inflammation; however, the translocated molecules and innate immune sensors that initiate inflammatory responses in the infected cell have yet to be identified.

Perspectives

Since the first discovery of the *Brucella* T4SS over ten years ago, much progress has been made on understanding the signals eliciting its expression, its function in the infected cell and the identification of the first substrates. However, key questions remain with regard to how the T4SS assembles and how its structural components, including pili, interact with host cells. Further, the mechanism of substrate secretion, the total number of T4SS substrates and the molecular interactions of the secreted proteins with component of the host cell remain to be determined. Considering the biology of *Brucella,* it could be predicted that the T4SS might play a role in evasion of degradation in the lysosome, interaction with ER exit sites or the ER-to-Golgi intermediate compartment, or modulation of host cell signalling pathways involved in host cell survival and inflammation. Well-characterized paradigms such as the *Agrobacterium* and *Legionella* T4SS have been instructive in the study of the *Brucella* VirB T4SS, yet it is probable that some key differences will emerge with regard to both the pathway of translocation from the *Brucella* cytoplasm into the infected cell, and the function of the effectors during infection.

Acknowledgements

The authors would like to thank J.J. Letesson and C. Baron for helpful discussions during the preparation of this chapter. Work in Renee Tsolis's laboratory is supported by US Public Health Service grant AI050553. Work in David O'Callaghan's laboratory is supported by INSERM, the Universite de Montpellier 1, the Region Languedoc Roussillon, La ville de Nimes and la Fondation Pour La Recherche Medicale.

References

Aguilar, J., Zupan, J., Cameron, T.A., and Zambryski, P.C. (2010). *Agrobacterium* type IV secretion system and its substrates form helical arrays around the circumference of virulence-induced cells. Proc. Natl. Acad. Sci. U.S.A. *107*, 3758–3763.

Alvarez-Martinez, C.E., and Christie, P.J. (2009). Biological diversity of prokaryotic type IV secretion systems. Microbiol. Mol. Biol. Rev. *73*, 775–808.

Aly, K.A., and Baron, C. (2007). The VirB5 protein localizes to the T-pilus tips in *Agrobacterium tumefaciens*. Microbiology *153*, 3766–3775.

Arocena, G.M., Sieira, R., Comerci, D.J., and Ugalde, R.A. (2010). Identification of the quorum-sensing target DNA sequence and N-Acyl homoserine lactone responsiveness of the *Brucella abortus virB* promoter. J. Bacteriol. *192*, 3434–3440.

Atmakuri, K., Ding, Z., and Christie, P.J. (2003). VirE2, a type IV secretion substrate, interacts with the VirD4 transfer protein at cell poles of *Agrobacterium tumefaciens*. Mol. Microbiol. *49*, 1699–1713.

Backert, S., Fronzes, R., and Waksman, G. (2008). VirB2 and VirB5 proteins: specialized adhesins in bacterial type-IV secretion systems? Trends Microbiol. *16*, 409–413.

Bailey, S., Ward, D., Middleton, R., Grossmann, J.G., and Zambryski, P.C. (2006). *Agrobacterium tumefaciens* VirB8 structure reveals potential protein–protein interaction sites. Proc. Natl. Acad. Sci. U.S.A. *103*, 2582–2587.

Bayliss, R., Harris, R., Coutte, L., Monier, A., Fronzes, R., Christie, P.J., Driscoll, P.C., and Waksman, G. (2007). NMR structure of a complex between the VirB9/VirB7 interaction domains of the pKM101 type IV secretion system. Proc. Natl. Acad. Sci. U.S.A. *104*, 1673–1678.

Boschiroli, M.L., Ouahrani-Bettache, S., Foulongne, V., Michaux-Charachon, S., Bourg, G., Allardet-Servent, A., Cazevieille, C., Liautard, J.P., Ramuz, M., and O'Callaghan, D. (2002). The *Brucella suis virB* operon is induced intracellularly in macrophages. Proc. Natl. Acad. Sci. U.S.A. *99*, 1544–1549.

Bourg, G., Sube, R., O'Callaghan, D., and Patey, G. (2009). Interactions between *Brucella suis* VirB8 and its homolog TraJ from the plasmid pSB102 underline the dynamic nature of type IV secretion systems. J. Bacteriol. *191*, 2985–2992.

Carle, A., Hoppner, C., Ahmed Aly, K., Yuan, Q., den Dulk-Ras, A., Vergunst, A., O'Callaghan, D., and Baron, C. (2006). The *Brucella suis* type IV secretion system assembles in the cell envelope of the heterologous host *Agrobacterium tumefaciens* and increases IncQ plasmid pLS1 recipient competence. Infect. Immun. *74*, 108–117.

Celli, J. (2006). Surviving inside a macrophage: the many ways of *Brucella*. Res. Microbiol. *157*, 93–98.

Celli, J., de, C.C., Franchini, D.M., Pizarro-Cerda, J., Moreno, E., and Gorvel, J.P. (2003). *Brucella* evades macrophage killing via VirB-dependent sustained interactions with the endoplasmic reticulum. J. Exp. Med. *198*, 545–556.

Celli, J., Salcedo, S.P., and Gorvel, J.P. (2005). *Brucella* coopts the small GTPase Sar1 for intracellular replication. Proc. Natl. Acad. Sci. U.S.A. *102*, 1673–1678.

Chandran, V., Fronzes, R., Duquerroy, S., Cronin, N., Navaza, J., and Waksman, G. (2009). Structure of the outer membrane complex of a type IV secretion system. Nature *462*, 1011–1015.

Comerci, D.J., Martinez-Lorenzo, M.J., Sieira, R., Gorvel, J.P., and Ugalde, R.A. (2001). Essential role of the VirB machinery in the maturation of the *Brucella abortus*-containing vacuole. Cell. Microbiol. *3*, 159–168.

Copin, R., De Baetselier, P., Carlier, Y., Letesson, J.J., and Muraille, E. (2007). MyD88-dependent activation of B220-CD11b+LY 6C+ dendritic cells during *Brucella melitensis* infection. J. Immunol. *178*, 5182–5191.

Covacci, A., and Rappuoli, R. (1993). Pertussis toxin export requires accessory genes located downstream from the pertussis toxin operon. Mol. Microbiol. *8*, 429–434.

Delrue, R.M., Martinez-Lorenzo, M., Lestrate, P., Danese, I., Bielarz, V., Mertens, P., De, B.X., Tibor, A., Gorvel, J.P., and Letesson, J.J. (2001). Identification of *Brucella* spp. genes involved in intracellular trafficking. Cell. Microbiol. *3*, 487–497.

Delrue, R.M., Deschamps, C., Leonard, S., Nijskens, C., Danese, I., Schaus, J.M., Bonnot, S., Ferooz, J., Tibor, A., De Bolle, X., *et al.* (2005). A quorum-sensing regulator controls expression of both the type IV secretion system and the flagellar apparatus of *Brucella melitensis*. Cell. Microbiol. *7*, 1151–1161.

Dozot, M., Boigegrain, R.A., Delrue, R.M., Hallez, R., Ouahrani-Bettache, S., Danese, I., Letesson, J.J., De Bolle, X., and Kohler, S. (2006). The stringent response mediator Rsh is required for *Brucella melitensis* and *Brucella suis* virulence, and for expression of the type IV secretion system virB. Cell. Microbiol. *8*, 1791–1802.

Dozot, M., Poncet, S., Nicolas, C., Copin, R., Bouraoui, H., Maze, A., Deutscher, J., De Bolle, X., and Letesson, J.J. (2010). Functional characterization of the incomplete phosphotransferase system (PTS) of the intracellular pathogen *Brucella melitensis*. PLoS One *5*, e12676.

Eisenbrandt, R., Kalkum, M., Lai, E.M., Lurz, R., Kado, C.I., and Lanka, E. (1999). Conjugative pili of IncP plasmids, and the Ti plasmid T pilus are composed of cyclic subunits. J. Biol. Chem. *274*, 22548–22555.

Fontes, P., varez-Martinez, M.T., Gross, A., Carnaud, C., Kohler, S., and Liautard, J.P. (2005). Absence of evidence for the participation of the macrophage cellular prion protein in infection with *Brucella suis*. Infect. Immun. *73*, 6229–6236.

Fronzes, R., Schafer, E., Wang, L., Saibil, H.R., Orlova, E.V., and Waksman, G. (2009). Structure of a type IV secretion system core complex. Science *323*, 266–268.

Gomis-Ruth, F.X., Moncalian, G., Perez-Luque, R., Gonzalez, A., Cabezon, E., de la Cruz, F., and Coll, M. (2001). The bacterial conjugation protein TrwB resembles ring helicases and F1-ATPase. Nature *409*, 637–641.

Guzman-Verri, C., Manterola, L., Sola-Landa, A., Parra, A., Cloeckaert, A., Garin, J., Gorvel, J.P., Moriyon, I., Moreno, E., and López-Goñi, I. (2002). The two-component system BvrR/BvrS essential for *Brucella abortus* virulence regulates the expression of outer membrane proteins with counterparts in members of the Rhizobiaceae. Proc. Natl. Acad. Sci. U.S.A. *99*, 12375–12380.

Haine, V., Sinon, A., Van Steen, F., Rousseau, S., Dozot, M., Lestrate, P., Lambert, C., Letesson, J.J., and De Bolle, X. (2005). Systematic targeted mutagenesis of *Brucella melitensis* 16M reveals a major role for GntR regulators in the control of virulence. Infect. Immun. *73*, 5578–5586.

Hare, S., Bayliss, R., Baron, C., and Waksman, G. (2006). A large domain swap in the VirB11 ATPase of *Brucella suis* leaves the hexameric assembly intact. J. Mol. Biol. *360*, 56–66.

den Hartigh, A.B., Rolan, H.G., de Jong, M.F., and Tsolis, R.M. (2008). VirB3 to VirB6 and VirB8 to VirB11, but not VirB7, are essential for mediating persistence of *Brucella* in the reticuloendothelial system. J. Bacteriol. *190*, 4427–4436.

Hilleringmann, M., Pansegrau, W., Doyle, M., Kaufman, S., MacKichan, M.L., Gianfaldoni, C., Ruggiero, P., and Covacci, A. (2006). Inhibitors of *Helicobacter pylori* ATPase Cagalpha block CagA transport and cag virulence. Microbiology *152*, 2919–2930.

Hong, P.C., Tsolis, R.M., and Ficht, T.A. (2000). Identification of genes required for chronic persistence of *Brucella abortus* in mice. Infect. Immun. *68*, 4102–4107.

Hubber, A., and Roy, C.R. (2010). Modulation of host cell function by *Legionella pneumophila* type IV effectors. Annu. Rev. Cell. Dev. Biol. *26*, 261–283.

Hwang, H.H., and Gelvin, S.B. (2004). Plant proteins that interact with VirB2, the *Agrobacterium tumefaciens* pilin protein, mediate plant transformation. Plant Cell. *16*, 3148–3167.

Isberg, R.R., O'Connor, T.J., and Heidtman, M. (2009). The *Legionella pneumophila* replication vacuole:

making a cosy niche inside host cells. Nat. Rev. Microbiol. *7*, 13–24.

Jimenez-Soto, L.F., Kutter, S., Sewald, X., Ertl, C., Weiss, E., Kapp, U., Rohde, M., Pirch, T., Jung, K., Retta, S.F., *et al.* (2009). *Helicobacter pylori* type IV secretion apparatus exploits beta1 integrin in a novel RGD-independent manner. PLoS Pathog. *5*, e1000684.

de Jong, M.F., Sun, Y.H., den Hartigh, A.B., van Dijl, J.M., and Tsolis, R.M. (2008). Identification of VceA and VceC, two members of the VjbR regulon that are translocated into macrophages by the *Brucella* type IV secretion system. Mol. Microbiol. *70*, 1378–1396.

Karimova, G., Dautin, N., and Ladant, D. (2005). Interaction network among *Escherichia coli* membrane proteins involved in cell division as revealed by bacterial two-hybrid analysis. J. Bacteriol. *187*, 2233–2243.

Kim, S., Watarai, M., Makino, S., and Shirahata, T. (2002). Membrane sorting during swimming internalization of *Brucella* is required for phagosome trafficking decisions. Microb. Pathog. *33*, 225–237.

Kohler, S., Foulongne, V., Ouahrani-Bettache, S., Bourg, G., Teyssier, J., Ramuz, M., and Liautard, J.P. (2002). The analysis of the intramacrophagic virulome of *Brucella suis* deciphers the environment encountered by the pathogen inside the macrophage host cell. Proc. Natl. Acad. Sci. U.S.A. *99*, 15711–15716.

Kumar, R.B., and Das, A. (2001). Functional analysis of the *Agrobacterium tumefaciens* T-DNA transport pore protein VirB8. J. Bacteriol. *183*, 3636–3641.

Kwok, T., Zabler, D., Urman, S., Rohde, M., Hartig, R., Wessler, S., Misselwitz, R., Berger, J., Sewald, N., Konig, W., *et al.* (2007). *Helicobacter* exploits integrin for type IV secretion and kinase activation. Nature *449*, 862–866.

Lai, E.M., Eisenbrandt, R., Kalkum, M., Lanka, E., and Kado, C.I. (2002). Biogenesis of T pilli in *Agrobacterium tumefaciens* requires precise VirB2 propillin cleavage and cyclization. J. Bacteriol. *184*, 327–330.

Lessl, M., and Lanka, E. (1994). Common mechanisms in bacterial conjugation and Ti-mediated T-DNA transfer to plant cells. Cell 77, 321–324.

Liang, L., Leng, D., Burk, C., Nakajima-Sasaki, R., Kayala, M.A., Atluri, V.L., Pablo, J., Unal, B., Ficht, T.A., Gotuzzo, E., *et al.* (2010). Large scale immune profiling of infected humans and goats reveals differential recognition of *Brucella melitensis* antigens. PLoS Negl. Trop. Dis. *4*, e673.

Mangan, M.W., Lucchini, S., Danino, V., Croinin, T.O., Hinton, J.C., and Dorman, C.J. (2006). The integration host factor (IHF) integrates stationary-phase and virulence gene expression in *Salmonella enterica* serovar Typhimurium. Mol. Microbiol. *59*, 1831–1847.

Marchesini, M.I, Hermann, C.K., Salcedo, S.P., Gorvel, J.-P., and Comerci, D.J. (2011). In search of *Brucella abortus* Type IV secretion substrates: screening and indentification of four proteins tranlocated into host cells through the VirB system. Cell. Microbiol. In press.

Martinez-Nunez, C., Altamirano-Silva, P., Alvarado-Guillen, F., Moreno, E., Guzman-Verri, C., and Chaves-Olarte, E. (2010). The two-component system BvrR/BvrS regulates the expression of the type IV secretion system VirB in *Brucella abortus*. J. Bacteriol. *192*, 5603–5608.

Nagai, H., Kagan, J.C., Zhu, X.J., Kahn, R.A., and Roy, C.R. (2002). A bacterial guanine nucleotide exchange factor activates ARF on *Legionella* phagosomes. Science *295*, 679–682.

O'Callaghan, D., Cazevieille, C., Allardet-Servent, A., Boschiroli, M.L., Bourg, G., Foulongne, V., Frutos, P., Kulakov, Y., and Ramuz, M. (1999). A homologue of the *Agrobacterium tumefaciens* VirB and *Bordetella pertussis* Ptl type IV secretion systems is essential for intracellular survival of *Brucella suis*. Mol. Microbiol. *33*, 1210–1220.

Paixao, T.A., Roux, C.M., den Hartigh, A.B., Sankaran-Walters, S., Dandekar, S., Santos, R.L., and Tsolis, R.M. (2009). Establishment of systemic *Brucella melitensis* infection through the digestive tract requires urease, the type IV secretion system, and lipopolysaccharide O antigen. Infect. Immun. 77, 4197–4208.

Paschos, A., Patey, G., Sivanesan, D., Gao, C., Bayliss, R., Waksman, G., O'Callaghan, D., and Baron, C. (2006). Dimerization and interactions of *Brucella suis* VirB8 with VirB4 and VirB10 are required for its biological activity. Proc. Natl. Acad. Sci. U. S. A *103*, 7252–7257.

Paschos, A., den, H.A., Smith, M.A., Atluri, V.L., Sivanesan, D., Tsolis, R.M., and Baron, C. (2011). An *in vivo* high-throughput screening approach targeting the type IV secretion system component VirB8 identified inhibitors of *Brucella abortus* 2308 proliferation. Infect. Immun. 79, 1033–1043.

Patey, G., Qi, Z., Bourg, G., Baron, C., and O'Callaghan, D. (2006). Swapping of periplasmic domains between *Brucella suis* VirB8 and a pSB102 VirB8 homologue allows heterologous complementation. Infect. Immun. *74*, 4945–4949.

Pei, J., Wu, Q., Kahl-McDonagh, M., and Ficht, T.A. (2008). Cytotoxicity in macrophages infected with rough *Brucella* mutants is type IV secretion system dependent. Infect. Immun. *76*, 30–37.

Pizarro-Cerdá, J., Meresse, S., Parton, R.G., van der, G.G., Sola-Landa, A., López-Goñi, I., Moreno, E., and Gorvel, J.P. (1998a). *Brucella abortus* transits through the autophagic pathway and replicates in the endoplasmic reticulum of nonprofessional phagocytes. Infect. Immun. *66*, 5711–5724.

Pizarro-Cerdá, J., Moreno, E., Sanguedolce, V., Mege, J.L., and Gorvel, J.P. (1998b). Virulent *Brucella abortus* prevents lysosome fusion and is distributed within autophagosome-like compartments. Infect. Immun. *66*, 2387–2392.

Porte, F., Liautard, J.P., and Kohler, S. (1999). Early acidification of phagosomes containing *Brucella suis* is essential for intracellular survival in murine macrophages. Infect. Immun. *67*, 4041–4047.

Rambow-Larsen, A.A., Rajashekara, G., Petersen, E., and Splitter, G. (2008). Putative quorum-sensing regulator BlxR of *Brucella melitensis* regulates virulence factors including the type IV secretion system and flagella. J. Bacteriol. *190*, 3274–3282.

Rohde, M., Puls, J., Buhrdorf, R., Fischer, W., and Haas, R. (2003). A novel sheathed surface organelle of the

Helicobacter pylori cag type IV secretion system. Mol. Microbiol. *49*, 219–234.

Rolan, H.G., and Tsolis, R.M. (2007). Mice lacking components of adaptive immunity show increased *Brucella abortus virB* mutant colonization. Infect. Immun. 75, 2965–2973.

Rolan, H.G., and Tsolis, R.M. (2008). Inactivation of the type IV secretion system reduces the Th1 polarization of the immune response to *Brucella abortus* infection. Infect. Immun. *76*, 3207–3213.

Rolan, H.G., den Hartigh, A.B., Kahl-McDonagh, M., Ficht, T., Adams, L.G., and Tsolis, R.M. (2008). VirB12 is a serological marker of *Brucella* infection in experimental and natural hosts. Clin. Vaccine Immunol. *15*, 208–214.

Rolan, H.G., Xavier, M.N., Santos, R.L., and Tsolis, R.M. (2009). Natural antibody contributes to host defense against an attenuated *Brucella abortus virB* mutant. Infect. Immun. *77*, 3004–3013.

Rosenberger, C.M., Scott, M.G., Gold, M.R., Hancock, R.E., and Finlay, B.B. (2000). *Salmonella typhimurium* infection and lipopolysaccharide stimulation induce similar changes in macrophage gene expression. J. Immunol. *164*, 5894–5904.

Rouot, B., Alvarez-Martinez, M.T., Marius, C., Menanteau, P., Guilloteau, L., Boigegrain, R.A., Zumbihl, R., O'Callaghan, D., Domke, N., and Baron, C. (2003). Production of the type IV secretion system differs among *Brucella* species as revealed with VirB5- and VirB8-specific antisera. Infect. Immun. *71*, 1075–1082.

Roux, C.M., Rolan, H.G., Santos, R.L., Beremand, P.D., Thomas, T.L., Adams, L.G., and Tsolis, R.M. (2007). *Brucella* requires a functional Type IV secretion system to elicit innate immune responses in mice. Cell. Microbiol. *9*, 1851–1869.

Savvides, S.N., Yeo, H.J., Beck, M.R., Blaesing, F., Lurz, R., Lanka, E., Buhrdorf, R., Fischer, W., Haas, R., and Waksman, G. (2003). VirB11 ATPases are dynamic hexameric assemblies: new insights into bacterial type IV secretion. EMBO J. *22*, 1969–1980.

Sieira, R., Comerci, D.J., Sanchez, D.O., and Ugalde, R.A. (2000). A homologue of an operon required for DNA transfer in *Agrobacterium* is required in *Brucella abortus* for virulence and intracellular multiplication. J. Bacteriol. *182*, 4849–4855.

Sieira, R., Comerci, D.J., Pietrasanta, L.I., and Ugalde, R.A. (2004). Integration host factor is involved in transcriptional regulation of the *Brucella abortus* virB operon. Mol. Microbiol. *54*, 808–822.

Sieira, R., Arocena, G.M., Bukata, L., Comerci, D.J., and Ugalde, R.A. (2010). Metabolic control of virulence genes in *Brucella abortus*: HutC coordinates virB expression and the histidine utilization pathway by direct binding to both promoters. J. Bacteriol. *192*, 217–224.

Sivanesan, D., Hancock, M.A., Villamil Giraldo, A.M., and Baron, C. (2010). Quantitative analysis of VirB8–VirB9-VirB10 interactions provides a dynamic model of type IV secretion system core complex assembly. Biochemistry *49*, 4483–4493.

Sivanesan, D., and Baron, C. (2011). The Dimer interface of *Agrobacterium tumefaciens* VirB8 is important for Type IV Secretion System function, stability and for association of VirB2 with the core complex. J. Bacteriol. *193*, 2097–2106.

Starr, T., Ng, T.W., Wehrly, T.D., Knodler, L.A., and Celli, J. (2008). *Brucella* intracellular replication requires trafficking through the late endosomal/lysosomal compartment. Traffic *9*, 678–694.

Sun, Y.H., Rolan, H.G., den Hartigh, A.B., Sondervan, D., and Tsolis, R.M. (2005). *Brucella abortus* virB12 is expressed during infection but is not an essential component of the type IV secretion system. Infect. Immun. 73, 6048–6054.

Taminiau, B., Daykin, M., Swift, S., Boschiroli, M.L., Tibor, A., Lestrate, P., De Bolle, X., O'Callaghan, D., Williams, P., and Letesson, J.J. (2002). Identification of a quorum-sensing signal molecule in the facultative intracellular pathogen *Brucella melitensis*. Infect. Immun. *70*, 3004–3011.

Terradot, L., Bayliss, R., Oomen, C., Leonard, G.A., Baron, C., and Waksman, G. (2005). Structures of two core subunits of the bacterial type IV secretion system, VirB8 from *Brucella suis* and ComB10 from *Helicobacter pylori*. Proc. Natl. Acad. Sci. U.S.A. *102*, 4596–4601.

Uzureau, S., Lemaire, J., Delaive, E., Dieu, M., Gaigneaux, A., Raes, M., De Bolle, X., and Letesson, J.J. (2010). Global analysis of quorum sensing targets in the intracellular pathogen *Brucella melitensis* 16 M. J. Proteome Res. *9*, 3200–3217.

Vergunst, A.C., van Lier, M.C., den Dulk-Ras, A., Stuve, T.A., Ouwehand, A., and Hooykaas, P.J. (2005). Positive charge is an important feature of the C-terminal transport signal of the VirB/D4-translocated proteins of *Agrobacterium*. Proc. Natl. Acad. Sci. U.S.A. *102*, 832–837.

Viadas, C., Rodríguez, M.C., Sangari, F.J., Gorvel, J.P., García-Lobo, J.M., and López-Goñi, I. (2010). Transcriptome analysis of the *Brucella abortus* BvrR/BvrS two-component regulatory system. PLoS One *5*, e10216.

Waksman, G., and Fronzes, R. (2010). Molecular architecture of bacterial type IV secretion systems. Trends Biochem. Sci. *35*, 691–698.

Watarai, M. (2004). Interaction between *Brucella abortus* and cellular prion protein in lipid raft microdomains. Microbes Infect. *6*, 93–100.

Watarai, M., Makino, S., Fujii, Y., Okamoto, K., and Shirahata, T. (2002a). Modulation of *Brucella*-induced macropinocytosis by lipid rafts mediates intracellular replication. Cell. Microbiol. *4*, 341–355.

Watarai, M., Makino, S., Michikawa, M., Yanagisawa, K., Murakami, S., and Shirahata, T. (2002b). Macrophage plasma membrane cholesterol contributes to *Brucella abortus* infection of mice. Infect. Immun. *70*, 4818–4825.

Watarai, M., Kim, S., Erdenebaatar, J., Makino, S., Horiuchi, M., Shirahata, T., Sakaguchi, S., and Katamine, S. (2003). Cellular prion protein promotes *Brucella* infection into macrophages. J. Exp. Med. *198*, 5–17.

Weeks, J.N., Galindo, C.L., Drake, K.L., Adams, G.L., Garner, H.R., and Ficht, T.A. (2010). *Brucella melitensis* VjbR and C12-HSL regulons: contributions of the N-dodecanoyl homoserine lactone signaling molecule and LuxR homologue VjbR to gene expression. BMC Microbiol. *10*, 167.

Yeo, H.J., Savvides, S.N., Herr, A.B., Lanka, E., and Waksman, G. (2000). Crystal structure of the hexameric traffic ATPase of the *Helicobacter pylori* type IV secretion system. Mol. Cell. *6*, 1461–1472.

Yeo, H.J., Yuan, Q., Beck, M.R., Baron, C., and Waksman, G. (2003). Structural and functional characterization of the VirB5 protein from the type IV secretion system encoded by the conjugative plasmid pKM101. Proc. Natl. Acad. Sci. U.S.A. *100*, 15947–15952.

Zygmunt, M.S., Hagius, S.D., Walker, J.V., and Elzer, P.H. (2006). Identification of *Brucella melitensis* 16M genes required for bacterial survival in the caprine host. Microbes Infect. *8*, 2849–2854.

Novel Targets for Antibacterial Agents in *Brucella* sp. 12

Christian Baron, Jean-Yves Winum and Stephan Köhler

Abstract

Novel strategies for the treatment of bacterial infectious diseases are urgently needed. In this chapter we will discuss two complementary approaches aimed at targeting *Brucella* species during the intracellular growth phase: depriving them of essential amino acid biosynthetic enzymes and disarmament by inhibiting type IV secretion system function that is essential for virulence. Small molecules targeting these functions that are essential exclusively during intracellular growth could serve as leads for the development of novel anti-infective and anti-virulence drugs, respectively. Since such drugs would not kill *Brucella* during extracellular growth, it is anticipated that the selection pressure for resistant mutants would be reduced. In addition, such specific molecules would not disrupt the human microbiome and thereby avoid side effects of conventional antibiotic treatments. Following similar approaches, anti-infective and anti-virulence drugs could be developed to treat infections caused by a wide variety of bacterial pathogens.

Introduction

Bacterial infections represent an alarming health problem worldwide. The widespread emergence of resistance and multiresistance to antibiotics among a growing number of bacterial pathogens represents a major threat and necessitates a permanent race for novel molecules with new mechanisms of action (Levy and Marshall, 2004). However, during the past 40 years, only two new classes of antibiotics have been discovered (Fischbach and Walsh, 2009) and strains resistant to these molecules appeared within a few years following the marketing of the drug. There are four possible approaches to reduce this trend: (i) to restrain use and prescription of antibiotics, (ii) to discover new drug targets, (iii) to design drugs with novel modes of action that will reduce the probability of resistance and (iv) to establish combination therapies hitting multiple targets, thereby reducing the likelihood of resistance development. The challenge at present is to identify and to validate novel pharmaceutical targets in bacteria, which would be the starting point for the discovery of new classes of antibacterial agents. With the constant accumulation of data from bacterial genomic sequencing projects, many potential new drug targets have been identified, but as of now, genomics approaches have not yielded the expected effective novel molecules. In the past few years, the growing information on the virulence mechanisms of various pathogens and the relationship with their hosts has led to increased interest among scientists for virulence factors as novel targets for anti-infectious agents, so-called anti-virulence drugs (Alksne, 2002; Clatworthy *et al.*, 2007; Baron, 2010). Other potential targets are metabolic enzymes (Liautard *et al.*, 2006) and in this review we will discuss the status and potential of these approaches for the development of novel anti-*Brucella* drugs.

Brucellosis is the most important bacterial zoonosis worldwide (Corbel, 1997; Godfroid *et al.*, 2005; Pappas *et al.*, 2005). In endemic areas (Mediterranean Europe, Middle East, Latin America), the incidence of human brucellosis may be as high as 200 per 100,000 habitants. Treatment of human brucellosis consists of antibiotic bitherapy (doxycycline and rifampicin) for 6 weeks, which is demanding on the patients (Ariza *et al.*, 2007). Despite the fact that this treatment is efficient, the isolated appearance of strains resistant to rifampicin has been reported, alerting to the risk of emergence of rifampicin resistance in regions where brucellosis is endemic (Memish *et al.*, 2000; Baykam *et al.*, 2004), and resistant strains could be engineered for abuse in bio-terror attacks or as biowarefare agents. Despite intense efforts, to date no effective vaccine has been developed for humans (Perkins *et al.*, 2011). It is therefore of interest to identify new targets for the development of novel anti-brucellosis agents.

As pathogenicity of intracellular bacteria such as brucellae is linked to their capacity to multiply within the host cell, the development of new antibacterials specifically active at the intracellular state targeting metabolic or virulence functions would block multiplication of the pathogen without affecting extracellular bacteria. The development of such anti-infectious or anti-virulence drugs has big potential to limit pressure for selection of resistant mutants and to reduce the secondary effects antibiotic treatments have on the commensal flora of the bacterial microbiome (Liautard *et al.*, 2006).

The virulence of *Brucella* spp. resides in its capability to enter, survive and replicate within phagocytic and non-phagocytic cells, among which macrophages are a major target in infected mammals. In the early phase of infection, the bacteria are located in a strongly acidified vacuole. This transient acidification of the vacuole is required for intracellular multiplication of the pathogen (Porte *et al.*, 1999). It allows the expression of a set of genes such as those encoding the major virulence factor VirB, essential for building of the replicative niche (Boschiroli *et al.*, 2002). Several genetic screens using transposon mutagenesis identified genes in *Brucella* that are required for survival, multiplication and adaptation to the environmental conditions encountered in the vacuole (for a review see Delrue *et al.*, 2004). The analysis of the function of these genes yielded information about the characteristic properties of the environment encountered by the pathogen within the *Brucella*-containing vacuole (Köhler *et al.*, 2003). This replicative niche is poor in certain nutritive elements, as numerous genes involved in the biosynthesis of amino acids and nucleotides are required for intramacrophagic multiplication (Köhler *et al.*, 2002). In addition, the essential role of the VirB system for *Brucella* virulence has been confirmed in different studies (O'Callaghan *et al.*, 1999; Hong *et al.*, 2000; Delrue *et al.*, 2001). The *virB* genes of *Brucella* species encode a type IV secretion system, a membrane-bound multiprotein complex that translocates virulence factors from *Brucella* and many other Gram-negative pathogens into eukaryotic cells (Christie *et al.*, 2005; de Jong *et al.*, 2008; Alvarez-Martínez and Christie, 2009). Structural and functional analyses have revealed many details on the individual protein components of type IV secretion systems and the data suggest that several of them may be suited targets for *in silico* or *in vivo* screening approaches towards the development of drugs that target *Brucella* virulence (Chandran *et al.*, 2009; Fronzes *et al.*, 2009a,b; Baron, 2010).

Based on the above described work, this chapter will focus on the exploration of the VirB type IV secretion system and of two metabolic enzymes (acetohydroxy acid synthase (AHAS) and histidinol dehydrogenase (HDH)) involved in branched amino acid and histidine biosynthesis, respectively, as novel targets for the development of drugs acting specifically on *Brucella* inside the host cell. The same targets may also be of interest for the development of drugs against other intracellular pathogens with similar systems necessary for virulence.

Acetohydroxy acid synthase (AHAS) and histidinol dehydrogenase (HDH) as targets for novel antibacterial agents

Both enzymes, acetohydroxy acid synthase (AHAS) and histidinol dehydrogenase (HDH), were identified as being encoded by genes essential in intramacrophagic replication of *B. suis*, hence belonging to the intramacrophagic virulome of the pathogen (Köhler *et al.*, 2002). AHAS, encoded by the genes *ilv* (loci BR1389 and BR1388), participates in the biosynthesis of isoleucine, leucine and valine, whereas HDH, encoded by *hisD* (BR0252), catalyses the last step of histidine biosynthesis. The essential character of both enzymes for intramacrophagic replication, together with other amino acid biosynthesis enzymes, led to the conclusion that the *Brucella*-containing vacuole is nutrient-poor (Köhler *et al.*, 2003). Another important aspect in the choice of both enzymes as potential targets of anti-infectious agents is the fact that these amino acid biosynthesis pathways are absent in mammals, therefore limiting the probability of secondary effects of enzyme inhibitors on the host. In contrast to classical antibiotic screening programmes, in which large collections of chemical products are systematically tested for an eventual effect on bacterial viability, here we describe an original, focused approach with defined biological targets known to be involved in the virulence of the pathogen of interest.

Acetohydroxy acid synthase as a target in *Brucella*

In contrast to mammals, bacteria, fungi, plants are able to synthesize all amino acids. Acetohydroxy acid synthase (AHAS; EC 2.2.1.6) has been studied in a wide range of organisms, such as *Escherichia coli*, yeast, and *Arabidopsis thaliana* with the aim of developing specific inhibitors (Chang and Duggleby, 1997; Hill *et al.*, 1997; Pang and Duggleby, 1999). Its activity is inhibited by sulfonylureas, which show low toxicity for mammals (Duggleby and Pang, 2000). These compounds have been commercialized and used efficiently as herbicides in the past. The observation by one of us (S.K.), that the *B. suis* 1330 *ilvI* (locus BR1389) encoding AHAS, as well as other genes of the biosynthesis pathway of branched amino acids are essential for intramacrophagic replication, led to the hypothesis that this enzyme might be a novel target for antimicrobial agents in intracellular bacteria (Liautard *et al.*, 2006).

Inhibition of AHAS activity in the presence of sulfonylureas

The activity of AHAS in *Brucella* was measured using the specific colorimetric end-point assay on *Brucella* lysates, according to established protocols (Epelbaum *et al.*, 1990; Hill *et al.*, 1997). Among the sulfonylureas investigated, chlorimuron ethyl (CE) and metsulfuron methyl (MSM) were the most effective in blocking AHAS activity in *B. suis*, with IC_{50} values in the nanomolar range (Fig. 12.1).

In minimal medium mimicking the presumably nutrient-poor *Brucella*-containing vacuole in the macrophage, wild-type brucellae lost their growth capacities in the presence of the sulfonylureas (Fig. 12.2). It is interesting to mention that the minimal inhibitory concentrations in brucellae were 1000-fold higher than the *in vitro* IC_{50} values (1–10 µmol/l for both sulfonylureas). One possible explanation for this observation is the fact that the bacterial envelope is a natural barrier to in- and efflux of ions and molecules, unless specific transport systems exist. In the case of these sulfonylureas, entry into the bacteria occurs most probably via passive diffusion only. As expected, these inhibitors do not have any effect on growth in complex tryptic soy broth in which amino acids are available, supporting the notion that the observed effects are not simply the consequence of general toxicity of these molecules.

Sulfonylureas block intramacrophagic replication of *Brucella* via their action on AHAS

A major aim of these studies was to evaluate the effectiveness of the sulfonylureas on blocking replication of *Brucella* in macrophage cells, as one of the authors (S.K.) published previously that a Tn5-mutant of *B. suis* affected in the AHAS-encoding gene *ilvI* was strongly attenuated in the classical macrophage infection model (Köhler *et al.*, 2002). Human THP-1 macrophage-like cells were used in the infection experiments (Caron *et al.*, 1994). The sulfonylureas CE or MSM were added to the macrophages infected with *B. suis* 1330 at 90 min post infection, applying concentrations of 1, 10, or 100 µmol/l. The lowest of the three concentrations resulted in 6- to 8-fold reduction of intracellular survival as compared with untreated cells. At 10 µmol/l, survival of *B. suis* was reduced 32- to 50-fold, and at the concentration of 100 µmol/l , the number of viable intracellular bacteria at 48 h post infection was identical to or lower than the number initially present at 90 min. Without inhibitor, the pathogen

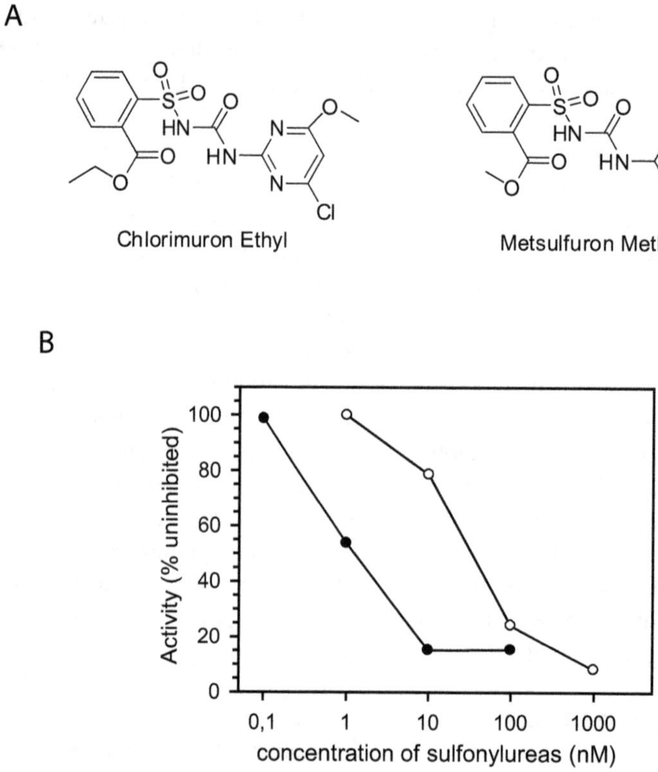

Figure 12.1 (A) Chemical structures of the sulfonylureas and inhibitors of AHAS chlorimuron ethyl (CE) and metsulfuron methyl (MSM). (B) Determination of the IC_{50}-values of CE (●) and MSM (O) using the endpoint assay for measurement of AHAS activity.

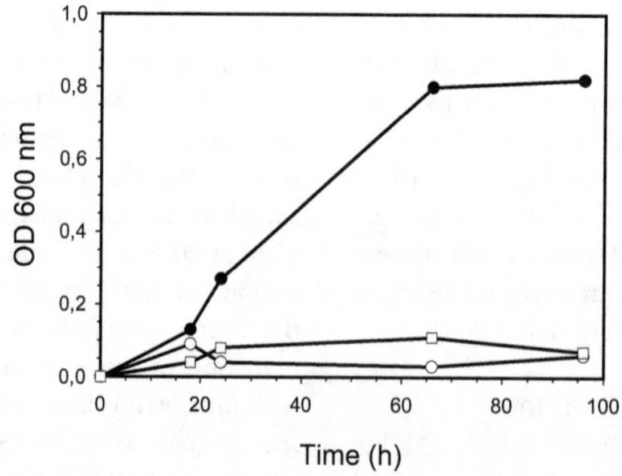

Figure 12.2 *In vitro* growth of *B. suis* 1330 in minimal medium without sulfonylureas (●), or in the presence of CE (□) or MSM (O) at a concentration of 10 µM.

multiplied 10^3-fold over the same period of time. Fig. 12.3 shows the effects of CE on intracellular *B. suis* at the concentration of 100 μmol/l. A direct toxic effect on the macrophages was excluded using viability staining with trypan blue.

Mutants spontaneously resistant to MSM were isolated on solid minimal medium containing MSM at a concentration of 10 μmol/l. As sulfonylurea resistance is linked to mutations of the active site of AHAS (Pang *et al.*, 2003), the question was raised of whether MSM-resistant mutants exhibited a reduced fitness in the macrophage. Three randomly chosen mutants had a significantly lower rate of intramacrophagic survival than the wild-type strain at 24h post infection (Boigegrain *et al.*, 2005). In addition, measurement of AHAS activity in the three mutants yielded significantly reduced activities as compared with the wild type. It was therefore concluded that sulfonylurea-resistant mutants are disadvantaged in their adaptation to the host cell, showing the self-limiting effect of these mutations (Boigegrain *et al.*, 2005).

These results clearly indicated that (i) *B. suis* needs to synthesize branched amino acids inside the macrophage vacuole in order to replicate within its niche, confirming results from Tn5-screening, (ii) sulfonylureas cross both the macrophage and the bacterial membranes, (iii) spontaneously resistant mutants are disadvantaged in host cell adaptation owing to impaired AHAS activity and (iv) AHAS may be a potentially interesting target for alternative antimicrobial therapy, aiming the pathogen at its intracellular state and limiting selective pressure.

Histidinol dehydrogenase as a target in *Brucella*

Histidinol dehydrogenase (HDH; EC 1.1.1.23) is a metalloenzyme, identified at the same time as AHAS as being essential for *Brucella* replication inside the macrophage (Köhler *et al.*, 2002). The essential character of histidine biosynthesis for virulence has been demonstrated in other pathogenic bacteria such as *S. typhimurium* (Fields *et al.*, 1986), *Burkholderia pseudomallei* (Pilatz *et al.*, 2006), and *Mycobacterium tuberculosis* (Parish, 2003). Moreover, HDH is absent from mammals and has been described only in bacteria, fungi and plants. Therefore, HDH has been proposed as a selective and promising target for the development of new antibacterial agents (Winum *et al.*, 2008). The use of HDH inhibitors as anti-brucellosis agents would inhibit the growth of the pathogen specifically in a histidine-poor environment such as the intramacrophagic niche. A major advantage of using these drugs would be the expected absence of side effects on the commensal floral of the host and the reduced rate of resistance development due to the reduced

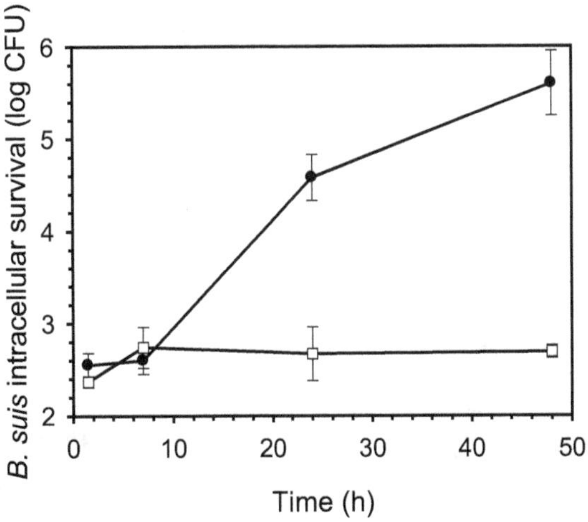

Figure 12.3 Intracellular replication of *B. suis* in human macrophage-like THP-1 cells in the absence (●) or presence (□) of 100 μM of the AHAS inhibitor CE.

selective pressure limited to intracellular pathogens.

We describe in this chapter the design, synthesis and evaluation of novel, highly potent and selective HDH inhibitors, and the assessment of their efficacy (1) on the purified target enzyme, (2) on the pathogen in minimal medium, and (3) on intracellular growth of *Brucella* in macrophage infection experiments.

Catalytic mechanism of HDH

Histidine biosynthesis is an ancient pathway found in bacteria, archaebacteria, fungi, and plants, and it is the biochemically and genetically most extensively studied pathway (Alifano *et al.*, 1996). This biosynthetic pathway converts 5-phosphorybosyl 1-pyrophosphate to l-histidine in ten enzymatic reactions. In *E. coli* and *S. typhimurium*, eight genes, organized in a single operon, encode enzymes involved in the ten steps of histidine biosynthesis. In *Brucella*, 11 genes have been annotated as being involved in histidine biosynthesis, and for three out of the ten enzymatic reactions, two genes have been assigned, respectively. Seven of the 11 genes are located in two distinct operons, one on each of the two chromosomes.

HDH catalyses the last two steps in the biosynthesis of l-histidine: sequential NAD-dependent oxidations of l-histidinol to l-histidinaldehyde and then to l-histidine. A catalytic mechanism has been proposed for the conversion of l-histidinol to l-histidine, which involves two consecutive oxidoreduction reactions accompanied by a reduction of two NAD^+ molecules (Barbosa *et al.*, 2002), as shown in Fig. 12.4.

HDH is a well-conserved metalloenzyme

HDH is a well-conserved metalloenzyme composed of two identical subunits with a molecular weight of 48 kDa each. Each subunit binds one Zn^{2+} cation as cofactor and removal of Zn^{2+} by chelation abolishes the enzymatic activity. Although the Zn^{2+} ion is essential for enzyme activity, Mn^{2+} or Cd^{2+} can replace Zn^{2+} in the formation of an active enzyme, indicating that the presence of divalent cations is essential for enzyme activity. Based on the three-dimensional structure of *E. coli* HDH in the apo state as well as complexed with Zn^{2+}, L-histidinol and NAD^+, it has been proposed recently that Zn^{2+} plays a crucial role in substrate binding, without, however, being directly involved in catalysis (Barbosa *et al.*, 2002). This structure provides detailed information of the active site, allows the assignment of catalytic roles to specific residues, and shows the mode of NAD^+ binding. HDH from cabbage, *Saccharomyces cerevisiae* and *E. coli* have amino acids identities of approximately 50%, and HDH from cabbage is biologically active in *E. coli*, showing high conservation (Nagai *et al.*, 1991). Although the pathway of histidine biosynthesis has been elucidated in various organisms, to date HDH was cloned and characterized only from three species of bacteria, *S. typhimurium*, *E. coli* and *B. suis* (Teng *et al.*, 1993; Teng and Grubmeyer, 1999; Barbosa *et al.*, 2002; Abdo *et al.*, 2007).

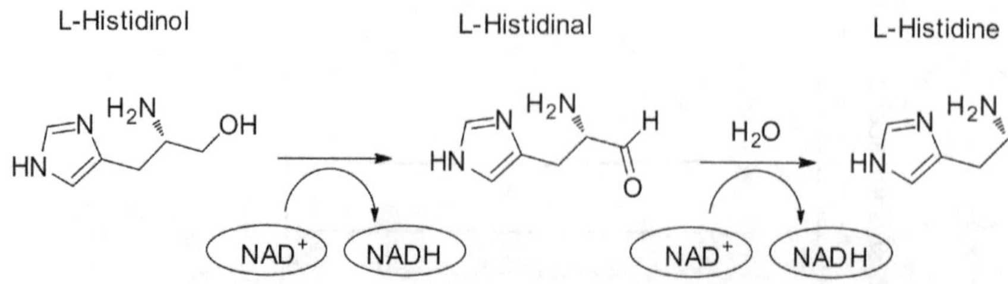

Figure 12.4 Representation of the last two steps of histidine biosynthesis, i.e. the oxydation of histidinol via histidinal to histidine. During this reaction, two molecules of NAD^+ per molecule of histidinol are reduced to NADH.

Inhibition of *B. suis* HDH by substituted benzylic ketones

Fourteen years ago, HDH was proposed as a suitable target for the development of herbicides. At that time, the approach developed focused on the design and synthesis of inhibitors blocking the activity of cabbage HDH. These inhibitors targeted the lipophilic binding pocket adjoining the active site of the enzyme (Dancer *et al.*, 1996).

Based on this previous report and with the aim to inhibit the HDH of *Brucella*, a series of fifteen substituted benzylic ketones derived from histidine whose structures are shown in Fig. 12.5, have been designed and synthesized recently. These molecules, which are analogues of the natural substrate histidinol, belong to the same class as the previously described cabbage HDH inhibitors (Dancer *et al.*, 1996), and their synthesis has been well described (Abdo *et al.*, 2007).

The potential inhibitory activity of these substituted benzylic ketones on *Brucella* HDH was tested at different levels: (i) on the purified HDH from *B. suis*, obtained as recombinant protein, (ii) on the *in vitro*-growth of *Brucella* in minimal medium and (iii) on the intramacrophagic multiplication of brucellae.

Inhibition of the enzymatic activity of B. suis HDH

The gene *hisD* (BR0252) encoding HDH of *B. suis* was cloned and over expressed in *E. coli* as an N-terminal hexahistidine-tagged fusion protein, then purified to further characterize its biological activity and to evaluate the inhibitory effect of the newly synthesized molecules. The results show that most of the compounds strongly inhibited *B. suis* HDH, the 50% inhibitory concentration (IC_{50}) being in the nanomolar range (Table 12.1).

The five HDH inhibitors previously described for cabbage HDH (1a, 1c, 1d, 1e, 1g, shown in Fig. 12.5) were between 2.5 and 25 times more active on *B. suis* HDH than on cabbage HDH (Abdo *et al.*, 2007). Among the 15 benzylic ketones, the most effective compound was the newly synthesized benzyloxy-1i (Fig. 12.6) with an IC_{50} of 3 nmol/l (Abdo *et al.*, 2007).

Several major observations could be made regarding the structure of the molecules: (i) the increase of the volume of substituent R on position 4 of the phenyl ring increases the inhibitory activity of the drug (1d > 1c > 1b), (ii) the increase of the lipophilic nature of R yields a better inhibitory activity (1e > 1a > 1g), (iii) the structure of compound 1i, which was characterized by a longer R-chain and which resulted in a stronger inhibitory effect than that observed with compound 1h, may give an indication of the distance between the active site and the lipophilic pocket, and (iv) the flexibility of the structure may play a key role in the adaptation to the catalytic site as compound 1i showed higher activity than compounds 1e and 1m.

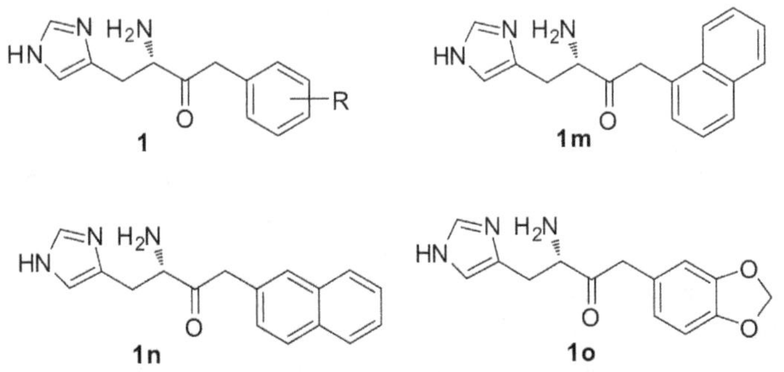

	R
1a	-H
1b	4-F
1c	4-Cl
1d	4-Br
1e	4-Ph
1f	4-CH_3
1g	4-OCH_3
1h	4-OPh
1i	4-OCH_2Ph
1j	2,4-F
1k	3,4-Cl
1l	Perfluoro

Figure 12.5 Structures of the substituted benzylic ketones derived from histidine and tested as potential inhibitors of histidinol dehydrogenase of *B. suis*.

Table 12.1 IC_{50} concentrations of the substituted benzylic ketones shown in Fig. 12.5, resulting in a 50% inhibition of *B. suis* HDH activity

Compounds	HDH (*Brucella suis*) IC_{50} (nM)
	40
1b	15
1c	12
1d	6
1e	12.5
1f	15
1g	150
1h	16
1i	3
1j	10
1k	20
1l	70
1m	200
1n	14.5
1o	25

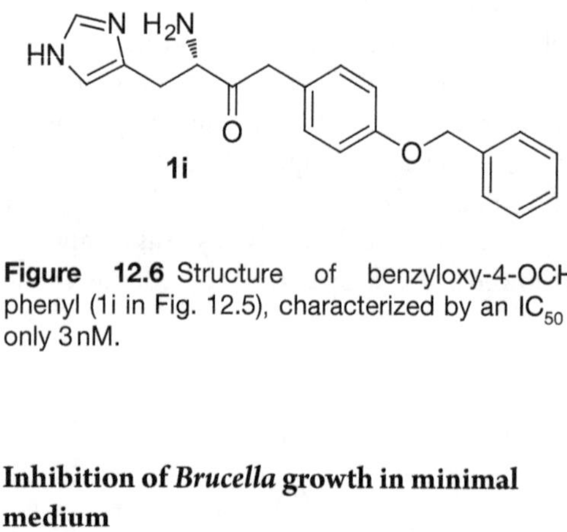

1i

Figure 12.6 Structure of benzyloxy-4-OCH_2-phenyl (1i in Fig. 12.5), characterized by an IC_{50} of only 3 nM.

Inhibition of *Brucella* growth in minimal medium

Following these enzymatic assays, the activity of the drugs on *Brucella* growth was investigated using a culture model with Gerhardt's modified minimal medium (Joseph *et al.*, 2007). Under these conditions, the inhibition of HDH by the drugs would be expected to abolish the capacity of the pathogen to grow, as an active histidine biosynthesis pathway is required in minimal medium devoid of histidine. The results showed that the inhibitors most active on the purified HDH (1b, 1c, 1d, 1e, 1n and 1i) were also most effective in blocking growth in minimal medium, but in the micromolar range at approximately 25 μmol/l (Fig. 12.7 and Joseph *et al.*, 2007). Thus, inhibition of the growth of the pathogen strongly correlates with the inhibition of HDH activity (Joseph *et al.*, 2007). A control experiment confirmed, as expected, that the growth of *Brucella* in rich medium (tryptic soy broth), which contains all amino acids, was not affected by these inhibitors. Under such conditions, the bacteria do not need an active histidine biosynthesis pathway and inhibition of HDH therefore remains without biological effect. The addition of histidine to the *Brucella* culture in minimal medium containing benzylic ketones relieves growth inhibition (Joseph *et al.*, 2007). This simple competition experiment demonstrates the inhibitory effect of the drugs on histidine biosynthesis *in vivo*, obviously targeting HDH.

Inhibition of intramacrophagic replication of *Brucella*

The effects of the most active drugs were measured on the intramacrophagic replication of *B. suis* by using an *in vitro* infection model with human macrophage-like THP1 cells. Overall, the inhibition of intramacrophagic growth was observed for inhibitor concentrations in the range of 10–25 μmol/l , and intracellular growth was reduced from 50- to 2500-fold as compared with *Brucella* replication in untreated cells (Joseph *et al.*, 2007). Following the previous observation of a specific biological effect of the drugs on the extracellular growth in minimal medium devoid of histidine, and because histidine biosynthesis is essential for the intramacrophagic replication of *B. suis*, the inhibition of intracellular replication is most probably due to the direct inhibition of HDH activity (Köhler *et al.*, 2002). The number of intracellular bacteria at the early stage of macrophage infection was drastically reduced following addition, for example, of compounds 1e or 1i, remaining at a low level until the end of the experiment at 24 h post infection (Fig. 12.8).

This result suggests that the active drug is stable and probably resistant to degradation by the bacteria under these experimental conditions. Moreover, it indicates that the drug efficiently crossed the macrophage membrane, the membrane forming the *Brucella*-containing vacuole, and the bacterial membranes, to reach

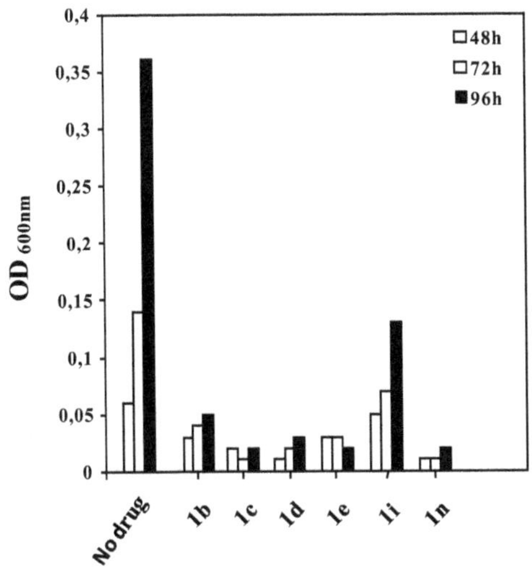

Figure 12.7 Inhibition of *Brucella* growth in minimal medium by HDH inhibitors1b, 1c, 1d, 1e, 1n and 1i, most active on the purified enzyme, at concentrations of 100 μM.

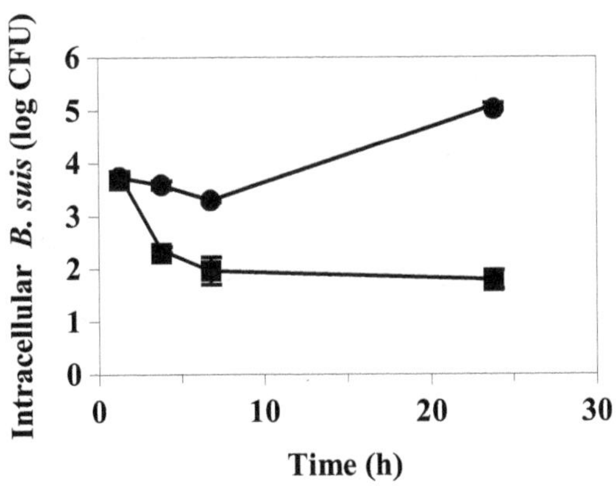

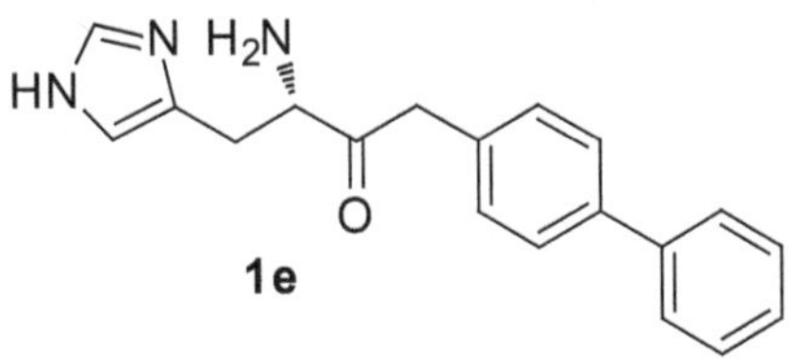

1e

Figure 12.8 Inhibition of intracellular replication of *B. suis* in human macrophage-like THP-1 cells following addition of the HDH inhibitor 1e (see also Fig. 12.5) at a concentration of 25 μM (■). *Brucella* growth within untreated cells is shown as control (●).

the cytoplasmic HDH target. The molecule 1i, which is also most effective on the purified HDH, therefore constitutes a promising anti-infectious lead molecule against brucellae.

The *Brucella* type IV secretion system as drug target

Secretions systems are essential for the virulence of many bacterial pathogens and they are increasingly being recognized as suited targets for the development of drugs that disarm bacteria. The *Brucella* type IV secretion system contains 12 cell envelope-associated VirB proteins (VirB1-VirB12) that form a conduit for the translocation of virulence factors into host cells (O'Callaghan *et al.*, 1999; Christie *et al.*, 2005; Fronzes *et al.*, 2009a) (see also Chapter 11). When individual *virB* genes were deleted it became apparent that most of them are essential for virulence and the gene products may therefore be suited as targets for drugs that inhibit intracellular survival (den Hartigh *et al.*, 2004, 2008) (Fig. 12.9). Whereas the knowledge on the individual components and structures of the different types of secretion systems has been growing steadily, work on their exploitation as drug targets has started more recently.

Previous work in other organisms and secretion systems

Type III secretion systems (T3SS) are the most advanced models for the approach of anti-virulence drugs. T3SS are widespread among Gram-negative pathogens such as *Chlamydia, Escherichia, Salmonella, Shigella* and *Yersinia* where they form needle-like structures on the bacterial surface that make contact with host cells and inject virulence factors (Mota *et al.*, 2005; Moraes *et al.*, 2008). Different laboratories have isolated promising T3SS inhibitors using *in vivo* screens aimed at identifying secretion defects (Keyser *et al.*, 2008). This approach takes advantage of the fact that under certain conditions many T3SS secrete virulence factors into the cell supernatant and that secretion is coupled to transcription, which can be exploited for the design of assays that screen for secretion system defects (Kauppi *et al.*, 2003; Gauthier *et al.*, 2005; Nordfelth *et al.*, 2005). Interestingly, many of these molecules belong to the chemical class of salicylidene acylhydrazides, or they were similar in structure, suggesting that they may hit common targets in *Yersinia* and *Escherichia* T3SS. Even more interestingly, some of these molecules were shown to be active against a broad spectrum of pathogens that depend on T3SS such as *Salmonella* and *Chlamydia* (Muschiol *et al.*, 2006; Negrea *et al.*, 2007). However, it is still not certain whether their primary target is a secretion system component or a regulator of transcription (Tree *et al.*, 2009; Veenendaal *et al.*, 2009). Clearly, despite the promising data on this interesting class of molecules, the absence of knowledge on their target(s) and binding site(s) somewhat limits the possibilities for their further development and of structure–activity relationship analysis by medicinal chemistry.

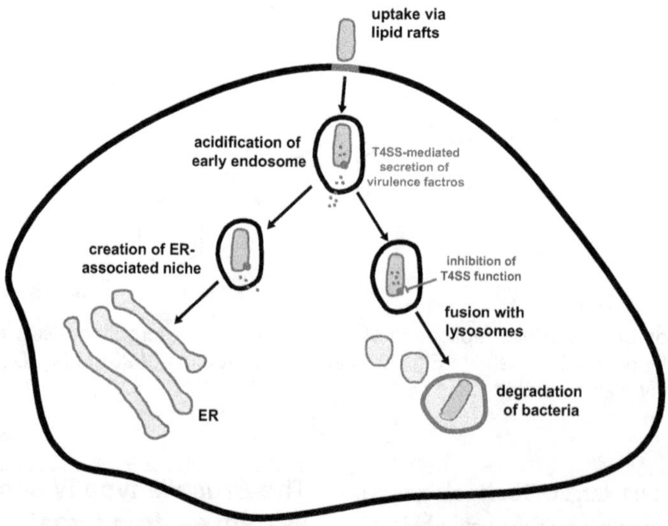

Figure 12.9 Targeting of the type IV secretion system with small molecules will inhibit intracellular survival of *Brucella*.

Similar to the approach chosen for type III secretion systems, an *in vivo* functional screen was used to isolate inhibitors of T4SS-mediated transfer of the antibiotic resistance plasmid R388 (Fernández-López *et al.*, 2005). The most active molecules identified from a library of 12,000 extracts of natural products were unsaturated fatty acids. It is not certain whether they have a specific impact on R388 plasmid transfer or whether their activity is due to a more general effect on membrane permeability. In contrast, a purified type IV secretion component (the *Helicobacter pylori* VirB11 ATPase) was used for high-throughput screening (HTS) of ATPase inhibitors from a chemical library of more than 500,000 small molecules. This industry-scale approach identified three molecules with promising inhibitory potential against the purified target (inhibitory constant lower than 1 μmol/l) (Hilleringmann *et al.*, 2006). At a concentration of 50 μmol/l, one of these molecules was active *in vivo* as measured by the inhibition of translocation of the CagA virulence factor from *H. pylori* to mammalian cells. Future work will show whether these promising molecules are specific for *Helicobacter* VirB11 or whether they have broader activity and whether their potential can be further increased by medicinal chemistry.

Target selection in the *Brucella* type IV secretion system

Based on previous work in many groups on the *Agrobacterium* model system (Das and Xie, 2000; Ward *et al.*, 2002; Judd *et al.*, 2005; Aguilar *et al.*, 2010; Sivanesan and Baron, 2011), the VirB8 protein was recognized as an essential T4SS assembly factor that interacts with many other VirB proteins. VirB8 may therefore be a druggable Achilles heel of the secretion system (Baron, 2006). The crystal structures of the periplasmic domains of VirB8 homologues from *Brucella* and from *Agrobacterium* were determined suggesting that the protein forms a homodimer (Terradot *et al.*, 2005; Bailey *et al.*, 2006). In the following, structure–function analyses of putative protein interaction interfaces revealed that amino acids at the putative dimerization interface of VirB8 are important for functionality of the *Agrobacterium* and *Brucella* secretion systems, respectively (Paschos *et al.*, 2006; Sivanesan and Baron, 2011). VirB8 interactions with other VirB proteins as well as its dimerization were next characterized *in vitro* with purified components and *in vivo* using the bacterial two-hybrid system. Both approaches supported the notion that VirB8 mediates dynamic interactions between the core components of the secretion system thus providing information on T4SS assembly (Sivanesan *et al.*, 2010). Previously, the yeast two-hybrid system was employed for the analysis of interactions between secretion system components and this gave important insights into their interactions (Kumar and Das, 2001; Ward *et al.*, 2002). However, the application of the bacterial two-hybrid system (Karimova *et al.*, 2005) permits analysis of these proteins in cellular compartments (periplasm of a Gram-negative cell) and membranes (cytoplasmic membrane of bacteria) that are very similar to the natural biological context. Therefore, the results are probably biologically more significant than those obtained with the yeast two-hybrid system, which place VirB proteins inside the nucleus of a eukaryotic cell.

Target-based screening for VirB8 interaction inhibitors

Once VirB8 was chosen as target, the next step was the design of a screen for the isolation of inhibitors. Traditionally, *in vitro* screens with purified components are considered to be the most robust by the pharmaceutical industry and following this approach, inhibitors of *H. pylori* VirB11 ATPase were identified (Hilleringmann *et al.*, 2006). In contrast, we exploited the bacterial two-hybrid system-based assay for inhibitor screens and this approach has several important advantages. First, this approach eliminates toxic compounds that kill bacteria and could therefore not serve as anti-virulence drugs. Second, active molecules isolated following this approach cross the bacterial outer membrane and therefore reach their target in the cell. In contrast, extensive chemical modification is often necessary to ensure that compounds identified by *in vitro* screening reach their target in the natural environment. Third, and most importantly, VirB8 is inhibited in a cellular compartment and while adopting a folding state that is probably very similar to that in its natural biological environment.

Using this approach, we isolated 20 potent and specific inhibitors of VirB8 interactions that belong to different chemical classes from a chemical library of 30,000 small molecules (Paschos *et al.*, 2011). One of these molecules, named B8I-2, inhibited the replication of *B. abortus* 2308 at low micromolar concentrations in a macrophage infection model and the time course of inhibitor action suggested that it works early in the infection process (Paschos *et al.*, 2011). Surprisingly, this molecule belongs to the chemical class of salicylidene acylhydrazides that were repeatedly identified as T3SS inhibitors. Considering that there is no similarity at the structural and sequence level between components of type III and type IV secretion systems, this discovery could constitute a mere coincidence, but it may also point to some common characteristics of inhibitor action. In support of this possibility, we found that B8I-2 and many other not structurally related VirB8 interaction inhibitors reduced the transcription of the *B. abortus virB* operon (Paschos *et al.*, 2011), which is reminiscent of the effects of T3SS inhibitors on transcription of genes encoding components of T3SS in other organisms (Tree *et al.*, 2009). One possibility that may explain these effects is that inhibition of the interactions between secretion system components during the assembly process may lead to misfolding and/or degradation of periplasmic proteins, thereby triggering a feedback loop that negatively regulates gene expression (Fig. 12.10). Alternatively, non-secreted substrates could induce a feedback loop blocking *virB* gene transcription and this may equally explain similar effects on T3SS and T4SS. Future work, e.g. using transcriptomics approaches, will identify the genes regulated by B8I-2 and other VirB8 interaction inhibitors and reveal how the inhibition of VirB8 interactions impacts the regulation of secretion system function(s).

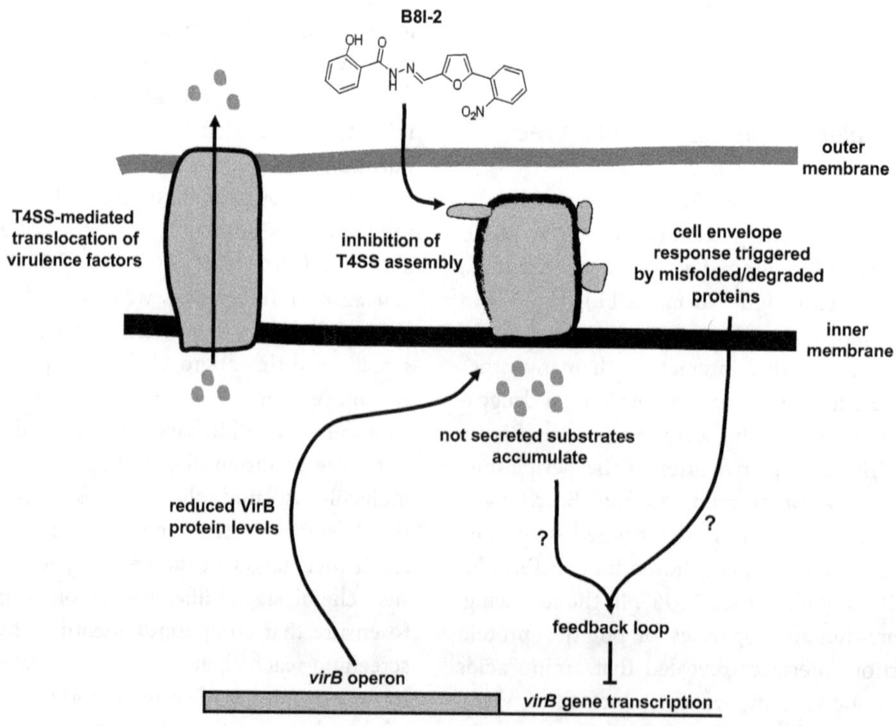

Figure 12.10 Models for the mode of action of VirB8 interaction inhibitors in *Brucella*.

Impact and next steps towards the development of anti-virulence drugs

The application of the bacterial two-hybrid system for inhibitor screens has led to the identification of the first small molecule hit that inhibits the protein–protein interactions of an essential T4SS component and that is active in an infection model (Paschos *et al.*, 2011). Whereas B8I-2 and some of the other inhibitors have interesting potential, they will need to undergo modification by medicinal chemistry to be active in the nanomolar concentration range before they can be considered as leads for drug development. To this end, *in vitro* interaction assays of inhibitor binding to VirB8 need to be established in order to correlate target binding with its activity in the BTH assay that may be impacted by membrane permeation and modifying enzymes. The X-ray structure of VirB8 is known and it will also be important to obtain co-crystal structures with B8I-2 and other inhibitors to determine the binding site on VirB8. With this information in hand, a structure–activity relationship (SAR) programme could be pursued in order to develop these molecules into potent leads to be assessed using tissue culture as well as mouse infection models. Once the proof of concept for their efficacy in animal models has been provided, clinical trials could be initiated to determine whether T4SS inhibitors are active as therapeutics. Alternatively, it would be interesting to determine whether they shorten the treatment time of a combination therapy of classical antibiotics of currently six weeks; this would significantly improve the quality of life of infected patients.

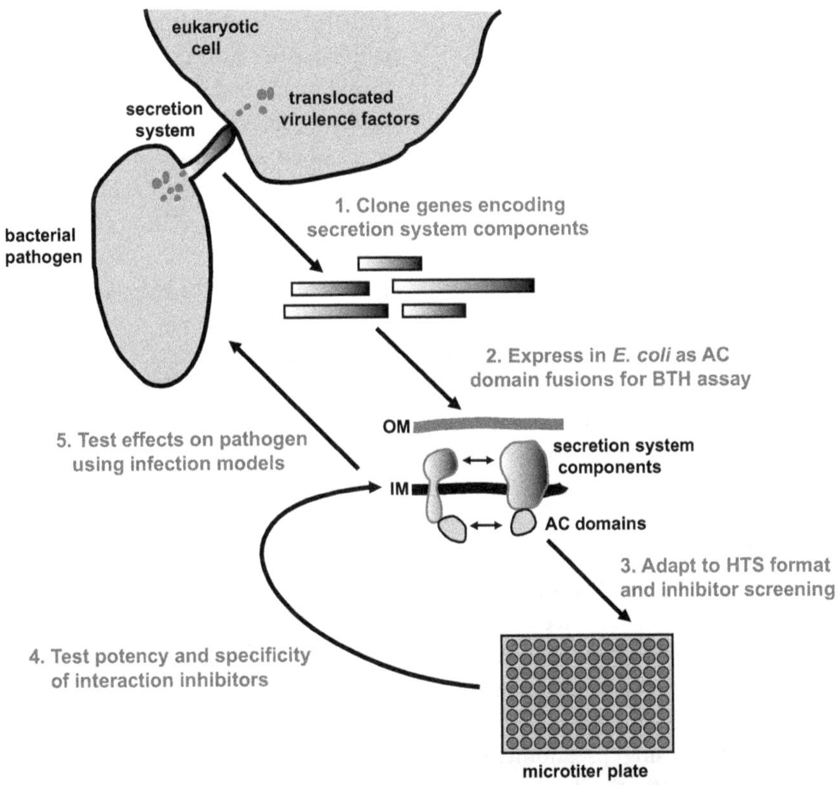

Figure 12.11 A generalized strategy for the development of anti-virulence drugs targeting bacterial secretion systems. In the BTH assay, the interaction between proteins fused to AC domains is monitored via restoration of cAMP formation leading to the induction of β-galactosidase activity in *E. coli* reporter strains (AC, adenylate cyclase; IM, inner membrane; OM, outer membrane; HTS, high-throughput screening; BTH, bacterial two-hybrid).

Since protein–protein interactions are important for all secretion systems, the application of the bacterial two-hybrid system for inhibitor screens may constitute a strategy that is generally applicable for the isolation of inhibitors of bacterial virulence functions from Gram-negative as well as from Gram-positive pathogens. This includes very important human pathogens such as *Mycobacterium tuberculosis* and *Staphylococcus aureus* that share the more recently discovered type VII secretion systems as virulence factors (Burts *et al.*, 2008; Simeone *et al.*, 2009; Chen *et al.*, 2010). Such a generalized strategy would require the establishment of a pipeline for expression of secretion system components, screening for strong and reproducible interactions using the BTH assay (alternatively: *in vitro* interaction assays), adaptation to the HTS format, screening including appropriate negative follow-up controls, followed by the assessment of inhibitor potency using suited biochemical as well as infection assays (Fig. 12.11).

Advantages of targeting specifically amino acid biosynthesis and virulence pathways of the pathogen at the intracellular state

One important advantage of using inhibitors of amino acid biosynthesis as well as of bacterial virulence functions is that they may limit the selective pressure for resistance to the intracellular niche where the targets are essential. As compared with classical antibiotics that kill bacteria or stop vital functions under most conditions, this should result in strongly reduced selection pressure for resistance. In addition, such molecules should cause little or no damage to the commensal bacterial flora. Disturbance of the balance of the intestinal microbiome is one of the most common side effect of antibiotic treatments and may even lead to superinfections with other pathogens such as *Clostridium difficile* causing pseudomembranous colitis (Rupnik *et al.*, 2009). Clearly, the use of pathogen-specific anti-virulence drugs would reduce such side-effects and it is an open question whether and when resistance will develop.

Efforts have been made to isolate spontaneously resistant mutants in minimal medium in the presence of high concentrations of various drugs affecting the activities of the amino acid biosynthetic enzymes (AHAS or HDH). Either no mutant was isolated, as it was the case with HDH inhibitors, indicating that enzymatic activity was incompatible with target resistance to these inhibitors (Joseph *et al.*, 2007). Alternatively, the mutants were characterized by reduced enzymatic activity and reduced fitness inside the macrophage, as observed with mutants resistant to AHAS inhibitors (Boigegrain *et al.*, 2005). Similar studies have not been conducted yet with the VirB8 protein, but it would be of scientific interest to isolate inhibitor-resistant variants, e.g. using the BTH assay for screening, followed by tests in infection models. A definite answer to this question will have to wait for the introduction of anti-virulence drugs into clinical practice.

In the future, the effectiveness of the above-described classes of inhibitors and their eventual toxicity on the host organism will have to be addressed in murine infection experiments, followed by clinical studies. The research described in this chapter may be the basis for the development of an alternative, non-classical antibacterial or anti-virulence therapy, notably against strains of brucellae or other intracellular pathogens with similar strategies of adaptation, resistant to conventional antibiotic treatments.

References

Abdo, M.R., Joseph, P., Boigegrain, R.A., Liautard, J.P., Montero, J.L., Köhler, S., and Winum, J.Y. (2007). *Brucella suis* histidinol dehydrogenase: synthesis and inhibition studies of a series of substituted benzylic ketones derived from histidine. Bioorg. Med. Chem. *15*, 4427–4433.

Aguilar, J., Zupan, J., Cameron, T.A., and Zambryski, P.C. (2010). *Agrobacterium* type IV secretion system and its substrates form helical arrays around the circumference of virulence-induced cells. Proc. Natl. Acad. Sci. U.S.A. *107*, 3758–3763.

Alifano, P., Fani, R., Lio, P., Lazcano, A., Bazzicalupo, M., Carlomagno, M.S., and Bruni, C.B. (1996). Histidine biosynthetic pathway and genes: structure, regulation, and evolution. Microbiol. Rev. *60*, 44–69.

Alksne, L.E. (2002). Virulence as a target for antimicrobial chemotherapy. Expert. Opin. Investig. Drugs *11*, 1149–1159.

Alvarez-Martínez, C.E., and Christie, P.J. (2009). Biological diversity of prokaryotic type IV secretion systems. Microbiol. Mol. Biol. Rev. 73, 775–808.

Ariza, J., Bosilkovski, M., Cascio, A., Colmenero, J.D., Corbel, M.J., Falagas, M.E., Memish, Z.A., Roushan, M.R., Rubinstein, E., Sipsas, N.V., *et al.* (2007). Perspectives for the treatment of brucellosis in the 21st century: the Ioannina recommendations. PLoS Med. *4*, e317.

Bailey, S., Ward, D., Middleton, R., Grossmann, J.G., and Zambryski, P. (2006). *Agrobacterium tumefaciens* VirB8 structure reveals potential protein–protein interactions sites. Proc. Natl. Acad. Sci. U.S.A. *103*, 2582–2587.

Barbosa, J.A., Sivaraman, J., Li, Y., Larocque, R., Matte, A., Schrag, J.D., and Cygler, M. (2002). Mechanism of action and NAD+-binding mode revealed by the crystal structure of L-histidinol dehydrogenase. Proc. Natl. Acad. Sci. U.S.A. *99*, 1859–1864.

Baron, C. (2006). VirB8, A conserved type IV secretion system assembly factor and drug target. Biochem. Cell. Biol. *84*, 890–899.

Baron, C. (2010). Antivirulence drugs to target bacterial secretion systems. Curr. Opin. Microbiol. *13*, 100–105.

Baykam, N., Esener, H., Ergonul, O., Eren, S., Celikbas, A.K., and Dokuzoguz, B. (2004). *In vitro* antimicrobial susceptibility of *Brucella* species. Int. J. Antimicrob. Agents *23*, 405–407.

Boigegrain, R.A., Liautard, J.P., and Köhler, S. (2005). Targeting of the virulence factor acetohydroxyacid synthase by sulfonylureas results in inhibition of intramacrophagic multiplication of *Brucella suis*. Antimicrob. Agents Chemother. *49*, 3922–3925.

Boschiroli, M.L., Ouahrani-Bettache, S., Foulongne, V., Michaux-Charachon, S., Bourg, G., Allardet-Servent, A., Cazevieille, C., Liautard, J.P., Ramuz, M., and O'Callaghan, D. (2002). The *Brucella suis virB* operon is induced intracellularly in macrophages. Proc. Natl. Acad. Sci. U.S.A. *99*, 1544–1549.

Burts, M.L., DeDent, A.C., and Missiakas, D.M. (2008). EsaC substrate for the ESAT-6 secretion pathway and its role in persistent infections of *Staphylococcus aureus*. Mol. Microbiol. *69*, 736–746.

Caron, E., Liautard, J.P., and Köhler, S. (1994). Differentiated U937 cells exhibit increased bactericidal activity upon LPS activation and discriminate between virulent and avirulent *Listeria* and *Brucella* species. J. Leukoc. Biol. *56*, 174–181.

Chandran, V., Fronzes, R., Duquerroy, S., Cronin, N., Navaza, J., and Waksman, G. (2009). Structure of the outer membrane complex of a type IV secretion system. Nature *462*, 1011–1015.

Chang, A.K., and Duggleby, R.G. (1997). Expression, purification and characterization of *Arabidopsis thaliana* acetohydroxyacid synthase. Biochem. J 327 *(Pt 1)*, 161–169.

Chen, J.M., Pojer, F., Blasco, B., and Cole, S.T. (2010). Towards anti-virulence drugs targeting ESX-1 mediated pathogenesis of *Mycobacterium tuberculosis*. Drug. Disc. Today: Disease Mech. 7, 25–31.

Christie, P.J., Atmakuri, K., Krishnamoorthy, V., Jakubowski, S., and Cascales, E. (2005). Biogenesis, architecture, and function of bacterial type IV secretion systems. Annu. Rev. Microbiol. *59*, 451–485.

Clatworthy, A.E., Pierson, E., and Hung, D.T. (2007). Targeting virulence: a new paradigm for antimicrobial therapy. Nat. Chem. Biol. *3*, 541–548.

Corbel, M.J. (1997). Brucellosis: an overview. Emerg. Infect. Dis. *3*, 213–221.

Dancer, J.E., Ford, J.M., Hamilton, K., Kilkelly, M., Lindell, S.D., O'Mahony, M.J., and Saville-Stones, E.A. (1996). Synthesis of potent inhibitors of histidinol dehydrogenase. Bioorg. Med. Chem. Lett. *6*, 2131–2136.

Das, A., and Xie, Y.-H. (2000). The *Agrobacterium* T-DNA transport pore proteins VirB8, VirB9 and VirB10 interact with one another. J. Bacteriol. *182*, 758–763.

Delrue, R.M., Martínez-Lorenzo, M., Lestrate, P., Danese, I., Bielarz, V., Mertens, P., De Bolle, X., Tibor, A., Gorvel, J.P., and Letesson, J.J. (2001). Identification of *Brucella* spp. genes involved in intracellular trafficking. Cell. Microbiol. *3*, 487–497.

Delrue, R.M., Lestrate, P., Tibor, A., Letesson, J.J., and De Bolle, X. (2004). *Brucella* pathogenesis, genes identified from random large-scale screens. FEMS Microbiol. Lett. *231*, 1–12.

Duggleby, R.G., and Pang, S.S. (2000). Acetohydroxyacid synthase. J. Biochem. Mol. Biol. *33*, 1–36.

Epelbaum, S., Chipman, D.M., and Barak, Z. (1990). Determination of products of acetohydroxy acid synthase by the colorimetric method, revisited. Anal. Biochem. *191*, 96–99.

Fernández-López, R., Machon, C., Longshaw, C.M., Martin, S., Molin, S., Zechner, E.L., Espinosa, M., Lanka, E., and de la Cruz, F. (2005). Unsaturated fatty acids are inhibitors of bacterial conjugation. Microbiology *151*, 3517–3526.

Fields, P.I., Swanson, R.V., Haidaris, C.G., and Heffron, F. (1986). Mutants of *Salmonella typhimurium* that cannot survive within the macrophage are avirulent. Proc. Natl. Acad. Sci. U.S.A. *83*, 5189–5193.

Fischbach, M.A., and Walsh, C.T. (2009). Antibiotics for emerging pathogens. Science *325*, 1089–1093.

Fronzes, R., Christie, P.J., and Waksman, G. (2009a). The structural biology of type IV secretion systems. Nat. Rev. Microbiol. *7*, 703–714.

Fronzes, R., Schafer, E., Wang, L., Saibil, H.R., Orlova, E.V., and Waksman, G. (2009b). Structure of a type IV secretion system core complex. Science *323*, 266–268.

Gauthier, A., Robertson, M.L., Lowden, M., Ibarra, J.A., Puente, J.L., and Finlay, B.B. (2005). Transcriptional inhibitor of virulence factors in enteropathogenic *Escherichia coli*. Antimicrob. Agents. Chemother. *49*, 4101–4109.

Godfroid, J., Cloeckaert, A., Liautard, J.P., Köhler, S., Fretin, D., Walravens, K., Garin-Bastuji, B., and Letesson, J.J. (2005). From the discovery of the Malta fever's agent to the discovery of a marine mammal reservoir, brucellosis has continuously been a re-emerging zoonosis. Vet. Res. *36*, 313–326.

den Hartigh, A.B., Sun, Y.H., Sondervan, D., Heuvelmans, N., Reinders, M.O., Ficht, T.A., and Tsolis, R.M. (2004). Differential requirements for VirB1 and VirB2 during *Brucella abortus* infection. Infect. Immun. *72*, 5143–5149.

den Hartigh, A.B., Rolan, H.G., de Jong, M.F., and Tsolis, R.M. (2008). VirB3 to VirB6 and VirB8 to VirB11, but not VirB7, are essential for mediating persistence of *Brucella* in the reticuloendothelial system. J. Bacteriol. *190*, 4427–4436.

Hill, C.M., Pang, S.S., and Duggleby, R.G. (1997). Purification of *Escherichia coli* acetohydroxyacid synthase isoenzyme II and reconstitution of active enzyme from its individual pure subunits. Biochem. J. *327 (Pt 3)*, 891–898.

Hilleringmann, M., Pansegrau, W., Doyle, M., Kaufman, S., MacKichan, M.L., Gianfaldoni, C., Ruggiero, P., and Covacci, A. (2006). Inhibitors of *Helicobacter pylori* ATPase Cagalpha block CagA transport and cag virulence. Microbiology *152*, 2919–2930.

Hong, P.C., Tsolis, R.M., and Ficht, T.A. (2000). Identification of genes required for chronic persistence of *Brucella abortus* in mice. Infect. Immun. *68*, 4102–4107.

de Jong, M.F., Sun, Y.H., den Hartigh, A.B., van Dijl, J.M., and Tsolis, R.M. (2008). Identification of VceA and VceC, two members of the VjbR regulon that are translocated into macrophages by the *Brucella* type IV secretion system. Mol. Microbiol. *70*, 1378–1396.

Joseph, P., Abdo, M.R., Boigegrain, R.A., Montero, J.L., Winum, J.Y., and Köhler, S. (2007). Targeting of the *Brucella suis* virulence factor histidinol dehydrogenase by histidinol analogues results in inhibition of intramacrophagic multiplication of the pathogen. Antimicrob. Agents. Chemother. *51*, 3752–3755.

Judd, P.K., Kumar, R.B., and Das, A. (2005). Spatial location and requirements for the assembly of the *Agrobacterium tumefaciens* type IV secretion apparatus. Proc. Natl. Acad. Sci. U.S.A. *102*, 11498–11503.

Karimova, G., Dautin, N., and Ladant, D. (2005). Interaction network among *Escherichia coli* membrane proteins involved in cell division as revealed by bacterial two-hybrid analysis. J. Bacteriol. *187*, 2233–2243.

Kauppi, A.M., Nordfelth, R., Uvell, H., Wolf-Watz, H., and Elofsson, M. (2003). Targeting bacterial virulence: inhibitors of type III secretion in *Yersinia*. Chem. Biol. *10*, 241–249.

Keyser, P., Elofsson, M., Rosell, S., and Wolf-Watz, H. (2008). Virulence blockers as alternatives to antibiotics: type III secretion inhibitors against Gram-negative bacteria. J. Intern. Med. *264*, 17–29.

Köhler, S., Foulongne, V., Ouahrani-Bettache, S., Bourg, G., Teyssier, J., Ramuz, M., and Liautard, J.P. (2002). The analysis of the intramacrophagic virulome of *Brucella suis* deciphers the environment encountered by the pathogen inside the macrophage host cell. Proc. Natl. Acad. Sci. U.S.A. *99*, 15711–15716.

Köhler, S., Michaux-Charachon, S., Porte, F., Ramuz, M., and Liautard, J.P. (2003). What is the nature of the replicative niche of a stealthy bug named *Brucella*? Trends Microbiol. *11*, 215–219.

Kumar, R.B., and Das, A. (2001). Functional analysis of the *Agrobacterium tumefaciens* T-DNA transport pore protein VirB8. J. Bacteriol. *183*, 3636–3641.

Levy, S.B., and Marshall, B. (2004). Antibacterial resistance worldwide: causes, challenges and responses. Nat. Med. *10*, S122–129.

Liautard, J.P., Jubier-Maurin, V., Boigegrain, R.A., and Köhler, S. (2006). Antimicrobials: targeting virulence genes necessary for intracellular multiplication. Trends Microbiol. *14*, 109–113.

Memish, Z., Mah, M.W., Al Mahmoud, S., Al Shaalan, M., and Khan, M.Y. (2000). *Brucella* bacteraemia: clinical and laboratory observations in 160 patients. J. Infect. *40*, 59–63.

Moraes, T.F., Spreter, T., and Strynadka, N.C. (2008). Piecing together the type III injectisome of bacterial pathogens. Curr. Opin. Struct. Biol. *18*, 258–266.

Mota, L.J., Sorg, I., and Cornelis, G.R. (2005). Type III secretion: the bacteria-eukaryotic cell express. FEMS Microbiol. Lett. *252*, 1–10.

Muschiol, S., Bailey, L., Gylfe, A., Sundin, C., Hultenby, K., Bergstrom, S., Elofsson, M., Wolf-Watz, H., Normark, S., and Henriques-Normark, B. (2006). A small-molecule inhibitor of type III secretion inhibits different stages of the infectious cycle of *Chlamydia trachomatis*. Proc. Natl. Acad. Sci. U.S.A. *103*, 14566–14571.

Nagai, A., Ward, E., Beck, J., Tada, S., Chang, J.Y., Scheidegger, A., and Ryals, J. (1991). Structural and functional conservation of histidinol dehydrogenase between plants and microbes. Proc. Natl. Acad. Sci. U.S.A. *88*, 4133–4137.

Negrea, A., Bjur, E., Ygberg, S.E., Elofsson, M., Wolf-Watz, H., and Rhen, M. (2007). Salicylidene acylhydrazides that affect type III protein secretion in *Salmonella enterica* serovar typhimurium. Antimicrob. Agents Chemother. *51*, 2867–2876.

Nordfelth, R., Kauppi, A.M., Norberg, H.A., Wolf-Watz, H., and Elofsson, M. (2005). Small-molecule inhibitors specifically targeting type III secretion. Infect. Immun. *73*, 3104–3114.

O'Callaghan, D., Cazevieille, C., Allardet-Servent, A., Boschiroli, M.L., Bourg, G., Foulongne, V., Frutos, P., Kulakov, Y., and Ramuz, M. (1999). A homologue of the *Agrobacterium tumefaciens* VirB and *Bordetella pertussis* Ptl type IV secretion systems is essential for intracellular survival of *Brucella suis*. Mol. Microbiol. *33*, 1210–1220.

Pang, S.S., and Duggleby, R.G. (1999). Expression, purification, characterization, and reconstitution of the large and small subunits of yeast acetohydroxyacid synthase. Biochemistry *38*, 5222–5231.

Pang, S.S., Guddat, L.W., and Duggleby, R.G. (2003). Molecular basis of sulfonylurea herbicide inhibition of acetohydroxyacid synthase. J. Biol. Chem. *278*, 7639–7644.

Pappas, G., Akritidis, N., Bosilkovski, M., and Tsianos, E. (2005). Brucellosis. N. Engl. J. Med. *352*, 2325–2336.

Parish, T. (2003). Starvation survival response of Mycobacterium tuberculosis. J. Bacteriol. *185*, 6702–6706.

Paschos, A., Patey, G., Sivanesan, D., Gao, C., Bayliss, R., Waksman, G., O'Callaghan, D., and Baron, C. (2006). Dimerization and interactions of *Brucella suis* VirB8 with VirB4 and VirB10 are required for its biological activity. Proc. Natl. Acad. Sci. U.S.A. *103*, 7252–7257.

Paschos, A., den Hartigh, A., Smith, M., Atluri, V.L., Sivanesan, D., Tsolis, R.M., and Baron, C. (2011). An *in vivo* high-throughput screening approach targeting the type IV secretion system component VirB8 identified inhibitors of *Brucella abortus* 2308 proliferation. Infect. Immun. *79*, 1033–1043.

Perkins, S.D., Smither, S.J., and Atkins, H.S. (2010). Towards a *Brucella* vaccine for humans. FEMS Microbiol. Rev. Epub ahead of print.

Pilatz, S., Breitbach, K., Hein, N., Fehlhaber, B., Schulze, J., Brenneke, B., Eberl, L., and Steinmetz, I. (2006). Identification of *Burkholderia pseudomallei* genes required for the intracellular life cycle and *in vivo* virulence. Infect. Immun. *74*, 3576–3586.

Porte, F., Liautard, J.P., and Köhler, S. (1999). Early acidification of phagosomes containing *Brucella suis* is essential for intracellular survival in murine macrophages. Infect. Immun. *67*, 4041–4047.

Rupnik, M., Wilcox, M.H., and Gerding, D.N. (2009). Clostridium difficile infection: new developments in epidemiology and pathogenesis. Nat. Rev. Microbiol. *7*, 526–536.

Simeone, R., Bottai, D., and Brosch, R. (2009). ESX/type VII secretion systems and their role in host–pathogen interaction. Curr. Opin. Microbiol. *12*, 4–10.

Sivanesan, D., and Baron, C. (2011). The dimer interace of *Agrobacterium tumefaciens* VirB8 is important for type IV secretion system function, stability and for association of VirB2 with the core complex. J. Bacteriol. *193*, 2097–2106.

Sivanesan, D., Hancock, M.A., Villamil Giraldo, A.M., and Baron, C. (2010). Quantitative analysis of VirB8–VirB9-VirB10 interactions provides a dynamic model of type IV secretion system core complex assembly. Biochemistry *49*, 4483–4493.

Teng, H., and Grubmeyer, C. (1999). Mutagenesis of histidinol dehydrogenase reveals roles for conserved histidine residues. Biochemistry *38*, 7363–7371.

Teng, H., Segura, E., and Grubmeyer, C. (1993). Conserved cysteine residues of histidinol dehydrogenase are not involved in catalysis. Novel chemistry required for enzymatic aldehyde oxidation. J. Biol. Chem. *268*, 14182–14188.

Terradot, L., Bayliss, R., Oomen, C., Leonard, G., Baron, C., and Waksman, G. (2005). Crystal structures of the periplasmic domains of two core subunits of the bacterial type IV secretion system, VirB8 from *Brucella suis* and ComB10 from *Helicobacter pylori*. Proc. Natl. Acad. Sci. U.S.A. *102*, 4596–4601.

Tree, J.J., Wang, D., McInally, C., Mahajan, A., Layton, A., Houghton, I., Elofsson, M., Stevens, M.P., Gally, D.L., and Roe, A.J. (2009). Characterization of the effects of salicylidene acylhydrazide compounds on type III secretion in *Escherichia coli* O157, H7. Infect. Immun. *77*, 4209–4220.

Veenendaal, A.K., Sundin, C., and Blocker, A.J. (2009). Small-molecule type III secretion system inhibitors block assembly of the *Shigella* type III secreton. J. Bacteriol. *191*, 563–570.

Ward, D.V., Draper, O., Zupan, J.R., and Zambryski, P.C. (2002). Peptide linkage mapping of the *A. tumefaciens vir*-encoded type IV secretion system reveals novel protein subassemblies. Proc. Natl. Acad. Sci. U.S.A. 99, 11493–11500.

Winum, J.Y., Köhler, S., Scozzafava, A., Montero, J.L., and Supuran, C.T. (2008). Targeting bacterial metalloenzymes: A new strategy for the developement of anti-infective agents. Anti-infect. Agents. Med. Chem. 7, 169–179.

Molecular and Chemical Approaches to *Brucella* Vaccine Development

13

Thomas A. Ficht and Allison C. Rice-Ficht

Abstract

Development of improved *Brucella* vaccines has focused on the identification of the genes that support intracellular survival and those encoding antigens capable of stimulating immune protection. Evaluation of live, attenuated vaccines and subunit vaccines expressed using numerous vector-based systems has provided two general results: (i) vaccines providing little protection with elevated safety, and (ii) vaccines exhibiting excellent protection that are of questionable safety. In order to address the limitations associated with *Brucella* vaccinology a fresh look at the interactions between host and pathogen may provide renewed insight. This chapter attempts to reconcile the varied results reported in the literature related to the host response to the invading pathogen and the mechanisms used by the organism to avoid both innate and adaptive immune responses. By reviewing and summarizing the current literature we expect to reveal focal points for new investigations and that inconsistencies limiting such development may be identified and dealt with constructively.

Introduction

Efforts to enhance protection of animals and humans against *Brucella* infections have now spanned a century. Over the course of the twentieth century the control of animal disease was shown to simultaneously limit human exposure. As a result, human intervention strategies lagged behind animal intervention, despite the compounding development of bioweapons in the 1940–1960s. *Brucella* species were initially developed as bioweapons in the USA in the 1950s. Several physical properties of the organism presumably made them preferable candidates for development over agents posing far greater risks to public health, such as smallpox and anthrax, at least according to the public record. Although President Richard M. Nixon officially terminated the development of bioweapons in the USA in 1969, the twenty-first century has brought renewed fear of bioweapon use warranting development of intervention strategies for the direct protection of public health. At the same time, it should not be concluded that current strategies to prevent animal disease are ideal. In fact, there are many problems related to the use of available vaccines including insufficient human safety, reduced efficacy in important target species, as well as a complete lack of efficacy in wild-life reservoirs that may now be corrected by taking advantage of recent advances in scientific knowledge. The potential for parallel development of improved animal and human vaccines may be enhanced by comparison of similarities and differences in immune response mechanisms and tissue distribution in various hosts affecting the outcome of infection. This chapter will attempt to incorporate recent advances in basic knowledge concerning the pathogenesis of *Brucella* infection to better define vaccination strategies based on the continued use of live, attenuated organisms rather than subunit or DNA vaccines. The premise for this is to take advantage of properties of live, attenuated organism to deliver antigens for processing and presentation to target cells that is difficult, if not impossible, to do with current subunit vaccine formulations.

As originally described by Ko and Splitter, the immune response to *Brucella* may be divided into the innate response designed to decrease the number of infecting organisms and to establish a population of functional memory cells, and an adaptive response resulting from recall of these memory cells (Ko and Splitter, 2003). The innate immune response is designed to non-specifically reduce the number of invading organisms until the more potent and specific adaptive response develops. Owing to the 'stealthy' nature of the organism, the innate immune response is poorly activated favouring establishment of infection, and potentially suppressing the development of the adaptive response. Clues have emerged concerning the identity of bacterial products, and their potential role in preventing clearance of the organism from the host. In contrast to previous publications describing such factors, an effort will be made here to describe their effect at selected stages of infection. This is based on current knowledge and does not rule out the possibility that some factors may have multiple effects, some of which may already be known while others remain to be described.

Initial pathogen host interaction and innate immunity

Evaluation of the state of the host following exposure to *Brucella* confirms the stealthy nature of the agent. First, there is no sign of sepsis or coagulation in the blood stream. Although polymorphonuclear leucocytes (PMN) exhibit excellent killing activity, cellular recruitment is not observed (Barquero-Calvo *et al.*, 2007), and little demonstrable reduction in bacterial number can be attributed to PMNs and/or complement activity. In addition, cytokine levels remain low and pathogen-associated molecular pattern (PAMP) recognition by toll-like receptors (TLRs) is muted most probably by the lipopolysaccharide (LPS) on the cell surface that both limits activation of macrophage and restricts uptake of the organism. Similar to macrophage, dendritic cells (DC) are readily infected and represent potential reservoirs for latent infection and or recrudescence of the organism (Billard *et al.*, 2007). *Brucella* infected DCs do not fully mature which is thought to contribute to the failure to establish a protective, adaptive immune response (Archambaud *et al.*, 2010; Billard *et al.*, 2007; Salcedo *et al.*, 2008). The failure to identify specific mechanisms responsible for the stealthy nature of the organism has suggested that this ability best reflects passive rather than active mechanisms designed to prevent immune recognition.

Perhaps the most obvious of the bacterial factors controlling the fate of the organism is the LPS on the cell surface. *Brucella* LPS contains a homopolymer of N-formylperosamine (O-polysaccharide), attached to a core polysaccharide composed of mannose, glucose, quinovosamine and KDO attached to lipid A containing much longer acyl side chains (C28:0 and C30:0) than enterobacterial LPS. These long acyl groups are responsible for reduced toxicity exhibited by *Brucella* lipid A and the reduced recognition by TLR4 that limits activation of phagocytic cells. Another feature of the LPS attributed to the O-polysaccharide portion is reduced interaction with and uptake by phagocytic cells. Rough organisms lacking O-polysaccharide are taken up at a reduced rate, but in greater number than smooth organisms expressing full-length O-polysaccharide (Pei *et al.*, 2008). The basis for this difference was originally thought to derive from differences in interaction at the cell surface involving lipid rafts (Porte *et al.*, 2003; Kim *et al.*, 2004; Olsson and Sundler, 2006). However, recent results suggest that naturally occurring rough organisms also interact with lipid rafts (Martín-Martín *et al.*, 2010). Whether this is a property unique to naturally occurring rough organisms remains to be determined. Results using smooth and rough variants of classical strains confirm significant differences in the outcome of infection as a result of early interactions. Primarily, rough variants are unable to cause disease resulting from enhanced early killing and the failure to prevent phagosome–lysosome fusion (Porte *et al.*, 2003). Enhanced early killing is associated with phagocytic cell activation, as demonstrated by enhanced levels of MAPK activity and translocation of NFκB, leading to the excretion of proinflammatory cytokines

TNFα and IFNγ (Jimenez de Bagues *et al.*, 2005; Pei *et al.*, 2008). In contrast, organisms expressing O-polysaccharide acquire a replicative niche associated with limited phagocyte activation. The elevated rate of transmission among exposed hosts is undoubtedly due to the low infectious dose and is consistent with the need to restrict target cell activation. The biological and chemical basis for restricted macrophage activation resides in the weak TLR agonist activity of *Brucella* cell surface components as has been described in detail elsewhere (Barquero-Calvo *et al.*, 2007). Suffice it to say that the poor activation results in a reduced innate immune response enhancing the capability of the organism to establish a replicative niche, as well as preventing the development of an effective adaptive immune response resulting in the systemic spread of infected cells throughout the host. In contrast, the enhanced immune response observed with rough mutants may be caused by increased exposure of cell surface proteins including the type IV secretion system (T4SS), consistent with the rapid clearance of the organism following inoculation (Turse *et al*, 2011).

Historically the organism has been evaluated in mice infected intraperitoneally, however, infection occurs naturally in primary hosts and in humans as a result of consumption of food (oral), aerosol exposure (intranasal or conjunctival) or on the surface of damaged skin. Despite the difference in initial interactions between host and pathogen the progress of disease differs only slightly based on the route of infection. Following natural inoculation, the organism may be expected to enter the host via M cells of the Peyer's patches located in the lymphoid tissue associated with the mucosa and other tissue resulting in systemic spread of the organism. *Brucella* has always been considered to be rapidly taken up prior to the oropharynx from which it is transferred to regional lymph nodes and the lymphoreticular system before ultimate detection in virtually any tissue/organ. Following the primary route of inoculation, initial interactions are expected to include exposure to complement, neutrophils, natural killer (NK) cells, macrophages and DC. The actual contributions of these factors to the ultimate outcome of infection has been estimated based on *in vitro* models and/or genetic or biochemical depletion of specific factors. Obviously, these approaches have their limitations, and the participation of host factors in the clearance of the organism may differ depending on the actual route of inoculation.

Despite minor differences in the sequence of exposure resulting from the route of challenge, *Brucella* is generally thought to encounter extracellular killing mechanisms prior to intracellular killing in macrophage and neutrophils. Although the role defensins play *in vivo* cannot be ruled out, resistance to polycationic defensins varies wildly in *in vitro* studies and is not always directly related to the presence of O-polysaccharide (Halling *et al.*, 2008). Among the three complement pathways only the classical and lectin-mediated pathways appear to play a role in reducing the number of organisms (Hoffmann and Houle, 1983, 1995; Fernández-Prada *et al.*, 2001). However, studies in cattle have revealed a competing effect of large numbers of IgG antibody on complement (Hoffmann and Houle, 1995). The dominant antigen on the *Brucella* cell surface is the O-polysaccharide portion of the LPS, also referred to as O-antigen. *Brucella* LPS protects against extracellular killing mechanisms. Organisms lacking complete O-antigen are reduced in number and eventually cleared from the host without causing disease. However, the antigenicity of *Brucella* O-antigen also appears to provide protection against complement-mediated lysis by depositing IgG on the bacterial cell surface preventing lysis of the organism and potentially enhancing uptake by phagocytic cells. Therefore, based on these results, the impact of extracellular killing on the outcome of *Brucella* infection appears not only to be ineffective at eliminating the organism, but quite possibly enhances infection by promoting phagocytic uptake via opsonization of the organism.

In the absence of extensive extracellular killing, the innate immune response must rely upon cellular killing mechanisms to reduce the number of infecting organisms, including neutrophils, resident macrophage and dendritic cells. However, survival within these cells has been well documented and evidence confirms the resistance of *Brucella* to numerous killing activities expressed by these cells. Several studies confirm the susceptibility of *Brucella* to killing activities present in neutrophils, i.e. myeloperoxidase activity (Riley and Robertson, 1984a,b; Gallego *et al.*, 1989). Although elevated in number and therefore more likely to be encountered by *Brucella*, the contribution of infected neutrophils to persistence and disease has not been fully elaborated. However, the shortened lifespan of the neutrophil does not promote persistence of the organism in a favourable replicative niche. In contrast, macrophage provides a longer-lived target cell for *Brucella* replication and persistence. The failure of either neutrophils or macrophage to control or prevent infection, may be attributable to the reduced endotoxic activity of the *Brucella* LPS, preventing complete activation of this important first line of host defence. In addition, the success of *Brucella* also results from the capacity to prevent apoptosis (Gross *et al.*, 2000; He *et al.*, 2006; Pei *et al.*, 2006). Reversal can be observed when host cells are treated with pro-apoptotic compound such as cisplatin (Erdogan *et al.*, 2010). The stealthy nature of *Brucella* organisms is defined by the reduced activation of cellular pathways resulting from poor recognition by the pattern recognition receptors PAMPs on the cell surface, the toll-like receptors TLRs. Changes occurring in these acyl groups in the rough strain RB51 have been cited as being responsible for differences in the capacity of these organisms to establish a successful infection (Campos *et al.*, 2004). Whether these changes are consistently observed in all rough *Brucella* remains to be demonstrated. The function of *Brucella* virulence factors must be considered against a backdrop in which 90% or more of invading organisms are destroyed, as revealed using macrophage in culture, suggesting that only 10% or less of incoming organisms are properly programmed for survival (see below).

Based on the fact that *Brucella* prevents apoptosis in parasitized host cells, enhancing the function of NK cells represents an obvious target to control early infection. Contact dependent association between human NK cells and macrophage was shown to result in restricted intracellular *in vitro* replication of *Brucella* (Dornand *et al.*, 2004). In addition, NK cell cytotoxicity was stimulated in response to *Brucella* fractions and IL-12 produced in response to heat-killed *Brucella* (Zaitseva *et al.*, 1996). The importance of NK cells has long been suspected based on its role in the production of IFNγ and TNFα. Secretion of cytokines appears to be of much less importance than granular release of perforin or other effectors, consistent with contact dependent reduction of *Brucella*. However, antibody depletion of NK cells in the mouse model revealed no demonstrable effect on the number of *Brucella* in the spleen or liver one week post infection (Fernandes *et al.*, 1995), although greatly reduced polyclonal antibody production suggests an important role for NK cell in protection against chronic infection (Gao *et al.*, 2010).

The role of different *Brucella* gene products in regulating the activity of NK cells remains unexplored, but could function directly on NK cells or via expression in macrophage, such as the reported reduction in expression of MHC class I (Dornand *et al.*, 2004). Dysfunctional NK cells have been noted in human patients infected with *Brucella*, leaving open the added possibility that the function of NK cells is different in humans than has been reported for the mouse model (Salmeron *et al.*, 1992). In support of this, the effect of IL-12 seen in the mouse model is independent of NK cell function (Fernandes *et al.*, 1995), while IL-12 has been shown to play a significant role in controlling infection via stimulation of human NK cells (Zaitseva *et al.*, 1996; Dornand *et al.*, 2004). Stimulation of NK function has taken a back seat to investigation of other arms of the innate immune response based on the discrepancies described above.

Macrophage invasion and intracellular trafficking

Uptake by macrophage and other cell types has been the primary focus of *Brucella* research, since it became obvious that survival of the organism in the host was controlled by intracellular survival within macrophage (Pomales-Lebron and Stinebring, 1957; Braun *et al.*, 1958). Early efforts evaluating the interactions resulting in uptake using antagonists to block receptor ligand interaction indicated that multiple interactions at the cell surface had significant impact on *Brucella* uptake by target cells (Campbell *et al.*, 1994). More recent reports describe scavenger receptor mediated uptake associated with lipid-rafts dependent on interaction with the T4SS (Kim *et al.*, 2004). Interaction and uptake via a lipid raft-associated pathway has been confirmed in several additional studies, along with an interaction with the class A scavenger receptor (Arellano-Reynoso *et al.*, 2005; Lapaque *et al.*, 2006a; Martin-Martin *et al.*, 2010). However, any interaction occurring at the cell surface may be presumed to be subtle or even passive in order to maintain a stealthy interaction between the organism and the target cells. Uptake may rely upon the natural phagocytic properties of the host cell to engulf associated particles that subsequently obtain an environment suitable for replication (Pei *et al.*, 2008a). These properties are consistent with the phagocytic function of the primary target cells of this organism, macrophages and the trophoblasts lining the placenta (Carvalho Neta *et al.*, 2008; Roop *et al.*, 2009; Watanabe *et al.*, 2009). The capacity to infect additional cell types presumably reflects their low level of shared phagocytic activity and a conservation of cell surface receptors.

Brucella uptake via lipid rafts suggests an immediate means by which *Brucella* may avoid destruction. Mycobacterial association with cholesterol rich rafts results in association with TACO, a host protein shown to prevent TLR-mediated activation of macrophage, NFκB translocation and phagosome–lysosome fusion (Tanigawa *et al.*, 2009). Recent results also indicate that naturally occurring rough species (*B. canis* and *B. ovis*) also interact with lipid rafts resulting in their successful infection (Martín-Martín *et al.*, 2010b). In contrast, this feature is not shared by rough derivatives of the smooth species (*B. melitensis, B. suis,* or *B. abortus*) suggesting that the mechanism(s) responsible for the success of *Brucella* infection may occur at the earliest stages of host–cell interaction, but does not necessarily rely upon the presence of O-polysaccharide (Porte *et al.*, 2003). Furthermore, since *in vitro* evaluation of natural rough species and engineered rough derivatives of smooth strains reveals little significant difference in susceptibility to intracellular killing mechanisms (Martín-Martín *et al.*, 2010a) and a normal rate of replication within macrophage (Porte *et al.*, 2003; Pei and Ficht, 2004), it appears that the fate of these organisms and their ability to cause disease is best attributed to events occurring at the earliest stages of infection including interactions at the cell surface. The presence of smooth LPS limits uptake of *Brucella* and activation of the host cell. Modification of the vacuolar compartment leads to changes in bacterial gene expression including expression of VirB, blocking entry into a fusogenic compartment, and promoting access into a reproductive niche. In contrast, the failure of rough derivatives of the smooth (Rittig *et al.*, 2003; Pei and Ficht, 2010) species to successfully infect host cells is presumably caused by a failure to correctly associate with lipid rafts resulting in increased interaction with host cells and enhanced uptake. Apparently, naturally occurring rough strains have altered these surface interactions so as to mimic infection by the naturally occurring smooth species. The stealthy nature of smooth organisms is lost in engineered or spontaneously occurring rough mutants. As a result, interaction with host cells induces high levels of proinflammatory cytokines that contribute to an inability of these organisms to establish a replicative niche (Rittig *et al.*, 2003; Pei and Ficht, 2010). One study documents changes in the fatty acyl content of the lipid A portion in the rough strain RB51, and hypothesizes that it is these changes rather than the loss of O-polysaccharide of the LPS that is the source of this difference (Campos *et al.*, 2004). This is a reasonable assumption since it is the presence of the longer fatty acid chains (C28:0 and C30:0) that is responsible for the reduced toxicity of *Brucella* LPS (Barquero-Calvo *et al.*, 2007). However, extrapolation based on the use of *B. abortus* RB51 may be problematic. Although derived from a model virulent organism *B. abortus* 2308, RB51 was extensively passaged *in vitro* during selection resulting in an accumulation of multiple mutational events. Until such time as each of these can be eliminated from consideration, this question may be best addressed using rough organisms that had been recently derived with minimal passage or using purified forms of the LPS in question (Barrio *et al.*, 2009).

Consideration of the interactions at the cell surface would not be complete without description of the role of the pathogen-associated pattern recognition proteins or TLRs. Although initially surprising, the observation that deletion of TLR2 or TLR4 provides little demonstrable difference in host cell activation or ultimate clearance of the organism is consistent with the stealthy nature of the organism (Campos *et al.*, 2004; Weiss *et al.*, 2005). In contrast, deletion of MyD88 reveals a significant difference in the ability to control infection and explains the significance of Tir containing proteins expressed by *Brucella* that interfere with this signalling via degradation of adaptor proteins to minimize activation of the host cells (Cirl *et al.*, 2008; Macedo *et al.*, 2008; Salcedo *et al.*, 2008; Lapaque *et al.*, 2009; Radhakrishnan *et al.*, 2009; Radhakrishnan and Splitter, 2010; Sengupta *et al.*, 2010). More recently, the potential role of additional receptors including TLR9, NOD and a type I interferon receptor have been examined. The results suggest the presence of another unidentified intracellular receptor that may be critical in the control of *Brucella* infection (Lapaque *et al.*, 2009). Although additional bacterial components including lipoproteins have been shown to activate target cells, proof of their activity during infection is more complex and has not been clearly defined (Giambartolomei *et al.*, 2004). The contribution of TLR expression to tissue-specific clearance of infection has not yet been evaluated (Behera *et al.*, 2006).

Following initial interactions with the host cell, *Brucella* organisms are taken up in endocytic vacuoles. *In vitro* experiments indicate that nine out of ten virulent organisms are inactivated within 8–10 h of infection. Extrapolation of *in vitro* studies in macrophage indicating that no more than one to two organisms are taken up per cell, suggests infection is successful in a only limited number of cells (Pei *et al.*, 2008a). A simple evaluation of virulence versus growth phase, *in vitro* or *in vivo* passage of *Brucella* has provided little additional insight, so that the basis for these observations remains undefined. Organisms that successfully infect ultimately reside in a *Brucella*-containing vacuole (BCV) associated with proteins derived from the endoplasmic reticulum. BCV typically accommodate a single organism throughout the course of infection suggesting that a massive amount of cellular membrane synthesis is necessary to support intracellular replication of *Brucella* (Starr *et al.*, 2008). Release of the organisms associated with successive rounds of infection and intracellular replication are suggested in the literature. While most researchers agree that the host cell ultimately succumbs to infection, this appears to contradict the observation that *Brucella* prevents apoptotic cell death in order to maintain a replicative niche. However, the mechanism(s) responsible for egress, be it cellular rupture of an immortalized cell, membrane budding or something different remain to be defined (Molmeret and Abu Kwaik, 2002; Hybiske and Stephens, 2008). Changes in gene expression, occurring late in infection, are one possible explanation. As the population of the organism increases changes in the cellular environment may serve as signal to alter gene expression. This may be as simple as a build-up in toxic intermediates having a direct effect on cellular structure or the capacity of *Brucella* to detect changing conditions serving as signal for altered gene expression. Quorum sensing systems could serve in this regard (discussed in Chapter 8), as well as two-component regulatory systems (Uzureau *et al.*, 2007; Godefroid *et al.*, 2010; Lavin *et al.*, 2010).

Apoptotic and necrotic cell death

Along with the inhibition of function of MyD88 described previously, inhibition of apoptosis is among the best-documented mechanisms for *Brucella* interference of innate immune function. By preventing macrophage apoptosis *Brucella* acquire a stable environment in which the organism may persist and replicate within the host, and evade the bacteriocidal effects of apoptosis and cytotoxic T-cell priming via antigen cross-presentation. As revealed in studies on unrelated intracellular pathogens, by inhibiting apoptosis of infected cells organisms may restrict development of protective immune responses (Behar *et al.*, 2010; Matheoud *et al.*, 2010). Identification and inactivation of the genetic loci in *Brucella* responsible for the inhibition of apoptosis may be expected to enhance immune protection. Comparisons of virulent and vaccine *Brucella* strains has not been performed to determine whether this is a feature that distinguishes such strains or whether this is a specific improvement that may be addressed in the development of improved vaccine strains to stimulate a more robust protective immune response.

In contrast to virulent smooth organisms, rough organisms lacking O-antigen are taken up by macrophage greater than ten-fold higher in number than smooth strains. Several reports suggest that rough organisms induce necrotic cell death (Pei *et al.*, 2008b; Zhong *et al.*, 2009). Rough organisms, replicate within the host cell, but fail to cause disease either from a failure to prevent phagosome–lysosome fusion or to achieve the elevated number of organisms necessary to perpetuate infection (Porte *et al.*, 2003; Moriyón *et al.*, 2004). In addition, numerous bacterial-surface changes may result from the loss of O-antigen expression (Cloeckaert *et al.*, 2002). Although the failure of rough strains to prevent apoptosis may suggest the capacity to stimulate an enhanced immune response, enhanced sensitivity to killing mechanisms, and the absence of antibody to LPS may compromise the ability of rough organisms to stimulate an immune response comparable to that provided by smooth strains. Targeted inactivation of anti-apoptosis genes represents a potential focus area for investigation in an effort to develop improved vaccine strains. Recent experiments have revealed inhibition of Bax-induced cell death resulting from expression of the *Brucella* Omp2b porin protein (Laloux *et al.*, 2010). As an essential protein, proof of a role for Omp2b in preventing apoptosis would require performance of gain of function assays with careful adherence to institutional biosafety and in the U.S. includes adherence to select agent regulations (Research Council, 2009). Alternatively, loss of function identification of bacterial factors responsible for preventing apoptotic cell death could be performed successfully with fewer restrictions using random mutagenesis approaches. Alternatively, the tir-containing proteins (Btp-1 and Tcp-1) expressed by *Brucella* represent potential candidates for inactivation in order to enhance apoptosis and the immune response.

Adaptive immune response

Brucella resides in macrophages and possibly dendritic cells within the host and resist destruction by inhibiting cellular apoptosis designed to eliminate the organism from the host. Clearance of the organism depends upon a cell mediated Th1 response including IL-12, IFNγ and TNFα expression that is limited by the non-classical structure of *Brucella* O-antigen (Lapaque *et al.*, 2006b). As a result, host phagocytic cells fail to become activated, reactive nitrogen and reactive oxygen compounds are formed in limited amounts, and proinflammatory cytokine expression and dendritic cells maturation is restricted (Billard *et al.*, 2007a,b; Macedo *et al.*, 2008). Among the factors that appear to be critical to preventing these responses the O-polysaccharide portion of *Brucella* LPS is the best characterized. Infection with rough organisms lacking O-polysaccharide results in enhanced expression of proinflammatory cytokines, dendritic cells maturation and an overall enhancement of the innate immune response, perhaps resulting from the enhanced uptake of the organism or exposure of *Brucella* lipoproteins (Zwerdling *et al.*, 2008; Surendran *et al.*, 2010, 2011). Persistence of *Brucella* vaccines has been clearly documented to enhance long-term protection (Grilló *et al.*, 2006; González *et al.*, 2008). However, questions concerning the long-term potency of rough vaccines persist. The rapid clearance of rough organisms, although enhancing safety potentially limits long-term efficacy.

Another gene of interest in the development of live attenuated vaccines Omp25 is directly associated with the prevention of TNFα production resulting in the impairment of innate immune function and the development of a Th1 response (Jubier-Maurin *et al.*, 2001). Although defects in the expression of Omp25 have no apparent effect on the intracellular survival of *Brucella* in macrophage culture, the mutants were attenuated for survival and provided a protective immune response in the mouse model (Cloeckaert *et al.*, 2002). Based on these results, removal of the gene encoding Omp25 from vaccine candidates remains a viable consideration for continued vaccine development (Edmonds *et al.*, 2002).

VirB expression appears to be critical to the stimulation of a proinflammatory response (Roux *et al.*, 2007; Rolan and Tsolis, 2008). In the murine model, comparison of wild-type and Δ*virB* mutants reveals a significant difference in expression of genes induced by type I and II interferons. Although, stimulation of the proinflammatory response by the T4SS is surprising, the differences had no demonstrable effect on bacterial persistence over the first four weeks of infection. The importance of the T4SS in eliciting the production of IL12, TNFα and IFNγ has been hypothesized to play a long term role in persistence resulting from granuloma formation and the capacity to cause disease (de Jong *et al.*, 2010). However, reduction in IL-12 was shown to enhance chronicity of infection while exogenous IL-12 has been shown to reduce chronicity in the mouse model (Zhan and Cheers, 1995; Zhan *et al.*, 1996; Pasquali *et al.*, 2002). VirB expression is also essential to the cytotoxic activity observed in macrophage infected with rough *Brucella*. Overexpression or VirB documented in these organisms is suspected to be the cause (Zhong *et al.*, 2009).

The roles of CD8 (MHC-I restricted) and CD4 (MHC-II restricted) cells has been investigated over the last two decades. The current consensus is that CD8 cells are critical to the control of *Brucella* infection (Oliveira and Splitter, 1995; Baldwin and Goenka, 2006), while additional evidence supports a role of CD4 T cells (Hu *et al.*, 2010). The critical feature for either cell type is the production of IFNγ to activate macrophage and cytotoxic-T-lymphocytes and to enhance IgG class switching (Baldwin and Goenka, 2006). γδ T cells also appear to play a role in infection that is especially prominent in humans and suspected in ruminants (Kunkle *et al.*, 1995; Oliaro *et al.*, 2005; Dudal *et al.*, 2006). The role of humoral immunity in *Brucella* infection has been repeatedly documented, both in mice and cattle. Antibody is presumed to play a direct role in the opsonization and intracellular killing of the organism rather than neutralization (Elzer *et al.*, 1994). However, the results of experimentation suggest that elevated levels of antibody prevent complement-mediated lysis and promote intracellular uptake that may in fact enhance infection (Bellaire *et al.*, 2005).

Improved vaccine design

In developing improved vaccines that are (i) safe and efficacious in humans, and/or (ii) safe and efficacious in animals there are several considerations that must be taken into account. First and foremost is vaccine safety. With this in mind there has been considerable effort towards the development of subunit and DNA vaccines. At this time subunit vaccines have, at best, a moderate level of success with *Brucella*, as well as unrelated intracellular organisms that continues to provide an active niche for development of live, attenuated vaccines. In addition, the low level of efficacy demonstrated by subunit vaccines requires the use of numerous supplements including potential vectors that will entail experimental proof of safe and effective use, as well as regulatory approval. All of these caveats require consideration among developmental costs. Difficulties persist in the development of subunit vaccines owing to limitations in the ability to target vaccines to appropriate tissues or cells or to traffic within cells in order to enhance stimulation of the various arms of the host immune response. In contrast, the strength of live attenuated vaccines is their built in capacity to target the appropriate tissues/cells and to traffic within these cells in a manner that is similar to virulent organisms. By manipulating the intracellular trafficking of the organism, persistence can be varied and the immune response enhanced. Current vaccines offer only a limited range in which the organism is either rapidly cleared owing to phagolysosomal fusion or persists too long spreading to other tissues (Pizarro-Cerda *et al.*, 1998). However, despite a long history of effective use and the capacity to manipulate the organisms genome, there remains lingering doubt about overall safety of live attenuated organisms (Lagrange *et al.*, 1985).

In order to address the questions raised, we will first examine the considerable amount of research over the last two decades focused on the identification of virulence factors required for *Brucella* to successfully cause disease (Sangari and Agüero, 1991; Foulongne *et al.*, 2000; Hong *et al.*, 2000; Lestrate *et al.*, 2000; Wu *et al.*, 2006; Zygmunt *et al.*, 2006). Knowledge of these genes and their effect on the organism enhances safe and effective use by crippling the organism while taking advantage of the organism's capacity to traffic within the host and to interact with the various components of the immune system. Furthermore, enhanced immune protection is possible by limiting the organism's capacity to prevent antigen presentation (Forestier *et al.*, 2000; Billard *et al.*, 2007a; Barrionuevo *et al.*, 2008). However, to do this care must be taken in combining the mutations so that the organism is not crippled to the extent that it is unable to successfully interact with the immune system. The critical aspects of safety that must be considered include the inability to cause symptoms associated with disease, as well as an inability to revert to virulent form (Moriyón *et al.*, 2004). Addressing the latter of the two requirements is a relatively easy task as recent work has identified the genes necessary for survival and approaches to remove them from the genome that precludes restoration of virulence. However, proven stability of phenotype and confidence in the use of such organisms generally acquires acceptance only after a great deal of time and experience with use. For example, past experience has shown that simple passage *in vitro*, i.e. without selection pressure, can result in undesirable outcomes (Chukwu, 1985). Yet, the examples of *B. abortus* strain 19 (S19) and *Mycobacterium bovis* BCG demonstrate that live, attenuated vaccines provide significant levels of protection while maintaining stability and safety. Recent genomic analysis of S19 has shown significant differences in sequence at 24 distinct genetic loci associated with virulence with additional changes in more than 200 loci (Crasta *et al.*, 2008). Restoration and/or exact reproduction of virulence of S19 have not been documented and would be near impossible spontaneously. In contrast, BCG can be mimicked by removing the 9.5 kb RD1 region from virulent organism (Lewis *et al.*, 2003). Similarly, virulence may be restored via the re-introduction of the RD1 DNA segments (Guinn *et al.*, 2004). Although recovery of virulence is possible under laboratory conditions, recovery of the 9.6 Kbp segment has not been documented despite more than 80 years of use and billions of vaccinations. As described above the genes required for virulence have been characterized in research spanning the last two decades. Based on the low probability of restoring virulence to deletion mutants, the greatest concern associated with the use of engineered vaccines is retention of low levels of virulence. Although such considerations may be explored using laboratory animal models, weaknesses in the models compromise the ability to make such assumptions and require testing directly in the target species or expensive model systems that more closely approximate the target species, i.e. non-human primates. Perhaps most important is organism attenuation while retaining the capacity to stimulate immune protection in target host species rather than laboratory animal models.

Considering the critical nature of vaccine safety, the first decision is the choice of gene to remove in order to attenuate the organism. This choice is predicated by specific requirements including stimulation of sufficient humoral and cellular immunity to prevent subsequent infection without evidence of disease. In the case of brucellosis, symptoms typically associated with early infection include splenomegaly and hepatomegaly caused by massive influx of macrophage into the reticuloendothelial tissues (Arenas-Gamboa *et al.*, 2009c). Numerous sequelae are observed as a result of the spread of infection and may be avoidable depending on reduced virulence (Fiori *et al.*, 2000). Efforts in several laboratories have indicated that rapid clearance of live, attenuated *Brucella* vaccines does not provide sufficient protective immunity and that persistence of the organism for some period of time in the host is necessary to stimulate an adaptive immune response (Young *et al.*, 1979; Hong *et al.*, 2000; Berchane *et al.*, 2006; Kahl-McDonagh and Ficht, 2006). In the absence of identification of genes that enhance persistence without enhancing virulence, alternative approaches were sought. Encapsulation has been used for several decades for the timed release of pharmaceuticals. Although most methods of capsule creation are not applicable to viable organisms, alginate capsules may be produced using aqueous solutions to provide gradual release of viable contents over time. Evaluation of this approach in conjunction with highly attenuated organisms that are otherwise rapidly cleared has proved to be a promising delivery platform that provides a natural boosting or depot effect which when coupled with the use of highly attenuated vaccine candidates provides an enhanced Th1 response and improved protection against infection by this intracellular pathogen (Arenas-Gamboa *et al.*, 2008, 2009a,b). Alterations of the capsule surface structure including differences in capsular material, cross-linker, surface charge, inclusion of adjuvants may all be evaluated in an effort to improve capsule performance (Rice-Ficht *et al.*, 2010). One drawback of this approach is that alginate formulations produce microcapsules (300 μm) that can only deliver their contents via diffusion and/or more probably degradation of the particles that are too large to be taken up by host cells. An alternative approach utilizes chitosan nanocapsules (0.4 μm) capable of penetrating the oral mucosa and taken up by individual cells to stimulate the caspase-1-mediated inflammasome activity to enhance macrophage function, cytokine release and ultimate protection against infection (Demento *et al.*, 2009; Eisenbarth and Flavell, 2009). Since bacterial particles themselves are not much smaller, this approach may be of applicability for use with recombinant viral vectors or subunit vaccines used as primary vaccine or as booster dose as follow up to primary vaccination with microcapsules. The use of encapsulation techniques may thus be applied to either live, attenuated candidates or subunit vaccines. It's use with live, attenuated mutants is expected to artificially extend the survival of mutants that would otherwise be rapidly cleared from the host.

Final considerations

The narrative above focused on the use of a novel delivery system capable of altering intracellular trafficking, antigen processing, and MHC presentation. Additional experiments must be performed to examine the route of delivery as well as the properties of particles to traffic differently based on altering the surface properties of particles including modifications using components of the pathogen. Improvement in vaccine design can take many forms and a review of the literature demonstrates that emphasis is often restricted to distinct silos focused on specific aspects for improving vaccination and immune protection that is consistent with reductionist approaches and emphasizes development of an individual researchers area of expertise. Since such approaches are most likely to receive the best scores in grant review, developments should be considered by a group of scientists having a broad range of expertise. Such 'meta-thinking' requires expertise of the multiple aspects affecting the outcome of development. In the case described, the authors have worked for the past 25 years to identify genes expressing factors important for the survival of the organism while a second author worked on the development of pharmaceutical delivery systems. Thanks to funding through the Western Regional Center of Excellence collaborating scientists having expertise in aerobiology, non-human primate models of infection, ruminant models of infection, host response and T cell immunology and good manufacturing production have broadened the overall research experience. Although investigation of the genes important for survival suggested a potential use as vaccine candidates, this idea was not novel having been tried with other pathogens with different levels of success. However, knowledge of the limitations of the *Brucella* vaccines and the risks and weaknesses of those under development led to cross-fertilization of thinking with researchers having complementing expertise. In the work described we have focused on one particular class of vaccine owing to the promising results observed, however, with enhanced knowledge of protective antigens and delivery methods including the variations described there is no reason that this approach may not be extended for broader investigation.

Summary

1. Provide improved protection over the use of rough organisms. Interference with common diagnostic tests will require adoption of new diagnostic tests. O-antigen is important for survival in the host including intracellular uptake and trafficking. Safety is a concern based on past use of genetically undefined variants and the inability to completely separate survival from pathology.
2. The choice of genetic background may be critical to the licensing of the vaccine. In the U.S., candidates derived from select biological agents require approvals from USDA and CDC prior to use and without previous

history of use to support renewed broader use. In contrast, previously employed vaccine strains have an established history of safe and stable use that provides greater confidence among regulators.

3 The failure of heat-killed vaccines suggests that similar approaches including subunit vaccines may have limited effectiveness unless delivery systems can provide some level of persistence to boost the immune response. Poor definition of protective immunogens and the inability to target appropriate tissues has hampered development.

4 DNA vaccines offer improved potential over other subunit vaccine approaches by providing a persistent source of immunogen and the ability to target specific tissues. However, the poor definition of protective immunogens hampers current development.

5 Delivery systems may be used to provide enhanced protection for live attenuated organisms including alginate capsules with or without the inclusion of additional adjuvants. These materials are already FDA approved for other uses in humans that will simplify approval in this context. There success with live, attenuated organisms suggests a potential for enhanced delivery of heat-killed or subunit vaccines. Particle delivery offers the potential to provide a long-lasting depot or boosting effect to enhance immune clearance, as well as providing independent activation of the host immune response.

6 Alternative materials include chitosan capable of nanoparticle formation and uptake by M cells and macrophage that stimulate the inflammasome in combination with TLR agonists including LPS to enhance the Th1 response.

7 Based on the threat of aerosol delivery and consumption of the organism stimulation of the mucosa is considered to be the optimal target to prevent against infection. The stealthy nature of *Brucella* organisms limits the host proinflammatory response and may be expected to provide stimulation without significant mucosal damage.

8 The intracellular trafficking of various vaccine candidates can have a dramatic effect on efficacy. Rough organisms are taken up in greater numbers and fail to prevent phagosome activation. Rough organisms do not prevent phagosome–lysosome fusion, stimulate a necrotic cell death and are rapidly cleared from the host. Smooth organisms prevent fusion and persist longer by preventing apoptotic cell death although a majority of incoming organisms are also cleared. Among currently available vaccine strains S19 fails to prevent phagosome–lysosome fusion.

9 Antigen processing and presentation is key to the development of an adaptive immune response. Protection against *Brucella* infection requires both MHCI and MHCII presentation, but a clear description of a protective immune response has not been obtained. Future manipulation of the *Brucella* organism must include the identification of factors responsible for preventing macrophage activation such as apoptosis in order to enhance development of a protective immune response.

10 There are numerous examples of attenuated mutations in the literature that may well make excellent vaccine candidates. Failure to provide a commercially viable product should not always be assumed to indicate unsuccessful safety or efficacy results. Continued efforts to bring vaccines to fruition is a daunting enterprise of which only 25–30% represents the scientific or discovery phase. Without continued developmental and scientific support from funding agencies, administrative and regulatory agencies and commercial developers success may be elusive.

Acknowledgements

ACR-F and TAF are supported by USAMRMC W81XWH-07-1-0304 and NIH 1 R41 AI068252-01A2 to the Western Regional Center of Excellence (David H. Walker, PI). This publication was also made possible by grants to TAF from NIAID/NIH 1U54AI057156-0100 and R01-048996-8. Its contents are solely the responsibility of the authors and do not necessarily represent the official views of the RCE Programs Office, NIAID or NIH. The authors thank Drs. L. Garry

Adams, Angela Arenas-Gamboa, Melissa M. Kahl-McDonagh, Jianwu Pei, Gabriel Gomez, Renée Tsolis and Priscilla Hong for thoughtful discussions, development of conceptual approaches, and diligent work that helped to formulate the concepts outlined in this manuscript.

References

Archambaud, C., Salcedo, S.P., Lelouard, H., Devilard, E., de Bovis, B., Van Rooijen, N., Gorvel, J.P., and Malissen, B. (2010). Contrasting roles of macrophages and dendritic cells in controlling initial pulmonary Brucella infection. Eur J Immunol *40*, 3458–3471.

Arellano-Reynoso, B., Lapaque, N., Salcedo, S., Briones, G., Ciocchini, A.E., Ugalde, R., Moreno, E., Moriyón, I., and Gorvel, J.P. (2005). Cyclic beta-1,2-glucan is a *Brucella* virulence factor required for intracellular survival. Nat. Immunol. *6*, 618–625.

Arenas-Gamboa, A.M., Ficht, T.A., Kahl-McDonagh, M.M., and Rice-Ficht, A.C. (2008). Immunization with a single dose of a microencapsulated *Brucella melitensis* mutant enhances protection against wild-type challenge. Infect. Immun. *76*, 2448–2455.

Arenas-Gamboa, A.M., Ficht, T.A., Davis, D.S., Elzer, P.H., Kahl-McDonagh, M., Wong-Gonzalez, A., and Rice-Ficht, A.C. (2009a). Oral vaccination with microencapsuled strain 19 vaccine confers enhanced protection against *Brucella abortus* strain 2308 challenge in red deer (*Cervus elaphus elaphus*). J. Wildl. Dis. *45*, 1021–1029.

Arenas-Gamboa, A.M., Ficht, T.A., Davis, D.S., Elzer, P.H., Wong-Gonzalez, A., and Rice-Ficht, A.C. (2009b). Enhanced immune response of red deer (*Cervus elaphus*) to live RB51 vaccine strain using composite microspheres. J. Wildl. Dis. *45*, 165–173.

Arenas-Gamboa, A.M., Ficht, T.A., Kahl-McDonagh, M.M., Gomez, G., and Rice-Ficht, A.C. (2009c). The *Brucella abortus* S19 DeltavjbR live vaccine candidate is safer than S19 and confers protection against wild-type challenge in BALB/c mice when delivered in a sustained-release vehicle. Infect. Immun. *77*, 877–884.

Baldwin, C.L., and Goenka, R. (2006). Host immune responses to the intracellular bacteria *Brucella*: does the bacteria instruct the host to facilitate chronic infection? Crit. Rev. Immunol. *26*, 407–442.

Barquero-Calvo, E., Chaves-Olarte, E., Weiss, D.S., Guzmán-Verri, C., Chacon-Diaz, C., Rucavado, A., Moriyón, I., and Moreno, E. (2007). *Brucella abortus* uses a stealthy strategy to avoid activation of the innate immune system during the onset of infection. PLoS ONE 2, e631.

Barrio, M.B., Grillo, M.J., Munoz, P.M., Jacques, I., Gonzalez, D., de Miguel, M.J., Marin, C.M., Barberan, M., Letesson, J.J., Gorvel, J.P., *et al.* (2009). Rough mutants defective in core and O-polysaccharide synthesis and export induce antibodies reacting in an indirect ELISA with smooth lipopolysaccharide and are less effective than Rev 1 vaccine against *Brucella melitensis* infection of sheep. Vaccine 27, 1741–1749.

Barrionuevo, P., Cassataro, J., Delpino, M.V., Zwerdling, A., Pasquevich, K.A., Garcia Samartino, C., Wallach, J.C., Fossati, C.A., and Giambartolomei, G.H. (2008). *Brucella abortus* inhibits major histocompatibility complex class II expression and antigen processing through interleukin-6 secretion via Toll-like receptor 2. Infect. Immun. *76*, 250–262.

Behar, S.M., Divangahi, M., and Remold, H.G. (2010). Evasion of innate immunity by *Mycobacterium tuberculosis*: is death an exit strategy? Nat. Rev. Microbiol. *8*, 668–674.

Behera, A.K., Hildebrand, E., Bronson, R.T., Perides, G., Uematsu, S., Akira, S., and Hu, L.T. (2006). MyD88 deficiency results in tissue-specific changes in cytokine induction and inflammation in interleukin-18-independent mice infected with *Borrelia burgdorferi*. Infect. Immun. 74, 1462–1470.

Bellaire, B.H., Roop, R.M., 2nd, and Cardelli, J.A. (2005). Opsonized virulent *Brucella abortus* replicates within nonacidic, endoplasmic reticulum-negative, LAMP-1-positive phagosomes in human monocytes. Infect. Immun. 73, 3702–3713.

Berchane, N.S., Jebrail, F.F., Carson, K.H., Rice-Ficht, A.C., and Andrews, M.J. (2006). About mean diameter and size distributions of poly(lactide-co-glycolide) (PLG) microspheres. J. Microencapsul. 23, 539–552.

Billard, E., Dornand, J., and Gross, A. (2007a). *Brucella suis* prevents human dendritic cell maturation and antigen presentation through regulation of tumor necrosis factor alpha secretion. Infect. Immun. 75, 4980–4989.

Billard, E., Dornand, J., and Gross, A. (2007b). Interaction of *Brucella suis* and *Brucella abortus* rough strains with human dendritic cells. Infect. Immun. 75, 5916–5923.

Braun, W., Pomales-Lebron, A., and Stinebring, W.R. (1958). Interactions between mononuclear phagocytes and *Brucella abortus* strains of different virulence. Proc. Soc. Exp. Biol. Med. 97, 393–397.

Campbell, G.A., Adams, L.G., and Sowa, B.A. (1994). Mechanisms of binding of *Brucella abortus* to mononuclear phagocytes from cows naturally resistant or susceptible to brucellosis. Vet. Immunol. Immumopathol. *41*, 295–306.

Campos, M.A., Rosinha, G.M., Almeida, I.C., Salgueiro, X.S., Jarvis, B.W., Splitter, G.A., Qureshi, N., Bruna-Romero, O., Gazzinelli, R.T., and Oliveira, S.C. (2004). Role of Toll-like receptor 4 in induction of cell-mediated immunity and resistance to *Brucella abortus* infection in mice. Infect. Immun. 72, 176–186.

Carvalho Neta, A.V., Stynen, A.P., Paixao, T.A., Miranda, K.L., Silva, F.L., Roux, C.M., Tsolis, R.M., Everts, R.E., Lewin, H.A., Adams, L.G., *et al.* (2008). Modulation of the bovine trophoblastic innate immune response by *Brucella abortus*. Infect. Immun. *76*, 1897–1907.

Chukwu, C.C. (1985). Serological investigation on cattle vaccinated with a killed *Brucella abortus* strain 45/20 adjuvant vaccine. Int. J. Zoonoses *12*, 14–21.

Cirl, C., Wieser, A., Yadav, M., Duerr, S., Schubert, S., Fischer, H., Stappert, D., Wantia, N., Rodriguez, N., Wagner, H., *et al.* (2008). Subversion of Toll-like receptor signaling by a unique family of bacterial Toll/

interleukin-1 receptor domain-containing proteins. Nat. Med. *14*, 399–406.

Cloeckaert, A., Vizcaíno, N., Paquet, J.Y., Bowden, R.A., and Elzer, P.H. (2002). Major outer membrane proteins of *Brucella* spp.: past, present and future. Vet. Microbiol. *90*, 229–247.

Crasta, O.R., Folkerts, O., Fei, Z., Mane, S.P., Evans, C., Martino-Catt, S., Bricker, B., Yu, G., Du, L., and Sobral, B.W. (2008). Genome sequence of *Brucella abortus* vaccine strain S19 compared to virulent strains yields candidate virulence genes. PLoS One 3, e2193.

Demento, S.L., Eisenbarth, S.C., Foellmer, H.G., Platt, C., Caplan, M.J., Mark Saltzman, W., Mellman, I., Ledizet, M., Fikrig, E., Flavell, R.A., *et al.* (2009). Inflammasome-activating nanoparticles as modular systems for optimizing vaccine efficacy. Vaccine 27, 3013–3021.

Dornand, J., Lafont, V., Oliaro, J., Terraza, A., Castaneda-Roldan, E., and Liautard, J.P. (2004). Impairment of intramacrophagic *Brucella suis* multiplication by human natural killer cells through a contact-dependent mechanism. Infect. Immun. 72, 2303–2311.

Dudal, S., Turriere, C., Bessoles, S., Fontes, P., Sanchez, F., Liautard, J., Liautard, J.P., and Lafont, V. (2006). Release of LL-37 by activated human Vgamma9Vdelta2 T cells: a microbicidal weapon against *Brucella suis*. J. Immunol. *177*, 5533–5539.

Edmonds, M.D., Cloeckaert, A., and Elzer, P.H. (2002). *Brucella* species lacking the major outer membrane protein Omp25 are attenuated in mice and protect against *Brucella melitensis* and *Brucella ovis*. Vet. Microbiol. *88*, 205–221.

Eisenbarth, S.C., and Flavell, R.A. (2009). Innate instruction of adaptive immunity revisited: the inflammasome. EMBO Mol. Med. *1*, 92–98.

Elzer, P.H., Jacobson, R.H., Jones, S.M., Nielsen, K.H., Douglas, J.T., and Winter, A.J. (1994). Antibody-mediated protection against *Brucella abortus* in BALB/c mice at successive periods after infection: variation between virulent strain 2308 and attenuated vaccine strain 19. Immunol. *82*, 651–658.

Erdogan, S., Tosyali, E., Duzguner, V., Kucukgul, A., Aslantas, O., and Celik, S. (2010). Cisplatin reduces *Brucella melitensis*-infected cell number by inducing apoptosis, oxidant and pro-inflammatory cytokine production. Res. Vet. Sci. *88*, 218–226.

Fernandes, D.M., Benson, R., and Baldwin, C.L. (1995). Lack of a role for natural killer cells in early control of *Brucella abortus* 2308 infections in mice. Infect. Immun. *63*, 4029–4033.

Fernandez-Prada, C.M., Nikolich, M., Vemulapalli, R., Sriranganathan, N., Boyle, S.M., Schurig, G.G., Hadfield, T.L., and Hoover, D.L. (2001). Deletion of *wboA* enhances activation of the lectin pathway of complement in *Brucella abortus* and *Brucella melitensis*. Infect. Immun. *69*, 4407–4416.

Fiori, P.L., Mastrandrea, S., Rappelli, P., and Cappuccinelli, P. (2000). *Brucella abortus* infection acquired in microbiology laboratories. J. Clin. Microbiol. *38*, 2005–2006.

Forestier, C., Deleuil, F., Lapaque, N., Moreno, E., and Gorvel, J.P. (2000). *Brucella abortus* lipopolysaccharide in murine peritoneal macrophages acts as a down-regulator of T cell activation. J. Immunol. *165*, 5202–5210.

Foulongne, V., Bourg, G., Cazevieille, C., Michaux-Charachon, S., and O'Callaghan, D. (2000). Identification of *Brucella suis* genes affecting intracellular survival in an *in vitro* human macrophage infection model by signature-tagged transposon mutagenesis. Infect. Immun. *68*, 1297–1303.

Gallego, M.C., Cuello, F., and Garrido, A. (1989). *In vitro* determination of phagocytosis and intracellular killing of Brucella melitensis by goat's polymorphonuclear phagocytes. Comp. Immunol. Microbiol. Infect. Dis. *12*, 9–15.

Gao, N., Jennings, P., Guo, Y., and Yuan, D. (2011). Regulatory role of natural killer (NK) cells on antibody responses to *Brucella abortus*. Innate Immun. *17*, 152–163.

Giambartolomei, G.H., Zwerdling, A., Cassataro, J., Bruno, L., Fossati, C.A., and Philipp, M.T. (2004). Lipoproteins, not lipopolysaccharide, are the key mediators of the proinflammatory response elicited by heat-killed *Brucella abortus*. J. Immunol. *173*, 4635–4642.

Godefroid, M., Svensson, M.V., Cambier, P., Uzureau, S., Mirabella, A., De Bolle, X., Van Cutsem, P., Widmalm, G., and Letesson, J.J. (2010). *Brucella melitensis* 16M produces a mannan and other extracellular matrix components typical of a biofilm. FEMS Immunol. Med. Microbiol. *59*, 364–377.

González, D., Grilló, M.J., De Miguel, M.J., Ali, T., Arce-Gorvel, V., Delrue, R.M., Conde-Alvarez, R., Muñoz, P., López-Goñii, I., Iriarte, M., *et al.* (2008). Brucellosis vaccines: assessment of *Brucella melitensis* lipopolysaccharide rough mutants defective in core and O-polysaccharide synthesis and export. PLoS One 3, e2760.

Grilló, M.J., Manterola, L., de Miguel, M.J., Muñoz, P.M., Blasco, J.M., Moriyón, I., and López-Goñi, I. (2006). Increases of efficacy as vaccine against *Brucella abortus* infection in mice by simultaneous inoculation with avirulent smooth *bvrS/bvrR* and rough *wbkA* mutants. Vaccine *24*, 2910–2916.

Gross, A., Terraza, A., Ouahrani-Bettache, S., Liautard, J.P., and Dornand, J. (2000). *In vitro Brucella suis* infection prevents the programmed cell death of human monocytic cells. Infect. Immun. *68*, 342–351.

Guinn, K.M., Hickey, M.J., Mathur, S.K., Zakel, K.L., Grotzke, J.E., Lewinsohn, D.M., Smith, S., and Sherman, D.R. (2004). Individual RD1-region genes are required for export of ESAT-6/CFP-10 and for virulence of *Mycobacterium tuberculosis*. Mol. Microbiol. *51*, 359–370.

Halling, S.M., Jensen, A.E., and Olsen, S.C. (2008). Defensin susceptibility and colonization in the mouse model of AJ100, a polymyxin B-resistant, *Brucella abortus* RB51 isolate. Curr. Microbiol. *56*, 274–278.

He, Y., Reichow, S., Ramamoorthy, S., Ding, X., Lathigra, R., Craig, J.C., Sobral, B.W., Schurig, G.G.,

Sriranganathan, N., and Boyle, S.M. (2006). *Brucella melitensis* triggers time-dependent modulation of apoptosis and down-regulation of mitochondrion-associated gene expression in mouse macrophages. Infect. Immun. *74*, 5035–5046.

Hoffmann, E.M., and Houle, J.J. (1983). Failure of *Brucella abortus* lipopolysaccharide (LPS) to activate the alternative pathway of complement. Vet. Immunol. Immunopathol. *5*, 65–76.

Hoffmann, E.M., and Houle, J.J. (1995). Contradictory roles for antibody and complement in the interaction of *Brucella abortus* with its host. Crit. Rev. Microbiol. *21*, 153–163.

Hong, P.C., Tsolis, R.M., and Ficht, T.A. (2000). Identification of genes required for chronic persistence of *Brucella abortus* in mice. Infect. Immun. *68*, 4102–4107.

Hu, X.D., Chen, S.T., Li, J.Y., Yu, D.H., Yi, Z., and Cai, H. (2010). An IL-15 adjuvant enhances the efficacy of a combined DNA vaccine against *Brucella* by increasing the CD8+ cytotoxic T cell response. Vaccine *28*, 2408–2415.

Hybiske, K., and Stephens, R.S. (2008). Exit strategies of intracellular pathogens. Nat. Rev. Microbiol. *6*, 99–110.

Jimenez de Bagues, M.P., Gross, A., Terraza, A., and Dornand, J. (2005). Regulation of the mitogen-activated protein kinases by *Brucella* spp. expressing a smooth and rough phenotype: relationship to pathogen invasiveness. Infect. Immun. *73*, 3178–3183.

de Jong, M.F., Rolan, H.G., and Tsolis, R.M. (2010). Innate immune encounters of the (Type) 4th kind: *Brucella*. Cell. Microbiol. *12*, 1195–1202.

Jubier-Maurin, V., Boigegrain, R.A., Cloeckaert, A., Gross, A., Alvarez-Martínez, M.T., Terraza, A., Liautard, J., Kohler, S., Rouot, B., Dornand, J., *et al.* (2001). Major outer membrane orotein Omp25 of *Brucella suis* is involved in inhibition of tumor necrosis factor alpha production during infection of human macrophages. Infect. Immun. *69*, 4823–4830.

Kahl-McDonagh, M.M., and Ficht, T.A. (2006). Evaluation of protection afforded by *Brucella abortus* and *Brucella melitensis* unmarked deletion mutants exhibiting different rates of clearance in BALB/c mice. Infect. Immun. *74*, 4048–4057.

Kim, S., Watarai, M., Suzuki, H., Makino, S., Kodama, T., and Shirahata, T. (2004). Lipid raft microdomains mediate class A scavenger receptor-dependent infection of *Brucella abortus*. Microb. Pathog. *37*, 11–19.

Ko, J., and Splitter, G.A. (2003). Molecular host–pathogen interaction in brucellosis: current understanding and future approaches to vaccine development for mice and humans. Clin. Microbiol. Rev. *16*, 65–78.

Kunkle, R.A., Steadham, E.M., and Cheville, N.F. (1995). Morphometric analysis of CD4+, CD8+, and gamma/delta+ T-lymphocytes in lymph nodes of cattle vaccinated with Brucella abortus strains RB51 and 19. Vet. Immunol. Immunopathol. *49*, 271–279.

Lagrange, P.H., Hurtrel, B., and Stach, J.L. (1985). Vaccines against mycobacteria and other intracellular multiplying bacteria. Ann. Inst. Pasteur Immunol. *136D*, 151–162.

Laloux, G., Deghelt, M., de Barsy, M., Letesson, J.J., and De Bolle, X. (2010). Identification of the essential *Brucella melitensis* porin Omp2b as a suppressor of Bax-induced cell death in yeast in a genome-wide screening. PLoS One *5*, e13274.

Lapaque, N., Forquet, F., de Chastellier, C., Mishal, Z., Jolly, G., Moreno, E., Moriyón, I., Heuser, J.E., He, H.T., and Gorvel, J.P. (2006a). Characterization of *Brucella abortus* lipopolysaccharide macrodomains as mega rafts. Cell. Microbiol. *8*, 197–206.

Lapaque, N., Takeuchi, O., Corrales, F., Akira, S., Moriyón, I., Howard, J.C., and Gorvel, J.P. (2006b). Differential inductions of TNF-alpha and IGTP, IIGP by structurally diverse classic and non-classic lipopolysaccharides. Cell. Microbiol. *8*, 401–413.

Lapaque, N., Muller, A., Alexopoulou, L., Howard, J.C., and Gorvel, J.P. (2009). *Brucella abortus* induces Irgm3 and Irga6 expression via type-I IFN by a MyD88-dependent pathway, without the requirement of TLR2, TLR4, TLR5 and TLR9. Microb. Pathog. *47*, 299–304.

Lavin, J.L., Binnewies, T.T., Pisabarro, A.G., Ussery, D.W., Garcia-Lobo, J.M., and Oguiza, J.A. (2010). Differences in two-component signal transduction proteins among the genus *Brucella*: implications for host preference and pathogenesis. Vet. Microbiol. *144*, 478–483.

Lestrate, P., Delrue, R.M., Danese, I., Didembourg, C., Taminiau, B., Mertens, P., De Bolle, X., Tibor, A., Tang, C.M., and Letesson, J.J. (2000). Identification and characterization of *in vivo* attenuated mutants of *Brucella melitensis*. Mol. Microbiol. *38*, 543–551.

Lewis, K.N., Liao, R., Guinn, K.M., Hickey, M.J., Smith, S., Behr, M.A., and Sherman, D.R. (2003). Deletion of RD1 from *Mycobacterium tuberculosis* mimics bacille Calmette-Guerin attenuation. J. Infect. Dis. *187*, 117–123.

Macedo, G.C., Magnani, D.M., Carvalho, N.B., Bruna-Romero, O., Gazzinelli, R.T., and Oliveira, S.C. (2008). Central role of MyD88-dependent dendritic cell maturation and proinflammatory cytokine production to control *Brucella abortus* infection. J. Immunol. *180*, 1080–1087.

Martín-Martín, A.I., Sancho, P., Tejedor, C., Fernández-Lago, L., and Vizcaíno, N. (2010a). Differences in the outer membrane-related properties of the six classical *Brucella* species. Vet. J. doi:10.1016/j.tvjl.2010.05.021.

Martín-Martín, A.I., Vizcaíno, N., and Fernández-Lago, L. (2010b). Cholesterol, ganglioside GM1 and class A scavenger receptor contribute to infection by *Brucella ovis* and *Brucella canis* in murine macrophages. Microbes Infect. *12*, 246–251.

Matheoud, D., Perie, L., Hoeffel, G., Vimeux, L., Parent, I., Maranon, C., Bourdoncle, P., Renia, L., Prevost-Blondel, A., Lucas, B., *et al.* (2010). Cross-presentation by dendritic cells from live cells induces protective immune responses *in vivo*. Blood *115*, 4412–4420.

Molmeret, M., and Abu Kwaik, Y. (2002). How does *Legionella pneumophila* exit the host cell? Trends Microbiol. *10*, 258–260.

Moriyón, I., Grilló, M.J., Monreal, D., González, D., Marín, C., López-Goñi, I., Mainar-Jaime, R.C., Moreno, E., and Blasco, J.M. (2004). Rough vaccines in animal brucellosis: structural and genetic basis and present status. Vet. Res. 35, 1–38.

Oliaro, J., Dudal, S., Liautard, J., Andrault, J.B., Liautard, J.P., and Lafont, V. (2005). Vgamma9Vdelta2 T cells use a combination of mechanisms to limit the spread of the pathogenic bacteria *Brucella*. J. Leukoc. Biol. 77, 652–660.

Oliveira, S.C., and Splitter, G.A. (1995). CD8+ type 1 CD44^{hi} CD45 RB^{lo} T lymphocytes control intracellular *Brucella abortus* infection as demonstrated in major histocompatibility complex class I- and class II-deficient mice. Eur. J. Immunol. 25, 2551–2557.

Olsson, S., and Sundler, R. (2006). The role of lipid rafts in LPS-induced signaling in a macrophage cell line. Mol. Immunol. 43, 607–612.

Pasquali, P., Adone, R., Gasbarre, L.C., Pistoia, C., and Ciuchini, F. (2002). Effect of exogenous interleukin-18 (IL-18) and IL-12 in the course of *Brucella abortus* 2308 infection in mice. Clin. Diagn. Lab. Immunol. 9, 491–492.

Pei, J., and Ficht, T.A. (2004). *Brucella abortus* rough mutants are cytopathic for macrophages in culture. Infect. Immun. 72, 440–450.

Pei, J., and Ficht, T.A. (2010). Lipopolysaccharide: A complex role in the pathogenesis of brucellosis. Vet. J. doi:10.1016/j.tvjl.2010.07.009.

Pei, J., Turse, J.E., Wu, Q., and Ficht, T.A. (2006). Brucella abortus rough mutants induce macrophage oncosis that requires bacterial protein synthesis and direct interaction with the macrophage. Infect. Immun. 74, 2667–2675.

Pei, J., Turse, J.E., and Ficht, T.A. (2008a). Evidence of *Brucella abortus* OPS dictating uptake and restricting NF-kappaB activation in murine macrophages. Microbes. Infect. 10, 582–590.

Pei, J., Wu, Q., Kahl-McDonagh, M., and Ficht, T.A. (2008b). Cytotoxicity in macrophages infected with rough *Brucella* mutants is type IV secretion system dependent. Infect. Immun. 76, 30–37.

Pizarro-Cerda, J., Meresse, S., Parton, R.G., Van Der Goot, G., Sola-Landa, A., Lopez-Goni, I., Moreno, E., and Gorvel, J.P. (1998). Brucella abortus transits through the autophagic pathway and replicates in the endoplasmic reticulum of nonprofessional phagocytes. Infect. Immun. 66, 5711–5724.

Pomales-Lebron, A., and Stinebring, W.R. (1957). Intracellular multiplication of *Brucella abortus* in normal and immune mononuclear phagocytes. Proc. Soc. Exp. Biol. Med. 97, 78–82.

Porte, F., Naroeni, A., Ouahrani-Bettache, S., and Liautard, J.P. (2003). Role of the *Brucella suis* lipopolysaccharide O antigen in phagosomal genesis and in inhibition of phagosome–lysosome fusion in murine macrophages. Infect. Immun. 71, 1481–1490.

Radhakrishnan, G.K., and Splitter, G.A. (2010). Biochemical and functional analysis of TIR domain containing protein from *Brucella melitensis*. Biochem. Biophys. Res. Commun. 397, 59–63.

Radhakrishnan, G.K., Yu, Q., Harms, J.S., and Splitter, G.A. (2009). *Brucella* TIR domain-containing protein mimics properties of the Toll-like receptor adaptor protein TIRAP. J. Biol. Chem. 284, 9892–9898.

Research Council (U.S.). (2009). Committee on Laboratory Security and Personnel Reliability Assurance Systems for Laboratories Conducting Research on Biological Select Agents and Toxins, (Academies Press (U.S.). Responsible research with biological select agents and toxins, p. 170.

Rice-Ficht, A.C., Arenas-Gamboa, A.M., Kahl-McDonagh, M.M., and Ficht, T.A. (2010). Polymeric particles in vaccine delivery. Curr. Opin. Microbiol. 13, 106–112.

Riley, L.K., and Robertson, D.C. (1984a). Brucellacidal activity of human and bovine polymorphonuclear leukocyte granule extracts against smooth and rough strains of *Brucella abortus*. Infect. Immun. 46, 231–236.

Riley, L.K., and Robertson, D.C. (1984b). Ingestion and intracellular survival of *Brucella abortus* in human and bovine polymorphonuclear leukocytes. Infect. Immun. 46, 224–230.

Rittig, M.G., Kaufmann, A., Robins, A., Shaw, B., Sprenger, H., Gemsa, D., Foulongne, V., Rouot, B., and Dornand, J. (2003). Smooth and rough lipopolysaccharide phenotypes of *Brucella* induce different intracellular trafficking and cytokine/chemokine release in human monocytes. J. Leukoc. Biol. 74, 1045–1055.

Rolan, H.G., and Tsolis, R.M. (2008). Inactivation of the type IV secretion system reduces the Th1 polarization of the immune response to *Brucella abortus* infection. Infect. Immun. 76, 3207–3213.

Roop, R.M., 2nd, Gaines, J.M., Anderson, E.S., Caswell, C.C., and Martin, D.W. (2009). Survival of the fittest: how *Brucella* strains adapt to their intracellular niche in the host. Med. Microbiol. Immunol. 198, 221–238.

Roux, C.M., Rolan, H.G., Santos, R.L., Beremand, P.D., Thomas, T.L., Adams, L.G., and Tsolis, R.M. (2007). *Brucella* requires a functional Type IV secretion system to elicit innate immune responses in mice. Cell. Microbiol. 9, 1851–1869.

Salcedo, S.P., Marchesini, M.I., Lelouard, H., Fugier, E., Jolly, G., Balor, S., Muller, A., Lapaque, N., Demaria, O., Alexopoulou, L., *et al.* (2008). *Brucella* control of dendritic cell maturation is dependent on the TIR-containing protein Btp1. PLoS Pathog. 4, e21.

Salmeron, I., Rodriguez-Zapata, M., Salmeron, O., Manzano, L., Vaquer, S., and Alvarez-Mon, M. (1992). Impaired activity of natural killer cells in patients with acute brucellosis. Clin. Infect. Dis. 15, 764–770.

Sangari, F., and Agüero, J. (1991). Mutagenesis of *Brucella abortus*: comparative efficiency of three transposon delivery systems. Microb. Pathog. 11, 443–446.

Sengupta, D., Koblansky, A., Gaines, J., Brown, T., West, A.P., Zhang, D., Nishikawa, T., Park, S.G., Roop, R.M., 2nd, and Ghosh, S. (2010). Subversion of innate immune responses by *Brucella* through the targeted degradation of the TLR signaling adapter, MAL. J. Immunol. 184, 956–964.

Starr, T., Ng, T.W., Wehrly, T.D., Knodler, L.A., and Celli, J. (2008). *Brucella* intracellular replication requires trafficking through the late endosomal/lysosomal compartment. Traffic 9, 678–694.

Surendran, N., Zimmerman, K., Seleem, M.N., Sriranganathan, N., Boyle, S.M., Hiltbold, E.M., Lawler, H., Heid, B., and Witonsky, S.G. (2010). Ability of *Brucella abortus* rough vaccine strains to elicit DC and innate immunity in lung using a murine respiratory model. Vaccine *28*, 7009–7015.

Surendran, N., Hiltbold, E.M., Heid, B., Sriranganathan, N., Boyle, S.M., Zimmerman, K.L., Makris, M.R., and Witonsky, S.G. (2011). Live *Brucella abortus* rough vaccine strain RB51 stimulates enhanced innate immune response *in vitro* compared to rough vaccine strain RB51SOD and virulent smooth strain 2308 in murine bone marrow-derived dendritic cells. Vet. Microbiol. *147*, 75–82.

Tanigawa, K., Suzuki, K., Kimura, H., Takeshita, F., Wu, H., Akama, T., Kawashima, A., and Ishii, N. (2009). Tryptophan aspartate-containing coat protein (CORO1A) suppresses Toll-like receptor signalling in *Mycobacterium leprae* infection. Clin. Exp. Immunol. *156*, 495–501.

Uzureau, S., Godefroid, M., Deschamps, C., Lemaire, J., De Bolle, X., and Letesson, J.J. (2007). Mutations of the quorum sensing-dependent regulator VjbR lead to drastic surface modifications in *Brucella melitensis*. J. Bacteriol. *189*, 6035–6047.

Watanabe, K., Tachibana, M., Kim, S., and Watarai, M. (2009). Participation of ezrin in bacterial uptake by trophoblast giant cells. Reprod. Biol. Endocrinol. *7*, 95.

Weiss, D.S., Takeda, K., Akira, S., Zychlinsky, A., and Moreno, E. (2005). MyD88, but not toll-like receptors 4 and 2, is required for efficient clearance of *Brucella abortus*. Infect. Immun. 73, 5137–5143.

Wu, Q., Pei, J., Turse, C., and Ficht, T.A. (2006). Mariner mutagenesis of *Brucella melitensis* reveals genes with previously uncharacterized roles in virulence and survival. BMC Microbiol. *6*, 102.

Young, E.J., Gomez, C.I., Yawn, D.H., and Musher, D.M. (1979). Comparison of *Brucella abortus* and *Brucella melitensis* infections of mice and their effect on acquired cellular resistance. Infect. Immun. *26*, 680–685.

Zaitseva, M., Golding, H., Manischewitz, J., Webb, D., and Golding, B. (1996). *Brucella abortus* as a potential vaccine candidate: induction of interleukin-12 secretion and enhanced B7.1 and B7.2 and intercellular adhesion molecule 1 surface expression in elutriated human monocytes stimulated by heat-inactivated *B. abortus*. Infect. Immun. *64*, 3109–3117.

Zhan, Y., Liu, Z., and Cheers, C. (1996). Tumor necrosis factor alpha and interleukin-12 contribute to resistance to the intracellular bacterium *Brucella abortus* by different mechanisms. Infect. Immun. *64*, 2782–2786.

Zhan, Y.F., and Cheers, C. (1995). Endogenous Interleukin-12 is involved in resistance to *Brucella abortus* infection. Infect. Immun. *63*, 1387–1390.

Zhong, Z., Wang, Y., Qiao, F., Wang, Z., Du, X., Xu, J., Zhao, J., Qu, Q., Dong, S., Sun, Y., *et al.* (2009). Cytotoxicity of *Brucella* smooth strains for macrophages is mediated by increased secretion of the type IV secretion system. Microbiology *155*, 3392–3402.

Zwerdling, A., Delpino, M.V., Barrionuevo, P., Cassataro, J., Pasquevich, K.A., Garcia Samartino, C., Fossati, C.A., and Giambartolomei, G.H. (2008). *Brucella* lipoproteins mimic dendritic cell maturation induced by *Brucella abortus*. Microbes Infect. *10*, 1346–1354.

Zygmunt, M.S., Hagius, S.D., Walker, J.V., and Elzer, P.H. (2006). Identification of *Brucella melitensis* 16M genes required for bacterial survival in the caprine host. Microbes Infect. *8*, 2849–2854.

Index

G

H

I

K

L

M

N

O

P

Q

R

Other Books of Interest

www.ingramcontent.com/pod-product-compliance
Lightning Source LLC
LaVergne TN
LVHW062121200926
848484LV00007B/1731
* 9 7 8 1 9 0 4 4 5 5 9 3 6 *